Improving Abiotic Stress Tolerance in Plants

Improving Abiotic Stress Tolerance in Plants

Edited by
M. Iqbal R. Khan, Amarjeet Singh and Péter Poór

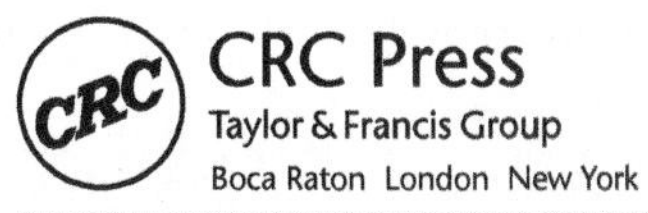

CRC Press is an imprint of the
Taylor & Francis Group, an **informa** business

CRC Press
Taylor & Francis Group
6000 Broken Sound Parkway NW, Suite 300
Boca Raton, FL 33487-2742

First issued in paperback 2022

ISBN-13: 978-0-367-13624-6 (hbk)
ISBN-13: 978-1-03-233617-6 (pbk)
DOI: 10.1201/9780429027505

Publisher's Note

The publisher has gone to great lengths to ensure the quality of this reprint but points out that some imperfections in the original copies may be apparent.

Visit the Taylor & Francis Web site at
http://www.taylorandfrancis.com

and the CRC Press Web site at
http://www.crcpress.com

Contents

Editors

Dr. M. Iqbal R. Khan is an Assistant Professor of Botany at Jamia Hamdard, New Delhi, India. His current research interests are elucidation of physiological and molecular mechanisms associated with abiotic stress tolerance and looking for suitable QTLs/genes/metabolites and/or germplasm for developing breeding or gene editing pipelines. Working on the metabolism of plants under different abiotic stresses, Dr. Khan has found a significant role of phytohormones in the regulation of plant growth and development and has suggested that phytohormones play an important role in controlling stress responses and interacting in coordination with each other for defense signal networking to fine-tune tolerance mechanisms. He is also exploring the regulatory role of signaling molecules and their impact on nutrient homeostasis and the source–sink relationship under abiotic stress. Dr. Khan has published more than 45 journal articles and 7 book chapters and has edited 4 books (including this). He has been recognized as Young Scientist of the Year by the Indian Society of Plant Physiology and Scientific and Environmental Research Institute, India, and Junior Scientist of the Year by the National Environmental Science Academy, New Delhi, India. He was guest editor of "Ethylene: A key regulatory molecule in plants" in *Frontiers in Plant Science* and is currently editing a special issue in the same journal on: "Plant Responses to the Dark Scenario".

Dr. Amarjeet Singh is a scientist and member of faculty at the National Institute of Plant Genome Research (NIPGR), New Delhi, India. He has undertaken several genomics and functional genomics studies to identify and understand the differential expression pattern at the whole genome level and functionally characterized crucial genes in rice and *Arabidopsis*. At NIPGR, his research group is interested in deciphering the signaling networks and molecular mechanisms of abiotic stress (drought, salinity, cold) tolerance, particularly those regulated by calcium and lipid signaling, in crop plants. Another major research focus of the group is to understand the molecular mechanism of nutrient (NPK) uptake, transport and homeostasis in crucial crop plants such as rice and chickpea. He has published about 20 research/review articles in highly reputed, peer-reviewed international journals. He has also published three book chapters in books by noted publishers and some short communications. For his significant contribution in the field of plant sciences and agriculture, he was awarded the prestigious Young Scientist Platinum Jubilee Award (2017) from the National Academy of Sciences (NASI), and the Pran Vohra Award (2018–19) from the Indian Science Congress Association (ISCA). He has also been conferred with several other national and international awards/honours/fellowships during his PhD and post-doctoral research including the SERB-DST Young Scientist award, Young Investigator Award-DBT, D.S. Kothari Post-Doctoral Fellowship, Travel Award by the American Society of Plant Biologists (ASPB), USA and Travel Award by CSIR-India.

Dr. Péter Poór is an assistant professor at the University of Szeged in Hungary. He is currently taking part in the education and research of the Department of Plant Biology and is a lecturer on the following courses: Plant Anatomy, Plant Cell Biology, Plant Physiology, Environmental Plant Biology, Plant Stress Physiology and Photosynthesis. Due to his excellent educational activities, he has won the Golden Chalk Award. He has published more than 40 peer-reviewed journal articles and contributed to several book chapters. Besides these, he has been awarded with various honors and scholarships. He is associate editor and manuscript reviewer for several plant biology journals. His current project is "Fine-tuning of plant defense in the dark: The role of salicylic acid, jasmonic acid, and ethylene".

Contributors

Mohammad Abass Ahanger
College of Life Sciences
North West A&F University
Shaanxi, China

Nazeer Ahmed
Transcriptomics Laboratory
Division of Plant Biotechnology
SKUAST-Kashmir
Shalimar, India

Insha Amin
Transcriptomics Laboratory
Division of Plant Biotechnology
SKUAST-Kashmir
Shalimar, India

Surendra Argal
School of Studies in Botany
Jiwaji University
Gwalior, India

Priya Arora
Department of Botanical and Environmental Sciences
Guru Nanak Dev University
Amritsar, Punjab, India

Farha Ashfaque
Department of Botany
Aligarh Muslim University
Aligarh, India

Pardeep Atri
Department of Botanical and Environmental Sciences
Guru Nanak Dev University
Amritsar, Punjab, India

Palak Bakshi
Department of Botanical and Environmental Sciences
Guru Nanak Dev University
Amritsar, Punjab, India

Ravinder Singh Bali
Department of Botanical and Environmental Sciences
Guru Nanak Dev University
Amritsar, Punjab, India

Shagun Bali
Department of Botanical and Environmental Sciences
Guru Nanak Dev University
Amritsar, Punjab, India

Aditya Banerjee
Department of Biotechnology
St. Xavier's College (Autonomous)
Kolkata, West Bengal, India

Sajid Ali Khan Bangash
Institute of Biotechnology and Genetic Engineering
University of Agriculture
Peshawar, Khyber Pakhtunkhwa, Pakistan

Krisztina Bela
Department of Plant Biology
University of Szeged
Szeged, Hungary

Renu Bhardwaj
Department of Botanical and Environmental Sciences
Guru Nanak Dev University
Amritsar, Punjab, India

Kaisar A. Bhat
School of Biosciences & Biotechnology
BGSB University
Rajouri, India
and
Division of Plant Biotechnology
Sher-e-Kashmir University of Agricultural Sciences & Technology of Kashmir
Shalimar, India

Pooja Bhatnagar-Mathur
International Crops Research Institute for the Semi-Arid Tropics (ICRISAT)
Patancheru, India

Cátia Brito
Centre for the Research and Technology of Agro-Environmental and Biological Sciences
Universidade de Trás-os-Montes e Alto Douro
Vila Real, Portugal

Himanshu Chhillar
Department of Botany
Jamia Hamdard
New Delhi, India

Priyanka Chopra
Department of Botany
Jamia Hamdard
New Delhi, India

Carlos Correia
Centre for the Research and Technology of Agro-Environmental and Biological Sciences
Universidade de Trás-os-Montes e Alto Douro
Vila Real, Portugal

Zalán Czékus
Department of Plant Biology
University of Szeged
Szeged, Hungary
and
Doctoral School in Biology
Faculty of Science and Informatics
University of Szeged
Szeged, Hungary

Riddhi Datta
Department of Botany
Dr. A.P.J. Abdul Kalam Government College
New Town, Kolkata
West Bengal, India

Loitongbam Lorinda Devi
National Institute of Plant Genome Research
New Delhi, India

Lia-Tânia Dinis
Centre for the Research and Technology of Agro-Environmental and Biological Sciences
Universidade de Trás-os-Montes e Alto Douro
Vila Real, Portugal

Samreena Farooq
Department of Botany
Jamia Hamdard
New Delhi, India

Gábor Feigl
Department of Plant Biology
University of Szeged
Szeged, Hungary

Vandana Gautum
Department of Botanical and Environmental Sciences
Guru Nanak Dev University
Amritsar, Punjab, India

Dhriti Kapoor
Department of Botany
School of Bioengineering and Biosciences
Lovely Professional University
Phagwara, Punjab, India

Parminder Kaur
Department of Botanical and Environmental Sciences
Guru Nanak Dev University
Amritsar, Punjab, India

Rupinder Kaur
Department of Biotechnology
DAV College
Amritsar, India

Kanika Khanna
Department of Botanical and Environmental Sciences
Guru Nanak Dev University
Amritsar, Punjab, India

Sukhmeen Kaur Kohli
Department of Botanical and Environmental Sciences
Guru Nanak Dev University
Amritsar, Punjab, India

Zsuzsanna Kolbert
Department of Plant Biology
University of Szeged
Szeged, Hungary

Shailesh Kumar
National Institute of Plant Genome Research (NIPGR)
Aruna Asaf Ali Marg
New Delhi, India

Khalid Z. Masoodi
Transcriptomics Laboratory
Division of Plant Biotechnology
SKUAST-Kashmir
Shalimar, India

Bilal Ahmad Mir
Department of Botany
School of Life Sciences
University of Kashmir
Jammu and Kashmir, India

Mudasir A. Mir
Transcriptomics Laboratory
Division of Plant Biotechnology
SKUAST-Kashmir
Shalimar, India

Rakeeb Ahmad Mir
School of Biosciences & Biotechnology
BGSB University
Rajouri, India

Rayees Ahmad Mir
School of Studies in Botany
Jiwaji University
Gwalior, India

Árpád Molnár
Department of Plant Biology
University of Szeged
Szeged, Hungary

José Moutinho-Pereira
Centre for the Research and Technology of Agro-Environmental and Biological Sciences
Universidade de Trás-os-Montes e Alto Douro
Vila Real, Portugal

Rahul B. Nitnavare
International Crops Research Institute for the Semi-Arid Tropics (ICRISAT)
Patancheru, India
and
Division of Plant and Crop Sciences
University of Nottingham, Sutton Bonington Campus
Loughborough, United Kingdom

Dóra Oláh
Department of Plant Biology
University of Szeged
Szeged, Hungary

Attila Ördög
Department of Plant Biology
University of Szeged
Szeged, Hungary

Anshika Pandey
National Institute of Plant Genome Research
New Delhi, India

Soumitra Paul
Department of Botany
University of Calcutta
Kolkata, West Bengal, India

Palakolanu Sudhakar Reddy
International Crops Research Institute for the Semi-Arid Tropics (ICRISAT)
Patancheru, India

Aryadeep Roychoudhury
Department of Biotechnology
St. Xavier's College (Autonomous)
Kolkata, West Bengal, India

Sushma Sagar
National Institute of Plant Genome Research
New Delhi, India

Salman Sahid
Department of Botany
University of Calcutta
Kolkata, West Bengal, India

A. A. Shah
School of Biosciences & Biotechnology
BGSB University
Rajouri, India

Aishwarya R. Shankhapal
International Crops Research Institute for the Semi-Arid Tropics (ICRISAT)
Patancheru, India

Momina Shanwaz
International Crops Research Institute for the Semi-Arid Tropics (ICRISAT)
Patancheru, India

Anket Sharma
State Key Laboratory of Subtropical Silviculture
Zhejiang A&F University
Hangzhou, China

Pooja Sharma
Department of Botanical and Environmental Sciences
Guru Nanak Dev University
Amritsar, Punjab, India
and
Department of Microbiology
DAV University
Jalandhar, Punjab, India

Amar Pal Singh
National Institute of Plant Genome Research
New Delhi, India

A.T. Vivek
National Institute of Plant Genome Research (NIPGR)
Aruna Asaf Ali Marg
New Delhi, India

Abhishek Walia
Department of Botany
DAV University
Jalandhar, Punjab, India

Shabir H. Wani
Mountain Research Centre for Field Crops
Khudwani Anantnag, Sher-e Kashmir
University of Agricultural Sciences and Technology of Kashmir
Jammu and Kashmir, India

Abbu Zaid
Plant Physiology and Biochemistry Section
Department of Botany
Aligarh Muslim University
Aligarh, India

Sajad Majeed Zargar
Division of Plant Biotechnology
Sher-e-Kashmir University of Agricultural
Sciences & Technology of Kashmir
Shalimar, Srinagar, India

1 Spectrum of Physiological and Molecular Responses in Plant Salinity Stress Tolerance

Insha Amin, Aditya Banerjee, Abbu Zaid, Mudasir A. Mir, Shabir H. Wani, Nazeer Ahmed, Aryadeep Roychoudhury, and Khalid Z. Masoodi

CONTENTS

1.1 INTRODUCTION

Salinity stress is regarded as one of the principal environment stresses that retard growth and productivity of crop plants, especially in arid and semi-arid regions of the world (Rozema and Flowers, 2008). According to Munns and Tester (2008), globally more than 800 million hectares of arable lands are severely affected by salinity stress, which corresponds to 50% of all irrigated lands (Sairam and Tyagi, 2004). Salt stress is a physiological condition characterized by increased concentrations of soluble salts inside the cells leading to an imbalance in the cell steady state (Joshi et al., 2016; Khan et al., 2017). Salt stress induces ion toxicity due to increased levels of ions like sodium (Na^+), chloride (Cl^-) and sulfate (SO_4^{2-}). Sodium chloride (NaCl) is the most widely present and most soluble salt, and therefore Na^+ accounts for the majority of the salt stress-related symptoms in the plants. There can be approximately 40 mM NaCl concentration and electrical conductivity (EC) of 4dS/m in the soils affected by salinity (Acosta-Motos et al., 2017). There is an increased concentration of Na^+ ions in the salt-rich soils with a concomitant increase in carbonate/bicarbonate levels making these soils highly alkaline (pH greater than 7). Salinity stress results in an imbalance of ion homeostasis due to an increase in the concentration of Na^+ ions as well as the simultaneous decrease in potassium (K^+) concentration (Liu et al., 2018). Many plants show a reduced growth, deteriorated quality and a significant decrease in productivity under such salt levels because salinity stress triggers complex signaling pathways to inhibit growth, development and plant physiological processes (Naeem et al., 2012). Salinity stress limits photosynthetic potential as a result of disorganized chloroplast thylakoids (Khan et al., 2014; Fatma et al., 2016) and impairment in the diffusion rate of carbon dioxide (CO_2) via decreasing conductance of stomata and mesophyll cells (Flexas et al., 2004). Rasool et al. (2013) conducted an experiment to evaluate the effect of salt stress on growth and some key antioxidants in eight chickpea genotypes which were grown in a hydroponic environment. Their results indicate that salt stress-induced oxidative stress by hampering the growth and physiology of the cells. Salinity stress results in oxidative damage through orchestrating the production of reactive oxygen species (ROS), which can cause cell death by damaging proteins, lipids, RNA and DNA (Gill and Tuteja, 2010; Anjum et al., 2015; Ahmad et al., 2016). Plants possess intricate mechanisms undergoing complex crosstalks for sensing of environmental stresses (Wani et al., 2013). To survive under such sub-optimal conditions, salt-tolerant plants like halophytes have evolved a well-integrated adaptive response at the molecular, cellular and physiological levels to ensure survival, distribution and productivity (Flowers and Muscolo, 2015). Various salt tolerance mechanisms have been comprehensively depicted in Figure 1.1. These adaptation mechanisms can be associated with detoxification, protein degradation, synthesis of osmoprotectant and antioxidants, overexpression of water and ion channels and accumulation of stress-responsive transcription factors (TFs) like *WRKY, NAC, bZIP, MYB, MYC*, etc. (Hiz et al., 2014; Banerjee and Roychoudhury, 2017). The TFs upregulate osmotic responsive (*OR*) genes encoding late embryogenesis abundant (LEA) proteins, heat shock proteins (HSPs) and antioxidant enzymes (Banerjee and Roychoudhury, 2016). The salt-tolerant genotypes accommodate the stress-mediated low water potential by maintaining a high relative water content (RWC) (Joshi and Karan, 2013). Salinity initiates multi-level regulation of gene expression through complex transcriptional networks (Singh and Laxmi, 2015). At the molecular level, the identification and characterization of candidate genes for the accumulation of ions and movement of water molecules are of paramount importance in dissecting underlying mechanisms of plants' salt stress tolerance. The salt overly sensitive (*SOS1*) gene, which encodes an antiporter Na^+/H^+ in plasma membrane, can play a significant role in deciphering mechanisms related to how Na^+ ions are excluded out of a salt-stressed cell and controlled via their long-distance transport from the roots to shoots in *Arabidopsis thaliana* (Shi et al., 2002). In plants, salt stress is known to increase the expression level of *SOS1*, which might confer salt stress tolerance (Gao et al., 2016). In a recent study, Liu et al. (2018) studied the growth, ionic response and gene expression analysis in ryegrass under salt stress conditions and observed that salinity tolerance is related to the decreased expressions of *SOS1*, *NHX1* and *TIP1* in the shoots, and increased expressions of *NHX1* and *PIP1* in the roots. These reports suggest that the coordination of genes for regulating the homeostasis of ions might prove beneficial for enhancing plant salinity stress tolerance. Nevertheless, the characterization and incorporation of selected salt-responsive genes like *DREB*,

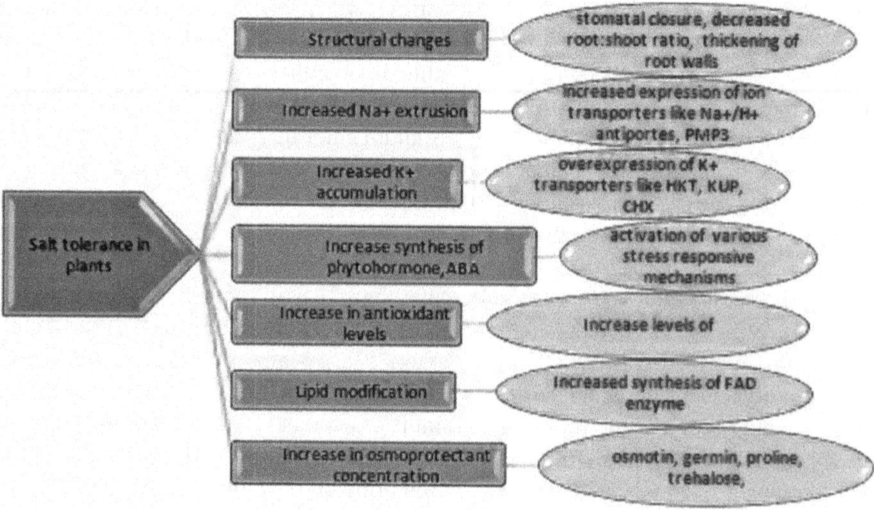

FIGURE 1.1 Various salt-tolerance mechanisms in plants.

SOS, HKT, NHX, PMP3, etc., using transgenic technology can help to design salt-tolerant lines and promote agricultural expansion.

1.2 MECHANISMS ADOPTED FOR SALT ADAPTATIONS

Ion toxicity and the hyperosmolar interior are the two important factors affecting the plants reared on soils with high salt concentrations. Adaptation to soil salinity is definitely one of the most complicated biological phenomena carried out by plants in order to maintain a steady state inside the cells. Plants adapt to many abiotic stresses at molecular, cellular, biochemical and physiological levels (Adem et al., 2014). Various adaptations include the regulation of ion transport and maintenance of water balance, vacuolar sequestration of Na^+ ions, retention of K^+ ions, accumulation of compatible solutes for osmotic adjustments and reactive oxygen species (ROS) scavenging at the gene and transcriptional levels.

1.3 EFFECT OF SALINITY ON PLANT GROWTH AND DEVELOPMENT

Ionic stress is predominant among the various abiotic stresses which act as limiting factors for plant growth and survival (Adem et al., 2014). There is a detrimental effect of salt excess on almost all the developmental parameters like seedling, flowering, chlorophyll content, internodal growth, etc. During salinity stress, there is a change in the concentration of ions in the soil around the root tip causing an imbalance in water potential leading to osmotic stress first and ending in ionic toxicity. This leads to shortening and swelling of the roots due to reduced cell division and proliferation in the root meristematic zone (Li et al., 2014b). It limits leaf extension, photosynthesis and biomass accumulation in plants (Rahnama et al., 2010). A significant reduction occurs in leaf elongation due to loosening of epidermal wall rigidity (Zörb et al., 2015). NaCl changes stem morphology by decreasing the number, diameter and length of internodes leading to stunted plant height (Nja et al., 2018). In olives, high salt levels cause significant reductions in the number and length of roots and an increased root turnover leading to restricted lifespan and development (Soda et al., 2017). Salt stress induces leaf discoloration, wilting, leaf bronzing and necrosis, thus derailing the aesthetic quality of plants (Valdez-Aguilar et al., 2011). Salt stress negatively affects flowering which can lead to drawbacks in the reproductive status by decreasing viable pollen grains (Yu et al., 2017). There is a decrease in root length when capsicum plants are exposed to 150 mM NaCl stress (Shivakumara et al., 2017). However, the transgenic plants overexpressing pea DNA Helicase 45 (*PDH45*) show a four-fold increase in root length. Meta-analyses of the responsive curves show that salinity markedly affects the leaf area and dry mass per unit area (Poorter et al., 2010).

Comparison of the total biomass of stressed and control plants has shown a relative decrease in plant biomass (RDPB) (Negrão et al., 2017). Dose-responsive curves revealed that growth rates of *Arabidopsis* plants decreased as a quadratic function of salt concentration by decreasing RGRs (relative growth rate) when NaCl concentrations increased above 25 mM (Claeys et al., 2014) Facilities like plant accelerators have been used to assess the ion-independent component of salt toxicity which inhibited shoot growth from the moment of salt imposition, i.e., even before the accumulation of Na^+ in the shoots (Berger et al., 2012; Campbell et al., 2015). Depending on plant adaptations like thick walls and space for ion sequestration, salt stress variably reduces the relative leaf area ratio (RLAR) (Negrão et al., 2017). Salinity affects water potential, hydraulic conductivity and transpiration use efficiency (TUE) in sensitive plants (Negrao et al., 2017). Salinity decreased relative water fraction (RWF) and leaf water fraction (WF) in the susceptible varieties. In spite of reduced water potential, the turgor pressure remains unaffected. This results in significant water losses from opened stomata (Boyer et al., 2008).

Essential physiological processes like photosynthesis, respiration and reproduction are negatively regulated during salinity. Rapid chlorosis resulted in lowering of the soil and plant analyzer development (SPAD) index in stressed plants (Adem et al., 2014). Reduction in the chlorophyll levels has been reported in barley exposed to salt stress wherein the chlorophyll fluorescence (F_v/F_m) value showed the highest correlation with the stress damage index (Chadchawan et al., 2017). Infra-red (IR) thermography has shown a strong genotype dependency between salinity, stomatal conductance and leaf temperature (Sirault et al., 2009). Similar phenomic approaches can be adopted to study salt-induced senescence (SIS) in each leaf rather than in the total shoot as a whole (Ward et al., 2014).

1.4 SALINITY AND THE ANTIOXIDANT SYSTEM

Reactive oxygen species (ROS) or free radicals like superoxides, hydroxyl ions, hydrogen peroxides and methylglyoxal are generated due to the degeneration of cell membranes as a result of any external stress stimuli, leading to the inhibition of plant growth and development (Saini et al., 2018; Mir et al., 2018). ROS stimulates peroxidation of membrane lipids producing toxic malondialdehyde (MDA) degeneration of nucleic acids and proteins, causing uncontrolled apoptosis (Rasouli et al., 2016). Plants increase the concentration of certain NAD(P)$^+$-dependent enzymes which act as 'aldehyde scavengers' (Zhu et al., 2014). Transgenic tall fescue plants overexpressing *Arabidopsis SOS* genes showed resistance to salt exposures and an increase in the activities of many antioxidative enzymes like superoxide dismutase, peroxidase and catalase (Ma et al., 2014). Although toxic by nature, ROS have now been recognized as vital signaling molecules in many biological processes like programmed cell death or apoptosis (Schmidt et al., 2013). Genes responsible for aldehyde dehydrogenase activity like *Aldh12A* from xerophytic grass, *Cleistogenes songorica* was found to improve tolerance to salinity and drought in transgenic *Arabidopsis* plants (Zhang et al., 2014a). Alfalfa plants transformed with the *CsALDH* gene from *Cleistogenes songorica*, a desert plant, showed improvement in the phenotypes when exposed to 200 mM of NaCl (Duan et al., 2015). This suggests that there exists a significant role played by aldehyde-scavenging enzymes in salt-combating mechanisms by destroying the aldehydes and free radicals generated in salinity stress. The cells of the lentil plant increase the concentration of antioxidative defense enzymes like superoxide dismutase (SOD), ascorbate peroxidase (APX), catalase (CAT) and glutathione reductase (GR) when subjected to salt concentrations of 100 and 200 mM (Cicerali, 2004). Therefore, the antioxidant defense mechanism helps in the maintenance of cellular integrity and cell homeostasis.

1.5 SALINITY AND OSMOTIC BALANCE

Salinity disrupts osmotic stability and cell homeostasis to confer osmotic stability; cells have an inherent capacity to undergo certain osmotic adjustments so as to maintain their water levels against the increased osmotic potentials. Thus, cells increase the concentration of certain osmolytes like glycine-betaine, β-alaninebetaine, proline-betaine, choline *O*-sulfate, hydroxyproline-betaine and pipecolate-betaine, polyamines, proline (Pro) and polyol (Roychoudhury et al., 2015; Roychoudhury and Banerjee, 2016); these compatible solutes act as osmoprotectants in the plants and help in maintaining water levels in the root cell (Table 1.1). Soluble carbohydrates like sucrose, trehalose and pinitol accumulate mostly in the cell under salt stress and confer membrane stability and maintenance of osmotic potentials under severe salinity (Slama et al., 2015). Rice varieties over-expressing the trehalose-6-phosphate synthase (TPS) gene responsible for trehalose biosynthesis show improved salt tolerance and also cause activation of various stress-related genes like HSP70 (Li et al., 2011). Free amino acid levels increase to equilibrate the cellular osmoticum. Certain other salt-induced proteins like 26 kDa protein osmotin, germin, etc., have been found to increase in certain plants under salt stress (Fatehi et al., 2012). Proline is an important amino acid found to increase in plants under salt stress (Surekha et al., 2014).

1.6 CELL MEMBRANE LIPID MODIFICATION

Lipids form an important component of the cell membrane and help in conferring membrane stability. Fatty acid desaturase (FAD) is a key enzyme for fatty acid synthesis and plays a role in inducing

TABLE 1.1
Osmoprotectants Involved in Salinity Tolerance

Solute Increased	Plant Species	Mechanism
Sugars (trehalose, mannitol)	Transgenic rice (Li et al., 2011)	Osmoprotectant Antiapoptotic
Proteins	Transgenic soybean (Weber et al.,	Structural modifications of cell wall
Osmotin	2014)	
Germin	Tobacco (Dani et al., 2005)	
Wheat LEA group 2 protein	Barley (Hurkman et al., 1991)	
(PMA80)	Transgenic rice (Cheng et al., 2002)	
Proline	Transgenic pigeon pea (Surekha	Regulates accumulation of osmotically
P5CSF129A	et al., 2014)	active nitrogen contributing to membrane
gene expression	Transgenic potato plants (Hmida-	stability
Δ1-pyrroline-5-carboxylate	Sayari et al., 2005)	
synthetase over-expression	Transgenic tobacco (Zhang et al.,	
OsP5CS1 and *OsP5CS2*	2014b)	
over-expression	Perennial rye grass (Li et al.,	
Pr5CS1 and *Pr5CS2* over-expression	2014a)	

resistance to various kinds of environmental stresses. *FAD2* is known to be required for salt resistance in *Arabidopsis* (Zhang et al., 2012a). There is an increase in salt sensitivity in *A. thaliana* fad2 mutant during seed germination and seedling stages and large amounts of Na^+ accumulate in the root cell cytoplasm. The FAD enzyme is required for maintaining plasma membrane-associated fatty acid desaturation which helps in maintaining Na^+/H^+ in exchangers required for Na^+ extrusion and maintenance of low cytosolic Na^+ concentrations. Transgenic tobacco plants over-expressing an antisense *FAD7* gene are more salt-sensitive, confirming the role of FAD7 in salt-tolerance mechanisms (Im et al., 2002). FAD6 is also involved in the acclimatization of *Arabidopsis* to salt overdose (Zhang et al., 2009). Therefore, regulation membrane composition by FAD enzymes can help in understanding various pathways involved in salt stress management.

1.7 SALINITY AND TRANSPORTERS

Plants have evolved three ways to alleviate Na^+ toxicity. Firstly, the influx of Na^+ into the cells via various influx transporters like non-selective cation channels (NSCC), high-affinity K^+ transporters (HKTs) and low-affinity cation transporters (LCT) in the root epidermal cells can be decreased by blocking these channels (Zhang et al., 2010). The genes encoding such transporters can be taken as strong gene candidates to study salt tolerance in plants. Secondly, the sequestration of excess Na^+ ions into vacuoles or apoplast lowers salt toxicity. Halophytes adapted to grow in high salt media are known to have high vacuolar Na^+ concentrations as Na^+ helps in osmotic adjustments inside the cell. *Arabidopsis* vacuolar Na^+/H^+ antiporter, *AtNH1*, present in the tonoplast, is responsible for the vacuolar compartmentalization of sodium (Shen et al., 2015). Thirdly, membrane channels like the Na^+/H^+ exporter and salt-overly sensitive 1 (*SOS1*) in association with *SOS2* and *SOS3* extrude Na^+ from the cytosol, leaving the cell free from Na^+ toxicities (Quintero et al., 2011). The manipulation of such ion exporters can help in the maintenance of salt-tolerance mechanisms in many crop plants. For example many transgenic lines of cotton have been generated. Cotton is an economically important fiber plant and is salt-sensitive. The overexpression of several candidate salt stress elated genes like *AtNHX1* from *Arabidopsis* (He et al., 2005), *AVP1* from *Arabidopsis* (Shen et al., 2015), *TsVP* from *Thellungiella halophila* (Lv et al., 2008), and *SNAC1* from rice (Liu et al., 2014a) enhanced salt tolerance significantly in the fiber-yielding crop, cotton. The transgenic cotton developed using all the above genes was more tolerant to salt shocks compared to the wild type due to proper sodium extrusion from cells. In maize, genes participating in ion homeostasis like *ZmCIPK16*, *ZmCBL4* and Na^+/H^+ Exchanger (*ZmNHX*) have been targeted to generate salt-tolerant lines (Zhao et al., 2009). Hence, using transgenics and novel omic techniques, salt-tolerant cultivars can be generated using the genes known for salt tolerance. Potassium retention is an essential physiological response of salt-tolerant plant species. A strong positive correlation is known to be present between K^+ retention and salt tolerance in many plants like wheat (Cuin et al., 2011) and barley (Chen et al., 2007). Thus, the tolerant cultivars possess the ability to retain K^+ or in some other way prevent K^+ loss under salt stress. K^+ is required for the maintenance of cell turgor pressure, osmotic adjustment, enzyme activations and cell homeostasis. So far, many K^+ transporters like cation/H^+ exchanger 17 (*CHX17*) (Sze and Chanroj, 2018), high affinity K^+ transporter 5 (*HAK5*) (Alnayef et al., 2018) and K^+ uptake transporter 1 (*KUP1*) (Saddiq et al., 2018) are known to play a role in K^+ transport into the cells during salt tolerance induction. There is an upregulation of the expression of K^+ transporter genes like *AtHAK5*, *AtCHX17* and *AtKUP1*, indicating that the K^+ retention property of cells is responsible for the induction of salinity tolerance in *Arabidopsis* (Sun et al., 2015).

1.8 SALINITY AND PHYTOHORMONE SYNTHESIS

Phytohormones are the chemical messengers released by plants that act locally or systematically and affect plant growth, development and environmental stress resistance (Javid et al., 2011). The universal stress phytohormone, abscisic acid (ABA), accumulates in stressed plants and modulates gene expression and adaptive physiological responses (Sreenivasulu et al., 2012). ABA accumulates in the vegetative tissues and triggers the upregulation of a set of ABA-responsive stress-related genes. Stress signals are perceived by the pyrabactin-resistance/pyrabactin-resistance-like/regula tory component of aba receptors (*PYR/PYL/RCAR*) complexed with ABA, which activate a mitogen-activated protein-kinase (MAPK) cascade to upregulate TFs like *ABF*, *bZIPs*, *MYB/MYCs*, *NACs*, etc. (Agarwal and Jha, 2010; Hossain et al., 2010) The ABA-independent pathway is less characterized and involves TFs like *DREB*s. Both the ABA-dependent and -independent pathways exhibit crosstalk while regulating response to dehydration 29A (*RD29A*) expression during stress. This gene promoter contains both ABA-responsive elements (ABREs) and drought-responsive elements (*DRE*s; Banerjee and Roychoudhury, 2017). Overexpression of the sulfurase gene, *AtLOS5*, upregulated ion homeostasis-associated genes like *ZmNHX1*, *ZmCBL4* and *ZmCIPK16*, thus highlighting the complexity of plant stress signaling (Zhang et al., 2016). Relative to desiccation A (*RD29A/COR78/LT178*) and *RD29B* are two important genes possessing ABA-responsive promoters and known to be over-expressed during salinity stress (Banerjee and Roychoudhury, 2017). In fava bean seedlings, Ahmad et al. (2018) reported that 100 mM NaCl toxicity decreased the endogenous concentrations of indole acetic acid (IAA) and indole butyric acid (IBA) but increased the contents of abscisic acid (ABA). However, the addition of salicylic acid to salt-stressed seedlings increased the levels of IAA and IBA but decreased the ABA concentration to an appreciable level.

1.9 FUTURE PERSPECTIVES

Salinity stress imposes a major setback on plant production and survival. Plants respond by alteration in the expression of several genes and their regulation, making them tolerant to adverse growth conditions. This has implications for the development and propagation of plant varieties resistant to salt stress. Plants have evolved in a number of ways to combat various stresses through many years of evolution. Salt stress due to toxic levels of Na^+ and Cl^- is one widely faced stress in plants. Mechanisms adopted to extrude Na^+ from cells and consequently sequester it inside the vacuoles help the plants to resist sodic niches. The novel way to achieve salt tolerance in plants is to know about the various pathways and transcription profiles related to salt stress tolerance. cDNA macro- and microarrays and the expressed sequence tags (ESTs) generated thereof can be used to formulate gene expression profiling in relation to stress tolerance in plants. Intricate physiological

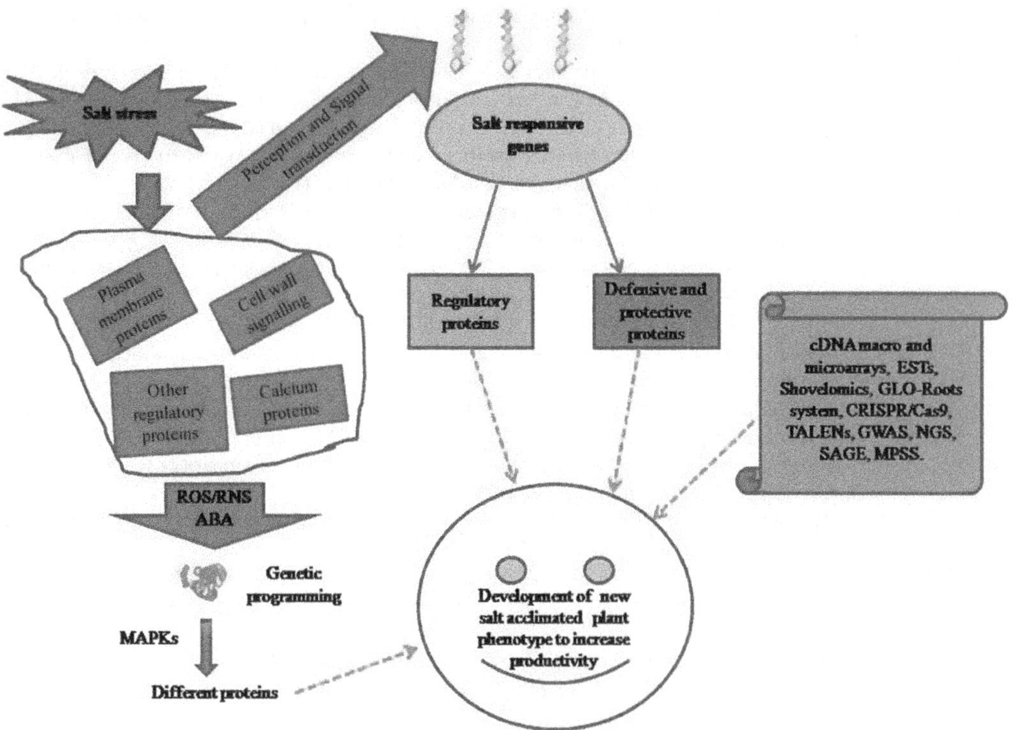

FIGURE 1.2 Technological advancements towards development of salt-tolerant plants.

studies for imaging and characterizing root responses against salt stress can further be performed using sophisticated technologies like shovelomics and the growth and luminescence observatory for roots (GLO-Roots) system. The subtle manipulation of target genes is being performed using genome editing technologies like clustered regularly interspaced short palindromic repeats (CRISPR)/CRISPR-associated system 9 (Cas9) and transcription activator-like effector nucleases (TALENs). Genome-wide associated studies (GWAS) based on next generation sequencing (NGS), and genomic approaches like EST-library screening, serial analysis of gene expression (SAGE) and massively parallel signature sequencing (MPSS) must be adopted to correlate plant genotype with the salt-tolerant phenotype (Figure 1.2). Therefore, by adopting the omics approaches like genomics, proteomics, phenomics, transcriptomics, ionomics, shovelomics, etc., many salt-tolerant varieties can be generated which can meet the growing needs of agricultural crops.

ACKNOWLEDGMENTS

The financial assistance of the Science & Engineering Research Board (SERB, DST), through the Early Career Research Award (EMR/2016/000025) and EMR/2016/005598 to Dr. Khalid Z. Masoodi is acknowledged. AZ is thankful to Aligarh Muslim University Aligarh and UGC-New Delhi India for financial assistance in the form of research fellowship 2015-BTM 04-GH-7403.

REFERENCES

Acosta-Motos, J. R., Ortuño, M. F., Bernal-Vicente, A., Diaz-Vivancos, P., Sanchez-Blanco, M. J., and Hernandez, J. 2017. Plant responses to salt stress: Adaptive mechanisms. *Agronomy* 7(1): 18.

Adem, G. D., Roy, S. J., Zhou, M., Bowman, J. P., and Shabala, S. 2014. Evaluating contribution of ionic, osmotic and oxidative stress components towards salinity tolerance in barley. *BMC Plant Biol* 14: 113.

Agarwal, P., and Jha, B. 2010. Transcription factors in plants and ABA dependent and independent abiotic stress signalling. *Biol Plant* 54(2): 201–212.

Ahmad, P., Abdel Latef, A. A., Hashem, A., Abd Allah, E. F., Gucel, S., and Tran, L.-S. P. 2016. Nitric oxide mitigates salt stress by regulating levels of osmolytes and antioxidant enzymes in chickpea. *Front Plant Sci* 7: 347.

Ahmad, P., Alyemeni, M. N., Ahanger, M. A., Egamberdieva, D., Wijaya, L., and Alam, P. 2018. Salicylic acid (SA) induced alterations in growth, biochemical attributes and antioxidant enzyme activity in Faba bean (*Vicia faba* L.) seedlings under NaCl toxicity. *Russ J Plant Physiol* 65(1): 104–114.

Alnayef, M., Bose, J., and Shabala, S. 2018. Potassium uptake and homeostasis in plants grown under hostile environmental conditions, and its regulation by CBL-interacting protein kinases. In: *Salinity Responses and Tolerance in Plants*, Volume 1, eds V. Kumar, S. H. Wani, P. Suprasanna, and L.-S. P. Tran. Springer, pp. 137–158.

Anjum, N. A., Sofo, A., Scopa, A., Roychoudhury, A., Gill, S. S., Iqbal, M., Lukatkin, A. S., Pereira, E., Duarte, A. C., and Ahmad, I. 2015. Lipids and proteins-major targets of oxidative modifications in abiotic stressed plants. *Environ Sci Pollut Res* 22(6): 4099–4121.

Banerjee, A., and Roychoudhury, A. 2016. Group II late embryogenesis abundant (LEA) proteins: Structural and functional aspects in plant abiotic stress. *Plant Growth Regul* 79(1): 1–17.

Banerjee, A., and Roychoudhury, A. 2017. Abscisic-acid-dependent basic leucine zipper (bZIP) transcription factors in plant abiotic stress. *Protoplasma* 254(1): 3–16.

Berger, B., de Regt, B., and Tester, M. 2012. Trait dissection of salinity tolerance with plant phenomics. In: *Plant Salt Tolerance*, eds S. Shabala and T. A. Cuin. Springer, pp. 399–413.

Boyer, J. S., James, R. A., Munns, R., Condon, T. A., and Passioura, J. B. 2008. Osmotic adjustment leads to anomalously low estimates of relative water content in wheat and barley. *Funct Plant Biol* 35: 1172–1182.

Campbell, M. T., Knecht, A. C., Berger, B., Brien, C. J., Wang, D., and Walia, H. 2015. Integrating image-based phenomics and association analysis to dissect the genetic architecture of temporal salinity responses in rice. *Plant Physiol.* 168(4): 1476–1489.

Chadchawan, S., Chokwiwatkul, R., Tantipirom, N., Khunpolwatatna, N., Imyim, A., Suriya-aroonroj, D., Buaboocha, T., Pongpanich, M., and Comai, L. 2017. Identification of genes involving in salt tolerance using GWAS data based on Na^+ content in local Thai rice leaves. *Genom Genet* 10(1&2): 27–37.

Chen, Z., Zhou, M., Newman, I. A., Mendham, N. J., Zhang, G., and Shabala, S. 2007. Potassium and sodium relations in salinised barley tissues as a basis of differential salt tolerance. *Funct Plant Biol* 34(2): 150–162.

Cheng, Z., Targolli, J., Huang, X., and Wu, R. 2002. Wheat LEA genes, PMA80 and PMA1959, enhance dehydration tolerance of transgenic rice (Oryza sativa L.). *Mol Breed* 10(1/2): 71–82.

Cicerali, I. 2004. Effect of salt stress on antioxidant defense systems of sensitive and resistant cultivars of lentil (*Lens Culinaris* M.). MSc Thesis, Middle East Technical University, Ankara, Turkey.

Claeys, H., Van Landeghem, S., Dubois, M., Maleux, K., and Inzé, D. 2014. What is stress? Dose-response effects in commonly used in vitro stress assays. *Plant Physiol* 165(2): 519–527.

Cuin, T. A., Bose, J., Stefano, G., Jha, D., Tester, M., Mancuso, S., and Shabala, S. 2011. Assessing the role of root plasma membrane and tonoplast Na^+/H^+ exchangers in salinity tolerance in wheat: In planta quantification methods. *Plant Cell Environ* 34(6): 947–961.

Dani, V., Simon, W. J., Duranti, M., and Croy, R. R. 2005. Changes in the tobacco leaf apoplast proteome in response to salt stress. *Proteomics* 5(3): 737–745.

Duan, Z., Zhang, D., Zhang, J., Di, H., Wu, F., Hu, X., Meng, X., Luo, K., Zhang, J., and Wang, Y. 2015. Co-transforming bar and CsALDH genes enhanced resistance to herbicide and drought and salt stress in transgenic alfalfa (*Medicago sativa* L.). *Front Plant Sci* 6: 1115.

Fatehi, F., Hosseinzadeh, A., Alizadeh, H., Brimavandi, T., and Struik, P. C. 2012. The proteome response of salt-resistant and salt-sensitive barley genotypes to long-term salinity stress. *Mol Biol Rep* 39(5): 6387–6397.

Fatma, M., Masood, A., Per, T. S., and Khan, N. A. 2016. Nitric oxide alleviates salt stress inhibited photosynthetic performance by interacting with sulfur assimilation in mustard. *Front Plant Sci* 7: 521.

Flexas, J., Bota, J., Loreto, F., Cornic, G., and Sharkey, T. D. 2004. Diffusive and metabolic limitations to photosynthesis under drought and salinity in C3 plants. *Plant Biol* 6(3): 269–279.

Flowers, T. J., and Muscolo, A. 2015. Introduction to the special issue: Halophytes in a changing world. *AoB Plants* 7: 15–20.

Gill, S. S., and Tuteja, N. 2010. Reactive oxygen species and antioxidant machinery in abiotic stress tolerance in crop plants. *Plant Physiol Biochem* 48(12): 909–930.

Gao, J., Sun, J., Cao, P., Ren, L., Liu, C., Chen, S., Chen, F., and Jiang, J. 2016. Variation in tissue Na+ content and the activity of SOS1 genes among two species and two related genera of Chrysanthemum. *BMC Plant Biol* 16: 98.

He, C., Yan, J., Shen, G., Fu, L., Holaday, A. S., Auld, D., Blumwald, E., and Zhang, H. 2005. Expression of an *Arabidopsis* vacuolar sodium/proton antiporter gene in cotton improves photosynthetic performance under salt conditions and increases fiber yield in the field. *Plant Cell Physiol* 46(11): 1848–1854.

Hiz, M. C., Canher, B., Niron, H., and Turet, M. 2014. Transcriptome analysis of salt tolerant common bean (*Phaseolus vulgaris* L.) under saline conditions. *PLOS ONE* 9(3): e92598.

Hmida-Sayari, A., Gargouri-Bouzid, R., Bidani, A., Jaoua, L., Savouré, A., and Jaoua, S. 2005. Overexpression of Δ1-pyrroline-5-carboxylate synthetase increases proline production and confers salt tolerance in transgenic potato plants. *Plant Sci* 169(4): 746–752.

Hossain, M. A., Cho, J. I., Han, M., Ahn, C. H., Jeon, J. S., An, G., and Park, P. B. 2010. The ABRE-binding bZIP transcription factor OsABF2 is a positive regulator of abiotic stress and ABA signaling in rice. *J Plant Physiol* 167(17): 1512–1520.

Hurkman, W. J., Tao, H. P., and Tanaka, C. K. 1991. Germin-like polypeptides increase in barley roots during salt stress. *Plant Physiol* 97(1): 366–374.

Im, Y. J., Han, O., Chung, G. C., and Cho, B. H. 2002. Antisense expression of an *Arabidopsis* omega-3 fatty acid desaturase gene reduces salt/drought tolerance in transgenic tobacco plants. *Mol Cells* 13(2): 264–271.

Javid, M. G., Sorooshzadeh, A., Moradi, F., ModarresSanavy, S. A. M., and Allahdadi, I. 2011. The role of phytohormones in alleviating salt stress in crop plants. *Aust J Crop Sci* 5: 726.

Joshi, R., and Karan, R. 2013. Physiological, biochemical and molecular mechanisms of drought tolerance in plants. In: *Molecular Approaches in Plant Abiotic Stress*, eds R. K. Gaur and P. Sharma. CRC Press, pp. 225–247.

Joshi, R., Singh, B., Bohra, A., and Chinnusamy, V. 2016. Salt stress signalling pathways: Specificity and crosstalk. In: *Manag Salinity Tolerance in Plants: Molecular and Genomic Perspectives*, pp. 51–78.

Khan, M. A., Shirazi, M. U., Shereen, A., Mujtaba, S. M., Khan, M. A., Mumtaz, S., and Mahboob, W. 2017. Identification of some wheat (*Triticum aestivum* L.) lines for salt tolerance on the basis of growth and physiological characters. *Pak J Bot* 49(2): 397–403.

Li, H., Guo, H., Zhang, X., and Fu, J. 2014a. Expression profiles of Pr5CS1 and Pr5CS2 genes and proline accumulation under salinity stress in perennial ryegrass (*Lolium perenne* L.). *Plant Breed* 133(2): 243–249.

Li, H., Yan, S., Zhao, L., Tan, J., Zhang, Q., Gao, F., Wang, P., Hou, H., and Li, L. 2014b. Histone acetylation associated up-regulation of the cell wall related genes is involved in salt stress induced maize root swelling. *BMC Plant Biol* 14(1): 105.

Li, H. W., Zang, B. S., Deng, X. W., and Wang, X. P. 2011. Overexpression of the trehalose-6-phosphate synthase gene OsTPS1 enhances abiotic stress tolerance in rice. *Planta* 234(5): 1007–1018.

Liu, G., Li, X., Jin, S., Liu, X., Zhu, L., Nie, Y., and Zhang, X. 2014. Overexpression of rice NAC gene SNAC1 improves drought and salt tolerance by enhancing root development and reducing transpiration rate in transgenic cotton. *PLOS ONE* 9(1): e86895.

Liu, M., Song, X., and Jiang, Y. 2018. Growth, ionic response, and gene expression of shoots and roots of perennial ryegrass under salinity stress. *Acta Physiol Plant* 40(6): 112.

Lv, S., Zhang, K., Gao, Q., Lian, L., Song, Y., and Zhang, J. 2008. Overexpression of an H+-PPase gene from Thellungiellahalophila in cotton enhances salt tolerance and improves growth and photosynthetic performance. *Plant Cell Physiol* 49(8): 1150–1164.

Ma, D. M., Xu, W. R., Li, H. W., Jin, F. X., Guo, L. N., Wang, J., Dai, H. J., and Xu, X. 2014. Co-expression of the *Arabidopsis* SOS genes enhances salt tolerance in transgenic tall fescue (*Festuca arundinacea* Schreb.). *Protoplasma* 251(1): 219–231.

Mir, M. A., John, R., Alyemeni, M. N., Alam, P., and Ahmad, P. 2018. Jasmonic acid ameliorates alkaline stress by improving growth performance, ascorbate glutathione cycle and glyoxylase system in maize seedlings. *Sci Rep* 8(1): 2831.

Munns, R., and Tester, M. 2008. Mechanisms of salinity tolerance. *Annu Rev Plant Biol* 59: 651–681.

Naeem, M. S., Warusawitharana, H., Liu, H., Liu, D., Ahmad, R., Waraich, E. A., *et al.* 2012. 5-aminolevulinic acid alleviates the salinity-induced changes in Brassica napus as revealed by the ultrastructural study of chloroplast. *Plant Physiol Biochem* 57: 84–92.

Negrão, S., Schmöckel, S., and Tester, M. 2017. Evaluating physiological responses of plants to salinity stress. *Ann Bot* 119(1): 1–11.

Nja, R. B., Merceron, B., Faucher, M., Fleurat-Lessard, P., and Béré, E. 2018. NaCl–Changes stem morphology, anatomy and phloem structure in Lucerne (*Medicago sativa* cv. Gabès): Comparison of upper and lower internodes. *Micron* 105: 70–81.

Poorter, H., Niinemets, Ü., Walter, A., Fiorani, F., and Schurr, U. 2010. A method to construct dose–response curves for a wide range of environmental factors and plant traits by means of a meta-analysis of phenotypic data. *J Exp Bot* 61(8): 2043–2055.

Quintero, F. J., Martinez-Atienza, J., Villalta, I., Jiang, X., Kim, W. Y., Ali, Z., Fujii, H., Mendoza, I., Yun, D. J., Zhu, J. K., and Pardo, J. M. 2011. Activation of the plasma membrane Na/H antiporter salt-overly-sensitive 1 (SOS1) by phosphorylation of an auto-inhibitory C-terminal domain. *PNAS* 108(6): 2611–2616.

Rahnama, A., James, R. A., Poustini, K., and Munns, R. 2010. Stomatal conductance as a screen for osmotic stress tolerance in durum wheat growing in saline soil. *Funct Plant Biol* 37(3): 255–263.

Rasool, S., Ahmad, A., Siddiqi, T. O., and Ahmad, P. 2013. Changes in growth, lipid peroxidation and some key antioxidant enzymes in chickpea genotypes under salt stress. *Acta Physiol Plant* 35(4): 1039–1050.

Rasouli, H., Farzaei, M. H., Mansouri, K., Mohammadzadeh, S., and Khodarahmi, R. 2016. Plant cell cancer: May natural phenolic compounds prevent onset and development of plant cell malignancy? A literature review. *Molecules* 21(9): 1104.

Roychoudhury, A., and Banerjee, A. 2016. Endogenous glycine betaine accumulation mediates abiotic stress tolerance in plants. *Trop Plant Res* 3: 105–111.

Roychoudhury, A., Banerjee, A., and Lahiri, V. 2015. Metabolic and molecular-genetic regulation of proline signaling and itscross-talk with major effectors mediates abiotic stress tolerance in plants. *Turk J Bot* 39: 887–910.

Rozema, J., and Flowers, T. 2008. Ecology. Crops for a salinized world. *Science* 322(5907): 1478–1480.

Saddiq, M. S., Afzal, I., Basra, S. M., Ali, Z., and Ibrahim, A. M. 2018. Sodium exclusion is a reliable trait for the improvement of salinity tolerance in bread wheat. *Arch Agron Soil Sci* 64(2): 272–284.

Saini, S., Kaur, N., and Pati, P. K. 2018. Reactive oxygen species dynamics in roots of salt sensitive and salt tolerant cultivars of rice. *Anal Biochem* 550: 99–108.

Sairam, R. K., and Tyagi, A. 2004. Physiology and molecular biology of salinity stress tolerance in plants. *Curr Sci* 86: 407–421.

Schmidt, R., Mieulet, D., Hubberten, H. M., Obata, T., Hoefgen, R., Fernie, A. R., Fisahn, J., San Segundo, B., Guiderdoni, E., Schippers, J. H., and Mueller-Roeber, B. 2013. Salt-responsive ERF1 regulates reactive oxygen species–dependent signaling during the initial response to salt stress in rice. *Plant Cell* 25(6): 2115–2131.

Shen, G., Wei, J., Qiu, X., Hu, R., Kuppu, S., Auld, D., Blumwald, E., Gaxiola, R., Payton, P., and Zhang, H. 2015. Co-overexpression of AVP1 and AtNHX1 in cotton further improves drought and salt tolerance in transgenic cotton plants. *Plant Mol Biol Rep* 33(2): 167–177.

Shi, H., Quintero, F. J., Pardo, J. M., and Zhu, J.-K. 2002. The putative plasma membrane Na+/H+ antiporter *SOS1* controls long distance Na+ transport in plants. *Plant Cell* 14(2): 465–477.

Shivakumara, T. N., Sreevathsa, R., Dash, P. K., Sheshshayee, M. S., Papolu, P. K., Rao, U., Tuteja, N., and UdayaKumar, M. 2017. Overexpression of Pea DNA helicase 45 (PDH45) imparts tolerance to multiple abiotic stresses in chili (*Capsicum annuum* L.). *Sci Rep* 7(1): 2760.

Singh, D., and Laxmi, A. 2015. Transcriptional regulation of drought response: A tortuous network of transcriptional factors. *Front Plant Sci* 6: 895.

Sirault, X. R., James, R. A., and Furbank, R. T. 2009. A new screening method for osmotic component of salinity tolerance in cereals using infrared thermography. *Funct Plant Biol* 36(11): 970–977.

Slama, I., Abdelly, C., Bouchereau, A., Flowers, T., and Savouré, A. 2015. Diversity, distribution and roles of osmoprotective compounds accumulated in halophytes under abiotic stress. *Ann Bot* 115(3): 433–447.

Soda, N., Ephrath, J. E., Dag, A., Beiersdorf, I., Presnov, E., Yermiyahu, U., and Ben-Gal, A. 2017. Root growth dynamics of olive (*Olea europaea* L.) affected by irrigation induced salinity. *Plant Soil* 411(1–2): 305–318.

Sreenivasulu, N., Harshavardhan, V. T., Govind, G., Seiler, C., and Kohli, A. 2012. Contrapuntal role of ABA: Does it mediate stress tolerance or plant growth retardation under long-term drought stress? *Gene* 506(2): 265–273.

Sun, Y., Kong, X., Li, C., Liu, Y., and Ding, Z. 2015. Potassium retention under salt stress is associated with natural variation in salinity tolerance among *Arabidopsis* accessions. *PLOS ONE* 10(5): e0124032.

Surekha, C. H., Kumari, K. N., Aruna, L. V., Suneetha, G., Arundhati, A., and Kishor, P. K. 2014. Expression of the Vignaaconitifolia P5CSF129A gene in transgenic pigeonpea enhances proline accumulation and salt tolerance. *Plant Cell Tissue Organ Cult* 116(1): 27–36.

Sze, H., and Chanroj, S. 2018. Plant endomembrane dynamics: Studies of K+/H+ antiporters provide insights on the effects of pH and ion homeostasis. *Plant Physiol* 177(3): 875–895.

Valdez-Aguilar, L. A., Grieve, C. M., Razak-Mahar, A., McGiffen, M. E., and Merhaut, D. J. 2011. Growth and ion distribution is affected by irrigation with saline water in selected landscape species grown in two consecutive growing seasons: Spring–summer and fall–winter. *Hort Sci* 46(4): 632–642.

Wani, S. H., Singh, N., Devi, T. R., Haribhushan, A., and Jeberson, S. 2013. Engineering abiotic stress tolerance in plants: Extricating regulatory gene complex. In: *Conventional and Non-Conventional Interventions in Crop Improvement*, eds C. P. Malik, G S. Sanghera, and S. H. Wani. CABI, New Delhi, pp. 1–19.

Ward, B., Bastian, J., van den Hengel, A., Pooley, D., Bari, R., Berger, B., and Tester, M. 2014. A model-based approach to recovering the structure of a plant from images. In *Proceedings of the European Conference on Computer Vision*. Springer, Cham, pp. 215–230.

Weber, R. L. M., Wiebke-Strohm, B., Bredemeier, C., Margis-Pinheiro, M., de Brito, G. G., Rechenmacher, C., *et al.* 2014. Expression of an osmotin-like protein from *Solanum nigrum* confers drought tolerance in transgenic soybean. *BMC Plant Biol* 14(1): 343.

Yu, Y., Wang, L., Chen, J., Liu, Z., Park, C. M., and Xiang, F. 2017. WRKY71 acts antagonistically against salt-delayed flowering in *Arabidopsis thaliana*. *Plant Cell Physiol* 59(2): 414–422.

Zhang, J., Duan, Z., Jahufer, Z., An, S., and Wang, Y. 2014a. Stress-inducible expression of a Cleistogenes songorica ALDH gene enhanced drought tolerance in transgenic Arabislopsis thaliana. *Plant OMICS* 7: 438.

Zhang, J., Liu, H., Sun, J., Li, B., Zhu, Q., Chen, S., and Zhang, H. 2012. *Arabidopsis* fatty acid desaturase FAD2 is required for salt tolerance during seed germination and early seedling growth. *PLOS ONE* 7(1): e30355.

Zhang, J., Yu, H., Zhang, Y., Wang, Y., Li, M., Zhang, J., Duan, L., Zhang, M., and Li, Z. 2016. Increased abscisic acid levels in transgenic maize overexpressing AtLOS5 mediated root ion fluxes and leaf water status under salt stress. *J Exp Bot* 67(5): 1339–1355.

Zhang, J. L., Flowers, T. J., and Wang, S. M. 2010. Mechanisms of sodium uptake by roots of higher plants. *Plant Soil* 326(1–2): 45.

Zhang, J. T., Zhu, J. Q., Zhu, Q., Liu, H., Gao, X. S., and Zhang, H. X. 2009. Fatty acid desaturase-6 (Fad6) is required for salt tolerance in *Arabidopsis thaliana*. *Biochem Biophys Res Commun* 390(3): 469–474.

Zhang, X. X., Tang, W. W., Liu, J., and Liu, Y. 2014b. Co-expression of rice *OsP5CS1* and *OsP5CS2* genes in transgenic tobacco resulted in elevated proline biosynthesis and enhanced abiotic stress tolerance. *Chin J App Environ Biol* 20: 717–722.

Zhao, J., Sun, Z., Zheng, J., Guo, X., Dong, Z., Huai, J., Gou, M., He, J., Jin, Y., Wang, J., and Wang, G. 2009. Cloning and characterization of a novel CBL-interacting protein kinase from maize. *Plant Mol Biol* 69(6): 661–674.

Zhu, C., Ming, C., Zhao-shi, X., Lian-cheng, L., Xue-ping, C., and You-zhi, M. 2014. Characteristics and expression patterns of the aldehyde dehydrogenase (ALDH) gene superfamily of foxtail millet (*Setaria italica* L.). *PLOS ONE* 9(7): e101136.

Zörb, C., Mühling, K. H., Kutschera, U., and Geilfus, C. M. 2015. Salinity stiffens the epidermal cell walls of salt-stressed maize leaves: Is the epidermis growth-restricting? *PLOS ONE* 10(3): e0118406.

2 Root Plasticity under Low Phosphate Availability

A Physiological and Molecular Approach to Plant Adaptation under Limited Phosphate Availability

Loitongbam Lorinda Devi, Anshika Pandey, and Amar Pal Singh

CONTENTS

2.1 INTRODUCTION

Inorganic phosphate (Pi) is a critical essential macronutrient for plant growth and productivity. In natural soils, the bioavailability of Pi is limited, thus affecting plant growth and productivity (Shang et al., 1996; Wang et al., 2017). Due to heterogeneous and differential availability of phosphate in the rhizosphere, plants evolved with different types of adaptive mechanisms, such as a change in the root system architecture (RSA) (Lynch, 1995; Hell & Hillebrand, 2001; Wang et al., 2017). In most soils, due to high reactivity, Pi makes complexes with organic and inorganic molecules, thus limiting its availability to the plants. The available Pi in most natural soils is approximately 10 μM (Schachtman et al., 1998), which is less than the endogenous pool of Pi concentration (5–20 mM) in the plants (Raghothama, 1999). The underground plant organ called root can uptake soluble Pi from the soils. Exogenous use of phosphorus-based fertilizers is the most favorable way to improve crop productivity in acidic and calcareous soils, however, it has a serious impact on the environment. The Pi diffusion rate in soils is very slow due to its high reactivity with organic and inorganic soil molecules, thus reducing the overall available Pi in the rhizosphere (Hinsinger, 2001). In acidic soils, generally, Pi forms complexes with calcium (Ca), aluminum (Al) and iron (Fe) such as gibbsite and hematite. The release of Pi in these soils depends on the ionic strength and pH of the rhizosphere. For example, increasing soil pH also increases the solubility of the Pi that is bound with the Al/Fe complexes (Hinsinger, 2001; Oelkers and Valsami-Jones, 2008; Arai and Sparks, 2007; Zhang et al., 2016; Shen et al., 2011). To increase Pi uptake, plants reprogram their RSA, for example, through increased lateral root density (LR), root hair length and number, to increase the root-to-soil surface area, thus promoting a shallow root system architecture (Sanchez-Calderon et al., 2005; Chiou and Lin, 2011; Abel, 2011; Zhang et al., 2016; Singh et al., 2014). The soil microbiome is an important factor in increasing the available Pi for plant uptake (Castrillo et al., 2017). Mycorrhizal symbiosis also increases the root Pi availability by the formation of mycorrhizal hyphae (Brundrett, 2009).

Pi availability in the rhizosphere regulates several biological and physiological processes such as energy metabolism, signal transduction, anthocyanin accumulation, organic acid formation and secretion (Ryan et al., 2011; Li et al., 2011; Mora-Macias et al., 2017). Cellular Pi homeostasis in plants is a very complex process and is linked with Pi acquisition efficiency (PAE) from the rhizosphere, phosphate utilization and translocation efficiency and internal Pi remobilization during low-Pi conditions (Ramaekers et al., 2010). Plants uptake Pi in the form of phosphate (PO_4^{3-}) through plasma membrane-localized low-Pi inducible high and low-affinity phosphate transporters (*PHTs*). *PHTs* have been well-characterized in several plants, such as *Arabidopsis*, *Glycine max*, *Oryza sativa*, *Zea mays*, *Medicago* and *Solanum lycopersicum* (Guo et al., 2008; Ai et al., 2009; Harrison et al., 2002; Nagy et al., 2005; Rausch et al., 2001; Bayuelo-Jimenez et al., 2011; Chang et al., 2019). Upon low Pi sensing by roots, plants transduce two interdependent dynamic responses, local and systemic response (Sanchez-Calderon et al., 2005; Thibaud et al., 2010; Rouached et al., 2011). In general, local Pi response is associated with the change in RSA to increase the Pi uptake from the rhizosphere while the systemic response is associated with long-distance Pi translocation and assimilation such as root-to-shoot internal Pi transport and distribution (Doerner, 2008). Apart from the *PHTs*, several genes have been characterized that are involved in low-Pi-dependent phosphate transport and signaling, such as *miRNA399b*, *INDUCED BY PHOSPHATE STARVATION1* (*IPS1*), *PHOSPHATE 1* (*PHO1*), *LOW PHOSPHATE ROOT 1* (*LPR1*), *PHOSPHATE DEFICIENCY RESPONSE 2* (*PDR2*), *PHOSPHATE STARVATION RESPONSE 1* (*PHR1*) and *small ubiquitin-like modifier* (*SUMO*) *E3 ligase* (*SIZ1*). (Hamburger et al., 2002; Arpat et al., 2012; Bari et al., 2006; Chiou et al., 2006; Miura et al., 2005).

The molecular basis of RSA changes under limited Pi availability is complex and involves the interdependent activity of nutrients and developmental signals (Lopez-Bucio et al., 2002, 2003). *Arabidopsis* roots undergo striking physiological and morphological changes under low Pi availability, such as inhibition in primary root elongation, cell expansion and division, increased lateral root and root hair density (Svistoonoff et al., 2007; Singh et al., 2014). Intrinsic factors such as growth hormones regulate these morphological and physiological changes associated with the RSA modification. For example, the plant hormones auxin and ethylene promote lateral root development and root hair elongation during Pi starvation, while gibberellic acid (GA) and brassinosteroid (BR) promote primary root (PR) elongation under similar conditions (Lopez-Bucio et al., 2005; Nacry et al., 2005; Chiou and Lin, 2011; Li et al., 2011; Jiang et al., 2007; Zhang et al., 2014; Singh et al., 2014, 2018). Thus, nutrient availability, particularly Pi availability, appears to have multiple effects on RSA change which is governed by a complicated regulatory network of environmental and developmental signals to regulate the Pi status of the plant.

2.2 PHOSPHATE (Pi) SENSING, REMOBILIZATION AND UTILIZATION

Plant responses to low Pi depend on the Pi levels in the rhizosphere and cellular Pi content. Root tip contact with the deficit Pi triggers interconnected systemic and local responses (Chiou and Lin, 2011). The systemic response regulates the cellular Pi homeostasis and specific physiological changes associated with the RSA modification, while the local response regulates the RSA morphology to increase Pi uptake. However, the mechanism of systemic and local Pi responses in plants is not well-defined, and further studies are required to decipher the interconnected responses of these signals. One of the important pathways that regulate Pi homeostasis and is involved in the loading of inorganic phosphate into the xylem of roots depends on the *Arabidopsis phosphate 1* (*PHO1*) gene. The *Arabidopsis* genome has 11 *PHO1* members. All the homologs of the *PHO1* protein contain an SPX tripartite domain (N-terminus) and an EXS domain in the C-terminus of the protein (Wang et al., 2004). Mutation in the *PHO1* gene (*pho1*) in *Arabidopsis* reduces Pi loading in the root xylem cells and causes low Pi content in the shoots of these plants (Poirier et al., 1991). In several other plants, the *PHO1* gene family has been characterized. Rice has three *PHO1* homologs (*OsPHO1;1, OsPHO1;2* and *OsPHO1;3)* and has both sense and antisense transcripts. The antisense transcript of *OsPHO1;2* is highly induced by low Pi as compared to other homologs. The genomes of *Brassica rapa*, soybean, *Brachypodium* and maize contain 23, 14, 2 and 2 homologs of *PHO1* genes, respectively (Secco et al., 2013; Wu et al., 2013). *PHO1* activity depends on *PHO2* levels. *PHO2* encodes a ubiquitin-conjugating E2 enzyme (UBC 24). The *pho2* mutant shows Pi toxicity and accumulates a high amount of Pi in the shoots due to increased Pi uptake. Further studies showed that *PHO1* degradation is *PHO2*-dependent during Pi stress (Delhaize and Randall, 1995; Dong et al., 1998; Hamburger et al., 2002; Stefanovic et al., 2011; Arpat et al., 2012; Lopez-Arredondo et al., 2014). In rice, a loss of function mutant *pho2* shows impaired Pi transport from root to shoot highlighting the role of *PHO2* in Pi homeostasis across the plant kingdom (Delhaize et al., 1995; Zhou et al., 2008; Wang et al., 2009; Hu et al., 2011). It has been proposed that *PHO2*-dependent systemic Pi response is controlled by *microRNA399* (*miRNA399*) (Bari et al., 2006; Chiou et al., 2006). *miRNA399* cleaves *PHO2* mRNA, thus reducing the *PHO2* activity. Low-Pi-induced *miRNA399* has been reported in several plants, including rice, soybean and oil rapeseed; therefore, the *PHO2*–miRNA399 module may be one important pathway that regulates systemic Pi response in plants (Sunkar et al., 2005; Liu et al., 2010; Pant et al., 2008; Desnos, 2008).

2.3 ENHANCING AVAILABLE Pi TO ROOTS

Pi is essential for the growth and development of the plant, and plant performance depends on the available Pi in the soil. More than 170 forms of phosphate exist in natural soils, and plants can uptake orthophosphates such as $H_2PO_4^-$ and HPO_4^{2-} (Cordell et al., 2009). However, due to the high reactivity of Pi with other minerals and metals such as Al, Ca and Fe, approximately 50–80% of Pi is not available for uptake by plant roots (Wang et al., 2009; Oelkers and Valsami-Jones, 2008; Hinsinger, 2001). Several factors such as acid phosphatases, organic acid exudation, soil microbiome, phosphate transporters and local changes in the RSA determine the availability of Pi to the roots.

2.3.1 Role of *Acid Phosphatases* (APases)

In natural soils, approximately 50–80% of P is in organic form and not available to roots for uptake (Wang et al., 2009; Hinsinger, 2001). Therefore, plants have evolved adaptive strategies to uptake the bound organic Pi from the soils, such as the secretion of acid phosphatases and nucleases into the rhizosphere during Pi deficiency (del Pozo et al., 1999; Brinch-Pedersen et al., 2002; Kuang et al., 2009; Tran et al., 2010; Mehra et al., 2017). Pi-deficiency-induced *APases* have been characterized in several plants such as *Arabidopsis*, rice, white lupin, soybean, tobacco and tomato (Miller et al., 2001; Bozzo et al., 2006; Liang et al., 2010; Lung et al., 2008; Wang et al., 2011). Intracellular *APases* are thought to be involved in regulating the endogenous reserve Pi pool of plants while external or secretory *APases* are believed to release Pi from bound organic Pi complexes in the rhizosphere, thus making Pi available to the roots (Wang et al., 2011). *Purple acid phosphatases* (*PAPs*) represent the largest group of *APases* (E.C. 3.1.3.2) due to their purple color in water solution. In *Arabidopsis*, out of 29 members of this group, 11 are highly induced by Pi limitation. *AtPAP10*, *AtPAP12*, *AtPAP15*, *AtPAP17*, *AtPAP25* and *AtPAP26* are characterized for their Pi-releasing activity (Kuang et al., 2009; Del Vecchio et al., 2014). *AtPAP15* gene overexpression in soybean improves the phosphate use efficiency (PUE), and transgenic plants accumulated more Pi with improved productivity. Similarly, in rice, 26 *PAPs* have been reported (Zhang et al., 2011). The overexpression of *OsPAP21b* improves Pi utilization efficiency in rice by increasing the root biomass, and its expression is controlled by a Pi-responsive transcription factor, *OsPHR2* (Mehra et al., 2017). The overexpression of Pi-deficiency-induced *PAPs*, *OsPAP10a* and *OsPAP10c* increased ATP hydrolysis (Lu et al., 2016; Tian et al., 2012). Thus, *PAPs* play an important role in promoting Pi use efficiency by hydrolyzing the bound Pi and making it available to roots.

2.3.2 Organic Acid Biosynthesis and Exudation under Low Pi

To cope with Pi deficiency, the plant produces organic acids, for example citrate and malate, and secretes them into the rhizosphere. It has been shown that the overexpression of the genes involved in malate and citrate biosynthesis, *malate dehydrogenase* and *citrate synthase*, enhance Pi acquisition and biomass by organic acid exudation. Citrate and malate are known to ameliorate aluminum toxicity and Pi scavenging from soil (Ryan et al., 2001; Vance et al., 2003). Recently, it was found that malate secretion is sufficient for increased iron accumulation in the root apical meristem and promotes cell differentiation under low-Pi conditions, thus promoting root meristem exhaustion. *ALUMINUM ACTIVATED MALATE TRANSPORTER 1* (*ALMT1*) promotes malate efflux in *Arabidopsis* roots and increases iron toxicity, thus inhibiting the primary root elongation (Mora-Macias et al., 2017).

2.3.3 Soil Microbiome

Soil microorganisms play an important role in solubilizing organic Pi content. Several species of bacteria and fungus have been identified for their phosphate-solubilizing property (Lopez-Arredondo et al., 2014; Nassal et al., 2018). Pi-solubilizing bacteria and fungus constitute 1–50% and 0.1–0.5%, respectively, to the total population (Zaidi et al., 2009). In tomatoes, inoculating soil with *Pseudomonas* sp. RU47 improves Pi uptake and plant biomass by increasing the microbial phosphatase activity in the rhizosphere (Spohn et al., 2018). In *Arabidopsis*, *in-vitro* colonization with a synthetic bacterial community under Pi limitation enhances the activity of the key phosphate-starvation-induced transcription factor, *PHR1*, thus providing a molecular link of the Pi-starvation-induced response with the soil microbiome (Castrillo et al., 2017). However, further studies are needed to decipher the role of the soil microbiome in improving nutrient use efficiency.

2.3.4 Phosphate Transporters

Pi uptake from soil and transport from root to shoot depend on phosphate transporters (*PHTs*). Phosphate transporters are plasma membrane-localized proteins and were first discovered by a homology search using a yeast, *Saccharomyces cerevisiae* phosphate transporter, *PHO84* (Bun-Ya et al., 1991). The role of *PHTs* in Pi uptake and transport has been well-characterized, and Pi uptake from soil largely depends on the *PHT1* gene family of transporters. Homologs of *PHT1* family transporters have been identified in several plants such as *Arabidopsis*, rice, tomato, *Medicago truncatula*, lotus, potato, wheat, tobacco, etc. (Mitsukawa et al., 1997; Paszkowski et al., 2002; Ming et al., 2005; Daram et al., 1998; Xiao et al., 2006; Tittarelli et al., 2007; Baek et al., 2001).

The transcriptional regulation of *PHT1* transporters mainly depends on the cellular Pi levels as well as Pi availability in the soils. Variation in Pi levels (0 to 1.25 μM) in the rhizosphere rapidly changes the expression of *PHTs* (Muchhal and Raghothama, 1999). The *Arabidopsis* genome has nine members of the *PHT1* family transporter with a differential affinity to Pi (Dunlop et al., 1997). Among the *PHT1* family of transporters, *PHT1;1* and *PHT1;4* show Pi uptake from the rhizosphere during optimal and deficit Pi conditions, while *PHT1;5* translocates Pi from the source to the organ

TABLE 2.1
Phosphate Transporters Involved in Pi Uptake and Transport in Different Plant Species

Name of Transporter	Plant Species	Reference
AtPHT1;1, *AtPHT1;4*, *AtPHT1;5*, *AtPHT1;9*	*Arabidopsis*	Shin et al. (2004); Nagarajan et al. (2011); Lapis-Gaza et al. (2014); Remy et al. (2012)
OsPht1;1, *OsPht1;4*, *OsPht1;8*	Rice	Seo et al. (2008); Ye et al. (2015); Jia et al. (2011)
HvPht1;1, *HvPht1;6*	Barley	Rae et al. (2003); Preuss et al. (2010)
LePT1	Tomato	Daram et al. (1998)
GmPT5	Soybean	Qin et al. (2012)

level. Genetic analysis with *pht1;9–1* and *pht1;8/pht1;9* as compared to the control plants suggested a role of *PHT1;8* and *PHT1;9* in Pi translocation from root to shoot during Pi deficiency. The overexpression of *PHT1;9* and *PHT1;5* improves Pi uptake and increases fresh weight, suggesting the crucial role of these transporters during Pi deficiency (Shin et al., 2004; Nagarajan et al., 2011; Lapis-Gaza et al., 2014; Remy et al., 2012). In agreement with the *Arabidopsis PHT1* transporter studies, the overexpression of rice *OsPT1* (*OsPht1;1*) increases the Pi uptake, shoot Pi levels as well as grain yield (Seo et al., 2008). Phosphate transporters that could be used to improve the Pi uptake and transport in different plants have been summarized in Table 2.1.

2.4 ROOT SYSTEM ARCHITECTURE (RSA) MODIFICATIONS UNDER LOW Pi AVAILABILITY

Nutrient availability in the rhizosphere reprograms the root system architecture (RSA) of plants. During Pi deficiency in *Arabidopsis*, the roots undergo several morphological and developmental changes such as increased root hair length and density, lateral root density and repression of primary root elongation (Figure 2.1). These adaptive changes in the RSA by plants allow the roots to

FIGURE 2.1 Root hair morphology in low-phosphate conditions. *Arabidopsis* seedlings grown on ½ MS medium for three days. Three-day post-germinated seedlings were transferred to low Pi (1 μM) or sufficient Pi (625 μM) for four days. Representative root images of seven-day-old seedling.

increase the root-to-soil area to increase the Pi uptake. In the past few years, several genes and transcription factors have been proposed as playing a role in mediating the root developmental changes in several plant species. Some of them are summarized in Table 2.2.

2.4.1 Regulation of Root Hairs

Root hairs are tubular structures and originate from asymmetric cell division from the epidermal precursor cell. Root hair increases the root-to-soil surface area, thus increasing the nutrient uptake in general and Pi uptake in particular. Root hair patterning and differentiation are controlled by several developmental and environmental cues (Kapulnik et al., 2011; Masucci and Schiefelbein., 1994; Salazar-Henao et al., 2016; Kirchner et al., 2018). Pi-deficiency promotes root hair elongation and density in *Arabidopsis*; thus longer root hair increases the root-to-soil surface area and

TABLE 2.2
Major Genes/Transcription Factors Involved in Root System Architecture Change (RSA) under Low-Pi Conditions in *Arabidopsis*

Gene/Transcription Factor Name	Putative Function	Reference
ZAT6	C2H2 zinc-finger transcription factor; involved in primary root growth and Pi uptake	Devaiah et al. (2007a)
PHL1/PHR1	MYB family transcription factor; root hair elongation	Bustos et al. (2010)
BHLH32	Transcription factor; involved in root hair formation under low Pi	Chen et al. (2007)
MYB62	Transcription factor; lateral root length	Devaiah et al. (2009)
WRKY75	WRKY domain; involved in lateral root development and root hair	Devaiah et al. (2007b)
MAX2-1/MAX4-1	Strigolactones signaling (*max2-1*) or biosynthesis (*max4-1*); root hair formation/development	Mayzlish-Gati et al. (2012)
PHR1	MYB family transcription factor; affects root-to-shoot ratio	Bustos et al. (2010)
LPR1/LPR2	Multicopper oxidase; primary root growth	Svistoonoff et al. (2007); Ticconi et al. (2009); Wang et al. (2010)
PDR2	A *P5-type ATPase*; involved in primary root growth and lateral root development	Ticconi et al. (2009)
SIZ1	*SUMO E3 ligase*; lateral root development	Miura et al. (2011)
BES1/BZR1	Brassinosteroid (BR) transcription factors; primary root growth	Singh et al. (2014); Singh et al. (2018); Kim et al. (2019)
PUB40	U-box protein; primary root growth	Kim et al. (2019)
IPK1	Inositol polyphosphate kinase; root hair development	Stevenson-Paulik et al. (2005)
RSL2/RSL4	Transcription factor; root hair development	Yi et al. (2010)
HSP2	Raf like kinase; involved in root hair development	Lei et al. (2011)
CLV2/PEPR2	Receptors; regulates root apical meristem in a CLE14-dependent manner under Pi deficiency	Gutierrez-Alanis et al. (2017)
Altered Phosphate Starvation Response1 (*APSR1*)	Involved in root meristem maintenance	Gonzalez-Mendoza et al. (2013)
TIR1	Component of the auxin perception complex *SCFTIR1/AuxIAA*; increases lateral root emergence	Pérez-Torres et al. (2008)

Pi acquisition (Figure 2.1). The increase in root hair density under low-Pi conditions is correlated with the reduced epidermal cell length, thereby increasing the overall root hair density (Jungk et al., 2001; Sanchez-Calderon et al., 2006). The determination of hair and non-hair epidermal cell fate depends on the key genes, including *TRANSPARENT TESTA GLABRA1 TTG1*, *WEREWOLF* (*WER*), *GLABRA 2* (*GL2*) and *GLABRA3* (*GL3*) (Grebe, 2012; Janes et al., 2018). In the past few years, several genes and transcription factors regulating root hair development and elongation during Pi deficiency have been discovered in *Arabidopsis*, rice and maize (Wang et al., 2013a; Wang et al., 2011; Zhou et al., 2008; Chen et al., 2007). In *Arabidopsis*, the basic helix–loop–helix transcription factor, *AtBHLH32*, negatively regulates root hair formation by interacting with TTG1 and GL3 during Pi deficiency (Chen et al., 2007). The basic helix–loop–helix transcription factor family is well-characterized for its role in root hair formation and elongation. It has been shown that *ROOT HAIR DEFECTIVE 6/ROOT HAIR DEFECTIVE 6 LIKE 1* (*RHD6/RSL1*) are required for root hair initiation while RSL2/RSL4 regulates root hair elongation. The overexpression of RSL4 promotes longer root hairs as compared to control plants (Menand et al., 2007; Yi et al., 2010; Bhosale et al., 2018). In rice, *OsPT1* (*OsPht1;1*) shows increased production of root hairs with a two-fold increase in shoot Pi content, suggesting the important function of root hairs in Pi uptake during limited Pi availability (Sun et al., 2012). The increased expression of maize and wheat orthologs of *PHR1*, *ZmPHR1* and *TaPHR1* increases root hair growth and Pi uptake in transgenic plants (Wang et al., 2013a; Wang et al., 2013b).

Growth hormones are the major regulator of RSA changes in response to developmental and environmental cues (Lopez-Bucio et al., 2002). Increased auxin sensitivity enhances root hair elongation and density. Reduced auxin synthesis and transport alter the root hair response to auxin. Recently it has been shown that gene involved in auxin transport, *AUX1*, regulates root hair elongation under low-Pi conditions in *Arabidopsis* (Bhosale et al., 2018). Genetic analysis showed that low Pi increases auxin sensitivity, thus promoting root hair elongation by modulating RSL2/RSL4 (Perez-Torres et al., 2008; Bhosale et al., 2018). Other phytohormones such as ethylene and strigolactones (SLs) influence root hair formation during Pi deficiency (Song et al., 2016; Kumar et al., 2015). Hypersensitive to phosphate starvation response mutant, *hps5* displays a constitutive ethylene response, and it has a mutation in the ethylene receptor gene, *ERS1* (Song et al., 2016). The *hps5* mutant shows increased root hair density with the longer root hairs independent of Pi levels. However, low Pi further increases the root hair density in *hps5* lines. SLs are known to regulate root hair and lateral root development (Kapulnik et al., 2011; Peret et al., 2011). SLs signaling and biosynthesis mutants *max2-1* and *max4-1* show reduced root hair density under Pi deficiency, suggesting the involvement of SLs in root hair formation and elongation under low-Pi conditions (Mayzlish-Gati et al., 2012; Koltai, 2013).

2.4.2 Primary Root Growth

Nutrient availability changes the biomass of the plant. The root-to-shoot ratio of biomass change is one of the characteristic features of nutrient deprivation. However, the differential availability of nutrients in the rhizosphere has a different impact on RSA. In the case of low phosphate availability, alteration in the RSA is well-studied, and several genetic components have been identified and characterized. Low Pi inhibits the primary root elongation, cell division and cell expansion in *Arabidopsis* (Figure 2.2). A reduction in cell division is associated with the loss of quiescent center (QC) identity as shown by using QC46 marker (Sanchez-Calderon et al., 2005).

Genetic screening for long root phenotypes under low-Pi conditions led to the identification of several low-phosphate-insensitive (*lpi*) mutants (Sanchez-Calderon et al., 2006). These mutants have a long primary root and improved cell division activity under low-Pi conditions as compared to the wild type. In a different approach, QTL analysis led to the identification of *LOW PHOSPHATE ROOT 1* and *2* (*LPR1* and 2). *LPR1* and *LPR2* encode *multicopper oxidases* (*MCO*), and *lpr1/lpr2* double mutants promote primary root elongation during Pi deficiency (Svistoonoff et al., 2007; Ticconi et al., 2009). *LPR1*-dependent root growth arrest depends on the physical contact of the root tip with the

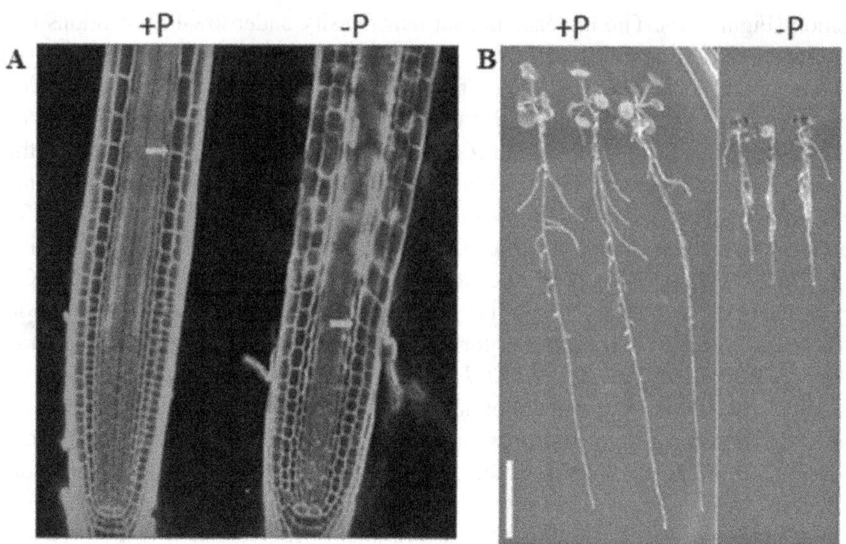

FIGURE 2.2 RSA change under low-Pi conditions. (A) Root meristem size of *Arabidopsis* roots treated with sufficient (1,250 μM) and low Pi (1 μM). Seedlings were grown in sufficient Pi conditions for 5 days and transferred to low Pi and sufficient Pi containing ½ MS medium for 48 hours. Representative images of propidium iodide (PI)-stained root meristem of seven-day-old plant. Arrows indicate the position of the meristem. (B) RSA change under the low-Pi conditions in 14-day-old plants. Three-day post-germinated seedlings were transferred to sufficient or deficient Pi conditions and allowed to grow for 11 days on an appropriate medium. Representative images of 14 days old seedlings. Scale: 1 cm.

low-Pi medium. Another gene, *PDR2*, functions with *LPR1* in maintaining root meristematic activity potentially by regulating SCARECROW (SCR) and SHORT-ROOT (SHR) protein levels. *PDR2* is a P5-type ATPase and the *pdr2* mutant is hypersensitive to low external Pi.

Recently it was found that Pi and iron levels affect the root growth and root apical meristem (RAM) activity. Pi-dependent apoplastic iron accumulation led to reduced RAM activity and cell elongation. The extent of low-Pi dependent root growth arrest depends on the iron levels in the medium. *LPR1* overexpression increases apoplastic iron accumulation in the root elongation zone, thus inhibiting cell elongation. It has also been shown that low Pi inhibits brassinosteroid (BR) biosynthesis, thus inhibiting the *BES1/BZR1* transcription factors of the brassinosteroid (BR) pathway. *BES1/BZR1* inhibits *LPR1* levels and represses its activity. Increased brassinosteroid signaling in a constitutive active BR signaling mutant, *bzr1*-1D, blocks the root response to external low phosphate by reducing the iron level in elongation zone cells, suggesting the role of BR and nutrient cross talk in reprogramming the root developmental plasticity under Pi and iron stress (Singh et al., 2014, 2018). Kim et al., 2019 identified a PUB40 protein that encodes an E3 ubiquitin ligase. PUB40 interacts with BZR1 and mediates its degradation. PUB triple mutants, *pub39pub40pub41*, show primary root elongation under low external Pi like *bzr1*-1D. Hence, PUB40-dependent BZR1 stability further suggests BR-dependent root growth under Pi deficiency. *BES1/BZR1* activity also regulates gibberellic acid (GA)-dependent cell elongation and growth (Tong et al., 2014; Unterholzner et al., 2015). The negative regulators of GA signaling, DELLA proteins, accumulated during Pi deficiency and DELLA-dependent inhibition of the GA pathway showed an increased root and shoot response to low external Pi. Thus, the intrinsic signals like BR and GA play an important role in modifying the RSA in general and primary root growth in particular under Pi deficiency.

2.4.3 Lateral Root Growth

Pi unavailability alters RSA across the plant species. Specialized tertiary lateral root structures also called cluster roots or proteoid roots are an important root trait and are produced in different plants in response to Pi depletion (Shane and Lambers, 2005; Lambers et al., 2006; Cheng et al., 2011). Low Pi promotes proteoid or cluster of lateral root in white lupin (*Lupinus albus*), thus promoting an adaptive response. The increased density of cluster roots increases the root volume and promotes Pi uptake (Lambers et al., 2006; Keerthisinghe et al., 1998; Neumann et al., 1999). In common bean (*Phaseolus vulgaris*), low Pi promotes shallow RSA by regulating the lateral root angle and perpendicular LR growth, thus increasing Pi foraging (Liao et al., 2004; Ho et al., 2005; Ramaekers et al., 2010; Lambers et al., 2011; Lynch, 2011; Funayama-Noguchi et al., 2015). In *Arabidopsis*, Pi deficiency inhibits primary root elongation and promotes LR elongation and density (Raghothama, 1999; Williamson et al., 2001; Lopez-Bucio et al., 2002; Franco-Zorrilla et al., 2007; Singh et al., 2014). Phytohormones are known to regulate several developmental aspects of root such as primary root elongation, lateral root development and root hair formation (Lopez-Bucio et al., 2002; Al Ghazi et al., 2003; Nacry et al., 2005; Jain et al., 2007; Singh and Savaldi-Goldstein, 2015). It is well-studied that auxin regulates LR development and formation. Low Pi increases auxin sensitivity and promotes LR formation (Al Ghazi et al., 2003; Mockaitis and Estelle, 2008). Exogenous auxin treatment promoted LR formation in the Pi-deprived roots as compared to the controlled conditions. Auxin *Transport Inhibitor Response 1* (*TIR1*), a component of the auxin perception complex (*SCFTIR1/AuxIAA*), and *AUXIN RESPONSE FACTOR 19* (*ARF19*)-dependent pathways have been shown to regulate the LR formation in Pi-deficient roots. Low Pi increases the expression of *TIR1*, thereby increasing the auxin sensitivity and correlating with the LR emergence in *Arabidopsis* (Perez-Torres et al., 2008). *Arabidopsis* SUMO E3 ligase, *AtSIZ1*, negatively regulates LR development in low-Pi conditions, and the mutant lacking functional *AtSIZ1* shows increased LR initiation and formation. In the *siz1* mutant, several auxin-induced genes were upregulated suggesting a molecular link of *AtSIZ1* with the auxin (Miura et al., 2005; Geiss-Friedlander and Melchior, 2007). Auxin-dependent RSA change has been reported in several plants during Pi limitation. In white lupin, reduced auxin sensitivity decreases the formation of cluster roots under limited Pi conditions (Gilbert et al., 2000; Vance et al., 2003; Yamagishi et al., 2011). In rice, the *arf16* mutant accumulated less auxin in LRs, suggesting auxin-dependent LR initiation and formation under low-Pi conditions (Shen et al., 2013). Signaling pathways such as ethylene, cytokinin and abscisic acid (ABA) are very complex and may function together to regulate the RSA under low-Pi conditions (Lopez-Bucio et al., 2002; Aloni et al., 2006; Beaudoin et al., 2000). Plant hormone ethylene and auxin interact to regulate its target genes in an auxin-dependent manner. It also regulates several of its targets independent of auxin (Stepanova et al., 2007). Under low-Pi conditions, the exogenous application of ethylene precursor 1-aminocyclopropane-1-carboxylic acid (ACC) in *Arabidopsis* inhibited primary root elongation and LR formation but overall LR density remained the same. Similarly, enhanced ethylene response mutants such as *ETHYLENE OVERPRODUCER 1* (*eto1*) and *CONSTITUTIVE TRIPLE RESPONSE 1* (*ctr1*) show reduced LR induction, suggesting that ethylene has a negative impact on LR initiation and formation under low-Pi conditions (Lopez-Bucio et al., 2002). Cytokinins regulate several developmental aspects of root growth such as root apical meristem (RAM) activity and LR development (Lai et al., 2007; Chapman and Estelle, 2009). The exogenous application of cytokinin reduces the expression of key phosphate-starvation-induced (PSI) marker genes such as *AtIPS1* (*TPSI1/Mt4* family) and *AtACP5*. Further, low Pi inhibits both cytokinin biosynthesis and signaling in *Arabidopsis* (Martin et al., 2000; Franco-Zorrilla et al., 2002). Lopez-Bucio et al., 2002 reported that cytokinins suppress LR formation during Pi deficiency in *Arabidopsis*, suggesting its negative effect on LR development.

2.5 CONCLUSIONS

The importance of Pi to the improvement of growth and productivity in plants has long been recognized worldwide. The exogenous use of phosphorus-based fertilizer is the most favorable method to improve the yield in several plant species. Due to the poor availability and high reactivity of Pi in natural soils, plants reprogram their RSA for optimal Pi acquisition. The adaptive root trait is the key factor that determines the Pi uptake and transport from the top soils, however, roots can explore only 20% of the top soils. A shallow RSA has been proposed for exploring the available Pi from the top soil layers. Increased root hairs, lateral roots and change in root angle in several plants such as *Arabidopsis*, white lupin, rice, wheat and maize have been shown to be important adaptive strategies to increase the root-to-soil surface area, thereby increasing Pi acquisition. Most of the studies on RSA modification have been performed under controlled laboratory conditions, however, phenotyping in a natural ecosystem is a major challenge and is required for developing plants with improved Pi use efficiency. Plant roots secrete several metabolites and enzymes in the rhizosphere during Pi stress. The secretion of organic acids such as citrate, malate and oxalate enhances Pi solubility in highly Pi-fixing soils, thus increasing the Pi availability. However, further research is required to understand the mechanistic way of organic acids exudation and the effect of soil composition such as pH, microbiome and differential availability of other soil minerals in their exudation. Intrinsic factors like hormones alter the RSA during Pi deficiency. The differential activity of hormones in roots promote an adaptive RSA from shallow to deep or vice versa depending on the availability of nutrients in the rhizosphere. Under low-Pi conditions, increased auxin sensitivity promotes the shallow over the deep RSA by increasing the LR and root hair density, while brassinosteroids (BRs) promote the deep RSA under similar conditions. The antagonistic effects of these hormones in regulating the root growth have been described in the past (Vert et al., 2008; Chaiwanon and Wang, 2015; Vragovic et al., 2015), however their interaction at the cellular level under limited Pi availability is yet to be explored for improving the root traits.

REFERENCES

Abel, S. 2011. Phosphate sensing in root development. *Current Opinion in Plant Biology* 14 (3):303–309.

Ai, P. H., S. B. Sun, J. N. Zhao, et al. 2009. Two rice phosphate transporters, OsPht1;2 and OsPht1;6, have different functions and kinetic properties in uptake and translocation. *Plant Journal* 57 (5):798–809.

Al-Ghazi, Y., B. Muller, S. Pinloche, et al. 2003. Temporal responses of *Arabidopsis* root architecture to phosphate starvation: Evidence for the involvement of auxin signalling. *Plant Cell and Environment* 26 (7):1053–1066.

Aloni, R., E. Aloni, M. Langhans, and C. I. Ullrich. 2006. Role of cytokinin and auxin in shaping root architecture: Regulating vascular differentiation, lateral root initiation, root apical dominance and root gravitropism. *Annals of Botany* 97 (5):883–893.

Arai, Y., and D. L. Sparks. 2007. Phosphate reaction dynamics in soils and soil components: A muiltiscale approach. *Advances in Agronomy* 94:135–179.

Arpat, A. B., P. Magliano, S. Wege, H. Rouached, A. Stefanovic, and Y. Poirier. 2012. Functional expression of PHO1 to the Golgi and trans-Golgi network and its role in export of inorganic phosphate. *Plant Journal* 71 (3):479–491.

Baek, S. H., I. M. Chung, and S. J. Yun. 2001. Molecular cloning and characterization of a tobacco leaf cDNA encoding a phosphate transporter. *Molecules and Cells* 11 (1):1–6.

Bari, R., B. D. Pant, M. Stitt, and W. R. Scheible. 2006. PHO2, microRNA399, and PHR1 define a phosphate-signaling pathway in plants. *Plant Physiology* 141 (3):988–999.

Bayuelo-Jimenez, J. S., M. Gallardo-Valdez, V. A. Perez-Decelis, L. Magdaleno-Armas, I. Ochoa, and J. P. Lynch. 2011. Genotypic variation for root traits of maize (*Zea mays* L.) from the Purhepecha Plateau under contrasting phosphorus availability. *Field Crops Research* 121 (3):350–362.

Beaudoin, N., C. Serizet, F. Gosti, and J. Giraudat. 2000. Interactions between abscisic acid and ethylene signaling cascades. *Plant Cell* 12 (7):1103–1115.

Bhosale, R., J. Giri, B. K. Pandey, et al. 2018. A mechanistic framework for auxin dependent *Arabidopsis* root hair elongation to low external phosphate. *Nature Communications* 9 (1):1409.

Bozzo, G. G., E. L. Dunn, and W. C. Plaxton. 2006. Differential synthesis of phosphate-starvation inducible purple acid phosphatase isozymes in tomato (*Lycopersicon esculentum*) suspension cells and seedlings. *Plant Cell and Environment* 29 (2):303–313.

Brinch-Pedersen, H., L. D. Sorensen, and P. B. Holm. 2002. Engineering crop plants: Getting a handle on phosphate. *Trends in Plant Science* 7 (3):118–125.

Brundrett, M. C. 2009. Mycorrhizal associations and other means of nutrition of vascular plants: Understanding the global diversity of host plants by resolving conflicting information and developing reliable means of diagnosis. *Plant and Soil* 320 (1–2):37–77.

Bun-Ya, M., M. Nishimura, S. Harashima, and Y. Oshima. 1991. The PHO84 gene of Saccharomyces cerevisiae encodes an inorganic phosphate transporter. *Molecular and Cellular Biology* 11 (6):3229–3238.

Bustos, R., G. Castrillo, F. Linhares, et al. 2010. A central regulatory system largely controls transcriptional activation and repression responses to phosphate starvation in *Arabidopsis*. *PLoS Genetics* 6 (9):e1001102.

Castrillo, G., P. J. Teixeira, S. H. Paredes, et al. 2017. Root microbiota drive direct integration of phosphate stress and immunity. *Nature* 543 (7646):513–518.

Chaiwanon, J., and Z. Y. Wang. 2015. Spatiotemporal brassinosteroid signaling and antagonism with auxin pattern stem cell dynamics in *Arabidopsis* roots. *Current Biology* 25 (8):1031–1042.

Chang, M. X., M. Gu, Y. W. Xia, et al. 2019. OsPHT1;3 mediates uptake, translocation, and remobilization of phosphate under extremely low phosphate regimes. *Plant Physiology* 179 (2):656–670.

Chapman, E. J., and M. Estelle. 2009. Cytokinin and auxin intersection in root meristems. *Genome Biology* 10:210.

Chen, Z. H., G. A. Nimmo, G. I. Jenkins, and H. G. Nimmo. 2007. BHLH32 modulates several biochemical and morphological processes that respond to Pi starvation in *Arabidopsis*. *Biochemical Journal* 405:191–198.

Cheng, L. Y., B. Bucciarelli, J. Q. Liu, et al. 2011. White lupin cluster root acclimation to phosphorus deficiency and root hair development involve unique glycerophosphodiester phosphodiesterases. *Plant Physiology* 156 (3):1131–1148.

Chiou, T. J., K. Aung, S. I. Lin, C. C. Wu, S. F. Chiang, and C. L. Su. 2006. Regulation of phosphate homeostasis by microRNA in *Arabidopsis*. *Plant Cell* 18 (2):412–421.

Chiou, T. J., and S. I. Lin. 2011. Signaling network in sensing phosphate availability in plants. *Annual Review of Plant Biology* 62:185–206.

Cordell, D., J. O. Drangert, and S. White. 2009. The story of phosphorus: Global food security and food for thought. *Global Environmental Change-Human and Policy Dimensions* 19 (2):292–305.

Daram, P., S. Brunner, B. L. Persson, N. Amrhein, and M. Bucher. 1998. Functional analysis and cell-specific expression of a phosphate transporter from tomato. *Planta* 206 (2):225–33.

del Pozo, J. C., I. Allona, V. Rubio, et al. 1999. A type 5 acid phosphatase gene from *Arabidopsis thaliana* is induced by phosphate starvation and by some other types of phosphate mobilising/oxidative stress conditions. *Plant Journal* 19 (5):579–589.

Del Vecchio, H. A., S. Ying, J. Park, et al. 2014. The cell wall-targeted purple acid phosphatase AtPAP25 is critical for acclimation of *Arabidopsis thaliana* to nutritional phosphorus deprivation. *Plant Journal* 80 (4):569–581.

Delhaize, E., and P. J. Randall. 1995. Characterization of a phosphate-accumulator mutant of *Arabidopsis thaliana*. *Plant Physiology* 107 (1):207–213.

Desnos, T. 2008. Root branching responses to phosphate and nitrate. *Current Opinion in Plant Biology* 11 (1):82–87.

Devaiah, B. N., A. S. Karthikeyan, and K. G. Raghothama. 2007a. WRKY75 transcription factor is a modulator of phosphate acquisition and root development in *Arabidopsis*. *Plant Physiology* 143 (4):1789–1801.

Devaiah, B. N., R. Madhuvanthi, A. S. Karthikeyan, and K. G. Raghothama. 2009. Phosphate starvation responses and gibberellic acid biosynthesis are regulated by the MYB62 transcription factor in *Arabidopsis*. *Molecular Plant* 2 (1):43–58.

Devaiah, B. N., V. K. Nagarajan, and K. G. Raghothama. 2007b. Phosphate homeostasis and root development in *Arabidopsis* are synchronized by the zinc finger transcription factor ZAT6. *Plant Physiology* 145 (1):147–159.

Doerner, P. 2008. Phosphate starvation signaling: A threesome controls systemic P(i) homeostasis. *Current Opinion in Plant Biology* 11 (5):536–40.

Dong, B., Z. Rengel, and E. Delhaize. 1998. Uptake and translocation of phosphate by pho2 mutant and wild-type seedlings of *Arabidopsis thaliana*. *Planta* 205 (2):251–6.

Dunlop J., Phung H. T., Meeking R., and White D. W. R. 1997. The kinetics associated with phosphate absorption by *Arabidopsis* and its regulation by phosphorus status. *Australian Journal of Plant Physiology* 24:623–629.

Franco-Zorrilla, J. M., A. C. Martin, R. Solano, V. Rubio, A. Leyva, and J. Paz-Ares. 2002. Mutations at CRE1 impair cytokinin-induced repression of phosphate starvation responses in *Arabidopsis*. *Plant Journal* 32 (3):353–360.

Franco-Zorrilla, J. M., A. Valli, M. Todesco, et al. 2007. Target mimicry provides a new mechanism for regulation of microRNA activity. *Nature Genetics* 39 (8):1033–1037.

Funayama-Noguchi, S., K. Noguchi, and I. Terashima. 2015. Comparison of the response to phosphorus deficiency in two lupin species, *Lupinus albus* and *L. angustifolius*, with contrasting root morphology. *Plant Cell and Environment* 38 (3):399–410.

Geiss-Friedlander, R., and F. Melchior. 2007. Concepts in sumoylation: A decade on. *Nature Reviews Molecular Cell Biology* 8 (12):947–956.

Gilbert, G. A., J. D. Knight, C. P. Vance, and D. L. Allan. 2000. Proteoid root development of phosphorus deficient lupin is mimicked by auxin and phosphonate. *Annals of Botany* 85 (6):921–928.

Giri, J., R. Bhosale, G. Q. Huang, et al. 2018. Rice auxin influx carrier OsAUX1 facilitates root hair elongation in response to low external phosphate. *Nature Communications* 9:1408.

Gonzalez-Mendoza, V., A. Zurita-Silva, L. Sanchez-Calderon, et al. 2013. APSR1, a novel gene required for meristem maintenance, is negatively regulated by low phosphate availability. *Plant Science* 205:2–12.

Grebe, M. 2012. The patterning of epidermal hairs in *Arabidopsis*: updated. *Current Opinion in Plant Biology* 15 (1):31–37.

Guo, B., Y. Jin, C. Wussler, E. B. Blancaflor, C. M. Motes, and W. K. Versaw. 2008. Functional analysis of the *Arabidopsis* PHT4 family of intracellular phosphate transporters. *New Phytologist* 177 (4):889–898.

Gutierrez-Alanis, D., L. Yong-Villalobos, P. Jimenez-Sandoval, et al. 2017. Phosphate starvation-dependent iron mobilization induces CLE14 expression to trigger root meristem differentiation through CLV2/PEPR2 signaling. *Developmental Cell* 41 (5):555–570.e3.

Hamburger, D., E. Rezzonico, J. M. C. Petetot, C. Somerville, and Y. Poirier. 2002. Identification and characterization of the *Arabidopsis* PHO1 gene involved in phosphate loading to the xylem. *Plant Cell* 14 (4):889–902.

Harrison, M. J., G. R. Dewbre, and J. Liu. 2002. A phosphate transporter from *Medicago truncatula* involved in the acquisition of phosphate released by arbuscular mycorrhizal fungi. *Plant Cell* 14 (10):2413–2429.

Hell, R., and H. Hillebrand. 2001. Plant concepts for mineral acquisition and allocation. *Current Opinion in Biotechnology* 12 (2):161–168.

Hinsinger, P. 2001. Bioavailability of soil inorganic P in the rhizosphere as affected by root-induced chemical changes: A review. *Plant and Soil* 237 (2):173–195.

Ho, M. D., J. C. Rosas, K. M. Brown, and J. P. Lynch. 2005. Root architectural tradeoffs for water and phosphorus acquisition. *Functional Plant Biology* 32 (8):737–748.

Hu, B., C. G. Zhu, F. Li, et al. 2011. LEAF TIP NECROSIS1 plays a pivotal role in the regulation of multiple phosphate starvation responses in rice. *Plant Physiology* 156 (3):1101–1115.

Huang, G. Q., W. Q. Liang, C. J. Sturrock, et al. 2018. Rice actin binding protein RMD controls crown root angle in response to external phosphate. *Nature Communications* 9:2346.

Jain, A., M. D. Poling, A. S. Karthikeyan, et al. 2007. Differential effects of sucrose and auxin on localized phosphate deficiency-induced modulation of different traits of root system architecture in *Arabidopsis*. *Plant Physiology* 144 (1):232–247.

Janes, George, Daniel von Wangenheim, Sophie Cowling, et al. 2018. Cellular patterning of *Arabidopsis* roots under low phosphate conditions. *Frontiers in Plant Science* 9:735.

Jia, H. F., H. Y. Ren, M. Gu, et al. 2011. The phosphate transporter gene OsPht1;8 is involved in phosphate homeostasis in rice. *Plant Physiology* 156 (3):1164–1175.

Jiang, C., X. Gao, L. Liao, N. P. Harberd, and X. Fu. 2007. Phosphate starvation root architecture and anthocyanin accumulation responses are modulated by the gibberellin-DELLA signaling pathway in *Arabidopsis*. *Plant Physiology* 145 (4):1460–1470.

Jungk, A. 2001. Root hairs and the acquisition of plant nutrients from soil. *Journal of Plant Nutrition and Soil Science* 164 (2):121–129.

Kapulnik, Y., P. M. Delaux, N. Resnick, et al. 2011. Strigolactones affect lateral root formation and root-hair elongation in *Arabidopsis*. *Planta* 233 (1):209–216.

Keerthisinghe, G., P. J. Hocking, P. R. Ryan, and E. Delhaize. 1998. Effect of phosphorus supply on the formation and function of proteoid roots of white lupin (*Lupinus albus* L.). *Plant Cell and Environment* 21 (5):467–478.

Kim, E. J., S. H. Lee, C. H. Park, et al. 2019. Plant U-box 40 mediates degradation of the brassinosteroid-responsive transcription factor BZR1 in *Arabidopsis* roots. *Plant Cell* 31:791–808.

Kirchner, T. W., M. Niehaus, K. L. Rossig, et al. 2018. Molecular background of Pi deficiency-induced root hair growth in *Brassica carinata*: A fasciclin-like arabinogalactan protein is involved. *Frontiers in Plant Science* 9:1372.

Koltai, H. 2013. Strigolactones activate different hormonal pathways for regulation of root development in response to phosphate growth conditions. *Annals of Botany* 112 (2):409–415.

Kuang, R. B., K. H. Chan, E. Yeung, and B. L. Lim. 2009. Molecular and biochemical characterization of AtPAP15, a purple acid phosphatase with phytase activity, in *Arabidopsis*. *Plant Physiology* 151 (1):199–209.

Kumar, M., N. Pandya-Kumar, A. Dam, et al. 2015. *Arabidopsis* response to low-phosphate conditions includes active changes in actin filaments and PIN2 polarization and is dependent on strigolactone signalling. *Journal of Experimental Botany* 66 (5):1499–1510.

Lai, F., J. Thacker, Y. Y. Li, and P. Doerner. 2007. Cell division activity determines the magnitude of phosphate starvation responses in *Arabidopsis*. *Plant Journal* 50 (3):545–556.

Lambers, H., J. C. Clements, and M. N. Nelson. 2013. How a phosphorus-acquisition strategy based on carboxylate exudation powers the success and agronomic potential of lupines (Lupinus, Fabaceae). *American Journal of Botany* 100 (2):263–288.

Lambers, H., P. M. Finnegan, E. Laliberte, et al. 2011. Phosphorus nutrition of Proteaceae in severely phosphorus-impoverished soils: Are there lessons to be learned for future crops? *Plant Physiology* 156 (3):1058–1066.

Lambers, H., M. W. Shane, M. D. Cramer, S. J. Pearse, and E. J. Veneklaas. 2006. Root structure and functioning for efficient acquisition of phosphorus: Matching morphological and physiological traits. *Annals of Botany* 98 (4):693–713.

Lapis-Gaza, Hazel R., Ricarda Jost, and Patrick M. Finnegan. 2014. *Arabidopsis* PHOSPHATE TRANSPORTER1 genes PHT1;8 and PHT1;9 are involved in root-to-shoot translocation of orthophosphate. *BMC Plant Biology* 14 (1):334.

Lei, M. G., C. M. Zhu, Y. D. Liu, et al. 2011. Ethylene signalling is involved in regulation of phosphate starvation-induced gene expression and production of acid phosphatases and anthocyanin in *Arabidopsis*. *New Phytologist* 189 (4):1084–1095.

Li, Y. S., Y. Gao, Q. Y. Tian, F. L. Shi, L. H. Li, and W. H. Zhang. 2011. Stimulation of root acid phosphatase by phosphorus deficiency is regulated by ethylene in *Medicago falcata*. *Environmental and Experimental Botany* 71 (1):114–120.

Liang, C. Y., J. Tian, H. M. Lam, B. L. Lim, X. L. Yan, and H. Liao. 2010. Biochemical and molecular characterization of PvPAP3, a novel purple acid phosphatase isolated from common bean enhancing extracellular ATP utilization. *Plant Physiology* 152 (2):854–865.

Liang, Y., D. M. Mitchell, and J. M. Harris. 2007. Abscisic acid rescues the root meristem defects of the *Medicago truncatula* latd mutant. *Developmental Biology* 304 (1):297–307.

Liao, H., X. L. Yan, G. Rubio, S. E. Beebe, M. W. Blair, and J. P. Lynch. 2004. Genetic mapping of basal root gravitropism and phosphorus acquisition efficiency in common bean. *Functional Plant Biology* 31 (10):959–970.

Liu, J. Q., D. L. Allan, and C. P. Vance. 2010. Systemic signaling and local sensing of phosphate in common bean: Cross-talk between photosynthate and microRNA399. *Molecular Plant* 3 (2):428–437.

Lopez-Arredondo, D. L., M. A. Leyva-Gonzalez, S. I. Gonzalez-Morales, J. Lopez-Bucio, and L. Herrera-Estrella. 2014. Phosphate nutrition: Improving low-phosphate tolerance in crops. *Annual Review of Plant Biology* 65:95–123.

Lopez-Bucio, J., A. Cruz-Ramirez, and L. Herrera-Estrella. 2003. The role of nutrient availability in regulating root architecture. *Current Opinion in Plant Biology* 6 (3):280–287.

Lopez-Bucio, J., E. Hernandez-Abreu, L. Sanchez-Calderon, M. F. Nieto-Jacobo, J. Simpson, and L. Herrera-Estrella. 2002. Phosphate availability alters architecture and causes changes in hormone sensitivity in the *Arabidopsis* root system. *Plant Physiology* 129 (1):244–256.

Lopez-Bucio, J., E. Hernandez-Abreu, L. Sanchez-Calderon, et al. 2005. An auxin transport independent pathway is involved in phosphate stress-induced root architectural alterations in *Arabidopsis*. Identification of BIG as a mediator of auxin in pericycle cell activation. *Plant Physiology* 137 (2):681–691.

Lu, L. H., W. M. Qiu, W. W. Gao, S. D. Tyerman, H. X. Shou, and C. Wang. 2016. OsPAP10c, a novel secreted acid phosphatase in rice, plays an important role in the utilization of external organic phosphorus. *Plant Cell and Environment* 39 (10):2247–2259.

Lung, S. C., A. Leung, R. Kuang, Y. Wang, P. Leung, and B. L. Lim. 2008. Phytase activity in tobacco (*Nicotiana tabacum*) root exudates is exhibited by a purple acid phosphatase. *Phytochemistry* 69 (2):365–373.

Lynch, J. 1995. Root architecture and plant productivity. *Plant Physiology* 109 (1):7–13.

Lynch, J. P. 2011. Root phenes for enhanced soil exploration and phosphorus acquisition: Tools for future crops. *Plant Physiology* 156 (3):1041–1049.

Martin, A. C., J. C. del Pozo, J. Iglesias, et al. 2000. Influence of cytokinins on the expression of phosphate starvation responsive genes in *Arabidopsis*. *Plant Journal* 24 (5):559–567.

Masucci, J. D., and J. W. Schiefelbein. 1994. The rhd6 mutation of *Arabidopsis thaliana* alters root-hair initiation through an auxin- and ethylene-associated process. *Plant Physiology* 106 (4):1335–1346.

Mayzlish-Gati, E., C. De-Cuyper, S. Goormachtig, et al. 2012. Strigolactones are involved in root response to low phosphate conditions in *Arabidopsis*. *Plant Physiology* 160 (3):1329–1341.

Mehra, P., B. K. Pandey, and J. Giri. 2017. Improvement in phosphate acquisition and utilization by a secretory purple acid phosphatase (OsPAP21b) in rice. *Plant Biotechnology Journal* 15 (8):1054–1067.

Menand, B., K. K. Yi, S. Jouannic, et al. 2007. An ancient mechanism controls the development of cells with a rooting function in land plants. *Science* 316 (5830):1477–1480.

Miller, S. S., J. Q. Liu, D. L. Allan, C. J. Menzhuber, M. Fedorova, and C. P. Vance. 2001. Molecular control of acid phosphatase secretion into the rhizosphere of proteoid roots from phosphorus-stressed white lupin. *Plant Physiology* 127 (2):594–606.

Ming, F., G. H. Mi, Q. Lu, et al. 2005. Cloning and characterization of cDNA for the *Oryza sativa* phosphate transporter. *Cellular and Molecular Biology Letters* 10 (3):401–411.

Mitsukawa, N., S. Okumura, Y. Shirano, et al. 1997. Overexpression of an *Arabidopsis thaliana* high-affinity phosphate transporter gene in tobacco cultured cells enhances cell growth under phosphate-limited conditions. *Proceedings of the National Academy of Sciences of the United States of America* 94 (13):7098–7102.

Miura, K., J. Lee, Q. Q. Gong, et al. 2011. SIZ1 regulation of phosphate starvation-induced root architecture remodeling involves the control of auxin accumulation. *Plant Physiology* 155 (2):1000–1012.

Miura, K., A. Rus, A. Sharkhuu, et al. 2005. The *Arabidopsis* SUMO E3 ligase SIZ1 controls phosphate deficiency responses (vol 102, pg 7760, 2005). *Proceedings of the National Academy of Sciences of the United States of America* 102 (27):9734–9734.

Mockaitis, K., and M. Estelle. 2008. Auxin receptors and plant development: A new signaling paradigm. *Annual Review of Cell and Developmental Biology* 24:55–80.

Mora-Macias, J., J. O. Ojeda-Rivera, D. Gutierrez-Alanis, et al. 2017. Malate-dependent Fe accumulation is a critical checkpoint in the root developmental response to low phosphate. *Proceedings of the National Academy of Sciences of the United States of America* 114 (17):E3563–E3572.

Muchhal, U. S., and K. G. Raghothama. 1999. Transcriptional regulation of plant phosphate transporters. *Proceedings of the National Academy of Sciences of the United States of America* 96 (10):5868–5872.

Nacry, P., G. Canivenc, B. Muller, et al. 2005. A role for auxin redistribution in the responses of the root system architecture to phosphate starvation in *Arabidopsis*. *Plant Physiology* 138 (4):2061–2074.

Nagarajan, V. K., A. Jain, M. D. Poling, A. J. Lewis, K. G. Raghothama, and A. P. Smith. 2011. *Arabidopsis* Pht1;5 mobilizes phosphate between source and sink organs and influences the interaction between phosphate homeostasis and ethylene signaling. *Plant Physiology* 156 (3):1149–1163.

Nagy, R., V. Karandashov, V. Chague, et al. 2005. The characterization of novel mycorrhiza-specific phosphate transporters from *Lycopersicon esculentum* and *Solanum tuberosum* uncovers functional redundancy in symbiotic phosphate transport in solanaceous species. *Plant Journal* 42 (2):236–250.

Nassal, D., M. Spohn, N. Eltlbany, et al. 2018. Effects of phosphorus-mobilizing bacteria on tomato growth and soil microbial activity. *Plant and Soil* 427 (1–2):17–37.

Neumann, G., A. Massonneau, E. Martinoia, and V. Romheld. 1999. Physiological adaptations to phosphorus deficiency during proteoid root development in white lupin. *Planta* 208 (3):373–382.

Oelkers, E. H., and E. Valsami-Jones. 2008. Phosphate mineral reactivity and global sustainability. *Elements* 4 (2):83–87.

Pant, B. D., A. Buhtz, J. Kehr, and W. R. Scheible. 2008. MicroRNA399 is a long-distance signal for the regulation of plant phosphate homeostasis. *Plant Journal* 53 (5):731–738.

Paszkowski, U., S. Kroken, C. Roux, and S. P. Briggs. 2002. Rice phosphate transporters include an evolutionarily divergent gene specifically activated in arbuscular mycorrhizal symbiosis. *Proceedings of the National Academy of Sciences of the United States of America* 99 (20):13324–13329.

Peret, B., M. Clement, L. Nussaume, and T. Desnos. 2011. Root developmental adaptation to phosphate starvation: Better safe than sorry. *Trends in Plant Science* 16 (8):442–450.

Perez-Torres, C. A., J. Lopez-Bucio, A. Cruz-Ramirez, et al. 2008. Phosphate availability alters lateral root development in *Arabidopsis* by modulating auxin sensitivity via a mechanism involving the TIR1 auxin receptor. *Plant Cell* 20 (12):3258–3272.

Poirier, Y., S. Thoma, C. Somerville, and J. Schiefelbein. 1991. Mutant of *Arabidopsis* deficient in xylem loading of phosphate. *Plant Physiology* 97 (3):1087–93.

Preuss, C. P., C. Y. Huang, M. Gilliham, and S. D. Tyerman. 2010. Channel-like characteristics of the low-affinity barley phosphate transporter PHT1;6 when expressed in Xenopus oocytes. *Plant Physiology* 152 (3):1431–1441.

Qin, L., J. Zhao, J. Tian, et al. 2012. The high-affinity phosphate transporter GmPT5 regulates phosphate transport to nodules and nodulation in soybean. *Plant Physiology* 159 (4):1634–1643.

Rae, A. L., D. H. Cybinski, J. M. Jarmey, and F. W. Smith. 2003. Characterization of two phosphate transporters from barley; evidence for diverse function and kinetic properties among members of the Pht1 family. *Plant Molecular Biology* 53 (1):27–36.

Raghothama, K. G. 1999. Phosphate acquisition. *Annual Review of Plant Physiology and Plant Molecular Biology* 50:665–693.

Raghothama, K. G., and A. S. Karthikeyan. 2005. Phosphate acquisition. *Plant and Soil* 274 (1–2):37–49.

Ramaekers, L., R. Remans, I. M. Rao, M. W. Blair, and J. Vanderleyden. 2010. Strategies for improving phosphorus acquisition efficiency of crop plants. *Field Crops Research* 117 (2–3):169–176.

Rausch, C., P. Daram, S. Brunner, et al. 2001. A phosphate transporter expressed in arbuscule-containing cells in potato. *Nature* 414 (6862):462–70.

Remy, E., T. R. Cabrito, R. A. Batista, M. C. Teixeira, I. Sa-Correia, and P. Duque. 2012. The Pht1;9 and Pht1;8 transporters mediate inorganic phosphate acquisition by the *Arabidopsis thaliana* root during phosphorus starvation. *New Phytologist* 195 (2):356–371.

Rouached, H., A. Stefanovic, D. Secco, et al. 2011. Uncoupling phosphate deficiency from its major effects on growth and transcriptome via PHO1 expression in *Arabidopsis*. *Plant Journal* 65 (4):557–570.

Ryan, P. R., E. Delhaize, and D. L. Jones. 2001. Function and mechanism of organic anion exudation from plant roots. *Annual Review of Plant Physiology and Plant Molecular Biology* 52:527–560.

Ryan, P. R., S. D. Tyerman, T. Sasaki, et al. 2011. The identification of aluminium-resistance genes provides opportunities for enhancing crop production on acid soils. *Journal of Experimental Botany* 62 (1):9–20.

Salazar-Henao, J. E., I. C. Velez-Bermudez, and W. Schmidt. 2016. The regulation and plasticity of root hair patterning and morphogenesis. *Development* 143 (11):1848–1858.

Sanchez-Calderon, L., J. Lopez-Bucio, A. Chacon-Lopez, et al. 2005. Phosphate starvation induces a determinate developmental program in the roots of *Arabidopsis thaliana*. *Plant and Cell Physiology* 46 (1):174–184.

Sanchez-Calderon, L., J. Lopez-Bucio, A. Chacon-Lopez, A. Gutierrez-Ortega, E. Hernandez-Abreu, and L. Herrera-Estrella. 2006. Characterization of low phosphorus insensitive mutants reveals a crosstalk between low phosphorus-induced determinate root development and the activation of genes involved in the adaptation of *Arabidopsis* to phosphorus deficiency. *Plant Physiology* 140 (3):879–889.

Schachtman, D. P., R. J. Reid, and S. M. Ayling. 1998. Phosphorus uptake by plants: From soil to cell. *Plant Physiology* 116 (2):447–453.

Secco, D., M. Jabnoune, H. Walker, et al. 2013. Spatio-temporal transcript profiling of rice roots and shoots in response to phosphate starvation and recovery. *Plant Cell* 25 (11):4285–4304.

Seo, H. M., Y. Jung, S. Song, et al. 2008. Increased expression of OsPT1, a high-affinity phosphate transporter, enhances phosphate acquisition in rice. *Biotechnology Letters* 30 (10):1833–1838.

Shane, M. W., and H. Lambers. 2005. Cluster roots: A curiosity in context. *Plant and Soil* 274 (1–2):101–125.

Shang, C., D. E. Caldwell, J. W. Stewart, H. Tiessen, and P. M. Huang. 1996. Bioavailability of organic and inorganic phosphates adsorbed on short-range ordered aluminum precipitate. *Microbial Ecology* 31 (1):29–39.

Shen, C. J., S. K. Wang, S. N. Zhang, et al. 2013. OsARF16, a transcription factor, is required for auxin and phosphate starvation response in rice (*Oryza sativa* L.). *Plant Cell and Environment* 36 (3):607–620.

Shen, J. B., L. X. Yuan, J. L. Zhang, et al. 2011. Phosphorus dynamics: From soil to plant. *Plant Physiology* 156 (3):997–1005.

Shin, H., H. S. Shin, G. R. Dewbre, and M. J. Harrison. 2004. Phosphate transport in *Arabidopsis*: Pht1;1 and Pht1;4 play a major role in phosphate acquisition from both low- and high-phosphate environments. *Plant Journal* 39 (4):629–642.

Singh, A. P., Y. Fridman, L. Friedlander-Shani, D. Tarkowska, M. Strnad, and S. Savaldi-Goldstein. 2014. Activity of the brassinosteroid transcription factors BRASSINAZOLE RESISTANT1 and BRASSINOSTEROID INSENSITIVE1-ETHYL METHANESULFONATE-SUPPRESSOR1/BRASSINAZOLE RESISTANT2 blocks developmental reprogramming in response to low phosphate availability. *Plant Physiology* 166 (2):678–688.

Singh, A. P., Y. Fridman, N. Holland, et al. 2018. Interdependent nutrient availability and steroid hormone signals facilitate root growth plasticity. *Developmental Cell* 46 (1):59–72.

Singh, A. P., and S. Savaldi-Goldstein. 2015. Growth control: Brassinosteroid activity gets context. *Journal of Experimental Botany* 66 (4):1123–1132.

Song, L., H. P. Yu, J. S. Dong, X. M. Che, Y. L. Jiao, and D. Liu. 2016. The molecular mechanism of ethylene-mediated root hair development induced by phosphate starvation. *Plos Genetics* 12 (2016):e1006194.

Spohn, M., A. Zavisic, P. Nassal, et al. 2018. Temporal variations of phosphorus uptake by soil microbial biomass and young beech trees in two forest soils with contrasting phosphorus stocks. *Soil Biology and Biochemistry* 117:191–202.

Stefanovic, A., A. B. Arpat, R. Bligny, et al. 2011. Over-expression of PHO1 in *Arabidopsis* leaves reveals its role in mediating phosphate efflux. *Plant Journal* 66 (4):689–699.

Stepanova, A. N., J. Yun, A. V. Likhacheva, and J. M. Alonso. 2007. Multilevel interactions between ethylene and auxin in *Arabidopsis* roots. *Plant Cell* 19 (7):2169–2185.

Stevenson-Paulik, J., R. J. Bastidas, S. T. Chiou, R. A. Frye, and J. D. York. 2005. Generation of phytate-free seeds in *Arabidopsis* through disruption of inositol polyphosphate kinases. *Proceedings of the National Academy of Sciences of the United States of America* 102 (35):12612–12617.

Sun, Huwei, Jinyuan Tao, Yang Bi, et al. 2018. OsPIN1b is involved in rice seminal root elongation by regulating root apical meristem activity in response to low nitrogen and phosphate. *Scientific Reports* 8 (1):13014.

Sun, S. B., M. A. Gu, Y. Cao, et al. 2012. A constitutive expressed phosphate transporter, OsPht1;1, modulates phosphate uptake and translocation in phosphate-replete rice. *Plant Physiology* 159 (4):1571–1581.

Sunkar, R., T. Girke, P. K. Jain, and J. K. Zhu. 2005. Cloning and characterization of MicroRNAs from rice. *Plant Cell* 17 (5):1397–1411.

Svistoonoff, S., A. Creff, M. Reymond, et al. 2007. Root tip contact with low-phosphate media reprograms plant root architecture. *Nature Genetics* 39 (6):792–796.

Thibaud, M. C., J. F. Arrighi, V. Bayle, et al. 2010. Dissection of local and systemic transcriptional responses to phosphate starvation in *Arabidopsis*. *Plant Journal* 64 (5):775–789.

Tian, J. L., C. Wang, Q. Zhang, X. W. He, J. Whelan, and H. X. Shou. 2012. Overexpression of OsPAP10a, a root-associated acid phosphatase, increased extracellular organic phosphorus utilization in rice. *Journal of Integrative Plant Biology* 54 (9):631–639.

Ticconi, C. A., R. D. Lucero, S. Sakhonwasee, et al. 2009. ER-resident proteins PDR2 and LPR1 mediate the developmental response of root meristems to phosphate availability. *Proceedings of the National Academy of Sciences of the United States of America* 106 (33):14174–14179.

Tittarelli, A., L. Milla, F. Vargas, et al. 2007. Isolation and comparative analysis of the wheat TaPT2 promoter: Identification in silico of new putative regulatory motifs conserved between monocots and dicots. *Journal of Experimental Botany* 58 (10):2573–2582.

Tong, H., Y. Xiao, D. Liu, et al. 2014. Brassinosteroid regulates cell elongation by modulating gibberellin metabolism in rice. *Plant Cell* 26 (11):4376–4393.

Tran, H. T., B. A. Hurley, and W. C. Plaxton. 2010. Feeding hungry plants: The role of purple acid phosphatases in phosphate nutrition. *Plant Science* 179 (1-2):14–27.

Unterholzner, S. J., W. Rozhon, M. Papacek, et al. 2015. Brassinosteroids are master regulators of gibberellin biosynthesis in *Arabidopsis*. *Plant Cell* 27 (8):2261–2272.

Vance, C. P., C. Uhde-Stone, and D. L. Allan. 2003. Phosphorus acquisition and use: Critical adaptations by plants for securing a nonrenewable resource. *New Phytologist* 157 (3):423–447.

Vert, G., C. L. Walcher, J. Chory, and J. L. Nemhauser. 2008. Integration of auxin and brassinosteroid pathways by auxin response factor 2. *Proceedings of the National Academy of Sciences of the United States of America* 105 (28):9829–34.

Vragovic, K., A. Sela, L. Friedlander-Shani, et al. 2015. Translatome analyses capture of opposing tissue-specific brassinosteroid signals orchestrating root meristem differentiation. *Proceedings of the National Academy of Sciences of the United States of America* 112 (3):923–928.

Wang D, S. Lv, P. Jiang, and Y. Li. 2017. Roles, regulation, and agricultural application of plant phosphate transporters. *Frontiers in Plant Science* 8:817.

Wang, J., J. H. Sun, J. Miao, et al. 2013a. A phosphate starvation response regulator Ta-PHR1 is involved in phosphate signalling and increases grain yield in wheat. *Annals of Botany* 111 (6):1139–1153.

Wang, X. H., J. R. Bai, H. M. Liu, Y. Sun, X. Y. Shi, and Z. Q. Ren. 2013b. Overexpression of a maize transcription factor ZmPHR1 improves shoot inorganic phosphate content and growth of *Arabidopsis* under low-phosphate conditions. *Plant Molecular Biology Reporter* 31 (3):665–677.

Wang, J. M., L. M. Pei, Z. Jin, K. W. Zhang, and J. R. Zhang. 2017. Overexpression of the protein phosphatase 2A regulatory subunit a gene ZmPP2AA1 improves low phosphate tolerance by remodeling the root system architecture of maize. *Plos One* 12 (4):e0176538.

Wang, L. S., Z. Li, W. Q. Qian, et al. 2011. The *Arabidopsis* purple acid phosphatase AtPAP10 is predominantly associated with the root surface and plays an important role in plant tolerance to phosphate limitation. *Plant Physiology* 157 (3):1283–1299.

Wang, X. M., G. K. Du, X. M. Wang, et al. 2010. The function of LPR1 is controlled by an element in the promoter and is independent of SUMO E3 ligase SIZ1 in response to low Pi stress in *Arabidopsis thaliana. Plant and Cell Physiology* 51 (3):380–394.

Wang, X. R., Y. X. Wang, J. Tian, B. L. Lim, X. L. Yan, and H. Liao. 2009. Overexpressing AtPAP15 enhances phosphorus efficiency in soybean. *Plant Physiology* 151 (1):233–240.

Wang, Y., C. Ribot, E. Rezzonico, and Y. Poirier. 2004. Structure and expression profile of the *Arabidopsis* PHO1 gene family indicates a broad role in inorganic phosphate homeostasis. *Plant Physiology* 135 (1):400–411.

Williamson, L. C., S. P. C. P. Ribrioux, A. H. Fitter, and H. M. O. Leyser. 2001. Phosphate availability regulates root system architecture in *Arabidopsis. Plant Physiology* 126 (2):875–882.

Wu, P., H. X. Shou, G. H. Xu, and X. M. Lian. 2013. Improvement of phosphorus efficiency in rice on the basis of understanding phosphate signaling and homeostasis. *Current Opinion in Plant Biology* 16 (2):205–212.

Xiao, K., H. Katagi, M. Harrison, and Z. Y. Wang. 2006. Improved phosphorus acquisition and biomass production in *Arabidopsis* by transgenic expression of a purple acid phosphatase gene from *M. truncatula. Plant Science* 170 (2):191–202.

Yamagishi, M., K. Q. Zhou, M. Osaki, S. S. Miller, and C. P. Vance. 2011. Real-time RT-PCR profiling of transcription factors including 34 MYBs and signaling components in white lupin reveals their P status dependent and organ-specific expression. *Plant and Soil* 342 (1–2):481–493.

Yang, W. T., D. Baek, D. J. Yun, et al. 2018. Rice OsMYB5P improves plant phosphate acquisition by regulation of phosphate transporter. *Plos One*13 (3):e0194628.

Ye, Y., J. Yuan, X. J. Chang, et al. 2015. The phosphate transporter gene OsPht1;4 is involved in phosphate homeostasis in rice. *Plos One* 10 (5):e0126186.

Yi, K., B. Menand, E. Bell, and L. Dolan. 2010. A basic helix-loop-helix transcription factor controls cell growth and size in root hairs. *Nature Genetics* 42 (3):264–267.

Yi, K. K., Z. C. Wu, J. Zhou, et al. 2005. OsPTF1, a novel transcription factor involved in tolerance to phosphate starvation in rice. *Plant Physiology* 138 (4):2087–2096.

Zaidi, A., M. S. Khan, M. Ahemad, M. Oves, and P. A. Wani. 2009. Recent advances in plant growth promotion by phosphate-solubilizing microbes. In M. S. Khan, A. Zaidi, J. Musarrat, Eds., *Microbial Strategies for Crop Improvement*: 23–50.

Zhang, D. S., C. C. Zhang, X. Y. Tang, et al. 2016. Increased soil phosphorus availability induced by faba bean root exudation stimulates root growth and phosphorus uptake in neighboring maize. *New Phytologist* 209 (2):823–831.

Zhang, Q., C. Wang, J. Tian, K. Li, and H. Shou. 2011. Identification of rice purple acid phosphatases related to posphate starvation signalling. *Plant Biology* 13 (1):7–15.

Zhang, Y. M., Y. S. Yan, L. N. Wang, et al. 2012. A novel rice gene, NRR responds to macronutrient deficiency and regulates root growth. *Molecular Plant* 5 (1):63–72.

Zhang, Z. L., H. Liao, and W. J. Lucas. 2014. Molecular mechanisms underlying phosphate sensing, signaling, and adaptation in plants. *Journal of Integrative Plant Biology* 56 (3):192–220.

Zhou, J., F. C. Jiao, Z. C. Wu, et al. 2008. OsPHR2 is involved in phosphate-starvation signaling and excessive phosphate accumulation in shoots of plants. *Plant Physiology* 146 (4):1673–1686.

3 The Role of Sugars in Improving Plant Abiotic Stress Tolerance

Pooja Sharma, Priya Arora, Dhriti Kapoor, Kanika Khanna, Pardeep Atri, Ravinder Singh Bali, Rupinder Kaur, Abhishek Walia, and Renu Bhardwaj

CONTENTS

3.1 INTRODUCTION

Sugars produced in plants by the process of photosynthesis play significant role in maintaining the metabolic and physiological functions of plants. They regulate the process of photosynthesis, carbon partitioning, carbohydrate and lipid metabolism, osmotic homeostasis, protein synthesis and gene expression under various abiotic stresses (Sami et al. 2016). Sucrose, trehalose, raffinose oligosaccharides (RFOs) and fructans are water-soluble carbohydrates involved in plant stress responses. Sucrose consists of the monosaccharides glucose and fructose and is widely distributed in nature. A change in concentration of glucose, sucrose and fructose increases plant tolerance to cold, drought and salinity stress. Soluble sugars play a critical role as osmoprotectants as well as nutrients and exert their positive effects to protect plant cells from the damage caused by cold stress (Yuanyuan et al. 2009). Sugars also have an emerging role as antioxidants and act as reactive oxygen species (ROS) scavengers. They protect the plant cells from the oxidative damage caused by an increased accumulation of ROS produced in response to abiotic stresses (Keunen et al. 2013) and also regulate a variety of processes such as embryogenesis, seed germination and early seedling development by interacting with phytohormones (Sami et al. 2016). They also act as signaling molecules which regulate many important processes throughout the life of the plant. Therefore, the perception and management of sugar levels by plants are important for their survival. Glucose and sucrose are the most important sugars involved with the process of signaling (Sakr et al. 2018). The objective of this review is to integrate the current understanding of the protective functions of sugars in plants against various abiotic stresses. It also sheds light on the antioxidant role of sugars, the sugar signaling network, the interaction of sugars with phytohormones and modulations in the expression patterns of sugar-responsive genes during abiotic stresses.

3.2 SUGARS AND ABIOTIC STRESS IN PLANTS

Plants subjected to different abiotic stresses undergo a number of changes that may affect their growth, metabolism, photosynthesis, respiration, seed germination and many other physiological processes (Arbona et al. 2017). Moreover, abiotic stresses are known to hinder the balance between carbon fixation and consumption through alteration in the intercellular sugar levels (Martinez-Noel and Tognetti, 2018). Different sugars such as glucose, fructose, raffinose, trehalose and sucrose accumulate in plants under abiotic stresses (Van den Ende and El-Esawe, 2014; Sami et al. 2016). This accumulation of sugars upon exposure to stresses such as salt, heavy metals, heat, cold, nutrient deficiency, etc., is the protective strategy adopted by the plants during initiation of the signaling cascade (Sami et al. 2016) (Figure 3.1).

3.2.1 Sugars and Salt Stress

The exposure of plants to high concentrations of salts leads to reduced nitrogen assimilation, growth inhibition, stomatal conductance and nutrient imbalance (Wani et al. 2013; Ahanger and Agarwal 2017a). The enhancement of glucose, fructose, galactose and sucrose in salt-treated plants plays an essential role in maintaining osmotic balance, membrane integrity and ROS scavenging (Rosa et al. 2009). It has been observed that salt stress leads to sucrose accumulation in tomato plants exposed to excess $CaCl_2$ stress (Saito and Matsukura 2015). The supplementation with glucose exogenously under salinity tends to improve the chlorophyll synthesis, enhance fresh and dry weight, maintain ionic balance, osmolyte accumulation and improve the antioxidative defense system of the plants (Hu et al. 2012). A study reported by Pattanagul and Thitisaksakul (2008) in rice found enhanced levels of osmoprotectants such as glucose and fructose along with the scavenging of free radicals generated under saline stress. The rate of germination under salinity was reduced in wheat seedlings and was reported to increase upon the exogenous application of glucose (Hu et al. 2012). On the other hand the germination rate was inhibited when glucose was applied in higher concentrations of 302 mM (Gibson 2005). It was reported by Nemati et al. (2011) that in wheat seedlings, supplementation with glucose regulates ionic homeostasis by delimiting sodium accumulation and enhancing

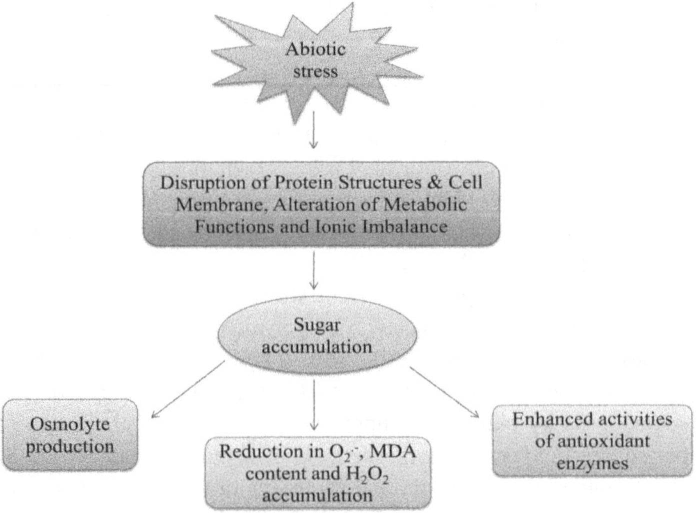

FIGURE 3.1 Soluble sugars, showing their role during abiotic stress.

the uptake of potassium during salinity. Trehalose is reported to induce the accumulation of soluble sugars in rice grown in saline conditions. The treatment of rice seeds with trehalose resulted in a significant increase in total carbohydrates (Abdallah et al. 2016). The treatment of wheat grains with 10 mM trehalose mitigated the damaging effects of salinity stress by lowering H_2O_2 and enhancing the content of sugars, proline, and phenolic compounds (Alla et al. 2019).

3.2.2 Sugars and Temperature Stress

Chilling or low temperature is another factor which affects plants through their membranes, ROS production and protein denaturation. In response to these effects, various carbohydrates are synthesized. These soluble sugars protect plant cells from the damage caused by cold stress by acting as osmoprotectants and function as primary messengers in signal transduction (Yuanyuan et al. 2009). They act as compatible solutes and regulate osmotic potential in plant cells. Apart from this, these sugars also contribute in providing protection to the cell membranes against high and low temperature stress. For example, trehalose provides stability to proteins and membranes and acts as an osmoprotectant (Eastmond and Graham, 2003). Trehalose is generally present in very minute doses, but gradually increases with stress. It was reported that the level of trehalose was found to increase during low temperature stress in rice plants (Garg et al. 2002, Fernandez et al. 2010). Fructans is the class of sugars with high water solubilities that not only provide resistance to crystallization during freezing but also lead to osmotic adjustment and membrane stabilization (Livingston et al. 2009; Krasensky and Jonak. 2012). Furthermore, the application of sucrose also improved responses to freezing stress in *Arabidopsis thaliana* plants (Li et al. 2006). The enhanced concentration of soluble sugars has also been reported in *Spinacia oleracea* exposed to low temperatures. Increased content of glucose, raffinose and maltose was observed in spinach exposed to low temperatures for 7 days (Yoon et al. 2017). High temperature stress is one of the most important constraints for agriculture. It affects plant growth, metabolism and productivity. In response to heat stress several heat-inducible genes are up-regulated. These "heat shock genes" (HSGs) encode "heat shock proteins" (HSPs) which are important for the survival of the plant under high temperatures. These proteins act as chaperones and prevent the denaturation of intracellular proteins and preserve their stability and function through protein folding. Plants produce sugars that maintain cellular structures, cell turgor by osmotic adjustment and modify the antioxidant system to re-establish the cellular redox balance and homeostasis (Hasanuzzaman et al. 2013). An increase in total sugar content was observed in 7-day-old seedlings of *Vigna aconitifolia* grown under normal conditions and then exposed to high temperatures for 1 hour (Harsh et al., 2016). The exogenous application of trehalose reduced the harmful effects of heat stress on photosynthesis in *Paeonea lactiflora* by reducing $O_2^{\bullet-}$, MDA content and H_2O_2 accumulation by improving the antioxidant enzyme activities (Zhao et al. 2019).

3.2.3 Sugars and Water Stress

Water is an important element for the growth and development of plants. In water-deficit (drought) conditions, the rate of photosynthesis decreases due to stomatal closure (Ashraf and Harris 2013). On exposure to drought stress, cell membrane oxidation can be inhibited by the increased accumulation of carbohydrates (Arabzadeh 2012). The sugars that get accumulated under these conditions provide stress alleviation in terms of free radicals scavenging, and protect the membrane structures from dehydration (Krasensky and Jonak 2012). It has been reported that different sugars such as verbascose, raffinose and stachyose are accumulated in desiccated seeds (Mohammadkhani and Heidari 2008). Moreover, they prevent the oxidation of membrane structures, proteins, turgidity and maintain ionic homeostasis in plants under drought conditions (Xu et al. 2007; Arabzadeh 2012). In flooding conditions gas diffusion is restricted between the plant and its environment, which results in oxygen deficiency inside the plants. Qin et al. (2013) reported that lower consumption and increased accumulation of soluble sugars are involved in flooding tolerance. The increased accumulation of sugars during water stress prevents the oxidative effects on membranes (Ahanger and Agarwal 2017b) and protects the functioning of key enzymes of nitrogen metabolism (Ahanger et al. 2017). Ibrahim and Abdellatif (2016) observed that water-stress tolerance in wheat plants was instigated with the foliar application of trehalose and maltose. Treatment with trehalose and maltose resulted in an increased concentration of total soluble and reducing sugars. It is reported that treating *Vigna unguiculata* with trehalose resulted in increased biomass, various solutes viz. proline, total soluble sugars, free amino acids, phenolic compounds and carbohydrates (Khater et al. 2018). Recently, Wang et al. (2019) documented that the concentration of soluble sugars increased in *Oryza sativa* grown in drought stress. The content of sugars increased with the severity of drought, but a decline in sugar concentration was observed when the plants were irrigated. Various reports showing the role of sugars in abiotic stress tolerance are mentioned in Table 3.1.

3.3 SUGARS SIGNALING NETWORKS IN PLANTS AND ABIOTIC STRESSES

Carbohydrates are synthesized in chloroplasts by the process of photosynthesis and are transferred to cytosol in the form of triose-phosphates, where they are converted to hexose-phosphates or sucrose for transport or storage in the vacuole. Sucrose is further converted into hexose in the presence of invertases and sucrose synthases (Ramon et al. 2008). These sugars are sensed by a composite mechanism in plants and show suitable responses to specific types of sugars (Sheen et al. 1999; Wind et al. 2010). In response to stress, sugars also act as a signaling molecule in plants and use various internal and external cues to achieve nutrient homeostasis (Li and Sheen 2016).

There are two types of sugar sensors in plants, i.e., intracellular and extracellular. Regulator of G-protein signaling 1 (RGS1), a seven-transmembrane-domain protein present on the plasma membrane, is an external sugar sensor, which responds to changes in glucose, fructose and sucrose levels (Phan et al. 2013; Grigston et al. 2008). G-proteins are deactivated by RGS1, and G-protein-mediated sugar signaling is activated by the endocytosis of RGS1. Sugars speed up the hydrolysis of GTP by Gα sub units and lead to RGS1 phosphorylation. With no lysine8 (K) [WINK] kinase phosphorylates RGS1, resulting in internalization of RGS1 (Sakr et al. 2018).

HXKs and other sugar kinases are the most ancient, conserved sugar sensors in many organisms (Rolland et al. 2006; Claeyssen and Rivoal, 2007). AtHXK1, an *Arabidopsis thaliana* mitochondrion-associated (type B) hexokinase, is an intracellular glucose sensor (Jang et al. 1997). It is observed that it has role in nuclear signaling. It forms a hetero-multimeric complex by interacting with vacuolar H^+ ATPase and 19 S regulatory particle of proteasome subunit for the recruitment of transcription factors. This complex binds to the promoters of glucose inducible genes to modulate the gene expression (Cho et al. 2006).

SNF-related serine/threonine-protein kinase (SnRK) family and Target of Rapamycin (TOR) kinases are two main signaling networks in plants which regulate metabolism and growth in a manner opposite to the response to sugars (Li and Sheen 2016). SnRKs are evolutionarily conserved proteins in plants. Sucrose non-fermenting 1 (SNF1) and AMP-activated protein kinase (AMPK) are their analogues in animals and yeast. In unfavorable conditions, SnRKs are energy sensors and are activated and regulate gene expression to deal with stress. These kinases maintain energy homeostasis in response to energy loss by the activation of catabolism and inhibition of anabolism (Rodrigues et al. 2013; Nukarinen et al. 2016).

In plants, SnRK1 plays a critical role in the metabolism, growth, development and plant responses to different stresses (Broeckx et al. 2016). SnRK1 signaling and activity is regulated by the phosphorylation of a highly conserved threonine residue near to the active site in the catalytic α-subunit. The dephosphorylation of SnRK1α by PP2C phosphatases deactivates the activation loop (Rodrigues et al. 2013). Glucose-6-phosphate (G6P) and glucose-1-phosphate (G1P) are inhibitors of SnRK1. Trehalose-6-phosphate (T6P) also represses SnRK1 at physiological concentrations

TABLE 3.1
The Role of Sugars during Abiotic Stress Tolerance in Plants

S.NO	Stress	Plant	Type of Sugars	Effect on Plants	References
1.	Drought	*Arachis hypogaea* L.	Sugar alcohols (inositol, mannitol, trehalose)	Increase in the rate of photosynthesis and rate of carbohydrate production as well as overall yield of the crop.	Chakraborty et al. (2016)
2.	Drought	*Solanum tuberosum*	Soluble sugars	Enhanced proline, soluble sugars, enzymatic antioxidants and photosynthetic pigments.	Farhad et al. (2011)
3.	Drought	*Selanginella tamariscina*	Soluble sugars	Up-regulation of genes associated with ABA biosynthesis and photosystem-associated genes along with the accumulation of fatty acids and sugars.	Liu et al. (2008)
4.	Water deficit	*Raphanus sativus* L.	Trehalose	Enhanced fresh and dry biomass, soluble sugars, chlorophyll, proline, water use efficiency, SOD activity, photosynthesis, K^+ and P contents.	Akram et al. (2016)
5.	Flooding	*Glycine max*	Sucrose, starch	Enhanced expression of O-fucosyl transferase and balanced starch-sugar synthesis and glycolysis.	Li et al. (2018)
6.	Water logging	*Mentha arvensis*	Soluble sugars	Induced expression of *MaRAP2-4* (water logging responsive group) and SWEET transporters (carbohydrates metabolism/ transport) in plants that regulate carbohydrate synthesis under stress conditions.	Phukan et al. (2018)
7.	Chilling	*Oryza sativa*	Fructans, glucose	Improved growth and other abnormalities in plants by accumulation of sugars.	Kawakami et al. 2008
8.	Chilling	*Miscanthus*	Raffinose	Increased in yield and biomass, higher accumulation of soluble sugars, growth rate and photosynthetic efficiency.	Fonteyne et al. (2016)
9.	Low temperature	*Lolium perenne*	Raffinose	Accumulation of different amino acids, sugars and other metabolites that enhanced photosynthetic intensities.	Bocian et al. (2015)
10.	Freezing	*Arabidopsis thaliana*	Raffinose	Stabilization of photosystem II, maximum quantum yield.	Knaupp et al. (2011)
11.	Low temperature	Grapevines	Starch, sucrose	Up-regulation of sugar- associated metabolizing enzymes, total chlorophyll and starch content.	Jiang et al. (2016)
12.	Salinity	*Brassica napus* L. (Canola)	Triacontanol	Improved fresh and dry weights, seed number, photosynthesis, transpiration, electron transport, proline, glycine betaine and K^+ contents.	Shahbaz et al. (2013)
13.	Salinity	*Triticum aestivum*	Glucose	Improved seed germination and yield of the crop.	Hu et al. (2012)
14.	Salinity	*Zea mays*	Trehalose	Elevated photosynthetic activities, nucleic acid, hydrolytic and protease enzyme activities. Stabilization of plasma membrane, reducing ion leakage and lipid peroxidation.	Zeid (2009)
15.	Heat stress	*Lycopersicum esculentum*	Soluble sugars	Decreased starch synthesis in anther walls and pollen grains and reduced anther viability.	Pressman et al. (2002)
16.	Heat stress	*Agrostis stolonifera*	Non-structural carbohydrates, fructans, glucans, glucose, sucrose	Reduced turf quality and increased root mortality along with the increased levels of different sugars under high temperature conditions.	Liu and Huang (2000)

(*Continued*)

TABLE 3.1 (CONTINUED)
The Role of Sugars during Abiotic Stress Tolerance in Plants

S.NO	Stress	Plant	Type of Sugars	Effect on Plants	References
17.	Heat stress	*Cocus nucifera* L.	Reducing sugars, soluble sugars	Reduced pollen germination under high temperatures and high synthesis of sugars reversed the negative effects.	Hebbar et al. (2018)
18.	Heat stress	*Oryza sativa*	Sucrose	Enhanced levels of soluble sugars, starch and non-structural proteins up-regulated the expression of heat shock proteins and genes encoding sugar metabolism (sucrose synthase, invertase, sucrose transporters).	Rezaul et al. (2018)

(Nunes et al. 2013). It has been found that in *Arabidopsis* and rice under hypoxic conditions, the stress-responsive expression of genes coding for alcohol dehydrogenase and pyruvate decarboxylase is induced by SnRK1 activity, which is involved in the tolerance of plants (Cho et al. 2012).

TOR is a kinase that acts as a regulator of growth in response to endogenous (nutrition and energy status) and exogenous (environmental factors) signals. It is highly conserved in eukaryotes (Dobrenel et al. 2016). TOR forms two complexes (TORC1 and TORC2, of which TORC1 was found in plants) with other proteins (Wullschleger et al. 2006; Cornu et al. 2013). The activity of TOR kinase is induced by sugars (Lastdrager et al. 2014). Xiong et al. (2013) reported that the activity of AtTOR was enhanced by glucose and resulted in the stimulation of root meristem activity. Due to TOR inhibition, plants show slower growth of roots, leaves and shoots, and TOR inhibition also results in delayed development, ultimately leading to low nutrient uptake and light energy utilization (Ren et al. 2012). TOR also appears to be involved in plant sensitivity to salt and osmotic stresses (Robaglia et al. 2012). Starch accumulation together with a decrease in biomass production by inhibiting the TORC1 activity has also been observed by Caldana et al. (2013). It is reported that AtSUT2/SUC3, a plasma membrane sucrose transporter, is a putative sucrose sensor (Barker et al. 2000). After the perception of sucrose by sensors, the signal is transmitted by calcium, calcium-dependent protein kinases (CDPKs) and protein phosphatases (Vitrac et al. 2000; Martinez-Noel et al. 2010).

3.4 INTERACTION OF SUGARS AND PHYTOHORMONES

Plant hormones play critical roles in governing various physiological and developmental processes. Crosstalk between phytohormones and the sugar-related signaling response regulates a diverse variety of processes, including tuberization, germination, seedling development and embryogenesis (Locascio et al. 2014). Numerous phytohormones such as auxins, gibberellins, brassinosteroids, ethylene and cytokinins interact with varied concentrations of sugars, mediating specific developmental and metabolic pathways in plants (Gibson 2004) (Figure 3.2).

Gibberellins (GAs), light and sugars have been found to play vital roles during bud outgrowth in roses through the augmented demand of sugar by upregulating the expression and activity of vacuolar invertases (Rabot et al. 2014). GAs have been thought to be associated with the remobilization of plant reserves. Loreti et al. (2008) are of the view that GAs hinder the repression of several genes (e.g., chalcone synthase) associated with the sucrose-mediated stimulation of the anthocyanin pathway. DELLA proteins, as suggested by Li et al. (2014), which act as negative regulators in gibberellin signaling, also act as significant positive regulators in sucrose signaling, thereby exerting control in anthocyanin biosynthesis.

Gibson (2004) reported that cytokinins promote tuberization in numerous species which otherwise don't form tubers. Levels of cytokinins correlate with the availability of sugars in plants,

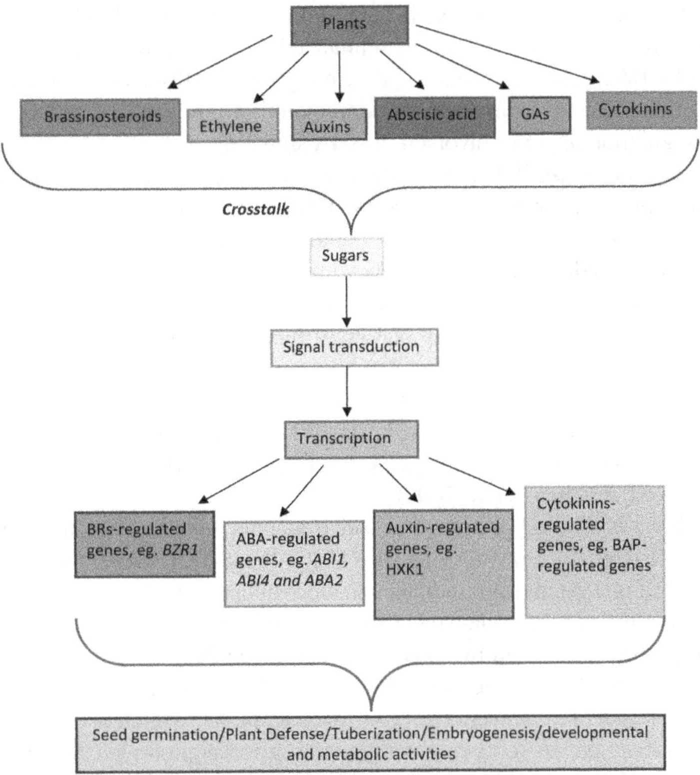

FIGURE 3.2 Crosstalk of phytohormones and sugars.

leading to increased photosynthesis via enhanced chlorophyll content. Kushwah and Laxmi (2014) have investigated the crosstalk between cytokinins and sugars by analyzing the effect of glucose and benzylaminopurine BAP (cytokinin) on gene expression patterns of *Arabidopsis.* They observed that 713 out of 941 BAP-regulated genes were influenced by glucose treatment, and 89% of these genes were agonistically controlled by either glucose or BAP, and the remaining 11% of genes were involved in stress responses. According to Riou-Khamlichi (2000), cytokinins and sugars interact with each other in the regulation of different phases of the cell cycle including the G1/S transition via the sucrose induction of *cycD3* (D-type cyclin gene) expression. Das et al. (2012) revealed that cytokinins enhanced sugar-induced anthocyanin biosynthesis via the redundant action of three *Arabidopsis* type-B response regulators (type-B ARRs), ARR1, ARR10 and ARR12.

In a study carried out on *Arabidopsis* seedlings, brassinosteroids and glucose were found to interact with each other in the formation of lateral roots (Singh et al. 2017). Sugars have been found to positively regulate the transcription of Brassinazole-resistant 1 (BZR1), thereby stabilizing the BZR1 protein (Zhang and He, 2015). The work of these authors also sheds light on the function of HXK1-BR-mediated signaling in the presence of glucose in *Arabidopsis*, revealing that BRs positively regulate the glucose-induced elongation of hypocotyl in darkness.

Abscisic acid (ABA) and sugars work in an orchestrated manner. Finkelstein and Gibson (2001) showed that ABA plays a pivotal role in helping plants to combat drought and other stresses. It has been stated that exogenously applied glucose increases the expression of ABA biosynthesis genes, leading to enhanced ABA levels in plants (Cheng et al. 2002). Leon and Sheen (2003) are of the view that ABA and glucose cause growth promotion only when they are present in lesser amounts and vice-versa. Increased concentrations of ABA and glucose hamper plant development (Dekkers et al. 2008). Zhu et al. (2009) revealed that treatment of *Oryza sativa* seeds with 6% glucose increased the expression levels of OsNCED1, OsNCED2, OsNCED3 and OsNCED4. It was found that the application of 167 mM glucose to *Arabidopsis* enhanced the expression of Abscisic acid insensitive (ABI1), ABI4 and ABA2 genes involved in ABA biosynthesis (Price et al. 2003). Carvalho et al. (2010) suggested that SR45 is involved in the suppression of the glucose-induced accumulation of ABA and down-regulation of genes involved in ABA signaling.

Various reports show crosstalk between ethylene and sugars. Jeong et al. (2010) stated that the expression levels of the sugar transporter gene (SUC1) in *Arabidopsis* were augmented upon treatment with an ethylene-binding inhibitor. These authors also showed that ethylene stalled the accumulation of anthocyanin induced by sucrose. They proposed that the repression of SUC1 expression by ethylene hampers sucrose-induced accumulation of anthocyanin. It was shown through inspection of the phenotype and genetic crossing that constitutive triple responsel (*ctrl*) is allelic to glucose-insensitive (*gin4*). It was also observed that glucose over-sensitivity in the ethylene insensitive 2 (*ein2*) mutant is epistatic to *ginl/aba2* in *ein2 ginl* double mutant plants and that the expression of the ethylene response gene plant defensin 1.2 (PDF1.2) is repressed by glucose in the wild type but not in the *ginl* mutant. These results suggest the antagonistic effects of ethylene and glucose are mediated in part by ABA (Cheng et al. 2002).

Auxin biosynthesis is directly promoted by sugar signals. Glucose and sucrose promote genes that encode auxin biosynthetic enzymes such as YUCCA8 and YUCCA9 (Ljung et al. 2015). These authors suggested that this mechanism indicates shade avoidance syndrome whereby shade detected mainly in cotyledons leads to the enhanced biosynthesis of auxins. HXK1 is involved in the activation of auxin biosynthesis mediated by sugars (Li and Sheen 2016). Baena-Gonzalez and Hanson (2017) are of the view that Snf1-related protein kinase-1 (SnRK1) and TOR networks are up-regulated by auxins, and target of rapamycin (TOR) is involved in the regulation of auxin signaling components. A study in *Arabidopsis* roots revealed that the expression of a WUSCHEL (WUS)-related homeobox gene, WOX5, is induced by auxin and turanose, a non-metabolizable sucrose analogue (Gonzali et al. 2005). Hexokinase (HXK)-mediated sugar signaling is thought to be involved in apical dominance. Since AtHXK1 over-expressors mitigated their apical dominance (Kelly et al. 2012), it was correlated with diminished expression levels of genes which encode crucial participants in auxin signaling, suggestive of the fact that glucose signals control downstream auxin signaling in *Arabidopsis*. Raya-González et al. (2017) reported that mediator genes (MED12) and MED13 can function as affirmative linkers of sugar sensing to the auxin response pathway. Also, MED12 acts upstream of Auxin Resistant 1 (AUX1), central for the spatio-temporal transport of auxin within the root tissue. The effect of sugar phytohormone crosstalk on various genes is shown in Table 3.2.

3.5 THE EXPRESSION OF GENES REGULATED BY SUGARS UNDER ABIOTIC STRESS

Soluble sugars (sucrose, glucose and fructose) play a significant role in maintaining the general structure and growth of plants (Rosa et al. 2009). Among the soluble sugars, sucrose acts as a transporter in plants, and plays an important role in the growth and development of the plant specifically in the vascular tissues. Preliminary analyses of the sucrose carrier (SUC) and sugars will eventually be exported transporter (SWEET) gene families have been conducted in many plant species (Aoki et al. 2004; Chen et al. 2012). The SWEET family is an innovative type of sugar transporter that has contributed to the regulation of plant growth, development and stress responses in many different plant species (Chen et al. 2010; Chong et al. 2014). Sucrose synthase (SUS) isozymes play important roles in plant metabolism, encoded by a small multigene family that contains at least three SUS genes in most plant species. SUS is also involved in the responses of plants to different stresses like low temperatures, drought, anoxia, high salinity and wounding (Baud et al. 2004; Barrero-Sicilia et al. 2011). The *Arabidopsis thaliana* sugar transporter 1 (AtSTP1) gene is regulated by light (Stadler et al. 2003). The AtSTP13 and AtSTP14 genes are also sensitive to light conditions. These genes show diurnal regulation, which may result from a direct light effect or from the sensing of sugars

TABLE 3.2
Up-Regulation/Down-Regulation of Genes/Proteins during Sugar–Phytohormones Crosstalk

S.No	Genes/Proteins	Plant	Effect of Sugar–Phytohormone Crosstalk on Gene Expression	References
1	ABI1, ABI4 and ABA2	*Arabidopsis*	Up-regulation	Price et al. (2003)
2	OsNCED1, OsNCED2, OsNCED3 and OsNCED4	*Oryza sativa*	Up-regulation	Zhu et al. (2009)
3	SUC1	*Arabidopsis*	Up-regulation	Jeong et al. (2010)
4	BZR1	*Arabidopsis*	Up-regulation	Zhang and He, (2015)
5	SnRK1	*Arabidopsis*	Up-regulation	Baena-Gonzalez and Hanson, (2017)
6	SR45	*Arabidopsis*	Down-regulation	Carvalho et al. (2010)
7	WOX5	*Arabidopsis*	Up-regulation	Gonzali et al. 2005
8	YUCCA8 and YUCCA9	*Arabidopsis*	Up-regulation	Ljung et al. (2015)
9	DELLA	*Arabidopsis*	Up-regulation	Li et al. (2014)
10	AtHXK1	*Arabidopsis thaliana*	Down-regulation	Kelly et al. (2012)
11	PDF1.2	*Arabidopsis*	Down-regulation	Cheng et al. (2002)
12	MED 12 and MED 13	*Arabidopsis*	Up-regulation	Raya-González et al. (2017)
13	cycD3	*Arabidopsis*	Up-regulation	Riou-Khamlichi et al. (2000)
14	ARR1, ARR10 and ARR12	*Arabidopsis*	Up-regulation	Das et al. (2012)

which are derived from photosynthesis (Büttner 2010). The tight connection between sugar signaling and the regulation of sugar transporters fits with the impact of SnRK1 alteration on the expression of several STPs (AtSTP1, AtSTP3, AtSTP4, AtSTP7 and AtSTP14) by transient expression in mesophyll protoplasts (Baena-Gonzalez et al. 2007). It has been reported that STP genes (STP1, STP4, STP13 and STP14) are down-regulated by glucose, but there are different pathways for the sugar control of STP expression. On the basis of the down-regulation of STP4 and STP10 in two independent hexokinase1 mutants, it has been reported that the glucose sensor HXK1 is required for their transcriptional control. On the other hand, STP1 is repressed by sugars, showing that glucose-dependent regulation is independent of HXK1 (Sakr et al. 2018). The expression of different sugar-regulated genes in plants in response to abiotic stresses are mentioned in Table 3.3.

3.6 SUGARS AS ANTIOXIDANTS IN ABIOTIC STRESS

Soluble sugars play a dual function as they are associated with both reactive oxygen species (ROS) anabolism and catabolism. Vacuolar carbohydrates are involved in vacuolar antioxidant processes, which are linked to the well-known cytosolic antioxidant processes. They play a vital role in regulatory mechanisms like signaling, scavenging of ROS generation and energy generation for various metabolic activities. However, small soluble sugars and their associated enzymes generate oxidative stress and induce signaling of ROS generation, but bigger soluble sugars like oligo-polysaccharides are involved in the scavenging of free radicals under various stress conditions (Van den and Valliuru 2009). In addition, endogenous soluble sugars enhance pentose phosphate cycle functioning, the production of reducing power and H_2O_2 assimilation (Couee et al. 2006). Sucrose and its derivatives trigger stress signaling at low concentration whereas at higher levels, these components directly enhance the antioxidant defense mechanism of plants in stressful

TABLE 3.3
Expression of Sugar-Regulated Gene in Plants under Abiotic Stress

Abiotic Stress	Sugar-Regulated Genes	Gene Expression	Response	Plant	References
Cold stress					
	AtSWEET 11/12	Increased	Increased tolerance by decreasing xylem number	*Arabidopsis thaliana*	Le Hir et al. (2015)
	AtSWEET16	Over-expression	Change in glucose fructose and soluble sugar	*Arabidopsis thaliana*	Klemens et al. (2013)
	CsSWEET16	Over-expression	Soluble sugar accumulation in vacuoles	*Arabidopsis thaliana*	Schulze et al. (2012)
	MaSWEET	Up-regulation in BX (BaXi Jiao) and FJ (Fen Jiao) tissues	Promotes early sugar transport to improve fruit quality and enhance stress resistance	*Musa acuminate*	Davey et al. (2013)
	SUS (sucrose synthase)	Up-regulation	SUS protein accumulation	*Triticum aestivum*	Crespi et al. (1991)
	AtSUC1	Up-regulation	Sucrose transport reduced	*Arabidopsis thaliana*	Lundmark et al. (2006)
Salt Stress					
	AtSWEET15	Over-expression	Increased tolerance by reducing root growth	*Arabidopsis thaliana*	Seo et al. (2011)
	AG-SUT1 (sucrose transporter)	Expressed in mature leaves and phloem of petioles and also in sink organs such as roots	Increased tolerance by decreasing root and all other organ growth	*Apium gaveolens* L.	Noiraud et al. (2000)
	OsSUT2	Up-regulation	Increase tolerance by facilitating transport of sucrose from photosynthetic cells	*Oryza sativa*	Ibraheem et al. (2011)
Oxidative Stress and Osmotic Stress					
	DsSWEET12	Over-expression	Increased tolerance by affecting the sugar metabolism	*Arabidopsis thaliana*	Zhou et al. (2018)
Flooding Stress					
	SUS	Induced expression (transcription increased)	SUS protein abundance	*Zea mays* *Arabidopsis thaliana* *Solanum tuberosum*	Subbaiah and Sachs, (2001) Déjardin et al. (1999) Biemelt et al. (1999)
	CsSUS3	Increase expression lateral roots	SUS protein abundance Membrane fraction is changed	*Cucumis sativus* L.	Wang et al. (2014)

(*Continued*)

TABLE 3.3 (CONTINUED)
Expression of Sugar-Regulated Gene in Plants under Abiotic Stress

Abiotic Stress	Sugar-Regulated Genes	Gene Expression	Response	Plant	References
	SBSS1 & SBSS2 (sugar beet sucrose synthase)	Transcription levels were raised in stressed root	Little change in SUS protein Enzyme activity relatively unchanged	*Beta vulgaris*	Klotz & Haagenson (2008)
Drought Stress					
	PtaSUT4	Expression decreased	Sucrose accumulation	*Populus tremula*	Frost et al. (2012)
	H-bSUS5	Induced expression in leaves and roots	Supplying energy for phloem loading during day time	*Hevea brasiliensis*	Xiao et al. (2014)

environments. Sucrose, fructose and trehalose act as osmoprotectants in regulating the osmotic adjustment and protect the membranes by scavenging toxic ROS generated under various kinds of stresses (Keunen et al. 2013). Furthermore, low concentrations of soluble sugars, such as glucose and sucrose, accumulate under salinity and stimulate the antioxidative enzymes (Boriboonkaset et al. 2012). Hu et al. (2012) have reported that the activity of antioxidant enzymes like peroxidase, catalase and superoxide dismutase is enhanced in wheat seedlings at low concentrations of glucose (0.1 mM and 0.5 mM).

3.7 CONCLUSION AND FUTURE PERSPECTIVES

Sugars regulate several metabolic events and also regulate the expression of various genes, especially those involved in photosynthesis, sucrose metabolism and the synthesis of osmoprotectants. Sugars like fructans, raffinose family oligosaccharides and trehalose are involved in protecting plants against abiotic stresses. An extensive amount of research is needed to understand the functioning of various enzymes that are regulated by sugars under stressful conditions. Substantial molecular and physiological research is warranted to understand the mechanism of sugar signaling in mediating various metabolic processes ranging from embryogenesis to senescence.

REFERENCES

Abdallah, M. S., Z. A. Abdelgawad, & H. M. S. El-Bassiouny. 2016. Alleviation of the adverse effects of salinity stress using trehalose in two rice varieties. *South African Journal of Botany* 103:275–282.

Ahanger, M. A., and R. M. Agarwal. 2017a. Salinity stress induced alterations in antioxidant metabolism and nitrogen assimilation in wheat (*Triticum aestivum* L.) as influenced by potassium supplementation. *Plant Physiology and Biochemistry* 115:449–460.

Ahanger, M. A., and R. M. Agarwal. 2017b. Potassium up-regulates antioxidant metabolism and alleviates growth inhibition under water and osmotic stress in wheat (*Triticum aestivum* L). *Protoplasma* 254(4):1471–1486.

Ahanger, M. A., M. Tittal, R. A. Mir, and R. M. Agarwal. 2017. Alleviation of water and osmotic stress-induced changes in nitrogen metabolizing enzymes in *Triticum aestivum* L. cultivars by potassium. *Protoplasma* 254(5):1953–1963.

Akram, N. A., M. Waseem, R. Ameen, and M. Ashraf. 2016. Trehalose pretreatment induces drought tolerance in radish (*Raphanus sativus* L.) plants: Some key physio-biochemical traits. *Acta Physiolgia Plantarum* 38:3.

Alla, M. N., E. Badran, and F. Mohammed. 2019. Exogenous trehalose alleviates the adverse effects of salinity stress in wheat. *Turkish Journal of Botany* 43(1):48–57.

Aoki, N., G. N. Scofield, X. D. Wang, J. W. Patrick, C. E. Offler, and R. T. Furbank. 2004. Expression and localisation analysis of the wheat sucrose transporter TaSUT1 in vegetative tissues. *Planta* 219:176–184.

Arabzadeh, N. 2012. The effect of drought stress on soluble carbohydrates (sugars) in two species of *Haloxylon persicum* and *Haloxylon aphyllum*. *Asian Journal of Plant Science* 11:44–51.

Arbona, V., M. Manzi, S. I. Zandalinas, V. Vives-Peris, R. M. Pérez-Clemente and A. Gomez-Cadenas. 2017. Physiological, metabolic, and molecular responses of plants to abiotic stress. In: *Stress Signaling in Plants: Genomics and Proteomics Perspective*, vol. 2. Eds. M. Sarwat, A. Ahmad, M. Z. Abdin, and M. M. Ibrahim. Springer, pp. 1–35.

Ashraf, M. H. P. J. C., and P. J. C. Harris. 2013. Photosynthesis under stressful environments: An overview. *Photosynthetica* 2:163–190.

Baena-Gonzalez, E., and J. Hanson. 2017. Shaping plant development through the SnRK1-TOR metabolic regulators. *Current Opinion in Plant Biology* 35:152–157.

Baena-Gonzalez, E., F. Rolland, J. M. Thevelein, and J. Sheen. 2007. A central integrator of transcription networks in plant stress and energy signaling. *Nature* 448:938.

Barker, L., C. Kühn, A. Weise, *et al.* 2000. SUT2, a putative sucrose sensor in sieve elements. *The Plant Cell* 12:1153–1164.

Barrero-Sicilia, C., S. Hernando-Amado, P. González-Melendi, and P. Carbonero. 2011. Structure, expression profile and subcellular localisation of four different sucrose synthase genes from barley. *Planta* 234:391–403.

Baud, S., M. N. Vaultier, and C. Rochat. 2004. Structure and expression profile of the sucrose synthase multigene family in *Arabidopsis*. *Journal of Experimental Botany* 55:397–409.

Biemelt, S., M. R. Hajirezaei, M. Melzer, G. Albrecht, and U. Sonnewald. 1999. Sucrose synthase activity does not restrict glycolysis in roots of transgenic potato plants under hypoxic conditions. *Planta* 210:41–49.

Bocian, A., Z. Zwierzykowski, M. Rapacz, G. Koczyk, D. Ciesiołka, and A. Kosmala. 2015. Metabolite profiling during cold acclimation of *Lolium perenne* genotypes distinct in the level of frost tolerance. *Journal of Applied Genetics* 56:439–449.

Boriboonkaset, T., C. Theerawitaya, A. Pichakum, S. Cha-Um, T. Takabe, and C. Kirdmanee. 2012. Expression levels of some starch metabolism related genes in flag leaf of two contrasting rice genotypes exposed to salt stress. *Australian Journal of Crop Science* 11:1579.

Broeckx, T., S. Hulsmans, and F. Rolland. 2016. The plant energy sensor: Evolutionary conservation and divergence of SnRK1 structure, regulation, and function. *Journal of Experimental Botany* 67:6215–6252.

Büttner, M. 2010. The *Arabidopsis* sugar transporter (AtSTP) family: An update. *Plant Biology* 12:35–41.

Caldana, C., Y. Li, A. Leisse, *et al.* 2013. Systemic analysis of inducible target of rapamycin mutants reveal a general metabolic switch controlling growth in *Arabidopsis thaliana*. *The Plant Journal* 73:897–909.

Carvalho, R. F., S. D. Carvalho, and P. Duque. 2010. The plant-specific SR45 protein negatively regulates glucose and ABA signaling during early seedling development in *Arabidopsis*. *Plant Physiology* 154:772–783.

Chakraborty K., M. K. Mahatma, L. K. Thawait, S. K. Bishi, K. A. Kalariya, and A. L. Singh. 2016. Water deficit stress affects photosynthesis and the sugar profile in source and sink tissues of groundnut (*Arachis hypogaea* L.) and impacts kernel quality. *Applied Botany and Food Quality* 89:98–104.

Chen, L. Q., Hou, B. H., S. Lalonde, *et al.* 2010. Sugar transporters for intercellular exchange and nutrition of pathogens. *Nature* 468:527.

Chen, L. Q., X. Q. Qu, B. H. Hou, *et al.* 2012. Sucrose efflux mediated by SWEET proteins as a key step for phloem transport. *Science* 335:207–211.

Cheng, W. H., A. Endo, L. Zhou, *et al.* 2002. A unique short-chain dehydrogenase/reductase in *Arabidopsis* abscisic acid biosynthesis and glucose signaling. *Plant Cell* 14:2723–2743.

Cho, Y. H., J. W. Hong, E. C. Kim, and S. D. Yoo. 2012. Regulatory functions of SnRK1 in stress-responsive gene expression and in plant growth and development. *Plant Physiology* 158:1955–1964.

Cho, Y. H., S. D. Yoo, and J. Sheen. 2006. Regulatory functions of nuclear hexokinase1 complex in glucose signaling. *Cell* 127:579–589.

Chong, J., M. C. Piron, S. Meyer, D. Merdinoglu, C. Bertsch, and P. Mestre. 2014. The SWEET family of sugar transporters in grapevine: VvSWEET4 is involved in the interaction with *Botrytis cinerea*. *Journal of Experimental Botany* 65:6589–6601.

Claeyssen, É., and J. Rivoal. 2007. Isozymes of plant hexokinase: Occurrence, properties and functions. *Phytochemistry* 68:709–731.

Cornu, M., V. Albert, and M. N. Hall. 2013. mTOR in aging, metabolism, and cancer. *Current Opinion in Genetics and Development* 23:53–62.

Couée I., C. Sulmon, G. Gouesbet, and A. El Amrani. 2006. Involvement of soluble sugars in reactive oxygen species balance and responses to oxidative stress in plants. *Journal of Experimental Botany* 57:449–459.

Crespi, M. D., E. J. Zabaleta, H. G. Pontis, and G. L. Salerno. 1991. Sucrose synthase expression during cold acclimation in wheat. *Plant Physiology* 96:887–891.

Das, P. K., D. H. Shin, S. B. Choi, S. D. Yoo, G. Choi, and Y. I. Park. 2012. Cytokinins enhance sugar-induced anthocyanin biosynthesis in *Arabidopsis. Molecular Cell* 34:93–101.

Davey, M. W., R. Gudimella, J. A. Harikrishna, L. W. Sin, N. Khalid, and J. Keulemans. 2013. A draft *Musa balbisiana* genome sequence for molecular genetics in polyploid, inter-and intra-specific Musa hybrids. *BioMed Central Genomics* 14:683.

Déjardin, A., L. N. Sokolov, and L. A. Kleczkowski. 1999. Sugar/osmoticum levels modulate differential abscisic acid-independent expression of two stress-responsive sucrose synthase genes in *Arabidopsis. Biochemical Journal* 344:503–509.

Dekkers, B., J. Schuurmans, and S. Smeekens. 2008. Interaction between sugar and abscisic acid signaling during early seedling development in *Arabidopsis. Plant Molecular Biology* 67:151–167.

Dobrenel, T., C. Caldana, J. Hanson, *et al.* 2016. TOR signaling and nutrient sensing. *Annual Review of Plant Biology* 67:261–285.

Eastmond, Peter J., and I. A. Graham. 2003. Trehalose metabolism: A regulatory role for trehalose-6-phosphate? *Current Opinion in Plant Biology* 3:231–235.

Farhad, M. S., A. M. Babak, Z. M. Reza, R. M. Hassan, and, T. Afshin. 2011. Response of proline, soluble sugars, photosynthetic pigments and antioxidant enzymes in potato (*Solanum tuberosum* L.) to different irrigation regimes in greenhouse condition. *Australian Journal of Crop Science* 5:55–60.

Fernandez, O., L. Bethencourt, A. Quero, R. Sangwan, and C. Clement. 2010. Trehalose and plant stress responses: Friend or foe? *Trends in Plant Sciences* 15:409–417.

Finkelstein, R., and S. Gibson. 2001. ABA and sugar interactions regulating development: Crosstalk or voices in a crowd? *Current Opinion in Plant Biology* 5:26–32.

Fonteyne, S., P. Lootens, H. Muylle, *et al.* 2016. Chilling tolerance and early vigour-related characteristics evaluated in two Miscanthus genotypes. *Photosynthetica* 54:295–306.

Frost, C. J., B. Nyamdari, C. J. Tsai, and S. A. Harding. 2012. The tonoplast-localized sucrose transporter in Populus (PtaSUT4) regulates whole-plant water relations, responses to water stress, and photosynthesis. *PLoS One* 7:44467.

Garg, A., J. Kim, T. Owens, A. Ranwala, Y. Choi, L. Kochian, and R. Wu. 2002. Trehalose accumulation in rice plants confers high tolerance levels to different abiotic stresses. *Proceedings in National Academy of Sciences* 99:15898–15903.

Gibson, S. 2004. Sugar and phytohormone response pathways: Navigating a signaling network. *The Journal of Experimental Botany* 55:253–264.

Gibson, S. 2005. Control of plant development and gene expression by sugar signaling. *Current Opinion in Plant Biology* 8:93–102.

Gonzali, S., G. Novi, E. Loreti, *et al.* 2005. A turanose-insensitive mutant suggests a role for WOX5 in auxin-homeostasis in *Arabidopsis thaliana. The Plant Journal* 44:633–645.

Grigston, J. C., D. Osuna, W. R. Scheible, C. Liu, M. Stitt, and A. M. Jones. 2008. D-Glucose sensing by a plasma membrane regulator of G signaling protein, AtRGS1. *Federation of European Biochemical Societies Letters* 582:3577–3584.

Harsh, A., Sharma, Y. K., Joshi, U., Rampuria, S., Singh, G., Kumar, S., and Sharma, R. 2016. Effect of short-term heat stress on total sugars, proline and some antioxidant enzymes in moth bean (*Vigna aconitifolia*). *Annals of Agricultural Sciences* 61(1):57–64.

Hasanuzzaman, M., K. Nahar, M. Alam, R. Roychowdhury, and M. Fujita. 2013. Physiological, biochemical, and molecular mechanisms of heat stress tolerance in plants. *International Journal of Molecular Sciences* 5:9643–9684.

Hebbar, K. B., H. M. Rose, A. R. Nair, *et al.* 2018. Differences in in vitro pollen germination and pollen tube growth of coconut (*Cocos nucifera* L.) cultivars in response to high temperature stress. *Environmental and Experimental Botany* 153:35–44.

Hu, M., Z. Shi, Z. Zhang, Y. Zhang, and H. Li. 2012. Effects of exogenous glucose on seed germination and antioxidant capacity in wheat seedlings under salt stress. *Plant Growth Regulation* 68:177–188.

Ibraheem, O., G. Dealtry, S. Roux, and G. Bradley. 2011. The effect of drought and salinity on the expressional levels of sucrose transporters in rice (*'Oryza sativa'Nipponbare*) cultivar plants. *Plant Omics* 4:68.

Ibrahim, H. A., and Abdellatif, Y. M. 2016. Effect of maltose and trehalose on growth, yield and some biochemical components of wheat plant under water stress. *Annals of Agricultural Sciences* 61(2):267–274.

Jang, J. C., P. León, L. Zhou, and J. Sheen. 1997. Hexokinase as a sugar sensor in higher plants. *The Plant Cell* 9:5–19.

Jeong, S. W., P. K. Das, S. C. Jeoung, *et al.* 2010. Ethylene suppression of sugar-induced anthocyanin pigmentation in *Arabidopsis. Plant Physiology* 154:1514–1531.

Jiang H. Y., Y. H. Gao, W. Tian, and W. B. J He. 2016. Evaluation of cold resistance in grapevines via photosynthetic characteristics, carbohydrate metabolism and gene expression levels. *Acta Physiolgia Plantarum* 38:251.

Kawakami, A., Y. Sato, and M. Yoshida. 2008. Genetic engineering of rice capable of synthesizing fructans and enhancing chilling tolerance. *Journal of Experimental Botany* 59:793–802.

Kelly, G., R. David-Schwartz, N. Sade, M. Moschelion, A. Levi, V. Alchanatis, *et al.* 2012. The pitfalls of transgenic selection and new roles of AtHXK1 expression uncouples hexokinase 1-dependent sugar signalling from exogenous sugar. *Plant Physiology* 159:47–51.

Keunen, E. L. S., D. Peshev, J. Vangronsveld, W. I. M. Van Den Ende, and A. N. N. Cuypers. 2013. Plant sugars are crucial players in the oxidative challenge during abiotic stress: Extending the traditional concept. *Plant, Cell and Environment* 7:1242–1255.

Khater, M. A., Dawood, M. G., Sadak, M. S., Shalaby, M. A., El-Awadi, M. E., and El-Din, K. G. 2018. Enhancement the performance of cowpea plants grown under drought conditions via trehalose application. *Middle East J* 7(3):782–800.

Klemens, P. A., K. Patzke, J. W. Deitmer, *et al.* 2013. Overexpression of the vacuolar sugar carrier AtSWEET16 modifies germination, growth and stress tolerance in *Arabidopsis thaliana. Plant Physiology* 163:1338–1352.

Klotz, K. L., and D. M. Haagenson. 2008. Wounding, anoxia and cold induce sugarbeet sucrose synthase transcriptional changes that are unrelated to protein expression and activity. *Journal of Plant Physiology* 165:423–434.

Knaupp, M., K. B. Mishra, L. Nedbal, and A. G. Heyer. 2011. Evidence for a role of raffinose in stabilizing photosystem II during freeze–thaw cycles. *Planta* 234(3):477–486.

Krasensky, J., and C. Jonak. 2012. Drought, salt and temperature stress-induced metabolic rearrangements and regulatory networks. *Journal of Experimental Botany* 63:1593e1608.

Kushwah, S., and A. Laxmi. 2014. The interaction between glucose and cytokinin signal transduction pathway in *Arabidopsis thaliana. Plant Cell and Environment* 37:235–253.

Lastdrager, J., J. Hanson, and S. Smeekens. 2014. Sugar signals and the control of plant growth and development. *Journal of Experimental Botany* 65:799–807.

Le Hir, R., L. Spinner, P. A. Klemens, *et al.* 2015. Disruption of the sugar transporters AtSWEET11 and AtSWEET12 affects vascular development and freezing tolerance in *Arabidopsis. Molecular Plant* 8:1687–1690.

Leon, P., and J. Sheen. 2003. Sugar and hormone connections. *Trends in Plant Science* 8:1360–1385.

Li, L., and J. Sheen. 2016. Dynamic and diverse sugar signaling. *Current Opinion in Plant Biology* 33:116–125.

Li, X., S. Rehman, H. Yamaguchi, *et al.* 2018. Proteomic analysis of the effect of plant-derived smoke on soybean during recovery from flooding stress. *Journal of Proteomics* 181:238–248.

Li, Y., K. Lee, S. Walsh, *et al.* 2006. Establishing glucose and ABA regulated transcription networks in *Arabidopsis* by microarray analysis and promoter classification using relevance vector machine. *Genome Research* 16:414–427.

Li, Y., W. Van den Ende, and F. Rolland. 2014. Sucrose induction of anthocyanin biosynthesis is mediated by DELLA. *Molecular Plant* 7:570–572.

Liu, M. S., C. T. Chien, and T. P. Lin. 2008. Constitutive components and induced gene expression are involved in the desiccation tolerance of *Selaginella tamariscina. Plant and Cell Physiology* 49:653–663.

Liu, X., and B. Huang. 2000. Carbohydrate accumulation in relation to heat stress tolerance in two creeping bentgrass cultivars. *Journal of the American Society for Horticultural Science* 125:442–447.

Livingston, D., D. Hincha, and A. Heyer. 2009. Fructan and its relationship to abiotic stress tolerance in plants. *Cellular and Molecular Life Sciences* 66:2007–2023.

Ljung, K., J. L. Nemhauser, and P. Perata. 2015. New mechanistic links between sugar and hormone signalling networks. *Current Opinion in Plant Biology* 25:130–137.

Locascio, A., A. Villanova, J. Bernardi, and S. Varotto. 2014. Current perspectives on the hormonal control of seed development in *Arabidopsis* and maize: A focus on auxin. *Frontiers in Plant Science* 5:412.

Loreti, E., G. Povero, G. Novi, C. Solfanelli, A. Alpi, and P. Perata. 2008. Gibberellins, jasmonate and abscisic acid modulate the sucrose-induced expression of anthocyanin biosynthetic genes in *Arabidopsis. New Phytologist* 179:1004–1016.

Lundmark, M., A. M. Cavaco, S. Trevanion, and V. Hurry. 2006. Carbon partitioning and export in transgenic *Arabidopsis thaliana* with altered capacity for sucrose synthesis grown at low temperature: A role for metabolite transporters. *Plant, Cell and Environment* 29:1703–1714.

Martínez-Noël, G. M. A, and J. A. Tognetti. 2018. Sugar signaling under abiotic stress in plants. In: *Plant Metabolites and Regulation Under Environmental Stress*, Eds. P. Ahmad, M. A. Ahanger, V. P. Singh, D. K. Tripathi, P. Alam, and M. N. Alyemeni. Elsevier, pp. 397–406.

Martinez-Noel, G. M. A, J. A. Tognetti, G. Salerno, and P. Horacio. 2010. Sugar signaling of fructan metabolism: New insights on protein phosphatases in sucrose-fed wheat leaves. *Plant Signaling and Behavior* 5:311–313.

Mohammadkhani, N., and R. Heidari. 2008. Drought-induced accumulation of soluble sugars and proline in two maize varieties. *World Applied Science Journal* 3:448–453.

Nemati, I., F. Moradi, S. Gholizadeh, M. A. Esmaeili, and M. R. Bihamta. 2011. The effect of salinity stress on ions and soluble sugars distribution in leaves, leaf sheaths and roots of rice (*Oryza sativa* L.) seedlings. *Plant Soil Environment* 57:26–33.

Noiraud, N., S. Delrot, and R. Lemoine. 2000. The sucrose transporter of celery. Identification and expression during salt stress. *Plant Physiology* 122:1447–1456.

Nukarinen, E., T. Nägele, L. Pedrotti, *et al.* 2016. Quantitative phosphoproteomics reveals the role of the AMPK plant ortholog SnRK1 as a metabolic master regulator under energy deprivation. *Scientific Reports* 6:31697.

Nunes, C., L. F. Primavesi, M. K. Patel, *et al.* 2013. Inhibition of SnRK1 by metabolites: Tissue-dependent effects and cooperative inhibition by glucose 1-phosphate in combination with trehalose 6-phosphate. *Plant Physiology and Biochemistry* 63:89–98.

Pattanagul, W., andM. Thitisaksakul. 2008. Effect of salinity stress on growth and carbohydrate metabolism in three rice (*Oryza sativa* L.) cultivars differing in salinity tolerance. *Indian Journal of Experimental Biology* 46:736–742.

Phan, N., D. Urano, M. Srba, L. Fischer, and A. M. Jones. 2013. Sugar-induced endocytosis of plant 7TM-RGS proteins. *Plant Signaling and Behavior* 8:22814.

Phukan, U. J., G. S. Jeena, V. Tripathi, and R. K. Shukla. 2018. MaRAP2-4, a waterlogging-responsive ERF from Mentha, regulates bidirectional sugar transporter AtSWEET 10 to modulate stress response in *Arabidopsis. Plant Biotechnology Journal* 16:221–233.

Pressman, E., M. M. Peet, and D. M. Pharr. 2002. The effect of heat stress on tomato pollen characteristics is associated with changes in carbohydrate concentration in the developing anthers. *Annals of Botany* 90:631–636.

Price, J., T. Li, S. G. Kang, J. K. Na, and J. C. Jang. 2003. Mechanisms of glucose signalling during germination of *Arabidopsis. Plant Physiology* 132:1424–1438.

Qin, X., F. Li, X. Chen, and Y. Xie. 2013. Growth responses and non-structural carbohydrates in three wetland macrophyte species following submergence and de-submergence. *Acta Physiologiae Plantarum* 7:2069–2074.

Rabot, A., V. Portemer, T. Peron, *et al.* 2014. Interplay of sugar, light and gibberellins in of *Rosa hybrida* vacuolar invertase 1 regulation. *Plant and Cell Physiology* 55:1734–1748.

Ramon, M., F. Rolland, and J. Sheen. 2008. Sugar sensing and signaling. *The Arabidopsis Book* (6):S185–S205.

Raya-González, J., J. S. López-Bucio, J. C. Prado-Rodríguez, L. F. Ruiz-Herrera, A. A. Guevara-García, and J. López-Bucio. 2017. The MEDIATOR genes *MED12* and *MED13* control *Arabidopsis* root system configuration influencing sugar and auxin responses. *Plant Molecular Biology* 95:141–156.

Ren, M., P. Venglat, S. Qiu, *et al.* 2012. Target of rapamycin signaling regulates metabolism, growth, and life span in *Arabidopsis. The Plant Cell* 24(12):4850–4874.

Rezaul, I. M., F. Baohua, C. Tingting, *et al.* 2018. Abscisic acid prevents pollen abortion under high-temperature stress by mediating sugar metabolism in rice spikelets. *Physiologia Plantarum* 165(3):644–663.

Riou-Khamlichi, C., M. Menges, J. S. Healy, and J. A. Murray. 2000. Sugar control of plant cell cycle: Differential regulation of *Arabidopsis* D-type cyclin gene expression. *Molecular and Cell Biology* 20:4513–4521.

Robaglia, C., M. Thomas, and C. Meyer. 2012. Sensing nutrient and energy status by SnRK1 and TOR kinases. *Current Opinion in Plant Biology* 15:301–307.

Rodrigues, A., M. Adamo, P. Crozet, *et al.* 2013. ABI1 and PP2CA phosphatases are negative regulators of Snf1-related protein kinase1 signaling in *Arabidopsis. The Plant Cell* 25(10):3871–3884.

Rolland, F., E. Baena-Gonzalez, and J. Sheen. 2006. Sugar sensing and signaling in plants: Conserved and novel mechanisms. *Annual Review of Plant Biology* 57:675–709.

Rosa, M., C. Prado, G. Podazza, *et al.* 2009. Soluble sugars: Metabolism, sensing and abiotic stress: A complex network in the life of plants. *Plant Signaling and Behavior* 4:388–393.

Saito, T, and C. Matsukura. 2015. Effect of salt stress on the growth and fruit quality of tomato plants. In: *Abiotic Stress Biology in Horticultural Plants*. Eds. Y. Kanayama and A. Kocheto. Springer, Japan, pp. 3–19.

Sakr, S., M. Wang, F. Dédaldéchamp, *et al.* 2018. The sugar-signaling hub: Overview of regulators and interaction with the hormonal and metabolic network. *International Journal of Molecular Sciences* 19:2506.

Sami, F., M. Yusuf, M. Faizan, A. Faraz, and S. Hayat. 2016. Role of sugars under abiotic stress. *Plant Physiology and Biochemistry* 109:54–61.

Schulze, W. X., T. Schneider, S. Starck, E. Martinoia, and O. Trentmann. 2012. Cold acclimation induces changes in *Arabidopsis* tonoplast protein abundance and activity and alters phosphorylation of tonoplast monosaccharide transporters. *The Plant Journal* 69:529–541.

Seo, P. J., J. M. Park, S. K. Kang, S. G. Kim, and C. M. Park. 2011. An *Arabidopsis* senescence-associated protein SAG29 regulates cell viability under high salinity. *Planta* 233:189–200.

Shahbaz M., N. Noreen, and S. Perveen. 2013. Triacontanol modulates photosynthesis and osmoprotectants in canola (*Brassica napus* L.) under saline stress. *Journal of Plant Interactions* 8:350–359.

Sheen, J., L. Zhou, and J. C. Jang. 1999. Sugars as signaling molecules. *Current Opinion in Plant Biology* 2:410–418.

Singh, M., A. Gupta, and A. Laxmi. 2017. Glucose and brassinosteroid signaling network in controlling plant growth and development under different environmental conditions. In: *Mechanism of Plant Hormone Signaling Under Stress*. Ed. G. K. Pandey. Wiley, pp. 443–469.

Stadler, R., M. Büttner, P. Ache, *et al.* 2003. Diurnal and light-regulated expression of AtSTP1 in guard cells of *Arabidopsis*. *Plant Physiology* 133:528–537.

Subbaiah, C. C., and M. M. Sachs. 2001. Altered patterns of sucrose synthase phosphorylation and localization precede callose induction and root tip death in anoxic maize seedlings. *Plant Physiology* 125:585–594.

Van den Ende, W., and S. K. El-Esawe. 2014. Sucrose signaling pathways leading to fructan and anthocyanin accumulation: A dual function in abiotic and biotic stress responses? *Environmental and Experimental Botany* 108:4–13.

Van den Ende, W., and R. Valluru. 2009. Sucrose, sucrosyl oligosaccharides, and oxidative stress: Scavenging and salvaging? *Journal of Experimental Botany* 60:9–18.

Vitrac, X., F. Larronde, S. Krisa, A. Decendit, G. Deffieux, and J. M. Mérillon. 2000. Sugar sensing and Ca^{2+}–calmodulin requirement in *Vitis vinifera* cells producing anthocyanins. *Phytochemistry* 53:659–665.

Wang, H., X. Sui, J. Guo, *et al.* 2014. Antisense suppression of cucumber (*Cucumis sativus L.*) sucrose synthase 3 (CsSUS3) reduces hypoxic stress tolerance. *Plant, Cell and Environment* 37:795–810.

Wang, X., Liu, H., Yu, F., Hu, B., Jia, Y., Sha, H., and Zhao, H. 2019. Differential activity of the antioxidant defence system and alterations in the accumulation of osmolyte and reactive oxygen species under drought stress and recovery in rice (*Oryza sativa* L.) tillering. *Scientific Reports* 9(1):8543.

Wani, A., A. Ahmad, S. Hayat, and Q. Fariduddin. 2013. Salt-induced modulation in growth, photosynthesis and antioxidant system in two varieties of *Brassica juncea*. *Saudi Journal of Biological Science* 20:183–193.

Wind, J., S. Smeekens, and J. Hanson. 2010. Sucrose: Metabolite and signaling molecule. *Phytochemistry* 71:610–1614.

Wullschleger, S., R. Loewith, and M. N. Hall. 2006. TOR signaling in growth and metabolism. *Cell* 124:471–484.

Xiao, X., C. Tang, Y. Fang, *et al.* 2014. Structure and expression profile of the sucrose synthase gene family in the rubber tree: Indicative of roles in stress response and sucrose utilization in the laticifers. *The Federation of European Biochemical Societies Journal* 281:291–305.

Xiong, Y., M. McCormack, L. Li, Q. Hall, C. Xiang, and J. Sheen. 2013. Glucose–TOR signaling reprograms the transcriptome and activates meristems. *Nature* 496:181–186.

Xu, S. M., L. X. Liu, K. C. Woo, and D. L. Wang. 2007. Changes in photosynthesis, xanthophyll cycle and sugar accumulation in two North Australia Tropical species differing in leaf angles. *Photosynthetica* 45:348–354.

Yoon, Y. E., Kuppusamy, S., Cho, K. M., Kim, P. J., Kwack, Y. B., and Lee, Y. B. 2017. Influence of cold stress on contents of soluble sugars, vitamin C and free amino acids including gamma-aminobutyric acid (GABA) in spinach (*Spinacia oleracea*). *Food Chemistry* 215:185–192.

Yuanyuan, M., Z. Yali, L. Jiang, and S. Hongbo. 2009. Roles of plant soluble sugars and their responses to plant cold stress. *African Journal of Biotechnology* 8:2004–2010.

Zeid, I. M. 2009. Trehalose as osmoprotectant for maize under salinity-induced stress. *Research Journal of Agriculture and Biological Sciences* 5:613–622.

Zhang, Y., and J. He. 2015. Sugar-induced plant growth is dependent on brassinosteroids. *Plant Signaling and Behavior* 10:1082700.

Zhao, D. Q., Li, T. T., Hao, Z. J., Cheng, M. L., and Tao, J. 2019. Exogenous trehalose confers high temperature stress tolerance to herbaceous peony by enhancing antioxidant systems, activating photosynthesis, and protecting cell structure. *Cell Stress and Chaperones* 24(1):247–257.

Zhou, A., H. Ma, S. Feng, S. Gong, and J. Wang. 2018. A novel sugar transporter from *Dianthus spiculifolius*, DsSWEET12, affects sugar metabolism and confers osmotic and oxidative stress tolerance in *Arabidopsis*. *International Journal of Molecular Sciences* 19:497.

Zhu, G., N. Ye, and J. Zhang. 2009. Glucose-induced delay of seed germination in rice is mediated by the suppression of ABA catabolism rather than an enhancement of ABA biosynthesis. *Plant and Cell Physiology* 50:644–651.

4 Glutathione Is a Key Component in Abiotic Stress Responses

Sajid Ali Khan Bangash and Krisztina Bela

CONTENTS

4.1 GLUTATHIONE STRUCTURE AND BIOSYNTHESIS

Glutathione is a low molecular weight thiol, present in almost all living organisms. It is built up from three amino acids: glutamate, cysteine and glycine (Noctor et al. 1998). The biosynthesis of glutathione is catalyzed by two ATP-dependent enzymes in two steps. In plants, the first step of glutathione biosynthesis takes place in plastids, and the second step in the cytosol. In contrast, in mammals, yeasts and bacteria, both steps take place in the cytosol (Galant et al. 2011). In the first reaction, L-cysteine and L-glutamate are combined by the action of the enzyme called glutamate–cysteine ligase (GSH1, EC 6.3.2.2), generating the intermediate γ-glutamylcysteine (γ-EC). In the second reaction, glycine is attached to the C-terminal of γ-EC and this reaction is catalyzed by glutathione synthase (GSH2, EC 6.3.2.3) forming the tripeptide reduced glutathione (GSH) (Meister 1988) (Figure 4.1).

GSH1 is the main regulatory enzyme in GSH biosynthesis, which localizes exclusively in plastids. This is the rate-limiting step of GSH biosynthesis; the feedback inhibition of GSH1 by glutathione is also an important regulatory point in glutathione homeostasis, both in plants and animals (Richman and Meister 1975; Meister 1988; Hell and Bergmann 1990). The enzyme of the second step, GSH2, localizes predominantly in the cytosol but also in the plastids (Wachter et al. 2005). Thus, glutathione biosynthesis occurs in both plastids and cytosol, and it requires an export of γ-EC from plastids to the cytosol (Pasternak et al. 2008). The relevance of the cytosolic production of GSH was characterized by complementing mutant plants deficient in GSH2 with a wild-type GSH2 exclusively targeted to the cytosol. The complementation restored GSH biosynthesis and hence rescued the phenotype (Pasternak et al. 2008).

As mentioned above, GSH is composed of three amino acids, L-glutamate, L-cysteine and glycine, with a γ-peptide bond between the amine group of cysteine and the γ-carboxyl group of the glutamate (Figure 4.1). The γ-peptide bond is important for not being degraded by most peptidases in the cell, with the exception of GGTs (gamma-glutamyl transferase) (Ohkama-Ohtsu et al. 2008).

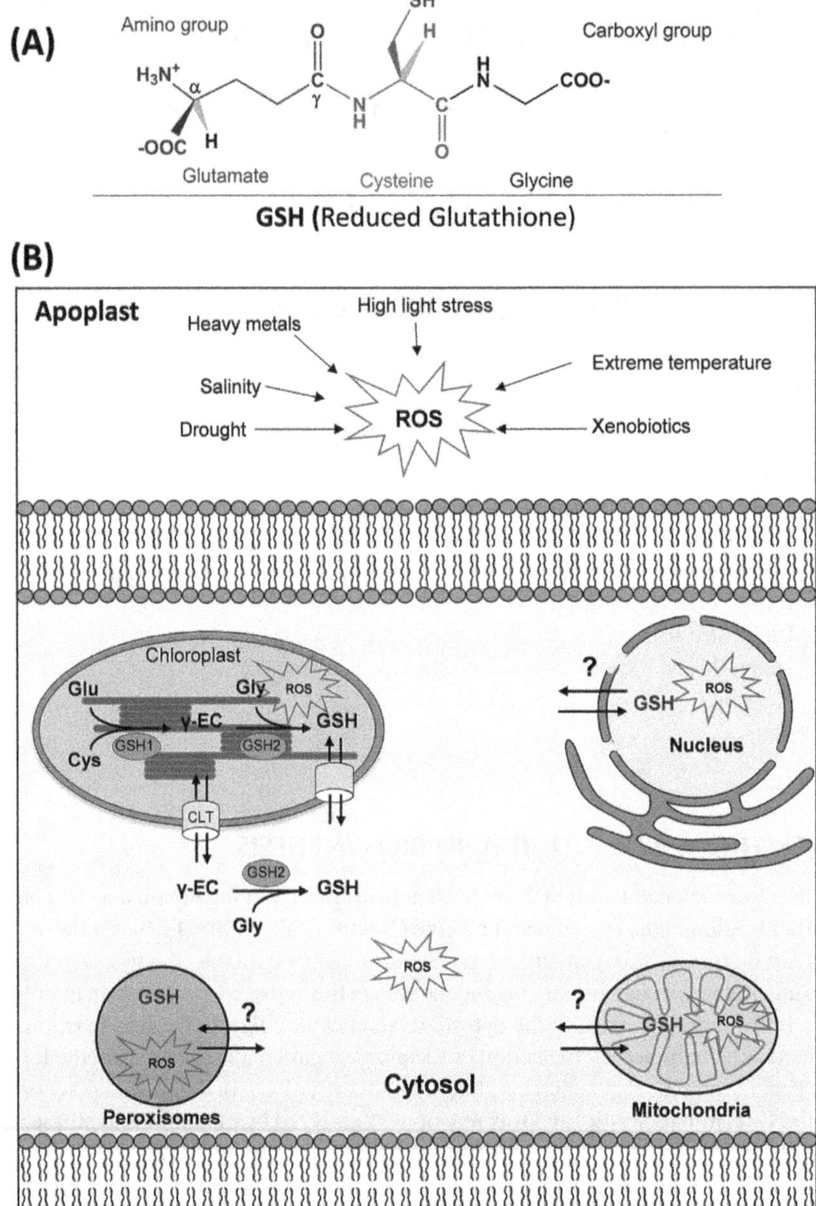

FIGURE 4.1 (A) 2D Structure of reduced glutathione (GSH). Tripeptide GSH consists of glutamate, cysteine and glycine. (B) Compartmentation of glutathione in plant cells. The figure reports the key roles of GSH in defense responses activated in plant cells against environmental stresses. Unfavorable environmental conditions result in the formation of reactive oxygen species (ROS) in different subcellular compartments, which may intracellularly activate anti-oxidants (e.g., key player GSH) and defense responses leading to plant acclimation.

Moreover, the reactivity of GSH depends on the thiol (-SH) group of cysteine, which is the key feature for the biological activities of GSH. The -SH group acts as an effective electron donor and acceptor for many biological reactions (Xiang et al. 2001). The glutathione reactivity, stability and high water solubility make it an ideal biomolecule to protect plants against abiotic stresses. It has several different functions in the cell such as detoxification of xenobiotics (Dixon et al. 1998), sequestration of heavy metals (Cobbett and Goldsbrough 2002) and defense against ROS (Foyer and Noctor 2005a). In addition, it is involved in numerous aspects of the life of plants: GSH is a substrate for different enzymes, which are also involved in the detoxification of ROS (Noctor et al. 2002), and GSH is one of the major forms of the long-distance transport of organic sulfur (Herschbach and Rennenberg 1994). Several different types of transporters may be important in the translocation of glutathione between subcellular compartments (Noctor et al. 2012).

4.2 GLUTATHIONE-RELATED ENZYMES

The destructive effects of elevated ROS levels in plants are controlled by different scavenging enzymatic and non-enzymatic mechanisms. The enzymatic scavenging system includes, for example, superoxide dismutase (SOD), catalase (CAT), ascorbate peroxidase (APX), monodehydroascorbate reductase (MDHAR), dehydroascorbate reductase (DHAR), glutathione reductase (GR), glutathione peroxidase (GPOX) and glutathione S-transferase (GST). The non-enzymatic system includes antioxidant molecules such as ascorbate (ASC), glutathione (GSH), tocopherols, proline, flavonoids, carotenoids, phenolic compounds, cytochromes and polyamines (Ahmad et al. 2010; Das and Roychoudhury 2014). Within the antioxidant system in plants, the ascorbate-glutathione pathway plays a central role in ROS homeostasis, which combines both enzymatic and non-enzymatic antioxidant components (Foyer and Noctor 2011). The pathway ensures the electron flow from NADPH to H_2O_2. In the first reaction, H_2O_2 is reduced to H_2O by APX, using ASC as the electron donor. Meanwhile, ASC is oxidized to monodehydroascorbate, which is reduced back to ASC by MDHAR or spontaneously converted to dehydroascorbate (DHA). Dehydroascorbate is reduced to ASC by DHAR, using the electron from GSH, which is therefore oxidized to glutathione disulfide (GSSG). As the final step of the cycle, the regeneration of GSSG back to GSH is catalyzed by GR, using NADPH as an electron donor (Foyer and Noctor 2011; Noctor et al. 2011).

GR is a flavoprotein belonging to the family of NADPH-dependent oxidoreductases. It catalyzes the reduction of GSSG to GSH and plays an essential role in cell defense against ROS. The catalytic activity of GR is a two-step process. In the first step, NADPH reduces flavin partition, then flavin is oxidized and a redox-active disulfide bridge is reduced, thereby producing a thiolate anion and a cysteine. In the second step, GSSG is reduced via the thiol–disulfide interchange reaction (Rao and Reddy 2008; Trivedi et al. 2013). GR activity was increased in many plant species under different abiotic stresses (Gill et al. 2013).

Dehydroascorbate reductase (EC 1.8.5.1), also known as glutathione dehydrogenase, is an oxidoreductase that recycles DHA to ASC, and so limits ROS-induced damage (Gallie 2012; Noshi et al. 2016). To one side it maintains the reduced ascorbate pool, and therefore also has roles in cell division, plant growth by cell wall metabolism and in different plant physiological processes, like photosynthesis (Smirnoff and Wheeler 2000; Chen and Gallie 2008).

Among the enzymes related to GSH metabolism, glutathione S-transferases (GSTs, EC 2.5.1.18) catalyze the binding of various xenobiotics (including numerous pesticides) and their electrophilic metabolites with GSH to produce less toxic and more water-soluble conjugates, so this isoenzyme family has a distinct role in plant detoxification reactions (Marrs 1996, Edwards et al. 2000). The GST enzyme subunits combine in homo- or heterodimeric form (Bartling et al. 1993). Numerous abiotic stress factors induce GST activity in plants (Dixon et al. 1998). Besides catalyzing the conjugation of electrophilic compounds to GSH, some GST isoenzymes also show peroxidase activity.

Glutathione peroxidases (GPOX, E.C. 1.11.1.9) are a family of isoenzymes that catalyze the reduction of H_2O_2 and organic hydroperoxides including lipid- and phospholipid hydroperoxides by using GSH. These isoenzymes help to protect cells against oxidative damage. Two decades ago, indications for the existence of GPOX enzyme activity from GST isoenzymes in plants were reported (Bartling et al. 1993, Edwards 1996, Eshdat et al. 1997, Dixon et al. 1998). There are glutathione peroxidase-like enzymes in plants also, however they prefer thioredoxins as a reducing substrate (Iqbal et al. 2006, Navrot et al. 2006).

The glyoxalase system is a GSH-dependent detoxification system in plants, where the glyoxalase I (Gly I; EC 4.4.1.5) and glyoxalase II (Gly II; EC 3.1.2.6) enzymes transform methylglyoxal to D-lactate in a two-step reaction (Hossain et al. 2011; Hoque et al. 2016). First methylglyoxal with GSH forms hemithioacetal, which is the substrate of Gly I. This enzyme forms the hemithiolacetal into S-D-lactoylglutathione, then Gly II converts the molecule further to D-lactate. Even though the glyoxalase system is involved in different plant physiological processes, one of its major functions is the role in plant abiotic stress tolerance (Hossain et al. 2009, 2014; Hossain and Fujita 2009; Hoque et al. 2016).

Phytochelatins (PCs) are the most important heavy metal chelators in living organisms, and have a high affinity toward heavy metals (Chia et al. 2013; Lee and Hwang 2015). PCs are synthesized by phytochelatin synthase (EC 2.3.2.15) in cytosol with GSH as the precursor. Both GSH and PCs chelate heavy metals and metalloids, and as the final result, deliver them into vacuoles (Cobbett and Goldsbrough 2002; Pilon-Smits 2005). Moreover, PCs have also important roles in tolerance to other abiotic stresses (Chaurasia et al. 2016).

Glutaredoxins (Grx) are small redox enzymes that are oxidized by their substrate and use GSH as a regenerator reducing substrate. They have a role in antioxidant defense, but on the other hand, Grxs are able to bind and deliver iron–sulfur clusters, also involved in the deglutathionylation process and in the regeneration of multiple enzymes, such as peroxiredoxins and methionine sulfoxide reductases (Rouhier et al. 2008; Rouhier 2010; Lillig and Berndt 2013).

4.3 THE ROLE OF GLUTATHIONE IN ABIOTIC STRESS RESPONSES

Plants are constantly exposed to a wide range of biotic and abiotic stresses such as high salinity, drought, extremes in temperature and heavy metals.

4.3.1 Drought

For several decades, significant increases in the extent and severity of drought have been estimated worldwide (Sheffield and Wood, 2008; Dai, 2011). It is one of the ongoing impacts of climate change. In agriculture, drought is the lack of an adequate amount of water required for normal plant growth/development to complete the life cycle (Manivannan et al. 2008). Water shortage is expected to lead to a drastic reduction in crop productivity by restricting the yield of many crops throughout the world. Therefore, it is a major challenge for agriculture in the context of climate change along with an increasing demand for food (Tardieu et al. 2018).

Under drought stress plants respond mainly by stomatal closure, thus reducing endogenous water loss. Drought-induced stomatal closure is regulated by a number of phytohormones such as jasmonic acid, brassinosteroids, cytokinins and ethylene, but mainly by abscisic acid (ABA) (Daszkowska-Golec and Szarejko 2013).

When plants experience drought stress, they accumulate ABA in the xylem sap and move to guard cells where the signaling network is activated; that causes shrinking of the guard cells which results in stomatal closure (Mittler and Blumwald 2015). Moreover, drought-induced impairment of stomatal conductance, membrane electron transport rate, CO_2 diffusion, carboxylation efficiency and photosynthesis leads to the generation of ROS in the apoplast, which causes oxidative damage. Ultimately, drought results in limited growth and developmental processes and lower final crop yield (Pinheiro and Chaves 2010). It has been shown that ROS generation mainly occurs by the activation of a plasma-membrane NADPH oxidase (Drerup et al. 2013; Nordzieke and Medraño-Fernandez 2018) and apoplastic SOD (Sirichandra et al. 2009). It then leads to the accumulation of H_2O_2 in the apoplast and intracellular compartments (Drerup et al. 2013). Moreover, apoplastic ROS produced by NADPH oxidases are essential for ABA-induced stomatal closure (Sierla et al. 2016). In addition, recently genetic evidence has shown the importance of ROS production by RBOHD, and RBOHF (the plasma membrane-localized NADPH oxidases), during sulfate-induced stomatal closure (Rajab et al. 2019).

Numerous studies have confirmed the role of GSH in drought stress via ABA-induced stomatal closure (Jahan et al. 2008; Okuma et al. 2011; Akter et al. 2012, 2013). It has been shown that glutathione was involved in stomatal closure in the *Arabidopsis cad2-1* mutant (impaired in glutathione biosynthesis). Reduced GSH in guard cells results in stomatal closure as it regulates ABA signaling (Okuma et al. 2011). In addition, chemical treatment (such as p-nitrobenzyl chloride, iodomethane) decreases GSH content in guard cells, in turn, increases the ROS accumulation significantly and thereby enhances guard cell sensitivity to ABA signaling (Okuma et al. 2011; Akter et al. 2012; Munemasa et al. 2013).

The signaling roles of GSH have been shown in *Arabidopsis thaliana*, Col-0 (wild type), and the GSH (*pad2-1*) and AsA (*vtc2-1*) deficient mutants under drought stress. It was observed that the GSH content strongly decreased in both mutants at the early stages of drought stress; consequently, both mutants showed lower activities of antioxidative enzymes, such as GR and DHAR in chloroplasts and peroxisomes. Impaired activity of these enzymes in mutants occurs due to the malfunctioning of the Calvin cycle and the occurrence of photorespiration under drought stress, which causes higher accumulation of ROS compared to Col-0 (Koffler et al. 2014).

Moreover, studies of different enzymes, which use GSH as a substrate, have highlighted their role in defense responses to drought. Under drought stress, increased GST levels have been shown in *Zea mays* and *Brassica napus* plants (Wang et al. 2016; Zhao et al. 2016). It has been revealed that GST17 is one of the most active GSTs in *Arabidopsis*. The *Arabidopsis* mutant with impaired GST17 (*atgst17*) shows higher GSH content compared to the wild type; therefore it has increased tolerance to drought. The increased GSH level activates ABA synthesis and in turn the ABA-protective effect against drought (Chen et al. 2012).

The higher glutathione content improved drought stress tolerance, and global translational changes have been shown in *Arabidopsis*. In addition, the transcriptomic study revealed the activation of abscisic acid, auxin, jasmonic acid biosynthesis and signaling genes during GSH treatment (Cheng et al. 2015).

Furthermore, treatment of the *Arabidopsis* mutant *chlorinal-1* (*chl-1*; defective in light-harvesting complexes and glutathione) with glutathione-monoethyl-ester restored the glutathione content of guard cells and the wild type phenotype, thus confirming the role of glutathione in controlling guard cell sensitivity to ABA (Jahan et al. 2008; Okuma et al. 2011). Jahan et al. (2016) suggested that defective glutathione biosynthesis in guard cells inhibits light-induced stomatal opening.

Bright et al. (2006) reported another reactive species, nitric oxide (NO), which is involved in the ABA-induced signaling pathway that promotes stomatal closure. Glutathione may play a role in this process since S-nitrosoglutathione (GSNO) represents storage NO compound in the cells, which can act as a trans-nitrosylating agent (de Pinto et al. 2013; Locato et al. 2016). GSNO is a nitric oxide-derived molecule, generated by the interaction of NO with GSH in a process called *S*-nitrosylation (Corpas et al. 2013). GSNO levels in cells are regulated by the enzyme GSNO reductase (GSNOR), which in turn controls the thiol levels of the cells (Locato et al. 2016). *Arabidopsis thaliana* GSNOR-defective mutants (*gsnor1-3*) over-accumulate S-nitroso-thiols (SNO) in guard cells and are insensitive to ABA-induced stomatal closure (Wang et al. 2015).

Drought stress regulates the sulfur assimilation pathway, which results in differential accumulation of the primary sulfur metabolism compounds (e.g., glutathione) and the secondary sulfur metabolism compounds (e.g., 3′-phosphoadenosine 5′-phosphate, PAP) (Estavillo et al. 2011; Ahmad et al. 2016). Therefore, glutathione acts as a redox buffer in different subcellular compartments during stress-induced accumulation of ROS (Meyer et al. 2007; Marty et al. 2009; Foyer et al. 2011; Noctor et al. 2012).

4.3.2 Salt Stress

Salt stress is the condition in soil characterized by a high concentration of soluble salts (approximately 40 mM) that is unfavorable for plant growth (Munns and Tester 2008). It is one of the most detrimental abiotic stresses, and critically damages agricultural crops and causes major reductions in their yield across the globe (Munns and Tester 2008; Shrivastava et al. 2015; Tanveer and Shah 2017; Chrysargyris et al. 2018). Plants can tolerate salt stress by maintaining the K^+/Na^+ ratio within a physiological range (Munns and Tester 2008). Hyperaccumulation of Na^+ in plant tissues results in ionic toxicity, which decreases cellular K^+ content since Na^+ competes with K^+ for intracellular transport. This decrease in cellular K^+ levels has a harmful effect on different metabolic pathways (Gupta and Huang 2014). Moreover, it has been shown that under salt stress, ROS activate guard cell outward rectifying potassium (GORK) channels, and as a consequence decrease K^+ levels in the cell. This ion leakage derived from the activation of GORK channels leads to programmed cell death (Demidchik et al. 2010). Chakraborty et al. (2016) confirmed that salt-tolerant *Brassica napus* has a higher ability to retain K^+ by reducing root K^+ permeable channel sensitivity to ROS. Salinity stress results in osmotic stress, ionic stress and oxidative stress, and together these effects reduce plant growth, development and survival rate (Hasegawa et al. 2000, Munns 2002). Among these effects, oxidative stress is considered to be the most harmful. Oxidative stress induces the production of different types of ROS at both intra- and extracellular levels. The toxic effects of the ROS are counteracted with a number of different enzymatic and non-enzymatic antioxidants (Bela et al. 2018).

Glutathione, which belongs to the non-enzymatic antioxidants, can aid in improving plant performance under salt stress. Several studies have shown the involvement of the GSH network in the defense response to salt. Enhanced GSH levels have been correlated with increased salt tolerance (Zagorchev et al. 2013). Under salt stress, glutathione helps to maintain cellular redox balance and performs signaling functions in plants. As an antioxidant, glutathione helps in reducing oxidative stress, preventing lipid peroxidation and protecting the plasma membrane, thus stabilizing the plasma membrane that in turn reduces passive Na^+ influx, which enhances plant salt tolerance (Foyer and Noctor 2005b). In addition, sulfur supplementation has been found to reduce salt-dependent oxidative stress in mustard by increasing GSH levels and increasing salt stress tolerance (Fatma et al. 2016). Moreover, wild type canola seedlings under salt stress showed a threefold increase in glutathione and cysteine synthesis compared to mutant plants that lacked the genes for glutathione and cysteine synthesis. In addition, the wild type also showed greater salinity tolerance than mutant plants, suggesting a protective role of GSH against salt stress (Ruiz and Blumwald 2002).

Roxas et al. (1997) proved that seedling growth in media containing salt is regulated by glutathione or the activities of glutathione-dependent and -regulating enzymes. They confirmed that, under salt stress, transgenic tobacco plants overexpressing both GST and GPX biosynthesis genes showed stimulated seedling growth. This effect might be due to the regulation of the glutathione pool (Roxas et al. 1997). Exogenous application of both proline and glycine betaine provide a protective function against NaCl-induced oxidative damage by reducing protein carbonylation, and an improved antioxidant defense and methylglyoxal detoxification systems. Furthermore, increased glutathione levels and activities of GPX, GST and Gly I were associated with salt stress tolerance in tobacco (Hoque et al. 2008).

Different types of GSTs play important roles in plant defense responses activated by NO signaling under salt stress. Moreover, under salt stress, expressions of GST genes are stimulated in soybean leaves. However, the regulation of GST enzymatic activity under stress conditions fluctuated throughout the treatment time (Dinler et al. 2014). The ectopic expression of tomato and rice GSTs in *Arabidopsis* seems to increase the tolerance to salt stress (Sharma et al. 2014; Xu et al. 2015). Xu et al. (2015) transformed *Arabidopsis* plants with tomato glutathione S-transferase (*LeGSTU2*). They found that the gene from tomato plays a positive role in improving tolerance to salinity and drought stresses in *Arabidopsis* by increasing the activity of antioxidant enzymes.

Jia et al. (2016) demonstrated the positive function of the *GsGSTU13* gene under salt by overexpressing it in the *Medicago sativa* plant. They found that under alkaline stress *GsGSTU13* transgenic lines showed better growth and physiological indicators compared to the wild type. They also suggested that *GsGSTU13* acts as a positive regulator in plant responses to salt stresses, and can be used as a good candidate for the generation of tolerant crops. In halophyte (*Eutrema salsugineum*), 300 mM NaCl caused a significant increase in hydrophilic antioxidants (ascorbate, total glutathione) (Wiciarz et al. 2018).

GPX is another player that plays an important role in plant response to salt stress (Islam et al. 2015; Pilarska et al. 2016; Bela et al. 2018). These enzymes are well-known for the protection of cells against ROS-dependent damage. Moreover, in plants, GPXs use glutathione and thioredoxin as a substrate to reduce hydroperoxides (Passaia and Margis-Pinheiro 2015). Islam et al. (2015) confirmed the role of GPX enzyme in increased salinity tolerance in rice, by expressing GPX from *Pennisetum glaucum*. El-Shabrawi et al. (2010) analyzed two different rice cultivars, salt-tolerant (Pokkali) and salt-sensitive (IR64), and showed that under salt stress the glutathione level was significantly higher in Pokkali compared to IR64. In addition, Pakkali compared to IR64 showed increased activity of different enzymatic antioxidants and higher GSH/GSSG and ascorbate (AsA)/dehydroascorbate (DHA) ratios.

In several studies it has been shown that the exogenous application of glutathione during salt stress regulates both enzymatic and non-enzymatic antioxidants (Foyer and Noctor 2005a; Khattab 2007; Aly-Salama and Al-Mutawa 2009). Furthermore, Khattab (2007) proved that under salt stress glutathione could prevent the loss of photosynthetic pigments in two canola cultivars. These responses indicate the fundamental mechanisms responsible for improved plant growth under salt stress induced by glutathione.

Furthermore, numerous studies verified the hormone-induced regulation of glutathione along with thecontribution of these hormones to abiotic stress tolerance. Bashandy et al. (2010) demonstrated a direct link between cellular redox status and auxin signaling by analyzing thioredoxin reductase and glutathione biosynthesis mutants of *Arabidopsis*. Translatome analysis in *Arabidopsis* also revealed that abscisic acid, auxin and jasmonic acid biosynthesis and signaling genes were activated during GSH treatment, which has not been reported in previously published transcriptomic data. In addition, they suggested that the increased glutathione level results in stress tolerance and global translational changes (Cheng et al. 2015). Furthermore, in the auxin autotrophic tobacco callus lines, the activities of antioxidant enzymes such as GST and GPX were higher, which were suggested to enhance salt stress tolerance. The higher activities of GST and GPX under NaCl stress (100 mmol/L NaCl) led to reduced levels of H_2O_2 and MDA (Csiszár et al. 2004). Exogenous cytokinin (6-benzyladenine) increased ascorbate, glutathione and proline contents in eggplant under salt stress (Wu et al. 2014). In mung bean (*Vigna radiata* L.), exogenous polyamines (such as spermine) improved glutathione levels and the GSH/GSSG ratio, and activities of GR and GPX which defended against the overproduction of ROS. In addition, polyamines application conferred salt tolerance by reducing sodium uptake, improving nutrient homeostasis, antioxidant defense and methylglyoxal detoxification systems (Nahar et al. 2016b).

Salt stress induces the catabolism of carbohydrates and amino acids, which in turn produce mainly a cytotoxic compound, called methylglyoxal (MG). Apart from salt stress tolerance, glutathione also contributes to MG detoxification via the glyoxalase system (Yadav et al. 2005). In rice, Gly II seems to be responsive to glutathione levels, and its activity appears to be correlated to salt tolerance (Singla-Pareek et al. 2008). Nahar et al. (2015a) suggested that under salt stress (200 mM NaCl, 24 and 48 h) exogenous glutathione effectively reduced oxidative damage and MG toxicity, and improved the physiological adaptation of mung bean (*Vigna radiata* L.). Furthermore, increased activities of Gly II enhanced glutathione levels, which contribute to diminishing toxic MG levels (Nahar et al. 2016a, b).

4.3.3 Extreme Temperatures

Extreme temperatures, either too high or too low, are one of the main reasons for reduced crop yield, or distribution limitation (Hasanuzzaman et al. 2013). As with other environmental stresses, heat or chilling also induces ROS production, and oxidative stress, in which the elements of the antioxidant system are essential to protect the cells. The non-enzymatic GSH pool and the redox state of glutathione all together have a pivotal role in stress responses and in the tolerance to extreme temperatures (Szalai et al. 2009).

Elevated glutathione content has been measured in heat-tolerant variants of wheat; the results indicate the positive correlation between the glutathione level and heat stress tolerance (Sairam et al. 2000; Dash and Mohanty 2002). Glutathione biosynthesis was also induced after applying high temperatures in different plant species, such as wheat (Kocsy et al. 2004), maize (Kocsy et al. 2002), rice (Kumar et al. 2012), tomato (Rivero et al. 2004), mustard (Dat et al. 1998), apple (Zhang et al. 2008) or in common beech (Peltzer et al. 2002). Exogenous pretreatment with GSH seems to be also an effective heat tolerance inducer in the case of mung bean, by stimulating the glyoxalase and other antioxidant systems (Nahar et al. 2015b). As mentioned before, not only the content but the redox state of the glutathione also affects the current stress response. Thus, induction of GR activity was commonly observed in short-term heat stress-exposed plants, like *Phalaenopsis* (Ali et al. 2005) and wheat (Wang et al. 2014). However long-term treatment often caused a relapse of GR activity or even reduction of glutathione levels (Sairam et al. 2000; Chaitanya et al. 2002; Ma et al. 2008; Nagesh Babu and Devaraj 2008).

Chilling causes oxidation of the glutathione pool by detoxification of the increased H_2O_2 level. This redox change is a signal toward stimulation of the defense mechanisms, like glutathione biosynthesis and activation of enzymes (Kocsy et al. 2001). Consequently, higher total glutathione levels and a chilling-triggered increase in the activity of GR have been reported in several plant species, including tomato (Walker and McKersie 1993), poplar (Foyer et al. 1995), strawberry (Luo et al. 2011), cucumber (Lee et al. 2000), cotton (Kornyeyev et al. 2003), wheat and maize (Kocsy et al. 2002). However, it is not a general rule, and exceptions always can be found. GR activity did not change in some wheat cultivars (Yordanova and Popova 2007); in rice it even declined (Huang and Guo 2005) after the application of low temperatures.

4.3.4 Light Stress

Plants are photosynthetic organisms that use light energy as an energy source to produce organic materials. However, unnecessary or extreme intensity of light causes damage in the cells by ROS production. To save the cellular compounds and molecules from this degradation, plants rapidly activate their antioxidant defense system (Foyer et al. 1994).

It was reported in many studies that plants respond with elevated glutathione levels to the applied abiotic stresses, but in case of excess light, it is a bit different. In *Arabidopsis* plants the glutathione pool did not differ between growing under normal and high light conditions (Müller-Moulé et al. 2004; Ogawa et al. 2004), however, shifting plants from low light to high light initially caused changes in the glutathione pool and also in the redox state (Müller-Moulé et al. 2004). But overexpression of the chloroplastic GR enhanced the tolerance to high light stress in tobacco (Aono et al. 1993), poplar (Foyer et al. 1995) and wheat (Melchiorre et al. 2009). Interestingly, overexpressing plastid-targeted *GSH2*, thereby increasing the plastidic glutathione pool, led to enlarged oxidative stress in high light conditions, which could be because of the damage of the redox-sensing process (Creissen et al. 1999). Plants have an extremely finely tuned signaling and protection system; sometimes overproduction of one element causes damage rather than an advantage.

4.3.5 Heavy Metal Exposure

Elements that have a density higher than 5 g cm^{-3} are called heavy metals. Exposure to these heavy metals causes oxidative stress by ROS formation in the Fenton reaction, leading to damage to membranes and proteins. Some plants try to avoid the uptake of these heavy metals (Hernández et al. 2015); others attempt to chelate metals in the cells and move them into the vacuoles (Anjum et al. 2015). Glutathione has the utmost importance in heavy metal detoxification processes as a metal chelator, a cellular antioxidant, a substrate of GSTs and a precursor of PCs (Cobbett and Goldsbrough 2002; Rouhier et al. 2008; Szalai et al. 2009; Hossain et al. 2012).

Cadmium, mercury, arsenic, chromium and lead are typical toxic heavy metals, causing serious problems in plant growth, development and cellular processes (Yadav 2010). The relation between glutathione content and cadmium stress is interesting. It has been proved that decreased glutathione content, either in biosynthesis mutants or by applying buthionine sulfoximine (BSO), led to increased cadmium toxicity in plants (Howden et al. 1995; Vernoux et al. 2000; Xiang et al. 2001; Schat et al. 2002; Ammar et al. 2008). However, elevated glutathione content by the overexpression of biosynthesis enzyme or by exogenously applied glutathione was not beneficial in cadmium toxicity in every species. Exogenously applied glutathione helped to maintain the growth of barley, tomato and mustard plants under cadmium stress (Zhu et al. 1999; Chen et al. 2010; Wang et al. 2011; Hasan et al. 2016), but in *Arabidopsis* it caused cadmium-hypersensitivity (Xiang et al. 2001; Lee et al. 2003; Peterson and Oliver 2006; Wójcik and Tukiendorf 2011). Lead accumulator *Sedum alfredii* ecotypes synthetized glutathione rather than PCs after lead treatment, which indicates the key role of GSH in heavy metal defense mechanisms (Gupta et al. 2010). Maintaining the glutathione content and its reduced state was also important in lead-treated rice plants, where the GR activity increased in a concentration-dependent manner (Verma and Dubey 2003). Exogenously applied glutathione was even able to amend the damaged germination of wheat, caused by mercury (Popa et al. 2007).

Copper, zinc, nickel, manganese and iron are essential elements for optimal plant development, however, in higher concentration, these also cause chlorosis and reduce plant growth (Yadav 2010). Applying a toxic concentration of copper to bean or *Arabidopsis* plants temporarily increased the glutathione content; longer treatments caused depletion of glutathione, probably because of the PCs synthesis (Cuypers et al. 2000; Drążkiewicz et al. 2003). The depletion of glutathione and homoglutathione by the application of BSO had an effect on the copper stress response of alfalfa; plants showed increased biomass reduction (Flores-Cáceres et al. 2015). In the case of zinc treatment, poplar plants responded with elevated total glutathione content (Di Baccio et al. 2005), and the protective role of this non-enzymatic antioxidant was proved with transgenic poplars with elevated glutathione level, but the balance between GSH and PCs is important (Bittsánszky et al. 2005). Maintenance of the GSH pool, and inducing the ascorbate-glutathione cycle-related enzymes were the key responses in nickel-treated wheat and rice (Gajewska et al. 2006; Maheshwari and Dubey 2009), and manganese-treated barley plants (Demirevska-Kepova et al. 2004).

4.3.6 Xenobiotics

In plants, xenobiotics, such as herbicides or insecticides, are converted into intermediates with reduced phytotoxicity by a three-phase process. During the first phase, oxidation, reduction or hydrolysis takes place. These initial reactions are followed by conjugation reactions in the second and third phases. The final result of these metabolic conversions is movement of the altered xenobiotics into the vacuole or cell wall (Sandermann 1992). Glutathione has a major rule in this process as one of the conjugating molecules. The glutathione reaction with toxic xenobiotics is catalyzed by GSTs (Edwards et al. 2000). Treatment with xenobiotics can lead to oxidative damage also, in which the GR enzyme has a pivotal role to play in maintaining the reduced glutathione pool.

2,4-dichlorophenoxyacetic acid (2,4-D) was the first successful selective herbicide; this synthetic auxin compound in higher concentrations kills dicots, therefore it is commercially used on wheat, maize and rice fields (Grossmann et al. 2000). Investigating the effect of this herbicide on the physiological response of pea plants, 2,4-D caused elevated total glutathione content in young and adult leaves; moreover, the GSH/GSSG ratio was higher in adult leaves. Meanwhile, GSTs transcript levels and enzyme activity also increased after treatment (Pazmino et al. 2011).

1,1′-dimethyl-4,4′-bipyridynium dichloride (paraquat) is a nonselective, soil-inactivated herbicide that catalyzes the production of superoxide from the water via photosynthetic electron transport, therefore leading to oxidative stress (Hawkes 2014). Treatment with paraquat caused elevated total glutathione content and activation of the GR enzyme in different plant species (Chiang et al. 2008; Ding et al. 2009; Gao et al. 2011; Hawkes 2014). In paraquat-tolerant rice plants, the reduced glutathione pool was maintained with significantly higher GR activity, thus dealing with oxidative damage (Tseng et al. 2013).

2-chloro-4-ethylamino-6-isopropylamino-s-triazine (atrazine) reduces electron flow from water to $NADP^+$ at the photochemical step in photosynthesis, which leads to oxidative damage. Increased activity of GR alleviates again this oxidative stress, maintaining the GSH/GSSG ratio. The activation of GR and GST enzymes was detected in atrazine-treated wheat plants (Del Buono et al. 2011).

Another nonselective, soil-inactivated herbicide is glyphosate, which inhibits the enolpyruvylshikimate-3-phosphate synthase enzyme, and so damages the synthesis of aromatic amino acids. Moreover, it can also harmfully affect other enzymes, and cause oxidative damage (Miteva et al. 2010). In pea and wheat plants this herbicide caused elevated glutathione levels and provoked the GR and GST activities (Uotila et al. 1995; Miteva et al. 2010), but in barley, tobacco, soybean and maize this has not been experienced (Smith 1985).

Having a look at the most commonly used herbicides we can conclude that glutathione and related enzymes (GR, GST) are the main factors to take the fight against the damage caused.

4.4 CONCLUSIONS

Abiotic stresses such as drought, salinity, heavy metal stress, xenobiotics, extreme temperature and high light affect global crop productivity. All these stress factors ultimately lead to ROS accumulation in the cells. To combat these stresses, plants already possess mechanisms to detoxify the elevated ROS by enzymatic and non-enzymatic antioxidant systems. A large number of studies have confirmed that glutathione, as a non-enzymatic redox player, occupies a central position in the antioxidant defense system and thus, plays a key role in plant protection against abiotic stresses. Glutathione has several different functions; it is involved in ROS signaling and detoxification, heavy metals sequestration, xenobiotic detoxification, MG signaling and detoxification, regulation of the expression of stress tolerance-related genes, protection of cellular structures and reproductive development. Glutathione performs all these functions by coordination with a number of antioxidant enzymes and phytohormones.

Thus, future research studies need to answer the key questions related to glutathione transport at the cellular and subcellular (cytosol, plastids, mitochondria and peroxisomes) levels in plants. Furthermore, there is a need to explore subcellular glutathione homeostasis under abiotic stresses. Several studies have also highlighted the importance of glutathione homeostasis among specific cell compartments. The variability in glutathione distribution within cells can modulate the redox state of specific compartments and may be part of the signaling pathways involved in defense responses. Nonetheless, the results with respect to overall stress responses are still extremely variable and, thus, need to be further explored. In addition, cross-talk of GSH with other phytohormones (such as jasmonic acid, salicylic acid and ethylene) and signaling compounds, like nitric oxide, H_2O_2 and Ca^{2+}, needs to be resolved. Besides the multiple functionalities of glutathione, its toxicity needs to be explored since a high amount of exogenous glutathione supplementation inhibits plant growth.

Moreover, the role of GSH is not clearly explained in several posttranslational modifications such as nitrosylation, hydroxylation and glutathionylation, which are involved in stress signaling. Additionally, more studies are needed to transform the current knowledge on the applied side to improve agriculture crops to combat climate change and abiotic stresses.

REFERENCES

Ahmad, Nisar, Mario Malagoli, Markus Wirtz, and Ruediger Hell. Drought stress in maize causes differential acclimation responses of glutathione and sulfur metabolism in leaves and roots. *BMC Plant Biology* 16, no. 1 (2016): 247.

Ahmad, Parvaiz, Shahid Umar, and Satyawati Sharma. Mechanism of free radical scavenging and role of phytohormones in plants under abiotic stresses. In: M. Ashraf, M. Ozturk, M. S. A. Ahmad, Eds., *Plant Adaptation and Phytoremediation*, pp. 99–118. Springer, Dordrecht, 2010.

Akter, Nasima, Muhammad Abdus Sobahan, Misugi Uraji, Wenxiu Ye, Mohammad Anowar Hossain, Izumi C. Mori, Yoshimasa Nakamura, and Yoshiyuki Murata. Effects of depletion of glutathione on abscisic acid-and methyl jasmonate-induced stomatal closure in *Arabidopsis thaliana. Bioscience, Biotechnology, and Biochemistry* 76, no. 11 (2012): 2032–2037.

Akter, Nasima, Eiji Okuma, Muhammad Abdus Sobahan, Misugi Uraji, Shintaro Munemasa, Yoshimasa Nakamura, Izumi C. Mori, and Yoshiyuki Murata. Negative regulation of methyl jasmonate-induced stomatal closure by glutathione in *Arabidopsis. Journal of Plant Growth Regulation* 32, no. 1 (2013): 208–215.

Ali, Mohammad Babar, Eun-Joo Hahn, and Kee-Yoeup Paek. Effects of temperature on oxidative stress defense systems, lipid peroxidation and lipoxygenase activity in *Phalaenopsis. Plant Physiology and Biochemistry* 43, no. 3 (2005): 213–223.

Aly-Salama, Karima Hamed, and M. M. Al-Mutawa. Glutathione-triggered mitigation in salt-induced alterations in plasmalemma of onion epidermal cells. *International Journal of Agriculture and Biology (Pakistan)* 11, no. 5 (2009): 639–642.

Ammar, W. Ben, C. Mediouni, B. Tray, M. H. Ghorbel, and F. Jemal. Glutathione and phytochelatin contents in tomato plants exposed to cadmium. *Biologia Plantarum* 52, no. 2 (2008): 314.

Anjum, Naser A., Mirza Hasanuzzaman, Mohammad A. Hossain, Palaniswamy Thangavel, Aryadeep Roychoudhury, Sarvajeet S. Gill, Miguel A. Merlos Rodrigo, et al. Jacks of metal/metalloid chelation trade in plants—an overview. *Frontiers in Plant Science* 6 (2015): 192.

Aono, Mitsuko, Akihiro Kubo, Hikaru Saji, Kiyoshi Tanaka, and Noriaki Kondo. Enhanced tolerance to photooxidative stress of transgenic *Nicotiana tabacum* with high chloroplastic glutathione reductase activity. *Plant and Cell Physiology* 34, no. 1 (1993): 129–135.

Bartling, Dieter, Renate Radzio, Ulrike Steiner, and Elmar W. Weiler. A glutathione S-transferase with glutathione-peroxidase activity from *Arabidopsis thaliana*: Molecular cloning and functional characterization. *European Journal of Biochemistry* 216, no. 2 (1993): 579–586.

Bashandy, Talaat, Jocelyne Guilleminot, Teva Vernoux, David Caparros-Ruiz, Karin Ljung, Yves Meyer, and Jean-Philippe Reichheld. Interplay between the NADP-linked thioredoxin and glutathione systems in *Arabidopsis* auxin signaling. *The Plant Cell* 22, no. 2 (2010): 376–391.

Bela, Krisztina, Riyazuddin Riyazuddin, Edit Horváth, Ágnes Hurton, Ágnes Gallé, Zoltán Takács, Laura Zsigmond, László Szabados, Irma Tari, and Jolán Csiszár. Comprehensive analysis of antioxidant mechanisms in *Arabidopsis* glutathione peroxidase-like mutants under salt-and osmotic stress reveals organ-specific significance of the AtGPXL's activities. *Environmental and Experimental Botany* 150 (2018): 127–140.

Bittsánszky, András, Tamás Kömives, Gábor Gullner, Gábor Gyulai, József Kiss, László Heszky, László Radimszky, and Heinz Rennenberg. Ability of transgenic poplars with elevated glutathione content to tolerate zinc (2+) stress. *Environment International* 31, no. 2 (2005): 251–254.

Bright, Jo, Radhika Desikan, John T. Hancock, Iain S. Weir, and Steven J. Neill. ABA-induced NO generation and stomatal closure in *Arabidopsis* are dependent on H_2O_2 synthesis. *The Plant Journal* 45, no. 1 (2006): 113–122.

Chaitanya, K. V., D. Sundar, S. Masilamani, and A. Ramachandra Reddy. Variation in heat stress-induced antioxidant enzyme activities among three mulberry cultivars. *Plant Growth Regulation* 36, no. 2 (2002): 175–180.

Chakraborty, Koushik, Jayakumar Bose, Lana Shabala, and Sergey Shabala. Difference in root K+ retention ability and reduced sensitivity of K+-permeable channels to reactive oxygen species confer differential salt tolerance in three *Brassica* species. *Journal of Experimental Botany* 67, no. 15 (2016): 4611–4625.

Chaurasia, Neha, Yogesh Mishra, Antra Chatterjee, Ruchi Rai, Shivam Yadav, and L. C. Rai. Overexpression of phytochelatin synthase (pcs) enhances abiotic stress tolerance by altering the proteome of transformed *Anabaena* sp. PCC 7120. *Protoplasma* 254, no. 4 (2017): 1715–1724.

Chen, Fei, Fang Wang, Feibo Wu, Weihua Mao, Guoping Zhang, and Meixue Zhou. Modulation of exogenous glutathione in antioxidant defense system against Cd stress in the two barley genotypes differing in Cd tolerance. *Plant Physiology and Biochemistry* 48, no. 8 (2010): 663–672.

Chen, Jui-Hung, Han-Wei Jiang, En-Jung Hsieh, Hsing-Yu Chen, Ching-Te Chien, Hsu-Liang Hsieh, and Tsan-Piao Lin. Drought and salt stress tolerance of an *Arabidopsis glutathione S-transferase U17* knockout mutant are attributed to the combined effect of glutathione and abscisic acid. *Plant Physiology* 158, no. 1 (2012): 340–351.

Chen, Zhong, and Daniel R. Gallie. Dehydroascorbate reductase affects non-photochemical quenching and photosynthetic performance. *Journal of Biological Chemistry* 283, no. 31 (2008): 21347–21361.

Cheng, Mei-Chun, Ko Ko, Wan-Ling Chang, Wen-Chieh Kuo, Guan-Hong Chen, and Tsan-Piao Lin. Increased glutathione contributes to stress tolerance and global translational changes in *Arabidopsis*. *The Plant Journal* 83, no. 5 (2015): 926–939.

Chia, Ju-Chen, Chien-Chih Yang, Yu-Ting Sui, Shin-Yu Lin, and Rong-Huay Juang. Tentative identification of the second substrate binding site in *Arabidopsis* phytochelatin synthase. *PloS One* 8, no. 12 (2013): e82675.

Chiang, Yeong-Jene, Yi-Xuan Wu, Mou-Yen Chiang, and Ching-Yuh Wang. Role of antioxidative system in paraquat resistance of tall fleabane (*Conyza sumatrensis*). *Weed Science* 56, no. 3 (2008): 350–355.

Chrysargyris, Antonios, Evgenia Michailidi, and Nikos Tzortzakis. Physiological and biochemical responses of *Lavandula angustifolia* to salinity under mineral foliar application. *Frontiers in Plant Science* 9 (2018): 489.

Cobbett, Christopher, and Peter Goldsbrough. Phytochelatins and metallothioneins: Roles in heavy metal detoxification and homeostasis. *Annual Review of Plant Biology* 53, no. 1 (2002): 159–182.

Corpas, Francisco Javier, Juan de Dios Alché, and Juan B. Barroso. Current overview of S-nitrosoglutathione (GSNO) in higher plants. *Frontiers in Plant Science* 4 (2013): 126.

Creissen, Gary, John Firmin, Michael Fryer, Baldeep Kular, Nicola Leyland, Helen Reynolds, Gabriela Pastori, et al. Elevated glutathione biosynthetic capacity in the chloroplasts of transgenic tobacco plants paradoxically causes increased oxidative stress. *The Plant Cell* 11, no. 7 (1999): 1277–1291.

Csiszár, Jolán, Margit Szabó, László Erdei, László Márton, Ferenc Horváth, and Irma Tari. Auxin autotrophic tobacco callus tissues resist oxidative stress: The importance of glutathione S-transferase and glutathione peroxidase activities in auxin heterotrophic and autotrophic calli. *Journal of Plant Physiology* 161, no. 6 (2004): 691–699.

Cuypers, Ann, Jaco Vangronsveld, and Herman Clijsters. Biphasic effect of copper on the ascorbate-glutathione pathway in primary leaves of *Phaseolus vulgaris* seedlings during the early stages of metal assimilation. *Physiologia Plantarum* 110, no. 4 (2000): 512–517.

Dai, Aiguo. Drought under global warming: A review. *Wiley Interdisciplinary Reviews: Climate Change* 2, no. 1 (2011): 45–65.

Das, Kaushik, and Aryadeep Roychoudhury. Reactive oxygen species (ROS) and response of antioxidants as ROS-scavengers during environmental stress in plants. *Frontiers in Environmental Science* 2 (2014): 53.

Dash, Sasmita, and Narendranath Mohanty. Response of seedlings to heat-stress in cultivars of wheat: Growth temperature-dependent differential modulation of photosystem 1 and 2 activity, and foliar antioxidant defense capacity. *Journal of Plant Physiology* 159, no. 1 (2002): 49–59.

Daszkowska-Golec, Agata, and Iwona Szarejko. Open or close the gate–stomata action under the control of phytohormones in drought stress conditions. *Frontiers in Plant Science* 4 (2013): 138.

Dat, James F., Humberto Lopez-Delgado, Christine H. Foyer, and Ian M. Scott. Parallel changes in H_2O_2 and catalase during thermotolerance induced by salicylic acid or heat acclimation in mustard seedlings. *Plant Physiology* 116, no. 4 (1998): 1351–1357.

de Pinto, Maria Concetta, Vittoria Locato, Alessandra Sgobba, Maria del Carmen Romero-Puertas, Cosimo Gadadeta, Massimo Delledonne, and Laura De Gara. S-Nitrosylation of ascorbate peroxidase is part of the programmed cell death signaling in tobacco By-2 cells. *Plant Physiology* 163, no. 4 (2013): 1766–1775.

Del Buono, Daniele, and Gerardina Ioli. Glutathione S-transferases of Italian ryegrass (*Lolium multiflorum*): Activity toward some chemicals, safener modulation and persistence of atrazine and fluorodifen in the shoots. *Journal of Agricultural and Food Chemistry* 59, no. 4 (2011): 1324–1329.

Demidchik, Vadim, Tracey A. Cuin, Dimitri Svistunenko, Susan J. Smith, Anthony J. Miller, Sergey Shabala, Anatoliy Sokolik, and Vladimir Yurin. *Arabidopsis* root K+-efflux conductance activated by hydroxyl radicals: Single-channel properties, genetic basis and involvement in stress-induced cell death. *Journal of Cell Science* 123, no. 9 (2010): 1468–1479.

Demirevska-Kepova, K., Lyudmila Simova-Stoilova, Z. Stoyanova, Regina Hölzer, and Urs Feller. Biochemical changes in barley plants after excessive supply of copper and manganese. *Environmental and Experimental Botany* 52, no. 3 (2004): 253–266.

Di Baccio, Daniela, Stanislav Kopriva, Luca Sebastiani, and Heinz Rennenberg. Does glutathione metabolism have a role in the defence of poplar against zinc excess? *New Phytologist* 167, no. 1 (2005): 73–80.

Ding, Hai-Dong, Xiao-Hua Zhang, Shu-Cheng Xu, Li-Li Sun, Ming-Yi Jiang, A.-Ying Zhang, and Yin-Gen Jin. Induction of protection against paraquat-induced oxidative damage by abscisic acid in maize leaves is mediated through mitogen-activated protein kinase. *Journal of Integrative Plant Biology* 51, no. 10 (2009): 961–972.

Dinler, Burcu Seckin, Chrystalla Antoniou, and Vasileios Fotopoulos. Interplay between GST and nitric oxide in the early response of soybean (*Glycine max* L.) plants to salinity stress. *Journal of Plant Physiology* 171, no. 18 (2014): 1740–1747.

Dixon, David P., David J. Cole, and Robert Edwards. Purification, regulation and cloning of a glutathione transferase (GST) from maize resembling the auxin-inducible type-III GSTs. *Plant Molecular Biology* 36, no. 1 (1998): 75–87.

Drążkiewicz, Maria, Ewa Skórzyńska-Polit, and Zbigniew Krupa. Response of the ascorbate–glutathione cycle to excess copper in *Arabidopsis thaliana* (L.). *Plant Science* 164, no. 2 (2003): 195–202.

Drerup, Maria Magdalena, Kathrin Schlücking, Kenji Hashimoto, Prabha Manishankar, Leonie Steinhorst, Kazuyuki Kuchitsu, and Jörg Kudla. The calcineurin B-like calcium sensors CBL1 and CBL9 together with their interacting protein kinase CIPK26 regulate the *Arabidopsis* NADPH oxidase RBOHF. *Molecular Plant* 6, no. 2 (2013): 559–569.

Edwards, Robert. Characterisation of glutathione transferases and glutathione peroxidases in pea (*Pisum sativum*). *Physiologia Plantarum* 98, no. 3 (1996): 594–604.

Edwards, Robert, David P. Dixon, and Virginia Walbot. Plant glutathione S-transferases: Enzymes with multiple functions in sickness and in health. *Trends in Plant Science* 5, no. 5 (2000): 193–198.

El-Shabrawi, Hattem, Bhumesh Kumar, Tanushri Kaul, Malireddy K. Reddy, Sneh L. Singla-Pareek, and Sudhir K. Sopory. Redox homeostasis, antioxidant defense, and methylglyoxal detoxification as markers for salt tolerance in Pokkali rice. *Protoplasma* 245, no. 1–4 (2010): 85–96.

Eshdat, Yuval, Doron Holland, Zehava Faltin, and Gozal Ben-Hayyim. Plant glutathione peroxidases. *Physiologia Plantarum* 100, no. 2 (1997): 234–240.

Estavillo, Gonzalo M., Peter A. Crisp, Wannarat Pornsiriwong, Markus Wirtz, Derek Collinge, Chris Carrie, Estelle Giraud, et al. Evidence for a SAL1-PAP chloroplast retrograde pathway that functions in drought and high light signaling in *Arabidopsis*. *The Plant Cell* 23, no. 11 (2011): 3992–4012.

Fatma, Mehar, Asim Masood, Tasir S. Per, and Nafees A. Khan. Nitric oxide alleviates salt stress inhibited photosynthetic performance by interacting with sulfur assimilation in mustard. *Frontiers in Plant Science* 7 (2016): 521.

Flores-Cáceres, María Laura, Sabrine Hattab, Sarra Hattab, Hamadi Boussetta, Mohammed Banni, and Luis E. Hernández. Specific mechanisms of tolerance to copper and cadmium are compromised by a limited concentration of glutathione in alfalfa plants. *Plant Science* 233 (2015): 165–173.

Foyer, Christine H., Maud Lelandais, and Karl J. Kunert. Photooxidative stress in plants. *Physiologia Plantarum* 92, no. 4 (1994): 696–717.

Foyer, Christine H., and Graham Noctor. Oxidant and antioxidant signalling in plants: A re-evaluation of the concept of oxidative stress in a physiological context. *Plant, Cell and Environment* 28, no. 8 (2005a): 1056–1071.

Foyer, Christine H., and Graham Noctor. Redox homeostasis and antioxidant signaling: A metabolic interface between stress perception and physiological responses. *The Plant Cell* 17, no. 7 (2005b): 1866–1875.

Foyer, Christine H., and Graham Noctor. Ascorbate and glutathione: The heart of the redox hub. *Plant Physiology* 155, no. 1 (2011): 2–18.

Foyer, Christine H., Nadège Souriau, Sophie Perret, Maud Lelandais, Karl-Josef Kunert, Christophe Pruvost, and Lise Jouanin. Overexpression of glutathione reductase but not glutathione synthetase leads to increases in antioxidant capacity and resistance to photoinhibition in poplar trees. *Plant Physiology* 109, no. 3 (1995): 1047–1057.

Gajewska, E., M. Skłodowska, M. Słaba, and J. Mazur. Effect of nickel on antioxidative enzyme activities, proline and chlorophyll contents in wheat shoots. *Biologia Plantarum* 50, no. 4 (2006): 653–659.

Galant, Ashley, Mary L. Preuss, Jeffrey Cameron, and Joseph M. Jez. Plant glutathione biosynthesis: Diversity in biochemical regulation and reaction products. *Frontiers in Plant Science* 2 (2011): 45.

Gallie, Daniel R. The role of L-ascorbic acid recycling in responding to environmental stress and in promoting plant growth. *Journal of Experimental Botany* 64, no. 2 (2012): 433–443.

Gao, Y., Y.-K. Guo, A.-H. Dai, W.-J. Sun, and J.-G. Bai. Paraquat pretreatment alters antioxidant enzyme activity and protects chloroplast ultrastructure in heat-stressed cucumber leaves. *Biologia Plantarum* 55, no. 4 (2011): 788.

Gill, Sarvajeet Singh, Naser A. Anjum, Mirza Hasanuzzaman, Ritu Gill, Dipesh Kumar Trivedi, Iqbal Ahmad, Eduarda Pereira, and Narendra Tuteja. Glutathione and glutathione reductase: A boon in disguise for plant abiotic stress defense operations. *Plant Physiology and Biochemistry* 70 (2013): 204–212.

Grossmann, Klaus, Cindy Rosenthal, and Jacek Kwiatkowski. Increases in jasmonic acid caused by indole-3-acetic acid and auxin herbicides in cleavers (*Galium aparine*). *Journal of Plant Physiology* 161, no. 7 (2004): 809–814.

Gupta, Bhaskar, and Bingru Huang. Mechanism of salinity tolerance in plants: Physiological, biochemical, and molecular characterization. *International Journal of Genomics* 2014 (2014): 701596.

Gupta, D. K., H. G. Huang, X. E. Yang, B. H. N. Razafindrabe, and M. Inouhe. The detoxification of lead in *Sedum alfredii* H. is not related to phytochelatins but the glutathione. *Journal of Hazardous Materials* 177, no. 1–3 (2010): 437–444.

Hasan, M. Kamrul, Congcong Liu, Fanan Wang, Golam Jalal Ahammed, Jie Zhou, Ming-Xing Xu, Jing-Quan Yu, and Xiao-Jian Xia. Glutathione-mediated regulation of nitric oxide, S-nitrosothiol and redox homeostasis confers cadmium tolerance by inducing transcription factors and stress response genes in tomato. *Chemosphere* 161 (2016): 536–545.

Hasanuzzaman, Mirza, Kamrun Nahar, and Masayuki Fujita. Extreme temperature responses, oxidative stress and antioxidant defense in plants. In: K. Vahdati, C. Leslie, Eds., *Abiotic Stress - Plant Responses and Applications in Agriculture*. InTech, 2013.

Hasegawa, Paul M., Ray A. Bressan, Jian-Kang Zhu, and Hans J. Bohnert. Plant cellular and molecular responses to high salinity. *Annual Review of Plant Biology* 51, no. 1 (2000): 463–499.

Hawkes, Timothy R. Mechanisms of resistance to paraquat in plants. *Pest Management Science* 70, no. 9 (2014): 1316–1323.

Hell, Rüdiger, and Ludwig Bergmann. λ-Glutamylcysteine synthetase in higher plants: Catalytic properties and subcellular localization. *Planta* 180, no. 4 (1990): 603.

Hernández, Luis E., Juan Sobrino-Plata, M. Belén Montero-Palmero, Sandra Carrasco-Gil, M. Laura Flores-Cáceres, Cristina Ortega-Villasanteand Carolina Escobar. Contribution of glutathione to the control of cellular redox homeostasis under toxic metal and metalloid stress. *Journal of Experimental Botany* 66, no. 10 (2015): 2901–2911.

Herschbach, Cornelia, and Heinz Rennenberg. Influence of glutathione (GSH) on net uptake of sulphate and sulphate transport in tobacco plants. *Journal of Experimental Botany* 45, no. 8 (1994): 1069–1076.

Hoque, Md Anamul, Mst Nasrin Akhter Banu, Yoshimasa Nakamura, Yasuaki Shimoishi, and Yoshiyuki Murata. Proline and glycinebetaine enhance antioxidant defense and methylglyoxal detoxification systems and reduce NaCl-induced damage in cultured tobacco cells. *Journal of Plant Physiology* 165, no. 8 (2008): 813–824.

Hoque, Tahsina S., Mohammad A. Hossain, Mohammad G. Mostofa, David J. Burritt, Masayuki Fujita, and Lam-Son P. Tran. Methylglyoxal: An emerging signaling molecule in plant abiotic stress responses and tolerance. *Frontiers in Plant Science* 7 (2016): 1341.

Hossain, Mohammad Anwar, and Masayuki Fujita. Purification of glyoxalase I from onion bulbs and molecular cloning of its cDNA. *Bioscience, Biotechnology, and Biochemistry* 73, no. 9 (2009): 2007–2013.

Hossain, Mohammad Anwar, Mohammad Zakir Hossain, and Masayuki Fujita. Stress-induced changes of methylglyoxal level and glyoxalase I activity in pumpkin seedlings and cDNA cloning of glyoxalase I gene. *Australian Journal of Crop Science* 3, no. 2 (2009): 53.

Hossain, Mohammad Anwar, Mohammad Golam Mostof, and Masayuki Fujita. Modulation of reactive oxygen species and methylglyoxal detoxification systems by exogenous glycinebetaine and proline improves drought tolerance in mustard (*Brassica juncea* L.). *International Journal of Plant Biology and Research* 2, no. 2 (2014): 1014.

Hossain, Mohammad Anwar, Pukclai Piyatida, Jaime A. Teixeira da Silva, and Masayuki Fujita. Molecular mechanism of heavy metal toxicity and tolerance in plants: Central role of glutathione in detoxification of reactive oxygen species and methylglyoxal and in heavy metal chelation. *Journal of Botany* 2012 (2012): 872875.

Hossain, Mohammad Anwar, Jaime A. Teixeira da Silva, and Masayuki Fujita. Glyoxalase system and reactive oxygen species detoxification system in plant abiotic stress response and tolerance: An intimate relationship. In: A. Shanker, B. Venkateswarlu, Eds., *Abiotic Stress in Plants - Mechanisms and Adaptations*, pp. 235–266. InTech, 2011.

Howden, Ross, Peter B. Goldsbrough, Chris R. Andersen, and Christopher S. Cobbett. Cadmium-sensitive, cad1 mutants of *Arabidopsis thaliana* are phytochelatin deficient. *Plant Physiology* 107, no. 4 (1995): 1059–1066.

Huang, M., and Z. Guo. Responses of antioxidative system to chilling stress in two rice cultivars differing in sensitivity. *Biologia Plantarum* 49, no. 1 (2005): 81–84.

Islam, Tahmina, Mrinalini Manna, and Malireddy K. Reddy. Glutathione peroxidase of *Pennisetum glaucum* (PgGPx) is a functional Cd2+ dependent peroxiredoxin that enhances tolerance against salinity and drought stress. *PLoS One* 10, no. 11 (2015): e0143344.

Iqbal, Aqib, Yukinori Yabuta, Toru Takeda, Yoshihisa Nakano, and Shigeru Shigeoka. Hydroperoxide reduction by thioredoxin-specific glutathione peroxidase isoenzymes of *Arabidopsis thaliana*. *The FEBS Journal* 273, no. 24 (2006): 5589–5597.

Jahan, Md Sarwar, Mohd Nozulaidi, Mohd Khairi, and Nashriyah Mat. Light-harvesting complexes in photosystem II regulate glutathione-induced sensitivity of *Arabidopsis* guard cells to abscisic acid. *Journal of Plant Physiology* 195 (2016): 1–8.

Jahan, Md Sarwar, Ken'ichiOgawa, Yoshimasa Nakamura, Yasuaki Shimoishi, Izumi C. Mori, and Yoshiyuki Murata. Deficient glutathione in guard cells facilitates abscisic acid-induced stomatal closure but does not affect light-induced stomatal opening. *Bioscience, Biotechnology, and Biochemistry* 72, no. 10 (2008): 2795–2798.

Jia, Bowei, Mingzhe Sun, Xiaoli Sun, Rongtian Li, Zhenyu Wang, Jing Wu, Zhengwei Wei, Huizi DuanMu, Jialei Xiao, and Yanming Zhu. Overexpression of GsGSTU13 and SCMRP in *Medicago sativa* confers increased salt–alkaline tolerance and methionine content. *Physiologia Plantarum* 156, no. 2 (2016): 176–189.

Khattab, H. Role of glutathione and polyadenylic acid on the oxidative defense systems of two different cultivars of canola seedlings grown under saline conditions. *Australian Journal of Basic and Applied Sciences* 1, no. 3 (2007): 323–334.

Kocsy, Gâbor, Gâbor Galiba, and Christian Brunold. Role of glutathione in adaptation and signalling during chilling and cold acclimation in plants. *Physiologia Plantarum* 113, no. 2 (2001): 158–164.

Kocsy, Gâbor, Gabriella Szalai, and Gâbor Galiba. Effect of heat stress on glutathione biosynthesis in wheat. *Acta Biologica Szegediensis* 46, no. 3–4 (2002): 71–72.

Kocsy, Gâbor, Gabriella Szalai, József Sutka, Emil Páldi, and Gâbor Galiba. Heat tolerance together with heat stress-induced changes in glutathione and hydroxymethylglutathione levels is affected by chromosome 5A of wheat. *Plant Science* 166, no. 2 (2004): 451–458.

Koffler, Barbara Eva, Nora Luschin-Ebengreuth, Edith Stabentheiner, Maria Müller, and Bernd Zechmann. Compartment specific response of antioxidants to drought stress in *Arabidopsis*. *Plant Science* 227 (2014): 133–144.

Kornyeyev, Dmytro, Barry A. Logan, Paxton R. Payton, Randy D. Allen, and A. Scott Holaday. Elevated chloroplastic glutathione reductase activities decrease chilling-induced photoinhibition by increasing rates of photochemistry, but not thermal energy dissipation, in transgenic cotton. *Functional Plant Biology* 30, no. 1 (2003): 101–110.

Kumar, Sanjeev, Deepti Gupta, and Harsh Nayyar. Comparative response of maize and rice genotypes to heat stress: Status of oxidative stress and antioxidants. *Acta Physiologiae Plantarum* 34, no. 1 (2012): 75–86.

Lee, Byoung Doo, and Seongbin Hwang. Tobacco phytochelatin synthase (NtPCS1) plays important roles in cadmium and arsenic tolerance and in early plant development in tobacco. *Plant Biotechnology Reports* 9, no. 3 (2015): 107–114.

Lee, Dong Hee, and Chin Bum Lee. Chilling stress-induced changes of antioxidant enzymes in the leaves of cucumber: In gel enzyme activity assays. *Plant Science* 159, no. 1 (2000): 75–85.

Lee, Sangman, Jae S. Moon, Tae-Seok Ko, David Petros, Peter B. Goldsbrough, and Schuyler S. Korban. Overexpression of *Arabidopsis* phytochelatin synthase paradoxically leads to hypersensitivity to cadmium stress. *Plant Physiology* 131, no. 2 (2003): 656–663.

Lillig, Christopher Horst, and Carsten Berndt. Glutaredoxins in thiol/disulfide exchange. *Antioxidants and Redox Signaling* 18, no. 13 (2013): 1654–1665.

Locato, Vittoria, Annalisa Paradiso, Wilma Sabetta, Laura De Gara, and Maria Concetta de Pinto. Nitric oxide and reactive oxygen species in PCD signaling. In: D. Wendehenne, Ed., *Advances in Botanical Research: Nitric Oxide and Signaling in Plants*, vol. 77, pp. 165–192. Academic Press, 2016.

Luo, Ya, Haoru Tang, and Yong Zhang. Production of reactive oxygen species and antioxidant metabolism about strawberry leaves to low temperatures. *Journal of Agricultural Science* 3, no. 2 (2011): 89.

Ma, Yu-Hua, Feng-Wang Ma, Jun-Ke Zhang, Ming-Jun Li, Yong-Hong Wang, and Dong Liang. Effects of high temperature on activities and gene expression of enzymes involved in ascorbate–glutathione cycle in apple leaves. *Plant Science* 175, no. 6 (2008): 761–766.

Maheshwari, Ruchi, and R. S. Dubey. Nickel-induced oxidative stress and the role of antioxidant defence in rice seedlings. *Plant Growth Regulation* 59, no. 1 (2009): 37–49.

Manivannan, Paramasivam, Cheruth Abdul Jaleel, Ramamurthy Somasundaram, and Rajaram Panneerselvam. Osmoregulation and antioxidant metabolism in drought-stressed *Helianthus annuus* under triadimefon drenching. *Comptes Rendus Biologies* 331, no. 6 (2008): 418–425.

Marrs, Kathleen A. The functions and regulation of glutathione S-transferases in plants. *Annual Review of Plant Biology* 47, no. 1 (1996): 127–158.

Marty, Laurent, Wafi Siala, Markus Schwarzländer, Mark D. Fricker, Markus Wirtz, Lee J. Sweetlove, Yves Meyer, Andreas J. Meyer, Jean-Philippe Reichheld, and Rüdiger Hell. The NADPH-dependent thioredoxin system constitutes a functional backup for cytosolic glutathione reductase in *Arabidopsis*. *Proceedings of the National Academy of Sciences* 106, no. 22 (2009): 9109–9114.

Meister, Alton. Glutathione metabolism and its selective modification. *Journal of Biological Chemistry* 263, no. 33 (1988): 17205–17208.

Melchiorre, Mariana, Germán Robert, Victorio Trippi, Roberto Racca, and H. Ramiro Lascano. Superoxide dismutase and glutathione reductase overexpression in wheat protoplast: Photooxidative stress tolerance and changes in cellular redox state. *Plant Growth Regulation* 57, no. 1 (2009): 57.

Meyer, Andreas J., Thorsten Brach, Laurent Marty, Susanne Kreye, Nicolas Rouhier, Jean-Pierre Jacquot, and Rüdiger Hell. Redox-sensitive GFP in *Arabidopsis thaliana* is a quantitative biosensor for the redox potential of the cellular glutathione redox buffer. *The Plant Journal* 52, no. 5 (2007): 973–986.

Miteva, L. P.-E., S. V. Ivanov, and V. S. Alexieva. Alterations in glutathione pool and some related enzymes in leaves and roots of pea plants treated with the herbicide glyphosate. *Russian Journal of Plant Physiology* 57, no. 1 (2010): 131–136.

Mittler, Ron, and Eduardo Blumwald. The roles of ROS and ABA in systemic acquired acclimation. *The Plant Cell* 27, no. 1 (2015): 64–70.

Müller-Moulé, Patricia, Talila Golan, and Krishna K. Niyogi. Ascorbate-deficient mutants of *Arabidopsis* grow in high light despite chronic photooxidative stress. *Plant Physiology* 134, no. 3 (2004): 1163–1172.

Munemasa, Shintaro, Daichi Muroyama, Hiroki Nagahashi, Yoshimasa Nakamura, Izumi C. Mori, and Yoshiyuki Murata. Regulation of reactive oxygen species-mediated abscisic acid signaling in guard cells and drought tolerance by glutathione. *Frontiers in Plant Science* 4 (2013): 472.

Munns, Rana. Comparative physiology of salt and water stress. *Plant, Cell and Environment* 25, no. 2 (2002): 239–250.

Munns, Rana, and Mark Tester. Mechanisms of salinity tolerance. *Annual Review of Plant Biology* 59 (2008): 651–681.

Nagesh Babu, R., and V. R. Devaraj. High temperature and salt stress response in French bean (*Phaseolus vulgaris*). *Australian Journal of Crop Science* 2, no. 2 (2008): 40–48.

Nahar, K., M. Hasanuzzaman, M. M. Alam, and M. Fujita. Roles of exogenous glutathione in antioxidant defense system and methylglyoxal detoxification during salt stress in mung bean. *Biologia Plantarum* 59, no. 4 (2015a): 745–756.

Nahar, K., M. Hasanuzzaman, M. M. Alam, and M. Fujita. Exogenous glutathione confers high temperature stress tolerance in mung bean (*Vigna radiata* L.) by modulating antioxidant defense and methylglyoxal detoxification system. *Environmental and Experimental Botany* 112 (2015b): 44–54.

Nahar, K., M. Hasanuzzaman, and M. Fujita. Physiological roles of glutathione in conferring abiotic stress tolerance to plants. *Abiotic Stress Response in Plants* (2016a): 151–180.

Nahar, K., Motiar Rahman, Mirza Hasanuzzaman, Md Mahabub Alam, Anisur Rahman, Toshisada Suzuki, and Masayuki Fujita. Physiological and biochemical mechanisms of spermine-induced cadmium stress tolerance in mung bean (*Vigna radiata* L.) seedlings. *Environmental Science and Pollution Research* 23, no. 21 (2016b): 21206–21218.

Navrot, Nicolas, Valérie Collin, José Gualberto, Eric Gelhaye, Masakazu Hirasawa, Pascal Rey, David B. Knaff, Emmanuelle Issakidis, Jean-Pierre Jacquot, and Nicolas Rouhier. Plant glutathione peroxidases are functional peroxiredoxins distributed in several subcellular compartments and regulated during biotic and abiotic stresses. *Plant Physiology* 142, no. 4 (2006): 1364–1379.

Noctor, Graham, Ana-Carolina M. Arisi, Lise Jouanin, Karl J. Kunert, Heinz Rennenberg, and Christine H. Foyer. Glutathione: Biosynthesis, metabolism and relationship to stress tolerance explored in transformed plants. *Journal of Experimental Botany* 49, no. 321 (1998): 623–647.

Noctor, Graham, Leonardo Gomez, Hélène Vanacker, and Christine H. Foyer. Interactions between biosynthesis, compartmentation and transport in the control of glutathione homeostasis and signalling. *Journal of Experimental Botany* 53, no. 372 (2002): 1283–1304.

Noctor, Graham, Amna Mhamdi, Sejir Chaouch, Y. I. Han, Jenny Neukermans, Belen. Marquez-Garcia, Guillaume Queval, and Christine H. Foyer. Glutathione in plants: An integrated overview. *Plant, Cell and Environment* 35, no. 2 (2012): 454–484.

Noctor, Graham, Guillaume Queval, Amna Mhamdi, Sejir Chaouch, and Christine H. Foyer. Glutathione. *The Arabidopsis Book/American Society of Plant Biologists* 9 (2011): e0142.

Nordzieke, Daniela, and Iria Medraño-Fernandez. The plasma membrane: A platform for intra-and intercellular redox signaling. *Antioxidants* 7, no. 11 (2018): 168.

Noshi, Masahiro, Risa Hatanaka, Noriaki Tanabe, Yusuke Terai, Takanori Maruta, and Shigeru Shigeoka. Redox regulation of ascorbate and glutathione by a chloroplastic dehydroascorbate reductase is required for high-light stress tolerance in *Arabidopsis*. *Bioscience, Biotechnology, and Biochemistry* 80, no. 5 (2016): 870–877.

Ogawa, Ken'ichi, Aya Hatano-Iwasaki, Mototsugu Yanagida, and Masaki Iwabuchi. Level of glutathione is regulated by ATP-dependent ligation of glutamate and cysteine through photosynthesis in *Arabidopsis thaliana*: Mechanism of strong interaction of light intensity with flowering. *Plant and Cell Physiology* 45, no. 1 (2004): 1–8.

Ohkama-Ohtsu, Naoko, Akira Oikawa, Ping Zhao, Chengbin Xiang, Kazuki Saito, and David J. Oliver. A γ-glutamyl transpeptidase-independent pathway of glutathione catabolism to glutamate via 5-oxoproline in *Arabidopsis*. *Plant Physiology* 148, no. 3 (2008): 1603–1613.

Okuma, Eiji, Md Sarwar Jahan, Shintaro Munemasa, Mohammad Anowar Hossain, Daichi Muroyama, Mohammad Mahbub Islam, Ken'ichi Ogawa, et al. Negative regulation of abscisic acid-induced stomatal closure by glutathione in *Arabidopsis*. *Journal of Plant Physiology* 168, no. 17 (2011): 2048–2055.

Passaia, Gisele, and Marcia Margis-Pinheiro. Glutathione peroxidases as redox sensor proteins in plant cells. *Plant Science* 234 (2015): 22–26.

Pasternak, Maciej, Benson Lim, Markus Wirtz, Rüdiger Hell, Christopher S. Cobbett, and Andreas J. Meyer. Restricting glutathione biosynthesis to the cytosol is sufficient for normal plant development. *The Plant Journal* 53, no. 6 (2008): 999–1012.

Pazmino, M. Diana, María Rodríguez-Serrano, María C. Romero-Puertas, Angustias Archilla-Ruiz, Luis A. Del Rio, and Luisa M. Sandalio. Differential response of young and adult leaves to herbicide 2, 4-dichlorophenoxyacetic acid in pea plants: Role of reactive oxygen species. *Plant, Cell and Environment* 34, no. 11 (2011): 1874–1889.

Peltzer, Detlef, Erwin Dreyer, and Andrea Polle. Differential temperature dependencies of antioxidative enzymes in two contrasting species: *Fagus sylvatica* and *Coleus blumei*. *Plant Physiology and Biochemistry* 40, no. 2 (2002): 141–150.

Peterson, Annita G., and David J. Oliver. Leaf-targeted phytochelatin synthase in *Arabidopsis thaliana*. *Plant Physiology and Biochemistry* 44, no. 11–12 (2006): 885–892.

Pilarska, Maria, Monika Wiciarz, Ivan Jajić, Małgorzata Kozieradzka-Kiszkurno, Petre Dobrev, Radomíra Vanková, and Ewa Niewiadomska. A different pattern of production and scavenging of reactive oxygen species in halophytic *Eutrema salsugineum* (*Thellungiella salsuginea*) plants in comparison to *Arabidopsis thaliana* and its relation to salt stress signaling. *Frontiers in Plant Science* 7 (2016): 1179.

Pilon-Smits, Elizabeth. Phytoremediation. *Annual Reviews in Plant Biology* 56 (2005): 15–39.

Pinheiro, C., and M. M. Chaves. Photosynthesis and drought: Can we make metabolic connections from available data? *Journal of Experimental Botany* 62, no. 3 (2010): 869–882.

Popa, Karin, Manuela Murariu, Ramona Molnar, Gitta Schlosser, Alexandru Cecal, and Gabi Drochioiu. Effect of radioactive and non-radioactive mercury on wheat germination and the anti-toxic role of glutathione. *Isotopes in Environmental and Health Studies* 43, no. 2 (2007): 105–116.

Rajab, Hala, Muhammad Sayyar Khan, Mario Malagoli, Rüdiger Hell, and Markus Wirtz. Sulfate-induced stomata closure requires the canonical aba signal transduction machinery. *Plants* 8, no. 1 (2019): 21.

Rao, A. S. V. Chalapathi, and Attipalli R. Reddy. Glutathione reductase: A putative redox regulatory system in plant cells. In: N. A. Khan, S. Singh, S. Umar, Eds., *Sulfur Assimilation and Abiotic Stress in Plants*, pp. 111–147. Springer, Berlin, Heidelberg, 2008.

Richman, P. G., and A. Meister. Regulation of gamma-glutamyl-cysteine synthetase by nonallosteric feedback inhibition by glutathione. *Journal of Biological Chemistry* 250, no. 4 (1975): 1422–1426.

Rivero, R. M., J. M. Ruiz, and L. Romero. Oxidative metabolism in tomato plants subjected to heat stress. *The Journal of Horticultural Science and Biotechnology* 79, no. 4 (2004): 560–564.

Rouhier, Nicolas. Plant glutaredoxins: Pivotal players in redox biology and iron–sulphur centre assembly. *New Phytologist* 186, no. 2 (2010): 365–372.

Rouhier, Nicolas, Stéphane D. Lemaire, and Jean-Pierre Jacquot. The role of glutathione in photosynthetic organisms: Emerging functions for glutaredoxins and glutathionylation. *Annual Reviews of Plant Biology* 59 (2008): 143–166.

Roxas, Virginia P., Roger K. Smith Jr, Eric R. Allen, and Randy D. Allen. Overexpression of glutathione S-transferase/glutathioneperoxidase enhances the growth of transgenic tobacco seedlings during stress. *Nature Biotechnology* 15, no. 10 (1997): 988.

Ruiz, J., and E. Blumwald. Salinity-induced glutathione synthesis in *Brassica napus*. *Planta* 214, no. 6 (2002): 965–969.

Sairam, R. K., G. C. Srivastava, and D. C. Saxena. Increased antioxidant activity under elevated temperatures: A mechanism of heat stress tolerance in wheat genotypes. *Biologia Plantarum* 43, no. 2 (2000): 245–251.

Sandermann Jr, Heinrich. Plant metabolism of xenobiotics. *Trends in Biochemical Sciences* 17, no. 2 (1992): 82–84.

Schat, Henk, Mercè Llugany, Riet Vooijs, Jeanette Hartley-Whitaker, and Petra M. Bleeker. The role of phytochelatins in constitutive and adaptive heavy metal tolerances in hyperaccumulator and non-hyperaccumulator metallophytes. *Journal of Experimental Botany* 53, no. 379 (2002): 2381–2392.

Sharma, Raghvendra, Annapurna Sahoo, Ragunathan Devendran, and Mukesh Jain. Over-expression of a rice tau class glutathione s-transferase gene improves tolerance to salinity and oxidative stresses in *Arabidopsis*. *PloS One* 9, no. 3 (2014): e92900.

Sheffield, Justin, and Eric F. Wood. Global trends and variability in soil moisture and drought characteristics, 1950–2000, from observation-driven simulations of the terrestrial hydrologic cycle. *Journal of Climate* 21, no. 3 (2008): 432–458.

Shrivastava, Pooja, and Rajesh Kumar. Soil salinity: A serious environmental issue and plant growth promoting bacteria as one of the tools for its alleviation. *Saudi Journal of Biological Sciences* 22, no. 2 (2015): 123–131.

Sierla, Maija, Cezary Waszczak, Triin Vahisalu, and Jaakko Kangasjärvi. Reactive oxygen species in the regulation of stomatal movements. *Plant Physiology* 171, no. 3 (2016): 1569–1580.

Singla-Pareek, Sneh L., Sudesh Kumar Yadav, Ashwani Pareek, M. K. Reddy, and S. K. SoporyEnhancing salt tolerance in a crop plant by overexpression of glyoxalase II. *Transgenic Research* 17, no. 2 (2008): 171–180.

Sirichandra, Caroline, Dan Gu, Heng-Cheng Hu, Marlène Davanture, Sangmee Lee, Michaël Djaoui, Benoît Valot, et al. Phosphorylation of the *Arabidopsis* AtrbohF NADPH oxidase by OST1 protein kinase. *Febs Letters* 583, no. 18 (2009): 2982–2986.

Smirnoff, Nicholas, and Glen L. Wheeler. Ascorbic acid in plants: Biosynthesis and function. *Critical Reviews in Plant Sciences* 19, no. 4 (2000): 267–290.

Smith, Ivan K. Stimulation of glutathione synthesis in photorespiring plants by catalase inhibitors. *Plant Physiology* 79, no. 4 (1985): 1044–1047.

Szalai, Gabriella, Tibor Kellős, Gábor Galiba, and Gábor Kocsy. Glutathione as an antioxidant and regulatory molecule in plants under abiotic stress conditions. *Journal of Plant Growth Regulation* 28, no. 1 (2009): 66–80.

Tanveer, Mohsin, and Adnan Noor Shah. An insight into salt stress tolerance mechanisms of *Chenopodium album*. *Environmental Science and Pollution Research* 24, no. 19 (2017): 16531–16535.

Tardieu, François, Thierry Simonneau, and Bertrand Muller. The physiological basis of drought tolerance in crop plants: A scenario-dependent probabilistic approach. *Annual Review of Plant Biology* 69 (2018): 733–759.

Trivedi, Dipesh Kumar, Sarvajeet Singh Gill, Sandep Yadav, and Narendra Tuteja. Genome-wide analysis of glutathione reductase (GR) genes from rice and *Arabidopsis*. *Plant Signaling and Behavior* 8, no. 2 (2013): e23021.

Tseng, Tai-You, Jen-Fu Ou, and Ching-Yuh Wang. Role of the ascorbate–glutathione cycle in paraquat tolerance of rice. *Weed Science* 61, no. 3 (2013): 361–373.

Uotila, M., Gábor Gullner, and Tibor Kömives. Induction of glutathione S-transferase activity and glutathione level in plants exposed to glyphosate. *Physiologia Plantarum* 93, no. 4 (1995): 689–694.

Verma, Shalini, and R. S. Dubey. Lead toxicity induces lipid peroxidation and alters the activities of antioxidant enzymes in growing rice plants. *Plant Science* 164, no. 4 (2003): 645–655.

Vernoux, Teva, Robert C. Wilson, Kevin A. Seeley, Jean-Philippe Reichheld, Sandra Muroy, Spencer Brown, Spencer C. Maughan, et al. The *ROOT MERISTEMLESS1/CADMIUM SENSITIVE2* gene defines a glutathione-dependent pathway involved in initiation and maintenance of cell division during postembryonic root development. *The Plant Cell* 12, no. 1 (2000): 97–109.

Wachter, Andreas, Sebastian Wolf, Heike Steininger, Jochen Bogs, and Thomas Rausch. Differential targeting of GSH1 and GSH2 is achieved by multiple transcription initiation: Implications for the compartmentation of glutathione biosynthesis in the *Brassicaceae*. *The Plant Journal* 41, no. 1 (2005): 15–30.

Walker, Mark A., and Bryan D. Mckersie. Role of the ascorbate-glutathione antioxidant system in chilling resistance of tomato. *Journal of Plant Physiology* 141, no. 2 (1993): 234–239.

Wang, Fang, Fei Chen, Yue Cai, Guoping Zhang, and Feibo Wu. Modulation of exogenous glutathione in ultrastructure and photosynthetic performance against Cd stress in the two barley genotypes differing in Cd tolerance. *Biological Trace Element Research* 144, no. 1–3 (2011): 1275–1288.

Wang, Limin, Xiang Jin, Qingbin Li, Xuchu Wang, Zaiyun Li, and Xiaoming Wu. Comparative proteomics reveals that phosphorylation of β carbonic anhydrase 1 might be important for adaptation to drought stress in *Brassica napus*. *Scientific Reports* 6 (2016): 39024.

Wang, Pengcheng, Yanyan Du, Yueh-Ju Hou, Yang Zhao, Chuan-Chih Hsu, Feijuan Yuan, Xiaohong Zhu, W. Andy Tao, Chun-Peng Song, and Jian-Kang Zhu. Nitric oxide negatively regulates abscisic acid signaling in guard cells by S-nitrosylation of OST1. *Proceedings of the National Academy of Sciences* 112, no. 2 (2015): 613–618.

Wang, Xiao, Jian Cai, Fulai Liu, Tingbo Dai, Weixing Cao, Bernd Wollenweber, and Dong Jiang. Multiple heat priming enhances thermo-tolerance to a later high temperature stress via improving subcellular antioxidant activities in wheat seedlings. *Plant Physiology and Biochemistry* 74 (2014): 185–192.

Wiciarz, Monika, E. Niewiadomska, and Jerzy Kruk. Effects of salt stress on low molecular antioxidants and redox state of plastoquinone and P700 in *Arabidopsis thaliana* (glycophyte) and *Eutrema salsugineum* (halophyte). *Photosynthetica* 56, no. 3 (2018): 811–819.

Wójcik, M., and A. Tukiendorf. Glutathione in adaptation of *Arabidopsis thaliana* to cadmium stress. *Biologia Plantarum* 55, no. 1 (2011): 125–132.

Wu, Xuexia, Jie He, Jianlin Chen, Shaojun Yang, and Dingshi Zha. Alleviation of exogenous 6-benzyladenine on two genotypes of eggplant (*Solanum melongena* Mill.) growth under salt stress. *Protoplasma* 251, no. 1 (2014): 169–176.

Xiang, Chengbin, Bonnie L. Werner, Christensen M. E'Lise, and David J. Oliver. The biological functions of glutathione revisited in *Arabidopsis* transgenic plants with altered glutathione levels. *Plant Physiology* 126, no. 2 (2001): 564–574.

Xu, Jing, Xiao-Juan Xing, Yong-Sheng Tian, Ri-He Peng, Yong Xue, Wei Zhao, and Quan-Hong Yao. Transgenic *Arabidopsis* plants expressing tomato glutathione S-transferase showed enhanced resistance to salt and drought stress. *PloS One* 10, no. 9 (2015): e0136960.

Yadav, S. K. Heavy metals toxicity in plants: An overview on the role of glutathione and phytochelatins in heavy metal stress tolerance of plants. *South African Journal of Botany* 76, no. 2 (2010): 167–179.

Yadav, S. K., Sneh L. Singla-Pareek, Manju Ray, M. K. Reddy, and S. K. Sopory. Methylglyoxal levels in plants under salinity stress are dependent on glyoxalase I and glutathione. *Biochemical and Biophysical Research Communications* 337, no. 1 (2005): 61–67.

Yordanova, R., and L. Popova. Effect of exogenous treatment with salicylic acid on photosynthetic activity and antioxidant capacity of chilled wheat plants. *General and Applied Plant Physiology* 33, no. 3–4 (2007): 155–170.

Zagorchev, Lyuben, Charlotte Seal, Ilse Kranner, and Mariela Odjakova. A central role for thiols in plant tolerance to abiotic stress. *International Journal of Molecular Sciences* 14, no. 4 (2013): 7405–7432.

Zhang, Jianguang, Shaochun Chen, Yingli Li, Bao Di, Jianqiang Zhang, and Yufang Liu. Effect of high temperature and excessive light on glutathione content in apple peel. *Frontiers of Agriculture in China* 2, no. 1 (2008): 97–102.

Zhao, Yulong, Yankai Wang, Hao Yang, Wei Wang, Jianyu Wu, and Xiuli Hu. Quantitative proteomic analyses identify ABA-related proteins and signal pathways in maize leaves under drought conditions. *Frontiers in Plant Science* 7 (2016): 1827.

Zhu, Yong Liang, Elizabeth A. H. Pilon-Smits, Lise Jouanin, and Norman Terry. Overexpression of glutathione synthetase in Indian mustard enhances cadmium accumulation and tolerance. *Plant Physiology* 119, no. 1 (1999): 73–80.

5 Improving Heavy Metal Tolerance through Plant Growth Regulators and Osmoprotectants in Plants

Farha Ashfaque, Samreena Farooq, Priyanka Chopra, Himanshu Chhillar, and M. Iqbal R. Khan

CONTENTS

5.1 INTRODUCTION

The rising rate of heavy metal contamination due to industrialization, urbanization, and modern agricultural practices has posed unparalleled problems for natural ecosystems and food chains either through direct ingestion or contact with environmental contamination (McLaughlin et al. 1999; Miransari 2011). These heavy metals are non-biodegradable components that frequently concentrate in the root environment due to excessive use of unplanned municipal and industrial wastewater, sewage sludge, pesticides, and chemical fertilizers (Pandey and Pandey 2009; El-Ramady et al. 2015). Some of these heavy metals such as Pb, Cd, Cr, As, and Hg do not take part in any known physiological functions and are very toxic even at very low concentrations (Ernst et al. 2008; Garzón et al. 2011; Chong-qing et al. 2013). However, the essential elements such as Fe, Mn, Ni, Zn, Co, and Cu are required for the normal growth and metabolic processes of plants, but they become toxic when their concentrations exceed their permissible limits (Blaylock and Huang 2000; Sebastiani et al. 2004). The most widespread effect of this metal toxicity on plants is reduction in growth including low biomass accumulation, leaf chlorosis, necrosis, inhibition of seed germination, photosynthesis, altered water balance and nutrient availability, and senescence which ultimately leads to plant death (Carrier et al. 2003; DalCorso et al. 2010). An excess of metals alters various physio-biochemical processes by the stimulation of ROS, enzyme inactivation, blocking functional groups of metabolically important molecules, disrupting membrane integrity, and displacing essential metal ions from specific binding sites (Rascio and Navari Izzo 2011; Villiers et al. 2012). Therefore, this review is focused on a physiological and molecular approach to metal toxicity, ROS production, avoidance, and scavenging in the light of heavy metal stress in plants, with special attention on the emrging role of osmoprotectants as antioxidants.

5.2 RESPONSES OF PLANTS UNDER HEAVY METAL STRESS

Firstly the heavy metals encounter the roots of the plant and thus this soil accumulated with metal adversely affects root growth resulting in hampering the overall growth performance of the plants (Kikui et al. 2005; Gangwar et al. 2010; Eleftheriou et al. 2012). The inhibition of root growth may be due to metal interference with cell division and irregular mitotic activity which can disturb seedling growth as root growth is the combination of cell division and elongation (Jiang et al. 2001; Liu et al. 2003). Prasad et al. (2001) reported that metal toxicity to new root primordia in *Salix viminalis* was in the order of Cd < Cr < Pb, while root length was more affected by Cr than other heavy metals studied. In this context, a study by Sundaramoorthy et al. (2010) showed that Cr (VI) caused a delay in the cell cycle that led to the inhibition in cell division, as a consequence reducing root growth. An excess amount of Cu affects both elongation and meristem zones by modifying auxin distribution through the PIN1 protein, and thus inhibits primary root elongation (Yuan et al. 2013). Correspondingly, Petö et al. (2011) have also observed that excess Cu inhibits root length by altering auxin levels. The primary cause of cell growth inhibition arises from a Pb-induced simulation of IAA oxidation. Prolonged exposure to higher concentrations of Cd and Cu in the roots of 3-week-old *Arabidopsis thaliana* seedlings reduced growth, combined with visible signs of chlorosis (Cuypers et al. 2011). The distinctive symptoms of Cd toxicity in plants are growth inhibition, progressive chlorosis, wilted leaves, and browned root systems, especially the root tips (Chugh and Sawhney 1999; Mohanpuria et al. 2007; Guo et al. 2008). Exposure to Cu adversely affected the growth of *Zea mays* (Ouzounidou et al. 1995) and exposure to Cd affected *Brassica juncea* (Singh and Tewari 2003). Furthermore, Talanova et al. (2001) and Liu et al. (2006) suggested that in maize, Cd also reduces plant growth. However, Zn toxicity also inhibits many plant metabolic functions, results in retarded growth and senescence, hinders the transfer of Mn and Cu from root to shoot, and consequently inhibits root and shoot growth (Ebbs and Kochian 1997; Fontes and Cox 1998). A high level of Zn in contaminated soil also induces abnormal morphology in plants including irregular radial thickening in roots, the cell wall of endodermis, and lignification of cortical parenchyma (Paivoke 1983). Pb also repressed root and stem growth and leaf expansion in onion (Gruenhage and Jager 1985) and barley (Juwarkar and Shende 1986). The degree to which inhibition of root elongation occurs depends upon the concentration of Pb and the pH of the medium. A recent study on the phytotoxicity of Co has shown an adverse effect on growth and biomass in barley (*Hordeum vulgare* L.), rape seed (*Brassica napus* L.), and tomato (*Lycopersicon esculentum* L.) (Li et al. 2009).

The exposure of plants to toxic heavy metals triggers alterations in a wide range of physiological and metabolic processes including the inhibition of proteins, cytoplasmic enzymes, substitution of essential metal ions from biomolecules and functional cellular units, blocking of functional groups of metabolically important molecules, conformational modifications and disturbance of membrane integrity, and damage to the cell structure due to oxidative stress (Assche and Clijsters 1990; Villiers et al. 2012), ultimately resulting in altered plant metabolism, activities of several key enzymes, and inhibition of photosynthesis and respiration (Sharma and Dubey 2007; Sharma and Dietz 2009; Tan et al. 2010). For instances, Cd impairs stomatal movement, photosynthetic machinery, and CO_2 fixation enzymes as well as activating NADPH oxidase activity and enhancing ROS production (Romero-Puertas et al. 2004).

Plasma membrane works as a checkpoint that hinders the entry of undesirable substances into the cell and protects the cell from different stress factors. An excessive amount of heavy metals can effectively bind to membranes through O_2 atoms or the tryptophan, tyrosine, and histidine groups of polypeptides (Maksymiec 1997). Cd^{2+} competes with either Zn^{2+} or Ca^{2+} leading to considerable changes in the membrane enzyme activity and lipid composition (Ouariti et al. 1997). The overall changes in membrane permeability and inhibition of membrane enzyme could shift the ionic balance in the cytoplasm. Kenderešová et al. (2012) reported that *Arabidopsis arenosa* and *Arabidopsis halleri* were more tolerant to heavy metal stress than *Arabidopsis thaliana* because the latter has the lowest membrane depolarization, indicating that rapid membrane voltage changes might be an effective tool for monitoring the effects of heavy metal toxicity. Growing plants in high Ni^{2+} disturbs the normal membrane function by distorting the lipid composition and H^+-ATPase activity of the plasma membrane as reported in *Oryza sativa* shoots (Ros et al. 1992). Similarly, others concluded that cell membrane disruption and ion leakage was the primary cause of Cu toxicity in the roots of *Silene vulgaris*, *Mimulus guttatus*, and *Triticum aestivum* (De Vos et al. 1991; Quartacci et al. 2001).

The photosynthetic apparatus is very sensitive in response to toxicity of heavy metals which invariably affects the photosynthetic functions by inhibiting the activities of different enzymes (Bertrand and Poirier 2005; Linger et al. 2005). Exposure to Cd stress in tomato seedlings had deleterious effects on various photosynthetic mechanisms such as the photosynthetic rate (P_N) and intracellular CO_2 concentration (Dong et al. 2005). In general it was accepted that Cd affected the water oxidizing system of PS II by replacing the Mn^{2+} and Ca^{2+} ions, thereby hindering the reaction of PS II leading to the uncoupling of the electron transport in the chlorophyll (Atal et al. 1991; Faller et al. 2005). Cd also inhibits chlorophyll biosynthesis which affects the inactivation of PS II (Siedlecka and Baszynski 1993). Moreover, PS II reaction centres and PS II electron transport are affected by an interaction of Cd impairing enzyme activity and protein structure. Cd treatments have been shown to reduce the ATPase activity of the plasma membrane fraction of wheat and sunflower roots (Fodor et al. 1995). The accumulation of excessive Zn in plants causes alterations in photosynthesis, chlorophyll biosynthesis, membrane integrity, and nutrient imbalance (Chaoui et al. 1997; De Vos et al. 1991; Doncheva et al. 2001). Khan and Khan (2014) have observed that Ni and Zn caused a significant decrease in the level of chlorophyll content, accompanied by a decline in the electron transport rate, photochemical quenching and efficiency of PS II, and photosynthetic nitrogen use efficiency in mustard. According to Rocchetta et al. (2006), Cr induced disorganization of the chloroplast ultrastructure and inhibition of excitation energy transfer that negatively affected the photosynthetic rate. Similarly, Ashfaque et al. (2017) have also reported that Cr reduced chlorophyll and carotenoids contents as well as the photosynthetic rate in mustard indicating that heavy metals have a negative impact on photosynthesis. It has also been hypothesized that the decrease in the chlorophyll a/b ratio due to Cr indicates that the size of the peripheral part of the antenna complex reduces due to Cr toxicity (Shanker 2003). In addition to the reaction to light, heavy metals diminish the photosynthetic enzymes, mainly RUBISCO and PEP carboxylase, bringing about decreased CO_2 assimilation (Mysliwa-Kurdziel et al. 2004). For example, Cd decreased the RUBISCO activity in *Erythrina variegate* (Muthuchelian et al. 2001); this decrease might be a reaction of Cd with the thiol group of RUBISCO. Similarly, Monnet et al. (2001) reported that in *Phaseolus vulgaris*, Zn application inhibits RUBISCO activity by replacing Zn^{2+} with Mg^{2+}. In another study Khan and Khan (2014) also observed that Ni and Zn reduce RUBISCO activity in mustard. Pb is also known to affect photosynthesis by inhibiting the activity of carboxylating enzymes by interacting with their -SH groups (Stiborova et al. 1987). The toxicity of Cd also damages the cell membrane and inactivates enzymes, possibly through reacting with the SH-group of proteins (Fuhrer 1988), which reflects the inhibitory effects of Pb^{2+}.

Furthermore, heavy metals can induce stress in plants through changes in water balance, for instance, Hg^{2+} can bind to water channel proteins, thus causing stomata closure and physical hindrance of water flow in plants (Zhang and Tyerman 1999). Conversely, the treatment of plants with Ni^{2+} induced a decline in water content, used as an indicator of the progression of Ni^{2+} toxicity in plants (Pandey and Sharma 2002; Gajewska et al. 2006). Moreover, water transport appears to be modulated by an impairment of aquaporin, which is one of the earliest responses to heavy metals in plants. It has been shown that the uptake and transport of water and elements in plants are altered by Cd (Das et al. 1997). Cd also reduced the absorption of nitrate and its transport from roots to shoots, by inhibiting the nitrate reductase activity in the shoots (Hernández et al. 1996).

In addition, one of the major consequences of heavy metals action is known to disturb redox homeostasis by the production of free radicals and reactive oxygen species (ROS) such as singlet oxygen ($_1O_2$), superoxide radicals ($O_2^{\bullet-}$), hydrogen peroxide (H_2O_2), and hydroxyl radicals (•OH) (Dietz et al. 1999; Dave et al. 2013), which usually damage the cellular components such as membranes, nucleic acids, and chloroplast pigments (Apel and Hirt 2004; Gunes et al. 2009; Petrov et al. 2015). The main sources of ROS due to toxic heavy metals are photosynthetic and respiratory electron transport processes. Chloroplasts are a major site of ROS generation in plants (Asada 2006). Mitochondria are also suggested to be an important source of stress-induced ROS generation. NADPH oxidase-dependent ROS stimulation has been observed in *Arabidopsis* in response to Cd and Cu (Remans et al. 2010), in *Pisum sativum* in response to Cd stress (Rodríguez-Serrano et al. 2006), in wheat in response to Ni stress (Hao et al. 2006), and in *Vicia faba* in response to Pb stress (Pourrut et al. 2008). Heavy metals usually form ROS either directly through Haber–Weiss reactions or through involvement in redox reaction such as interaction with the antioxidant defence system, disruption of the electron transport chain, or induction of lipid peroxidation (Shi and Dalal 1989; Mithofer et al. 2004). The latter can be due to a heavy metal-induced increase in lipoxygenase (LOX) activity. In addition, heavy metal excess may stimulate the formation of free radicals; ROS can react with macromolecules such as pigments, lipids, proteins, DNA, and other cellular molecules, ultimately leading to oxidative stress (Mittler 2002; Sharma and Dietz 2009; Hossain et al. 2012). Conversely, ROS are also involved in the regulation of many key physiological processes by acting as signalling molecules, such as cell growth, root hair growth, stomatal movement, and cell differentiation (Foreman et al. 2003; Tsukagoshi et al. 2010). Practically all amino acids can serve as targets for oxidative attack by ROS, although some amino acids such as tryptophan, tyrosine, histidine, and cysteine are particularly sensitive to ROS (Stadtman 1993). Exposure of plants to excess Cu causes oxidative stress through ROS generation, leading to the disturbance of metabolic pathways and damage to macromolecules (Stadtman and Oliver 1991; Hegedus et al. 2001). Increased production of ROS in plants triggered oxidative stress also induced by higher Pb concentration (Reddy et al. 2005). Although Cd does not take part in Fenton-type reactions (Stohs and Bagchi 1995), it can indirectly favor the formation of different ROS, such as hydrogen peroxide (H_2O_2), superoxide ($O_2^{\bullet-}$), and hydroxyl radical (•OH), by unknown mechanisms, giving rise to an oxidative burst (Romero-Puertas et al. 2004; Garnier et al. 2006). One of the most destructive consequences of heavy metals exposure in plants is lipid peroxidation, which may lead to bio-membrane dysfunction and ultimately oxidative stress. Polyunsaturated fatty acids of the membrane disintegrate to produce malondialdehyde (MDA), a reliable indicator of oxidative stress (Demiral and Türkan 2005). A study by Pandolfini et al. (1992) observed that the exposure of wheat to high levels of Ni^{2+} enhanced MDA concentration. Such changes may possibly disturb membrane functions and ion balance in the cytoplasm and consequently result in leakage of K^+ mobile ion which is mobile across the plant cell membrane. Similarly, Cd exposure alters the membrane's functionality by enhancing lipid peroxidation (Fodor et al. 1995). Additionally, heavy metal-induced H_2O_2 accumulation in cellular compartments, usually connected with alterations in the cellular redox status, alerts the plant cell counter to abiotic stresses (Lamb and Dixon 1997; Orozco-Cárdenas et al. 2001; Foyer and Noctor 2003). H_2O_2 is produced primarily in chloroplasts, peroxisomes, and glyoxysomes, however, the mitochondria, cytosol, endomembrane system, and nucleus are also involved in H_2O_2 synthesis (Gechev et al. 2006; Ashtamker et al. 2007). The accumulation of H_2O_2 usually occurred in Mn-treated barley (Demirevska-Kepova et al. 2004), upon Cu and Cd exposure in *Arabidopsis* (Romero-Puertas et al. 2004; Maksymiec and Krupa 2006), and Hg treatment in tomato plants (Cho and Park 2000). It is possible that increased H_2O_2 formation, especially through increased activity of NADPH-oxidase, can decrease cell wall extensibility as indicated by Foreman et al. (2003). Recently, many studies have demonstrated that methylglyoxal (MG), a cytosolic compound, was increased in response to various stresses in plants, including heavy metals (Yadav et al. 2005; Singla-Pareek et al. 2006; Hossain et al. 2009). A rapid increase in MG levels further strengthens the generation of ROS which inactivates the antioxidant defence system, interfering with physiological and metabolic processes such as photosynthesis (Hoque et al. 2010; Saito et al. 2011). Singla-Pareek et al. (2006) reported a significant increase in

MG levels in tobacco in response to Zn and in pumpkin (*Cucurbita maxima* Duch.) in response to Cd (Hossain et al. 2009). This enhancement in ROS and MG induces oxidative stress, leading to membrane dismantling, biomolecule deterioration, lipid peroxidation, ion leakage, and DNA-strand cleavage and ultimately plant death (Navari-Izzo 1998; Barconi et al. 2011). However, plants use several strategies to overcome and repair the damage caused by ROS. These ROS are normally scavenged instantly at their sites of synthesis by nearby antioxidants. Besides, antioxidants play important roles in scavenging ROS and minimizing the oxidative stress triggered by heavy metals, which have also been studied (Hirschi et al. 2000; Tseng et al. 2007). The antioxidant defence system contains several antioxidant enzymes like superoxide dismutase (SOD), catalase (CAT), peroxidase (POX), ascorbate peroxidase (APX), glutathione reductase (GR), glutathione S-transferase (GST), monodehydroascorbate reductase (MDHAR), and dehydroascorbate reductase (DHAR) as well as some non-enzymatic antioxidants such as phenolic compounds, ascorbate (AsA) glutathione (GSH), non-protein amino acids, and α-tocopherols (Noctor and Foyer 1998; Hasanuzzaman et al. 2012; Rahman et al. 2016). The enzymes SOD, CAT, and POX participate in the detoxification of $O2^{\bullet-}$ and H_2O_2 respectively, thus preventing the formation of •OH radicals. Moreover, APX and GR through the ascorbate-glutathione cycle exclude H_2O_2 in different cellular compartments (Jiménez et al. 1998; Noctor et al. 1998). See Table 5.1.

5.3 IMPROVING HEAVY METAL TOLERANCE IN PLANTS THROUGH OSMOPROTECTANTS

Plants are frequently exposed to environmental stresses both due to some natural causes and rough agricultural practices. Plant stresses like oxidative, metals toxicity, drought and salinity, and extreme temperatures along with the attack of insects, pests, and plant pathogens result in limited crop productivity by adversely affecting various physiological and molecular activities of plants which is a serious threat to agriculture (Jamil et al. 2005; Panuccio et al. 2009; Hassan et al. 2017). The capacity of plants to cope with unfavorable environments is known as stress resistance. Plant adaptations to tolerate stress depend upon genetically modified resistance genes that improve resistance as a result of prior exposure of a plant to stress. Plants have evolved numerous defence mechanisms viz. reduced heavy metal uptake, sequestration and compartmentalization of metal into vacuoles by ion-selective metal transporter, binding to phytochelatins/metallothioneins, and activation of various antioxidants (Cohen et al. 1998; Guerinot 2000; Hu et al. 2001; Shahid et al. 2015). One mechanism by which many plants respond to and apparently detoxify toxic heavy metals is the production of osmolytes (Delauney and Verma 1993; Schat et al. 1997; Mehta and Gaur 1999). In response to different stresses plants accumulate large quantities of different types of osmolytes/compatible solutes (Serraj and Sinclair 2002). Osmolytes are low molecular weight compatible solutes that are usually non-toxic at high cellular concentrations. These osmolytes provide protection to plants from stress via different mechanisms including adjustment of cellular osmoticum, ROS detoxification, maintenance of membrane integrity, and enzymes/protein stabilization (Zhu 2001; Ashraf and Foolad 2007). These osmolytes include proline, sugars (glucose, fructose, sucrose, trehalose, raffinose), amino acids, and quaternary ammonium compounds (QACs) such as glycinebetaine, alanine betaine, proline betaine, and polyamines which play a vital role in adjusting the osmotic potential and stabilizing of plant cells and tissues (Hasegawa et al. 2000; Mudgal et al. 2010). See Table 5.2.

5.4 PROLINE

Proline is a nitrogen containing five-carbon -amino acid that acts as an osmoprotectant, free radical scavenger, antioxidant, and protein stabilizer and metal chelator (Mishra and Dubey 2006; Zarei et al. 2012). In addition, proline also plays a role as a molecular chaperon, stabilizing the 3D structure of protein, and it maintains the cytosolic pH which aids in balancing the redox reaction status of the cell (Paleg et al. 1981). Rustgi et al. (1977) have revealed that C5 of Pro can react with hydroxyl radicals (·OH). Moreover, it reacts directly with singlet oxygen ($^1O_2$), forming reversible charge-transfer

TABLE 5.1
Metal Transporter Genes Involved in Heavy Metal Uptake and Homeostasis in Different Plant Species

Family	Gene	Metal Transported	Plant	Reference
Heavy metal associated (HMA)	*HMA1*	Cd	*Populus trichocarpa*	Li et al. (2015)
	HMA5	Cu	*Oryza sativa*	Deng et al. (2013)
	HMA3	Cd	*Arabidopsis*	Chao et al. (2012)
	HMA2 and *HMA4*	Zn and Cd	*Arabidopsis*	Wong et al. (2009); Hussain et al. (2004)
	HMA1	Zn	*Arabidopsis thaliana*	Moreno et al. (2008)
	HMA9	Cu, Zn, and Cd	*Oryza sativa*	Lee et al. (2007)
	HMA8	Cu	*Glycine max*	Bernal et al. (2007)
	HMA5	Cu	*Arabidopsis thaliana*	Andrés-Colás et al. (2006)
Natural resistance-associated macrophage (NRAMP)	*NRAMP5*	Mn and Cd	*Oryza sativa*	Sasaki et al. (2012)
	NRAT1	Al	*Saccharomyces cerevisae*	Xia et al. (2011)
	NRAMP3 and *NRAMP4*	Fe	*Arabidopsis thaliana*	Lanquar et al. (2005)
	NRAMP4	Ni	*Thlapsi japonicum*	Mizuno et al. (2005)
	NRAMP1-3	Fe	*Lycopersicon esculentum*	Bereczky et al. (2003)
Cation diffusion facilitator (CDF)	*MTP1*	Zn	*Oryza sativa*	Menguer et al. (2013)
	MTP3	Zn	*Arabidopsis thaliana*	Arrivault et al. (2006)
	MTP1	Zn, Co	*Nicotiana tabacum*	Shingu et al. (2005)
	MTP1	Zn	*Arabidopsis halleri*	Dräger et al. (2004)
	MTP8 and *MTP11*	Mn	*Arabidopsis thaliana*	Delhaize et al. (2003)
ZRT-IRT like protein (ZIP)	*IRT1*, *ZIP1*, and *ZIP3*	Mg	*Oryza sativa*	Chou et al. (2011)
	ZIP1-12	Zn	*Arabidopsis thaliana*	Roosens et al. (2008); Weber et al. (2004)
	ZNT1	Zn	*Thlapsi caerulescence*	Assunção et al. (2001); Pence et al. (2000)
ABC transporter	*ABCC*	As	*Oryza sativa*	Song et al. (2014)
	PDR8	Cd and Pb	*Arabidopsis thaliana*	Kim et al. (2007)
	MRP1 and *MRP2*	PC–Cd complex	*Arabidopsis thaliana*	Lu et al. (1998;1997)
	HMT1	PC–Cd complex	*Saccharomyces pombe*	Ortiz et al. (1995)
Arsenate reductase (ACR)	*HAC1*	As	*Arabidopsis thaliana*	Chao et al. (2014)
	ACR2	As	*Arabidopsis thaliana*	Duan et al. (2007)
Yellow stripe like transporter (YSL)	*YSL3*	Fe, Ni	*Thlapsi caerulescence*	Gendre et al. (2006)
	YSL1	Fe	*Arabidopsis thaliana*	Le Jean et al. (2005)
	YSL2	Zn	*Arabidopsis thaliana*	Schaaf et al. (2005)
	YS1	Fe	*Zea mays*	Curie et al. (2001)
Copper transporter family (CTR)	*COPT1*	Cu	*Arabidopsis thaliana*	Andrés-Colás et al. (2010); Sancenón et al. (2004); Kampfenkel et al. (1995)
CaCA transporter	*CAX2* and *CAX4*	Cd	*Arabidopsis thaliana*	Korenkov et al. (2007)

TABLE 5.2
Responses of Different Osmoprotectants under Heavy Metal Stress in Plants

Omoprotectants	Metals	Plants	Response	References
Proline (Pro)	As	*Glycine max*	Scavenged free radicals and ROS, alleviated lipid and protein oxidation	Chandrakar et al. (2018)
	Ni	*Lolium perenne*	Lipid peroxidation and electrolyte leakage ↓, biosynthetic enzyme activities ↑	Shahid et al. (2014)
	Cd	*Cicer arietinum*	Maintains intracellular redox homeostasis potential, photosynthetic activities ↑	Hayat et al. (2013)
	Se	*Phaseolus vulgaris*	Oxidative stress↓, endogenous proline level ↓,	Aggarwal et al. (2011)
	Cd	*Vigna radiata*	High GSH/GSSG ratio GSH maintaining enzymes	Hossain et al. (2010)
	Hg	*Oryza sativa*	ROS level ↓, water potential (ψ)↓	Wang et al. (2009)
	Cd	*Nicotian tabacum*	Mitigated inhibitory effect on the growth of BY-2 cells	Islam et al. (2009)
	Cd	*Solanum nigrum*	ROS level ↓, protects plasma membrane	Xu et al. (2009)
	Cr	*Ocimum tenuiflorum*	NR activity, photosynthesis and antioxidant activity ↑	Rai et al. (2004)
	Cu, Cr, Ni and Zn	*Chlorella vulgaris*	Counteract lipid peroxidation, K^+ efflux ↑	Mehta and Gaur (1999)
	Cu	*Anacystis nidulans*	K^+ ion efflux ↓	Wu et al. (1995)
Glycinebetaine (GB)	Cd	*Amarnathus* sp.	Soluble sugars and dissolved organic carbon ↑, enhanced photosynthetic activity	Yao et al. (2018)
	Cd	*Spinacea oleracea*	Metal accumulation ↓, photosynthesis and antioxidant activity ↑	Aamer et al. (2018)
	Cd	*Gossypium* sp.	Oxidative stress ↓, enzymatic activity ↑	Farooq et al. (2016a)
	Cr	*Triticum aestivum*	Metal concentration followed by oxidative stress ↓	Ali et al. (2015)
	Pb	*Gossypium* L.	Electrolyte leakage, MDA and H_2O_2 ↓, positively regulates antioxidant system	Bharwana et al. (2014)
	Cd	*Oryza sativa*	Maintains nutrient balance, photosynthesis ↑	Cao et al. (2013)
	Cd	*Vigna radiata*	Increased APX activity and decreased ROS levels	Hossain et al. (2010)
	Cd	*Nicotiana tabacum*	GB accumulation ↑, reduction in lipid peroxidation level	Islam et al (2009)

(*Continued*)

TABLE 5.2 (CONTINUED)
Responses of Different Osmoprotectants under Heavy Metal Stress in Plants

Omoprotectants		Metals	Plants	Response	References
Polyamines (PAs)	Put, Spd, Spm	Cd	*Inula chrithmoides*	Modulate NH_3/NO_2 ratio, stabilizes cellular structure	Ghabriche et al. (2017)
	Spm and Spd	Cd, Pb, Cu, Ni, Zn	*Triticum aestivum*	RWC and WUE ↑, reduced heavy metals translocation from root to shoot and leaf	Aldesuquy (2016)
	Put	Cd	*Vigna radiata*	Lowers biological accumulation co-efficient (BAC) and translocation factor (TF)	Nahar et al. (2016)
	Put, Spd, Spm	Pb	*Triticum aestivum*	Maintains membrane stability and nutrients availability	Rady et al. (2016)
	Spd	Cu	*Raphanus sativus*	Improved antioxidant system and Cu homeostasis	Choudhary et al. (2012)
	Spm	Cd and Cu	*Triticum aestivum*	H_2O_2 ↓, restore GR activity	Groppa et al. (2007)
	Spd	Cd	*Typha latifolia*	Inhibits MDA and ROS production	Tang et al. (2005)
Soluble sugars	Tre	Cu	*Oryza sativa*	Lowers ROS and MDA, Glyoxalase I and II ↑, and improved redox status	Mostofa et al. (2015)
	Tre	Cd and Cu	*Nicotiana tabacum*	Scavenge ROS production	Martins et al. (2014)
	Tre	Cd	*Lemna gibba*	Lipid peroxidation ↓, photosynthetic pigments and antioxidant ↑	Duman et al. (2010)

complexes, which effectively quench free radicals. Pro is synthesized from glutamate involving two consecutive reductions catalyzed by the enzymes D1-pyrroline-5-carboxylate (P5C) synthetase (P5CS) and P5C reductase (P5CR) respectively (Sekhar et al. 2007). Numerous genes can be expressed under heavy metals stress which activate the particular enzymes to cope with negative response. In this way, Siripornadulsil et al. (2002) observed that proline reduces free radical damage and enhances the Cd tolerance capability of a transgenic alga *Chlamydomonos reinhardtii*. In this research, a gene encoding moth bean D1-pyrroline-5-carboxylate synthetase (P5CS) starts the proline production and it is introduced into the nuclear genome of *C. reinhardtii*. Furthermore, Wojas et al. (2008) have shown that the phytochelatins (PC) level along with the detoxification potential of As- and Cd-stressed tobacco plant increased by the overexpression of CePCS and AtPCS1. Likewise, in *Arabidopsis* it has been observed that the overexpression of an antisense proline dehydrogenase cDNA speeds up the proline addition which increases the resistance potential of the plant against freezing and salinity conditions (Nanjo et al. 2003). Similarly, Su and Wu (2004) showed that overexpression of the P5CS gene increased the tolerance potential of the plant against salt and drought stress by the increase of P5CS mRNA and proline levels in transgenic rice. The induction of Pro in response to HM-induced plants is to a great extent concentration-dependent, and organ- and metal-specific. Gohari et al. (2012) observed that Pro concentration in the roots of rape seed increased linearly as the plant was exposed to rising concentrations of Pb^{2+} (100 to 400 M). Similarly, the organ-specific root-to-shoot ratio of Pro accumulation was higher, as exhibited in different plants including wheat subjected to Cd (Leskó and Simon-Sarkadi 2002), mustard subjected to Pb and Cd stress (John et al. 2009), *Cymbopogon flexuosus* Stapf subjected to Hg and Cd stress (Handique and Handique 2009), and *Solanum nigrum* L. exposed to Cu stress (Fidalgo et al. 2013).

Similarly, Nikolić et al. (2008) have shown that Pro accumulation in roots was almost two-fold higher than in leaves in response to Cd stress in hybrid poplar (*Populus trichocarpa*×*deltoides*). Furthermore, Zengin and Kirbag (2007) reported that Pro content in sunflower seedlings subjected to various amounts of HMs was strongly induced in the order of Hg > Cd > Cu > Pb. Therefore, studies showed that the proline accumulation may depend on the concentration and specificity of HMs, their toxicity threshold, and the plant species.

5.5 GLYCINEBETAINE

Glycinebetaine (GB), an ammonium compound, is one of the main organic osmoregulators of plants (Chen and Murata 2011). GB is environmentally friendly, soluble in water, and it has no toxic effects (Makela et al. 1996). GB alleviates oxidative stress and enhances the resistance of the plants under various kinds of abiotic stress such as drought (Anjum et al. 2011; Raza et al. 2014), salinity (Korkmaz et al. 2012; Wei et al. 2017), heat (Li et al. 2011; Kanechi et al. 2013), cold (Park et al. 2006), and heavy metals (Chen and Murata 2011; Cao et al. 2013; Jabeen et al. 2015). Glycinebetaine (GB) is synthesized abundantly in the chloroplast from choline by a two-step oxidation reaction in plants (Chen and Murata 2002). The first oxidation is catalyzed by choline monooxygenase (CMO) which converts choline into betaine aldehyde; then a NAD^+-dependent enzyme betaine aldehyde dehydrogenase (BADH) produces GB from betaine aldehyde. Among the different GB biosynthetic genes, choline oxidase (*codA*) from *A. globiformis* has been widely used for GB production in transgenic plants (Nuccio et al. 1998). Though in some plants, GB accumulates naturally at levels that can reduce the adverse effects of various environmental stresses. The proposed mechanisms for GB-mediated abiotic stress tolerance include stabilization of proteins and enzymes, membrane integrity, osmoregulation, photochemical quenching, and detoxification of ROS that induces stress tolerance through the signal transduction (Yancey 1994; Allakhverdieva et al. 2001; Subbarao et al. 2001). GB efficiently stabilizes the 3D structures proteins and enzymes and induces stress tolerance by protecting the components of photosynthetic machinery, such as RUBISCO and photosystem II (PS II) and maintains the turgor pressure (Allakhverdieva et al. 1996), thereby sustaining the fixation of CO_2, which, in turn, reduces ROS production. Furthermore, GB activates the expression of genes for ROS-scavenging enzymes, which decrease the ROS levels in cells (Chen and Murata 2011). The ascorbate-glutathione (ASC-GSH) cycle is one of the crucial mechanisms scavenging ROS (Noctor and Foyer 1998). Islam et al. (2009a) found that exogenous application of GB increased the activities of ASC-GSH cycle enzymes, resulting in the suppression of ROS production and considerably restoring the membrane integrity under Cd stress in tobacco bright yellow-2 cells. However, the effect of exogenously applied GB on heavy metal tolerance in plants remains unclear.

5.6 SOLUBLE SUGARS

Soluble sugars such as glucose, sucrose, trehalose, and related oligosaccharides have an active role in growth regulation, photosynthesis, carbon partitioning, osmotic homeostasis, membrane stabilization, and gene expression under various abiotic stresses (Hoekstra et al. 2001; Rosa et al. 2009; Hieu et al. 2015). Moreover, they also control transcriptional and translational processes, acting as signalling molecules in plants (Koch 1996; Muller et al. 2011). They can protect the plant from various stresses viz. drought, salt, and oxidative stress through direct or indirect triggering of the production of ROS scavengers and repair enzymes (Almeida et al. 2005, 2007; Van den Ende and Valluru 2009; Hasegawa 2000). However, the beneficial role of sugars and the associated mechanisms involved in protecting plants against heavy metal toxicity remain elusive. The effects of soluble sugar on gene expression result from sugar-specific signalling cascades (Couée et al. 2006). However, hexokinase (HXK), Snf1-related kinase 1, and INV are currently identified as conserved sugar signalling components (Valluru and Van den Ende 2011). They have the capability to modify the superoxide dismutase (Koch 1996), heat shock proteins, and glutathione-S-transferases gene expression (Price et al. 2004) against abiotic stress responses. An increase in other saccharides (galactose, myoinositol, trehalose, and raffinose) was observed after Cd-treatment of *Arabidopsis* (Sun et al. 2010b). Rahoui et al. (2015) found that Cd-tolerant lines of *Medicago truncatula* showed higher mobilization of total soluble sugars (glucose, fructose, and sucrose), as well as overconsumption of glucose compared to the Cd- susceptible lines. Furthermore, Ibrahim and Abdellatif (2016) also observed that trehalose (Tre) led to a significant increase in phenolic compounds in wheat plants under water stress. These compounds scavenge ROS production through the antioxidant enzymes that use polyphenols as substrates (Sgherri et al. 2003). Tre up-regulated the expression of AOX and concomitantly retarded the accumulation of lipid peroxidation and H_2O_2 as well as increasing phenolics content resulting in overcoming the oxidative stress (Shalaby and Horwitz 2015). Finally, mannitol is a sugar alcohol able to retain ROS scavenging capacity, thereby protecting against oxidation by °OH radicals (Stoyanova et al. 2011). It also protects the function of TRX, ferredoxin, and GSH in genetically modified *Nicotiana tabacum* by increasing their own concentration in the chloroplast as compared to wild-type seedlings. However, the capacity to scavenge these radicals did increase in transgenic seedlings without any negative influence on photosynthesis and productivity (Shen et al. 1997; Bolouri-Moghaddam et al. 2010).

5.7 ROLE OF PLANT GROWTH REGULATORS IN ALLEVIATION OF METALS STRESS

Plants have the ability to mitigate toxic effects of heavy metals stress, which seems to be a very important determinant of their tolerance to various abiotic and biotic stresses. An alternative strategy to alleviate the effects of abiotic stress factors, including heavy metals, on plant growth by the use of exogenous application of plant growth hormones by manipulating their endogenous level has recently been gaining importance (Kováčik et al. 2010; Ali et al. 2013). Plant hormones as a signalling molecule may participate in the signal-transduction pathway, or they may stimulate reactions in response to different stresses (Argueso et al. 2010; Qin et al. 2011). There are five classical phytohormones: auxins, cytokinins, gibberellins, ethylene, and abscisic acid (ABA). Recently, a class of "new plant hormones" which includes salicylic acid, brassinosteroids, jasmonates, and stringolactones has also been identified. Moreover, polyamines, systemin, and nitric oxide have been known to possess activities resembling phytohormones (Chen et al. 2009; Peleg and Blumwald 2011). Phytohormones play a fundamental role in regulating and synchronizing growth, and also take part in all developmental processes, including abiotic stress responses (Spoel and Dong 2008). The phytohormones play an important role in the adaptation of plants to stressful environments by improving heavy metal tolerance by increasing the biomass, and enhance the efficacy of antioxidant systems in plants, consequently facilitating the acquisition and accumulation of toxic elements (Liphadzi et al. 2006; Israr et al. 2011). In this context, this review is focused on several phytohormones involved in the regulation of heavy metal tolerance.

5.8 ACTION OF AUXIN AGAINST METAL STRESS

Auxin is considered to be the most important plant growth regulator involved in cell division and differentiation. In addition to this, it also takes part in the protection and regulation of the plant metabolism under stress conditions (Parker et al. 1992; Mockaitis and Estelle 2008). It has been reported that As stress alters the level of endogenous auxin (IAA, NAA, and IBA) in mustard (Srivastava et al. 2013). As-induction leads to the repression of auxin influx (AUX1) and efflux (PIN5) carriers. Surprisingly, Cd increased AUX1 expression, also extending it to more cells, but strongly inhibited PIN5 expression (Ronzan et al. 2018). The AUX1 controls many aspects of root development in rice in response to Cd stresses (Zhao et al. 2015). Furthermore, Sun et al. (2010a) showed that Al-induced auxin redistribution is modulated by AUX1 and PIN2. In another case, IAA homeostasis in barley root tips was disturbed after exposure to Cd (Zelinová et al. 2015). It is known that *OsPIN5b* is associated with the modulation of IAA homeostasis, transport, and distribution (Lu et al. 2015). Similarly, Bücker-Neto et al. (2017) recently reviewed that heavy metal stress leads to a decrease in the endogenous levels of auxins by elevating IAA oxidase activity and auxin degradation. Studies have shown that auxins play a critical role in enhancing metal tolerance in plants. Auxin interacting with ethylene leads plant tolerance to Al toxicity through altered auxin distribution via AUX1 and PIN2 auxin transporters (Sun et al. 2010a). It has been suggested that the auxin efflux transport mutants *aux1*, *pin1*, and *pin2* were highly sensitive to As(III); thereby, auxin transport inhibitors reduced plant tolerance to As(III) due to the increased levels of H_2O_2 (Krishnamurthy and Rathinasabapathi 2013). However, an exogenous IAA application may improve As(III) tolerance of *aux1* suggesting a positive role for auxin transport through AUX1 in As(III) tolerance to plant through ROS-mediated signalling.

It is quite possible that the exogenous application of auxin may rescue the level and distribution of endogenous auxin. RD64, an IAA overproducing strain, accumulated higher levels of endogenous osmolyte in *Medicago truncatula* showing increased tolerance to several stress conditions, thus indicating the role of auxin in imparting tolerance against stress (Bianco and Defez 2009). A study by Ghorbanli et al. (2000) indicated that plant growth regulators reduced the opposing effect of Cd by increasing plant growth, photosynthetic pigment, CO_2 consumption, net assimilation rate, and leaf area ratio in heavy metal-stressed plants. According to Srivastava et al. (2013), the exogenous application of IAA improved As-induced plant growth by modulating the expression of miR167, miR319, and miR854, suggesting a protective role of IAA in enhancing metal tolerance. In a similar way, Farooq et al. (2015) further reported that the application of L-TRP (an auxin precursor) to rice seedlings enhanced plant growth and yield compared to seedlings without auxin precursor in Cd-contaminated soil. In pea seedlings, the exogenous application of IAA protected several metabolic processes against Cr(VI) and Mn toxicity by regulating the antioxidant defence system (Gangwar and Singh 2011). The exogenous application of phytohormones including auxin inhibits oxidative burst through the stimulation of antioxidant machinery (SOD, CAT, and peroxidase) in wheat (Szechynska-Hebdam et al. 2007) and in *Chlorella vulgaris* (Piotrowska-Niczyporuk et al. 2012) in response to heavy metal stress. In another study, Ostrowski et al. (2016) demonstrated that an auxin conjugate, IAA-Asp, modulates catalase and peroxidase activity in Cd-induced pea seedlings. It also induces carbonylated protein and reduced H_2O_2 concentration. Conversely, Hac-Wydro et al. (2016) indicate that IAA or NAA can promote membrane organization by alleviating the toxic effect of Pb exposure. It has also been suggested that auxin increased the hemicellulose 1 level in *Arabidopsis*, consequently fixing Cd^{2+} in the roots, thus reducing the translocation of Cd^{2+} from the root to aboveground parts (Zhu et al. 2013). Some recent approaches showed that in addition to protecting plants against different stresses, exogenously auxin can be used as a tool in phytoremediation programs for the detoxification of heavy metal-contaminated soil. Sanjaya et al. (2008) described that the Cd tolerance ability of *Arabidopsis* and tomato plants increased by the exogenous application of tryptophan (Trp). In another study, Fässler et al. (2010) showed that exogenously applied IAA may protect plants as well as increase the phytoextraction potential of Pb and Zn. Similar observations were also reported indicating that IAA improved phytoextraction of Pb in *Medicago sativa* (López et al. 2005), maize (Hadi et al. 2010), and IBA as well as NAA enhanced Pb and Zn extraction in maize (Fuentes et al. 2000).

5.9 ACTION OF CYTOKININ AGAINST METAL STRESS

Cytokinins (CKs) participate in various physiological processes such as seed germination, cell expansion and differentiation, and leaf and chloroplast senescence (Rashotte et al. 2006; Dong et al. 2009). Several biotic and abiotic factors including metals, temperature, high light, drought, and salinity, lower the production as well as transport of CKs from roots. The decrease in CKs might be due to the oxidation of CKs which is an important pathway for their inactivation (Kaminek et al. 1997). However the effect of Cd on the activity of cytokinin oxidase suggests a more direct effect of this element on the process of hormone metabolism. In agreement with this, cadmium acetate increased the activity of cytokinin oxidase (Somashekaraiah et al. 1992). However, in *Arabidopsis*, it may be ascribed that up-regulation of cytokinin oxidase (*AtCKX*) is activated only above a certain threshold level of Cd toxicity (Vitti et al. 2013). The regulatory role of CKs in controlling development seems feasible given their involvement in responses to adverse environmental conditions. It has been suggested that CKs play an important role in the regulation of environmental stress responses and are involved in intense interactions and crosstalk with other hormones (Jeon et al. 2010; Ha et al. 2012; Chang et al. 2016). Studies have shown that the exogenous application of CKs increases the stress-tolerance capacity of plants, indicating a beneficial effect of CKs in the regulation of plants against environmental stresses. Thus, these findings show the potential for the application of cytokinins in agricultural biotechnology. For instance, CKs can act as a radical scavenger against damage caused by ROS through down-regulation of lipoxygenase activity as well as modulation of the activities of antioxidant enzymes such as SOD, CAT, and peroxidase (Prakash et al. 1990; Gidrol et al. 1994; Chalupkova and Smart 1994; Petit-Paley et al. 1999). In this respect, Gangwar et al. (2010) stated that the application of kinetin enhanced Mn tolerance as well as seedling growth by increasing ammonium assimilation and the antioxidant defence system in pea. In green algae, exogenous CK mitigated the stress induced by heavy metals (Cd, Pb, and Cu) via decreasing metal absorption and stimulating the antioxidant defence system (Piotrowska-Niczyporuk et al. 2012). Recently, Wang et al. (2015) suggested that kinetin could alleviate the destructive effects of As-induced maize plant by enhancing antioxidant enzyme activities, photosynthesis, and reducing lipid peroxidation levels. It has been reported that zeatin mitigated the lead-induced inhibition of root growth and the electron transport activity in *Picea abies*, indicating that cytokinin in the rhizosphere might be an important factor affecting Pb toxicity (Vodnik et al. 1999). Gemrotová et al. (2013) suggested that an inhibitor of cytokinin degradation, INCYDE, mitigated the negative effects of Cd in *Bulbine natalensis* and suggested that the degradation of inhibitors of cytokinin controls the cytokinin status which may be useful in protecting plants against the adverse effects of Cd toxicity. Moreover, Pospíšilová (2003) observed that phytoextraction efficiency can be enhanced by the application of CKs as it regulates stomatal opening. As the stomata open, the greater the rate of transpiration, the more contaminants will be transferred from soil to plants. In addition, the levels of ethylene and ABA enhanced in plant tissues in response to environmental stresses causes a decline in stomatal conductance while the application of cytokinins may counteract these changes (Lipiec 2012). Similarly, Thomas et al. (2005) observed that the cytokinin-synthesizing gene (*ipt*) stimulates the synthesis of metallothionein by expressing the metallothionein gene (MT-L2) in *Nicotiana plumbaginifolia* during Cu stress. Tassi et al. (2008) found that kinetin increases Pb and Zn accumulation in sunflower leaves, leading to enhanced phytoextraction efficiency. In another study, it has been reported that cytokine treatment reduced Hg contamination in various parts of *Helianthus annus* and mustard (Cassina et al. 2012).

5.10 ROLE OF ETHYLENE DURING HEAVY METAL STRESS

Ethylene is a gaseous phytohormone with multiple roles in the regulation of metabolism at the cellular, molecular, and whole plant levels (Pierik et al. 2006; Lin et al. 2009; Schaller 2012; Khan N. A. and Khan M. I. R. 2014; Khan N.A. et al. 2017). It is designated as the first identified plant hormone known to regulate growth and developmental responses to biotic and abiotic stresses. Ethylene is considered as an ageing hormone and, being the most influential plant growth regulator, it is required for processes such as ripening, senescence, and abscission (Schaller 2012). Ethylene production and signalling is one of the most prominent networks of hormonal crosstalk and signalling triggered by plants to manifest stress-induced symptoms in acclimation processes. Among the most commonly occurring abiotic stresses observed in agricultural production, such as drought, submerging, and extreme temperatures (Thao and Tran 2012; Xia et al. 2015; Thao et al. 2015), heavy metal (HM) stress has arisen as a new pervasive threat (Srivastava et al. 2016; Ahmad et al. 2015; Thao et al. 2015). Progressive studies in ethylene biology have paved the way in establishing ethylene as not only the phytohormone which regulates the plant's physiological processes but also as a regulatory hormone during heavy metal stress. 1-Aminocyclopropane-1-carboxylate synthase (ACS) and 1-aminocyclopropane-1-carboxylate oxidase (ACO) are the two major regulatory enzymes involved in ethylene biosynthesis, and are encoded by multigene families, which are also the primary regulation points in the ethylene biosynthetic pathway (Xu and Zhang 2015; Thao et al. 2015). HM stress increases the activity of these two enzymes, resulting in increased ethylene production (Schellingen et al. 2014; Khan et al. 2015b; Thao et al. 2015) During unstressed conditions, methionine is converted into S- adenosyl-methionine (SAM) in the presence of enzyme SAM synthetase. Furthermore, using SAM as a substrate, ACC is produced by ACC synthase (ACS) (Keunen et al. 2016). Under severe stress conditions, ethylene production might be simply increased by tissue damage and necrosis (Lynch and Brown 1997; Keunen et al. 2016). Stress severity will in turn effect the activation of specific signal transduction pathways, for example those related to ethylene (Kacperska 2004; Keunen et al. 2016). Autocatalytic stress-related ethylene synthesis is triggered as the response to stress-induced effects on plants. Strong lines of evidence have shown the multiple facets of ethylene in plant responses to different abiotic stresses, including excessive HM, depending upon endogenous ethylene concentration (Thao et al. 2015) and ethylene sensitivities that differ according to the developmental stage, plant species, and culture systems (Pierik et al. 2006; Kim et al. 2007; Khan and Khan 2014; Thao et al. 2015).

Various external and internal stimuli (biotic and abiotic stress) are known to regulate the production of ethylene at the level of ACS gene expression (Tsuchisaka et al. 2009; Van de Poel and Van Der Straeten 2014; Keunen et al. 2016). An illustration was given by Sandmann and Böger (1980), describing the regulatory role of ethylene in the condition of heavy metal stress, in which they demonstrated the inhibition of photosynthetic electron transport and enhanced synthesis of ethylene in spinach (*Spinacia oleracea*) chloroplasts under high concentrations of Cu. More evidence illustrated that Ni and Zn did not stimulate ethylene production in *Arabidopsis* (Arteca and Arteca 2007; Thao et al. 2015), however, ethylene activity was reportedly increased in mustard plants by enhancing ACS activity (Khan and Khan 2014; Thao et al. 2015). Relative changes in PS II activity and the reciprocal mechanism of photosynthetic inhibition by Ni and Zn stress with alterations in photosynthetic-NUE and antioxidant capacity within mustard due to ethylene were confirmed using the thylene action inhibitor, norbornadiene (Khan and Khan 2014).

5.11 ACTION OF JASMONATES AGAINST METAL STRESS

Jasmonic acid (JA) is a lipid-derived naturally occurring phytohormone that is involved in the regulation of the physio-biochemical as well as antioxidant activities of plants in response to biotic and abiotic stresses (Wasternack 2014; Huang et al. 2017). It also functions as a signal for the stimulation of detoxification enzymes and related gene expression under stressful conditions (Shan and Liang 2010). Several studies suggested that JA plays a significant role in alleviating heavy metal stress by increasing the endogenous JA level in plants, which seems to be a promising method of providing protection against stress (Piotrowska et al. 2009; Keramat et al. 2010). Endogenous JA content increases under different metal stresses were also observed in *Arabidopsis*, *Phaseolus coccineus*, rice, and *Kandelia obovata* (Koeduka et al. 2005; Maksymiec et al. 2005; Chen et al. 2014). Additionally, it has been shown that Cu or Cd can induce some JA-responsive events, such as VSP2 transcripts in *Arabidopsis* (Mira et al. 2002) and MAPK in rice (Agrawal et al. 2003; Kim et al. 2003). The stimulatory influence of JA was found to ameliorate the stress imposed by Pb in *Wolfia arrhiza* at low concentrations (Piotrowska et al. 2009), Cd in *Capsicum frutescens* and rape seed (Yan and Tam 2013 Ali et al. 2018), Cu in *Cajanus cajan* (Poonam et al. 2013), Ni in *Glycine max* (Sirhindi et al. 2016), and As in rape seed (Farooq et al. 2016b, 2018). This may be due to the fact that exogenous JA causes a rapid induction of defensive proteins called jasmonate inducible proteins (JIPs) (Wasternack and Parthier 1997; Farmer et al. 2003), which helps in resistance to particular stress factors (Kessler and Baldwin 2002). Similarly, Chen et al. (2014) also reported that the exogenous application of 0.1–1 mM methyl jasmonate (MJ) improves ion homeostasis and supresses oxidative damage through enhanced antioxidant activities in *Kandelia obovata* under Cd toxicity. In another study, it has been stated that MJ application increased biomass production in soybean plants under Cd stress (Keramat et al. 2010). The application of MJ induced plant tolerance against metal stress by maintaining the antioxidant system which plays a key role in scavenging the H_2O_2 contents

(Farooq et al. 2016b). Furthermore, an elevation of antioxidant capacity by several fold in SOD, POD activity, and proline accumulation, which help to scavenge free radicals, was also observed after JA application (Poonam et al. 2013). Evidence suggested that JA modified the fatty acid composition of the membrane which may have protective effects on the cell membrane lipid by reducing lipid peroxidation levels (Wang 1999). The benefit of exogenous JA application in terms of minimizing oxidative stress was evidenced by the reduced MDA levels in plants under heavy metal stress (Singh and Shah 2014; Hanaka et al. 2016). Additionally, the accumulation of compatible solutes, proline, and GB has been reported to be a good indicator of stress tolerance. It has been reported that JA induces mRNAs encoding proline biosynthetic enzymes during stress conditions and protects the plants from oxidative burst by scavenging ROS (Creelman and Mullet 1991; Chen and Kao 1993), and a greater accumulation of osmolytes results from the up-regulation and down-regulation of biosynthetic genes of osmolyte-generating pathways (Ahanger et al. 2017). Exogenously applied JA enhanced the proline and GB content in *Glycine max* and pear plants respectively (Gao et al. 2004; Sirhindi et al. 2016). In addition to this, JA is also involved in the formation of secondary metabolites that play a crucial role in reducing oxidative stress (Erbilgin et al. 2006; Wang et al. 2011). It has been also suggested that exogenous JA increased phenolic compounds that have an important role in plant defence response (Kim et al. 2007). Furthermore, JA is also involved in the accumulation of phytochelatins in *Arabidopsis* against Cu and Cd toxicity (Maksymiec and Krupa 2007).

5.12 ACTION OF POLYAMINES (PAS) AGAINST METAL TOXICITY

Polyamines (PAs) are a low molecular weight aliphatic compound, important modulators of biological processes that play a vital role in regulating plant physiology and development, as well as stress management (Kakkar and Sawhney 2002; Nahar et al. 2016). The common PAs in plants are spermidine (Spd), spermine (Spm), and putrescine (Put). They are positively charged, therefore, they can bind with negatively charged molecules such as DNA, RNA, proteins, and phospholipids, resulting in the modulation of replication, translation, membrane stabilization, cell division, and differentiation (Kuzentsov et al. 2006; Minois 2014). In addition, PAs can mitigate the overproduction of ROS by reducing free radicals (Benavides et al. 2018) and play vital roles in regulating the plant defence response to diverse environmental stresses including heavy metal toxicity (Groppa et al. 2003). Yiu et al. (2009) reported that exogenous Put reduces the oxidative damage in *Allium fistulosum* by reducing the superoxide radical ($O_{2\bullet-}$) and H_2O_2 contents and thereby, decreasing oxidative stress in plant cells. Similarly, exogenous PAs (Spd, Spm) protect the tissues from oxidative damage induced by copper and cadmium as reported by Groppa et al. (2007) and Wang et al. (2007). However, PAs synthesized endogenously in plants further enhance antioxidant defence mechanisms, including stimulating antioxidants, ROS scavenging, membrane stability, and metal chelation (Lomozik et al. 2005; Alcazar et al. 2010; Chen et al. 2013). In stress condition, PAs act as signalling molecules and consequently regulate ion homeostasis and control ion transportation through interaction with ion channels (Podlešáková et al. 2019). Andrea et al. (2004) have shown that PAs play an important role in the alleviation of Cd toxicity in tobacco cells by the activation of stress-defending mechanisms which protect the cells from programmed cell death (PCD). It has been observed from several studies that PAs are not only involved in numerous cellular and molecular processes but are also involved in alleviating abiotic stresses when plants are exposed to different stress factors like salinity, oxidative, drought, chilling, water, and heavy metals and paraquat (Shevyakova et al. 2006; Zhao et al. 2003; Yamaguchi et al. 2006; Hausman et al. 2000; Nayyar and Chander 2004; Walters 2003; Ye et al. 1997; Xing et al. 2000; Weinstein et al. 1986; Groppa et al. 2001; Wang et al. 2004; Taulavuori et al. 2005; Shevyakova et al. 2010; Benavides et al. 2000). Rady and Hemida (2015) observed that treatment with Spm or Spd enhanced the Cd tolerance in wheat seedlings. Recently, Taie et al. (2019) also found that application of Spm, Spd, or Put considerably improved wheat plant growth and yield under Cd and Pb stress by increasing tolerance mechanisms. Furthermore, it has been noted that gene encoding polyamine biosynthetic enzymes such as arginine decarboxylase (ADC), ornithine decarboxylase (ODC), *S*-adenosylmethionine decarboxylase (SAMDC), and Spd synthase (SPDS) improved environmental stress tolerance in various plant species (Bagni and Tassoni 2001; Liu et al. 2007). Little information is known about this biosynthetic gene regulation in response to heavy metals (Weinstein et al. 1986; Groppa et al. 2003; Balestrasse et al. 2005).

REFERENCES

Aamer, M., Muhammad, U. H., Li, Z., Abid, A., Su, Q., Liu, Y., Adnan, R., Muhammad, A. U. K., Tahir, A. K. and Huang, G. 2018. Foliar application of glycinebetaine (GB) alleviates the cadmium toxicity in spinach through reducing Cd uptake and improving the activity of anti-oxidant system. *Appl Ecol Environ Res* 16:7575–7583.

Aggarwal, M., Sharma, S., Kaur, N., Pathania, D. and Bhandhari, K. 2011. Exogenous proline application reduces phytotoxic effects of selenium by minimising oxidative stress and improves growth in bean (*Phaseolus vulgaris* L.) seedlings. *Biol Trace Elem Res* 140:354–367.

Agrawal, G. K., Iwahashi, H. and Rakwal, R. 2003. Rice MAPKs. *Biochem Biophys Res Commun* 302:171–180.

Ahanger, M. A., Akram, N. A., Ashraf, M., Alyemeni, M. N., Wijaya, L. and Ahmad, P. 2017. Plant responses to environmental stresses-from gene to biotechnology. *AoB Plants* 9:4.

Ahmad, P., Sarwat, M., Bhat, N. A., Wani, M. R., Kazi, A. G. and Tran, L. S. 2015. Alleviation of cadmium toxicity in *Brassica juncea* L. (Czern. & Coss.) by calcium application involves various physiological and biochemical strategies. *PLoS One* 10:e0114571.

Alcázar, R., Altabella, T., Marco, F., Bortolotti, C., Reymond, M., Koncz, C., et al. 2010. Polyamines: Molecules with regulatory functions in plant abiotic stress tolerance. *Planta* 231:1237–1249.

Aldesuquy, H. S. 2016. Polyamines in relation to metal concentration, distribution, relative water content and abscisic acid in wheat plants irrigated with waste water heavily polluted with heavy metals. *Int J Bioassays* 5(5):4534–4546.

Ali, E., Hussain, N., Shamsi, I. H., Jabeen, Z., Siddiqui, M. H. and Jiang, L. 2018. Role of jasmonic acid in improving tolerance of rapeseed (*Brassica napus* L.) to Cd toxicity. *J Zhejiang Univ-Sci B (Biomed and Biotechnol)* 19:130–146.

Ali, S., Chaudhary, A., Rizwan, M., Anwar, H. T., Adrees, M., Farid, M., Irshad, M. K., Hayat, T. and Anjum, S. A. 2015. Alleviation of chromium toxicity by glycinebetaine is related to elevated antioxidant enzymes and suppressed chromium uptake and oxidative stress in wheat (*Triticum aestivum* L.). *Environ Sci Pollut Res* 22:10669–10678.

Ali, T., Mahmood, S., Khan, M., Aslam, A., Hussain, M. B., Asghar, H. N. and Akhtar, M. J. 2013. Phytoremediation of cadmium contaminated soil by auxin assisted bacterial inoculation. *Asian J Agric Biol* 1:79–84.

Allakhverdiev, S. I., Feyziev, Y. M., Ahmed, A., Hayashi, H., Alie, J. A., Klimov, V. V., Murata, N. and Carpentier, R. 1996. Stabilization of oxygen evolution and primary electron transport reactions in photosystem II against heat stress with glycine betaine and sucrose. *Photochem Photobiol* 34:149–157.

Allakhverdieva, M. Y., Mamedov, D. M. and Gasanov, R. A. 2001. The effect of glycinebetaine on the heat stability of photosynthetic reactions in thylakoid membranes. *Turk J Bot* 25:11–17.

Almeida, A. M., Cardoso, L. A., Santos, D. M., Torne, J. M. and Fevereiro, P. S. 2007. Trehalose and its applications in plant biotechnology. *Vitro Cell Dev Biol Plant* 43:167–177.

Almeida, A. M., Villalobos, E., Arau´jo, S. S., Leyman, B. V. D. P., Alfaro-Cardoso, L., Fevereiro, P. S., Torné, J. M. and Santos, D. M. 2005. Transformation of tobacco with an *Arabidopsis thaliana* gene involved in trehalose biosynthesis increases tolerance to several abiotic stresses. *Euphytica* 146:165–176.

Andrea, K., Lenka, G., Sylva, Z., Josef, E., Ivana, M., Denek, O. Z. and Milena, C. 2004. Cytological changes and alterations in polyamine contents induced by cadmium in tobacco BY-2 cells. *J Plant Physiol Biochem* 42:149–156.

Andrés-Colás, N., Perea-Garcia, A., Puig, S. and Penarrubia, L. 2010. Deregulated copper transport affects *Arabidopsis* development especially in the absence of environmental cycles. *Plant Physiol* 153:170–184.

Andrés-Colás, N., Sancenón, V., Rodríguez-Navarro, S., Mayo, S., Thiele, D. J., Ecker, J. R., Puig, S. and Peñarrubia, L. 2006. The *Arabidopsis* heavy metal P-type ATPase HMA5 interacts with metallochaperones and functions in copper detoxification of roots. *Plant J* 45:225–236.

Anjum, S. A., Farooq, M., Wang, L. C., Xue, L. L., Wang, S. G., Wang, L., Zhang, S. and Chen, M. 2011. Gas exchange and chlorophyll synthesis of maize cultivars are enhanced by exogenously-applied glycinebetaine under drought conditions. *Plant Soil Environ* 57:326–331.

Apel, K. and Hirt, H. 2004. Reactive oxygen species: Metabolism, oxidative stress, and signal transduction. *Ann Rev Plant Biol* 5:373–399.

Argueso, C. T., Raines, T. and Kieber, J. J. 2010. Cytokinin signaling and transcriptional networks. *Curr Opin Plant Biol* 13:533–539.

Arrivault, S., Senger, T. and Krämer, U. 2006. The *Arabidopsis* metal tolerance protein AtMTP3 maintains metal homeostasis by mediating Zn exclusion from the shoot under Fe deficiency and Zn oversupply. *Plant J* 46:861–879.

Arteca, R. N. and Arteca, J. M. 2007. Heavy-metal-induced ethylene production in *Arabidopsis thaliana. J Plant Physiol* 164:1480–1488.

Asada, K. 2006. Production and scavenging of reactive oxygen species in chloroplasts and their functions. *Plant Physiol* 141:391–396.

Ashfaque, F., Inam, A., Inam, A., Iqbal, S. and Sahay, S. 2017. Response of silicon on metal accumulation, photosynthetic inhibition and oxidative stress in chromium-induced mustard (*Brassica juncea* L.). *S Afr J Bot* 111:153–160.

Ashraf, M. and Foolad, M. R. 2007. Roles of glycine betaine and proline in improving plant abiotic stress resistance. *Environ Exp Bot* 59:206–216.

Ashtamker, C., Kiss, V., Sagi, M., Davydov, O. and Fluhr, R. 2007. Diverse subcellular locations of cryptogein-induced reactive oxygen species production in tobacco Bright Yellow-2 cells. *Plant Physiol* 143:1817–1826.

Assche, F. and Clijsters, H. 1990. Effects of metals on enzyme activity in plants. *Plant Cell Environ* 24:1–15.

Assunção, A. G. L., Da Costa Martins, P., De Folter, S., Vooijs, R., Schat, H. and Aarts, M. G. M. 2001. Elevated expression of metal transporter genes in three accessions of the metal hyperaccumulator *Thlaspi caerulescens. Plant Cell Environ* 24:217–226.

Atal, N., Saradhi, P. P. and Mohanty, P. 1991. Inhibition of the chloroplast photochemical reactions by treatment of wheat seedlings with low concentrations of cadmium: Analysis of electron transport activities and changes in fluorescence yield. *Plant Cell Physiol* 32:943–951.

Bagni, N. and Tassoni, A. 2001. Biosynthesis, oxidation and conjugation of aliphatic polyamines in higher plants. *Amino Acids* 20:301–317.

Balestrasse, K. B., Gallego, S. M., Benavides, M. P. and Tomaro, M. L. 2005. Polyamines and proline are affected by cadmium stress in nodules and roots of soybean plants. *Plant Soil* 270:343–353.

Barconi, D., Bernardini, G. and Santucci, A. 2011. Linking protein oxidation to environmental pollutants: Redox proteome approaches. *J Proteomics* 74:2324–2337.

Benavides, M. P., Gallego, S. M., Comba, M. E. and Tomaro, M. L. 2000. Relationship between polyamines and paraquat toxicity in sunflower leaf discs. *Plant Growth Regul* 31:215–224.

Benavides, M. P., Groppa, M. D., Recalde, L. and Verstraeten, S. V. 2018. Effects of polyamines on cadmium- and copper-mediated alterations in wheat (*Triticum aestivum* L.) and sunflower (*Helianthus annuus* L.) seedling membrane fluidity. *Arch Biochem Biophys* 654:27–39.

Bereczky, Z., Wang, H. Y., Schubert, V., Ganal, M. and Bauer, P. 2003. Differential regulation of Nramp and IRT metal transporter genes in wild type and iron uptake mutants of tomato. *J Biol Chem* 278:24697–24704.

Bernal, M., Testillano, P. S., Alfonso, M., Del Carmen, R. M., Picorel, R. and Yruela, I. 2007. Identification and subcellular localization of the soybean copper P1B-ATPase GmHMA8 transporter. *J Struct Biol* 158:146–158.

Bertrand, M. and Poirier, I. 2005. Photosynthetic organisms and excess of metals. *Photosynthetica* 43:345–353.

Bharwana, S. A., Ali, S., Farooq, M. A., Iqbal, N., Hameed, A., Abbas, F. and Ahmad, M. S. A. 2014. Gycinebetaine-induced lead toxicity tolerance related to elevated photosynthesis, antioxidant enzymes suppressed lead uptake and oxidative stress in cotton. *Turkish J Bot* 38:281–292.

Bianco, C. and Defez, R. 2009. *Medicago truncatula* improves salt tolerance when nodulated by an indole-3-acetic acid-overproducing *Sinorhizobium meliloti* strain. *J Exp Bot* 60:3097–3107.

Blaylock, J. M. and Huang, J. W. 2000. Phytoextraction of metals. In: Raskin, I. and Ensley, B. D. (eds), *Phytoremediation of Toxic Metals: Using Plants to Clean Up the Environment*, pp. 53–70. Wiley, New York.

Bolouri-Moghaddam, M. R., Roy, K. L., Xiang, L., Rolland, F. and Van den Ende, W. 2010. Sugar signalling and antioxidant network connections in plant cells. *FEBS Journal* 277:2022–2037.

Bücker-Neto, L., Paiva, A. L. S., Machado, R. D., Arenhart, R. A. and Margis-Pinheiro, M. 2017. Interactions between plant hormones and heavy metals responses. *Genet Mol Biol* 40:373–386. doi: 10.1590/1678-4685-GMB-2016-0087

Cao, F., Liu, L., Ibrahim, W., Cai, Y. and Wu, F. 2013. Alleviating effects of exogenous glutathione, glycinebetaine, brassinosteroids and salicylic acid on cadmium toxicity in rice seedlings (*Oryza sativa*). *Agrotechnology* 2:107.

Carrier, P., Baryla, A. and Havaux, M. 2003. Cadmium distribution and microlocalization in oilseed rape (*Brassica napus*) after long-term growth on cadmium-contaminated soil. *Planta* 216:939–950.

Cassina, L., Tassi, E., Pedron, F., Petruzzelli, G., Ambrosini, P. and Barbafieri, M. 2012. Using a plant hormone and a thioligand to improve phytoremediation of Hg-contaminated soil from a petrochemical plant. *J Hazard Mater* 15:36–42.

Chalupkova, K. and Smart, C. C. 1994. The abscisic acid induction of novel peroxidase is antagonized by cytokinin in *Spirodela polyrhiza* L. *Plant Physiol* 105:497–504.

Chandrakar, V., Dubey, A. and Keshavkant, S. 2018. Modulation of arsenic-induced oxidative stress and protein metabolism by diphenyleneiodonium, 24-epibrassinolide and proline in *Glycine max* L. *Acta Bot Croat* 77:51–61. doi: 10.2478/botcro-2018-0004

Chang, Z., Liu, Y., Dong, H., et al. 2016. Effects of cytokinin and nitrogen on drought tolerance of creeping *Bentgrass*. *PLoS One* 11:e0154005.

Chao, D. Y., Chen, Y., Chen, J., Shi, S., Chen, Z., Wang, C., et al. 2014. Genome-wide association mapping identifies a new arsenate reductase enzyme critical for limiting arsenic accumulation in plants. *PLoS Biol* 12:e1002009. doi: 10.1371/journal.pbio.1002009

Chao, D. Y., Silva, A., Baxter, I., Huang, Y. S., Nordborg, M., Danku, J., Lahner, B., Yakubova, E. and Salt, D. E. 2012. Genome-wide association studies identify heavy metal ATPase3 as the primary determinant of natural variation in leaf cadmium in *Arabidopsis thaliana*. *PLoS Genet* 8:e1002923.

Chaoui, A., Mazhoudi, S., Ghorbal, M. N. and Ferjani, E. E. 1997. Cadmium and zinc induction of lipid peroxidation and effects on antioxidant enzymes activities in bean (*Phaseolus vulgaris* L.). *Plant Sci* 127:139–147.

Chen, C. T. and Kao, C. H. 1993. Osmotic stress and water stress have opposite effects on putrescine and proline production in excised rice leaves. *Plant Growth Regul* 13:197–202.

Chen, C. Y., Zou, J. H., Zhang, S. Y., Zaitlin, D. and Zhu, L. H. 2009. Stringolactones are a new-defined class of plant hormones which inhibit shoot branching and mediate the interaction of plant-AM fungi and plant-parasitic weeds. *Sci China Ser C* 52:693–700.

Chen, J., Yan, Z. and Li, X. 2014. Effect of methyl jasmonate on cadmium uptake and antioxidative capacity in *Kandelia obovata* seedlings under cadmium stress. *Ecotoxicol Environ Saf* 104:349–356. doi: 10.1016/j.ecoenv.2014.01.022

Chen, L., Wang, L., Chen, F., Korpelainen, H. and Li, C. 2013. The effects of exogenous putrescine on sex-specific responses of *Populus cathayana* to copper stress. *Ecotoxicol Environ Saf* 97:94–102.

Chen, T. H. and Murata, N. 2002. Enhancement of tolerance to abiotic stress by metabolic engineering of betaines and other compatible solutes. *Curr Opin Plant Biol* 5:250–257.

Chen, T. H. and Murata, N. 2011. Glycinebetaine protects plants against abiotic stress: Mechanisms and biotechnological applications. *Plant Cell Environ* 34:1–20.

Cho, U. H. and Park, J. O. 2000. Mercury-induced oxidative stress in tomato seedlings. *Plant Sci* 156:1–9.

Chong-qing, W., Tao, W., Ping, M., Zi-chao, L. and Ling, Y. 2013. Quantitative trait loci for mercury tolerance in rice seedlings. *Rice Sci* 20:238–242. doi: 10.1016/S1672-6308(13)60124-9

Chou, T. S., Chao, Y. Y., Huang, W. D., Hong, C. Y. and Kao, C. H. 2011. Effect of magnesium deficiency on antioxidant status and cadmium toxicity in rice seedlings. *J Plant Physiol* 168:1021–1030.

Choudhary, S. P., Oral, H. V., Bhardwaj, R., Yu, J. Q. and Tran, L. S. 2012. Interaction of brassinosteroids and polyamines enhances copper stress tolerance in *Raphanus sativus*. *J Exp Bot* 63:5659–5675.

Chugh, L. K. and Sawhney, S. K. 1999. Photosynthetic activities of *Pisum sativum* seedlings grown in the presence of cadmium. *Plant Physiol Biochem* 37:297–303.

Cohen, C. K., Fox, T. C., Garvin, D. F. and Kochian, L. V. 1998. The role of iron-deficiency stress responses in stimulating heavy-metal transport in plants. *Plant Physiol* 116:1063–1072.

Couée, I., Sulmon, C., Gouesbet, G. and El Amrani, A. 2006. Involvement of soluble sugars in reactive oxygen species balance and responses to oxidative stress in plants. *J Exp Bot* 57:449–459.

Creelman, R. A. and Mullet, J. E. 1991. Water deficit modulates gene expression in growing zones of soybean seedlings. Analysis of differentially expressed cDNAs, a new β-tubulin gene, and expression of genes encoding cell wall proteins. *Plant Mol Biol* 17:591–608.

Curie, C., Panaviene, Z., Loulergue, C., Dellaporta, S. L., Briat, J. F. and Walker, E. L. 2001. Maize yellow stripe1 encodes a membrane protein directly involved in Fe(III) uptake. *Nature* 409:346–349.

Cuypers, A., Karen, S., Jos, R., Kelly, O., Els, K., Tony, R., Nele, H., Nathalie, V., Yves, G., Jan, C. and Jaco, V. 2011. The cellular redox state as a modulator in cadmium and copper responses in *Arabidopsis thaliana* seedlings. *J Plant Physiol* 168:309–316.

DalCorso, G., Farinati, S. and Furini, A. 2010. Regulatory networks of cadmium stress in plants. *Plant Signal Behav* 5:663–667.

Das, P., Samantaray, S. and Rout, G. R. 1997. Studies on cadmium toxicity in plants: A review. *Environ Pollut* 98:29–36.

Dave, R., Tripathi, R. D., Dwivedi, S., Tripathi, P., Dixit, G., Sharma, Y. K., Trivedi, P. K., Corpas, F. J., Barroso, J. B. and Chakrabarty, D. 2013. Arsenate and arsenite exposure modulate antioxidants and amino acids in contrasting arsenic accumulating rice (*Oryza sativa* L.) genotypes. *J Hazard Mater* 15:1123–1131. doi: 10.1016/j.jhazmat.2012.06.049

De Vos, C. H. R., Schat, H., de Waal, M. A. M., Voojs, R. and Ernst, W. H. O. 1991. Increased resistance to Cu-induced damage of root cell plasmalemma in Cu tolerant *Silene cucubalus. Physiol Plant* 82:523–528.

Delauney, A. J. and Verma, D. P. S. 1993. Proline biosynthesis and osmoregulation in plants. *Plant J* 4:215–223. doi: 10.1046/j.1365-313X.1993.04020215.x

Delhaize, E., Kataoka, T., Hebb, D. M., White, R. G. and Ryan, P. R. 2003. Genes encoding proteins of the cation diffusion facilitator family that confer manganese tolerance. *Plant Cell* 15:1131–1142. doi: 10.1105/tpc.009134

Demiral, T. and Türkan, I. 2005. Comparative lipid peroxidation, antioxidant defense systems and proline content in roots of two rice cultivars differing in salt tolerance. *Environ Exp Bot* 53:247–257.

Demirevska-Kepova, K., Simova-Stoilova, L., Stoyanova, Z., Holzer, R. and Feller, U. 2004. Biochemical changes in barley plants after excessive supply of copper and manganese. *Environ Exp Bot* 52:253–266.

Deng, F., Yamaji, N., Xia, J. and Ma, J. F. 2013. A member of the heavy metal P-type ATPase OsHMA5 is involved in xylem loading of copper in rice. *Plant Physiol* 163:1353–1362.

Dietz, K. J., Bair, M. and Krämer, U. 1999. Free radical and reactive oxygen species as mediators of heavymetal toxicity in plants. In: Prasad, M. N. V. and Hagemeyer, J. (eds), *Heavy Metal Stress in Plants from Molecules to Ecosystems*, pp. 73–79. Spinger-Verlag, Berlin, Germany.

Doncheva, S., Stoynova, Z. and Velikova, V. 2001. Influence of succinate on zinc toxicity of pea plants. *J Plant Nutr* 24:789–804.

Dong, H. Z., Niu, Y. H., Kong, X. Q., et al. 2009. Effects of early-fruit removal on endogenous cytokinins and abscisic acid in relation to leaf senescence in cotton. *Plant Growth Regul* 59:93–101.

Dong, J., Wu, F. and Zhang, G. 2005. Effect of cadmium on growth and photosynthesis of tomato seedlings. *J Zhejiang Univ Sci* 10:974–980.

Dräger, D. B., Desbrosses-Fonrouge, A. G., Krach, C., Chardonnens, A. N., Meyer, R. C., Saumitou-Laprade, P., et al. 2004. Two genes encoding *Arabidopsis halleri* MTP1 metal transport proteins co-segregate with zinc tolerance and account for high MTP1 transcript levels. *Plant J* 39:425–439. doi: 10.1111/j.1365-313X.2004.02143.x

Duan, G. L., Zhou, Y., Tong, Y. P., Mukhopadhyay, R., Rosen, B. P. and Zhu, Y. G. 2007. A CDC25 homologue from rice functions as an arsenate reductase. *New Phytol* 174:311–321.

Duman, F., Aksoy, A., Aydin, Z. and Temizgul, R. 2010. Effects of exogenous glycinebetaine and trehalose on cadmium accumulation and biological responses of an aquatic plant (*Lemna gibba* L). *Water Air Soil Pollut* 217:545–556.

Ebbs, S. D. and Kochian, L. V. 1997. Toxicity of zinc and copper to Brassica species: Implications for phytoremediation. *J Environ Qual* 26:776–781.

Eleftheriou, E. P., Adamakis, I. D. and Melissa, P. 2012. Effects of hexavalent chromium on microtubule organization, ER distribution and callose deposition in root tip cells of *Allium cepa* L. *Protoplasma* 249:401–416. doi: 10.1007/s00709-011-0292-3

El-Ramady, H., Abdalla, N., Alshaal, T., Domokos-Szabolcsy, É., Elhawat, N., Prokisch, J., Sztrik, A., Fári, M., El-Marsafawy, S. and Shams, M. S. 2015. Selenium in soils under climate change, implication for human health. *Environ Chem Lett* 13:1–19.

Erbilgin, N., Krokene, P., Christiansen, E., Zeneli, G. and Gershenzon, J. 2006. Exogenous application of methyl jasmonate elicits defenses in Norway spruce (*Picea abies*) and reduces host colonization by the bark beetle Ips typographus. *Oecologia* 148:426–436.

Ernst, W. H. O., Krauss, G. J., Verkleij, J. A. C. and Wesenberg, D. 2008. Interaction of heavy metal with the sulphur metabolism in angiosperms from an ecological point of view. *Plant Cell Environ* 31:123–143. doi:10.1111/j.1365-3040.2007.01746.x

Faller, P., Kienzler, K. and Krieger-Liszkay, A. 2005. Mechanism of Cd_2^+ toxicity: Cd_2^+ inhibits photoactivation of Photosystem II by competitive binding to the essential Ca_2^+ site. *Biochim Biophys Acta* 1706:158–164.

Farmer, E. E., Almeras, E. and Krishnamurthy, V. 2003. Jasmonates and related oxylipins in plant responses to pathogenesis and herbivory. *Curr Opin Plant Biol* 6:372–378.

Farooq, H., Asghar, H. N., Khan, M. Y., Saleem, Z. A. 2015. Auxin-mediated growth of rice in cadmium-contaminated soil. *Turkish J Agric For* 39:272–276.

Farooq, M. A., Ali, S., Hameed, A., Bharwana, S. A., Rizwan, M., Ishaque, W., Farid, M., Mahmood, K. and Iqbal, Z. 2016a. Cadmium stress in cotton seedlings: Physiological, photosynthesis and oxidative damages alleviated by glycinebetaine. *S Afr J Bot* 104:61–68.

Farooq, M. A., Gill, R. A., Islam, F., Ali, B., Liu, H., Xu, J. and Zhou, W. 2016b. Methyl jasmonate regulates antioxidant defense and suppresses arsenic uptake in *Brassica napus* L. *Front Plant Sci.* 7:468 doi: 10.3389/fpls.2016.00468

Farooq, M. A., Islam, F., Yang, C., Nawaz, A., Gill, R. A., Ali, B. and Zhou, W. 2018. Methyl jasmonate alleviates arsenic-induced oxidative damage and modulates the ascorbate–glutathione cycle in oilseed rape roots. *Plant Growth Regul* 84:135–148.

Fässler, E., Evangelou, M. W., Robinson, B. H. and Schulin, R. 2010. Effects of indole-3-acetic acid (IAA) on sunflower growth and heavy metal uptake in combination with ethylene diamine disuccinic acid (EDDS). *Chemosphere* 80:901–907.

Fidalgo, F., Azenha, M., Silva, A. F., et al. 2013. Copper-induced stress in *Solanum nigrum* L. and antioxidant defense systemresponse. *Food and Energy Security* 2:70–80.

Fodor, A., Szabo-Nagy, A. and Erdei, L. 1995. The effects of cadmium on the fluidity and H^+-ATPase activity of plasma membrane from sunflower and wheat roots. *J Plant Physiol* 14:787–792.

Fontes, R. L. S. and Cox, F. R. 1998. Zinc toxicity in soybean grown at high iron concentration in nutrient solution. *J Plant Nutr* 21:1723–1730.

Foreman, J., Demidchik, V., Bothwell, J. H., Mylona, P., Miedema, H., Torres, M. A., et al. 2003. Reactive oxygen species produced by NADPH oxidase regulate plant cell growth. *Nature* 422:442–446.

Foyer, C. H. and Noctor, G. 2003. Redox sensing and signalling associated with reactive oxygen in chloroplasts, peroxisomes and mitochondria. *Physiol Plant* 119:355–364.

Fuentes, H. D., Khoo, C. S., Pe, T., Muir, S. and Khan, A. G. 2000. Phytoremediation of a contaminated mine site using plant growth regulators to increase heavy metal uptake. In: Sánchez, M. (ed.), *Proceedings of the 5th International Conference on Clean Technologies for the Mining Industry, Vol. 1*, pp. 427–435. Universidad de Concepción, Santiago.

Fuhrer, J. 1988. Ethylene biosynthesis and cadmium toxicity in leaf tissue of beans *Phaseolus vuglaris* L. *Plant Physiol* 70:162–167.

Gajewska, E., Sklodowska, M., Slaba, M. and Mazur, J. 2006. Effect of nickel on antioxidative enzyme activities, proline and chlorophyll contents in wheat shoots. *Biol Planta* 50:653–659.

Gangwar, S. and Singh, V. P. 2011. Indole acetic acid differently changes growth and nitrogen metabolism in *Pisum sativum* L. seedlings under chromium (VI) phytotoxicity: Implication of oxidative stress. *Sci Hortic* 129:321–328.

Gangwar, S., Singh, V. P., Prasad, S. M., et al. 2010. Modulation of manganese toxicity in *Pisum sativum* L. seedlings by kinetin. *Sci Hortic* 126:467–474.

Gao, X. P., Wang, X. F., Lu, Y. F., Zhang, L. Y., Shen, Y. Y., Liang, Z. and Zhang, D. P. 2004. Jasmonic acid is involved in the water-stress-induced betaine accumulation in pear leaves. *Plant Cell Environ* 27:497–507.

Garnier, L., Simon-Plas, F., Thuleau, P., Agnel, J. P., Blein, J. P., Ranjeva, R., et al. 2006. Cadmium affects tobacco cells by a series of three waves of reactive oxygen species that contribute to cytotoxicity. *Plant Cell Environ* 29:1956–1969. doi: 10.1111/j.1365-3040.2006.01571.x

Garzón, T., Gunsé, B., Moreno, A. R., Tomos, A. D., Barceló, J. and Poschenrieder, C. 2011. Aluminium-induced alteration of ion homeostasis in root tip vacuoles of two maize varieties differing in Al tolerance. *Plant Sci* 180:709–715. doi: 10.1016/j.plantsci.2011.01.022

Gechev, T. S., Van Breusegem, F., Stone, J. M., Denev, I. and Laloi, C. 2006. Reactive oxygen species as signals that modulate plant stress responses and programmed cell death. *Bio Essays* 28:1091–1101. doi: 10.1002/bies.20493

Gemrotová, M., Kulkarni, M. G., Stirk, W. A., Strnad, M., Van Staden, J. and Spichal, L. 2013. Seedlings of medicinal plants treated with either a cytokinin antagonist (PI-55) or an inhibitor of cytokinin degradation (INCYDE) are protected against the negative effects of cadmium. *Plant Growth Regul* 71:137–145.

Gendre, D., Czernic, P., Conéjéro, G., Pianelli, K., Briat, J. F., Lebrun, M. and Mari, S. 2006. TcYSL3, a member of the YSL gene family from the hyperaccumulator *Thlaspi caerulescens*, encodes a nicotianamine-Ni/Fe transporter. *Plant J* 49:1–15.

Ghabriche, R., Ghnaya, T., Zaier, H., Baioui, R., Vromman, D., Abdelly, C. and Lutts, S. 2017. Polyamine and tyramine involvement in NaCl-induced improvement of Cd resistance in the halophyte *Inula chrithmoides* L. *J Plant Physiol* 216:136–144.

Ghorbanli, M., Kaveh, S. H. and Sepehr, M. F. 2000. Effects of cadmium and gibberellin on growth and photosynthesis of *Glycine max*. *Photosynthetica* 37:627–631.

Gidrol, X., Lin, W. S., Dégousée, N., et al. 1994. Accumulation of reactive oxygen species and oxidation of cytokinin in germinating soybean seeds. *Eur J Biochem* 224:21–28.

Gohari, M., Habib-Zadeh, A. R., and Khayat, M. 2012. Assessing the intensity of tolerance to lead and its effect on amount of protein and proline in root and aerial parts of two varieties of rape seed (*Brassica napus* L.). *J Basic Appl Sci Res* 2:935–938.

Groppa, M. D., Benavides, P. and Tomaro, L. 2003. Polyamine metabolism in sunflower and wheat leaf discs under cadmium or copper stress. *Plant Science* 164:293–299.

Groppa, M. D., Tomaro, M. L. and Benavides, M. P. 2001. Polyamines as protectors against cadmium or copper-induced oxidative damage in sunflower leaf discs. *Plant Sci* 161:481–488.

Groppa, M. D., Tomaro, M. L. and Benavides, M. P. 2007. Polyamines and heavy metal stress: The antioxidant behavior of spermine in cadmiumand copper-treated wheat leaves. *Bio Metals* 20:185–195.

Gruenhage, L. and Jager, I. I. J. 1985. Effect of heavy metals on growth and heavy metals content of *Allium Porrum* and *Pisum sativum*. *Angew Bot* 59:11–28.

Guerinot, M. L. 2000. The ZIP family of metal transporters. *Biochim Biophys Acta* 1465:190–198.

Gunes, A., Pilbeam, J. D. and Inal, A. 2009. Effect of arsenic–phosphorus interaction on arsenic-induced oxidative stress in chickpea plants. *Plant Soil* 314:211–220.

Guo, J., Dai, X., Xu, W. and Ma, M. 2008. Overexpressing GSH1 and AsPCS1 simultaneously increases the tolerance and accumulation of cadmium and arsenic in *Arabidopsis thaliana*. *Chemosphere* 72:1020–1026.

Ha, S., Vankova, R., Yamaguchi-Shinozaki, K., Shinozaki, K. and Phan Tran, L. S. 2012. Cytokinins: Metabolism and function in plant adaptation to environmental stresses. *Trends Plant Sci* 17:172–179.

Hac-Wydro, K., Sroka, A. and Jablo, K. 2016. The impact of auxins used in assisted phytoextraction of metals from the contaminated environment on the alterations caused by lead (II) ions in the organization of model lipid membranes. *Colloids Surfaces B Biointerfaces* 143:124–130.

Hadi, F., Bano, A. and Fuller, M. P. 2010. The improved phytoextraction of lead (Pb) and the growth of maize (*Zea mays* L.): The role of plant growth regulators (GA_3 and IAA) and EDTA alone and in combinations. *Chemosphere* 80:457–462.

Hanaka, A., Wójcik, M., Dresler, S., Mroczek-Zdyrska, M. and Maksymiec, W. 2016. Does methyl jasmonate modify the oxidative stress response in *Phaseolus coccineus* treated with Cu? *Ecotoxicol Environ Saf* 124:480–488. doi: 10.1016/j.ecoenv.2015.11.024

Handique, G. K. and Handique, A. K. 2009. Proline accumulation in lemongrass (*Cymbopogon flexuosus* Stapf.) due to heavy metal stress. *J Environ Biol* 30:299–302.

Hao, F., Wang, X. and Chen, J. 2006. Involvement of plasma membrane NADPH oxidase in nickel-induced oxidative stress in roots of wheat seedlings. *Plant Sci* 170:151–158.

Hasanuzzaman, M., Hossain, M. A., da Silva, J. A. T. and Fujita, M. 2012. Plant response and tolerance to abiotic oxidative stress: Antioxidant defense is a key factor. In: Bandi, V., Shanker, A. K., Shanker, C. and Mandapaka, M. (eds), *Crop Stress and Its Management: Perspectives and Strategies*, pp. 261–316. Springer, Berlin.

Hasegawa, P. M., Bressan, R. A., Zhu, J. K. and Bohnert, H. J. 2000. Plant cellular and molecular responses to high salinity. *Annu Rev Plant Phys* 51:463–499.

Hassan, T. U., Bano, A. and Naz, I. 2017. Alleviation of heavy metals toxicity by the application of plant growth promoting rhizobacteria and effects on wheat grown in saline sodic field. *Int J Phytoremediation* 19:522–529. doi: 10.1080/15226514.2016.1267696

Hausman, J. F., Evers, D., Thiellement, H. and Jouve, L. 2000. Compared responses of poplar cuttings and in vitro raised shoots to short-term chilling treatments. *Plant Cell Rep* 19:954–60.

Hayat, S., Hayat, Q., Alyemeni, M. N. and Ahmad, A. 2013. Proline enhances antioxidative enzyme activity, photosynthesis and yield of *Cicer arietinum* L. exposed to cadmium stress. *Acta Botanica Croatica* 72:323–335.

Hegedus, A., Erdei, S. and Horvath, G. 2001. Comparative studies of H_2O_2 detoxifying enzymes in green and greening barley seedings under cadmium stress. *Plant Sci* 160:1085–1093.

Hernández, L. E., Carpena-Rutz, R. and Garate, A. 1996. Alterations in the mineral nutrition of pea seedlings exposed to cadmium. *J Plant Nutr* 19:1581–1598.

Hieu, H. C., Li, H., Miyauchi, Y., Mizutani, G., Fujita, N. and Nakamura, Y. 2015. Wetting effect on optical sum frequency generation (SFG) spectra of D-glucose, D-fructose, and sucrose. *Spectrochim Acta A Mol Biomol Spectrosc* 138:834–839.

Hirschi, K. D., Korenkov, V. D., Wilganowski, N. L. and Wagner, G. J. 2000. Expression of *Arabidopsis* CAX2 in tobacco. Altered metal accumulation and increased manganese tolerance. *Plant Physiol* 124:125–133. doi: 10.1104/pp.124.1.125

Hoekstra, F., Golovina, E. and Buitink, J. 2001. Mechanisms of plant dessication tolerance. *Trends Plant Sci* 6:431–438.

Hoque, M. A., Uraji, M., Banu, M. N., Mori, I. C., Nakamura, Y. and Murata, Y. 2010. The effects of methylglyoxal on glutathione S-transferase from *Nicotiana tabacum*. *Biosci Biotechnol Biochem* 74:2124–2126.

Hossain, M. A., Hasanuzzaman, M. and Fujita, M. 2010. Up-regulation of antioxidant and glyoxalase systems by exogenous glycinebetaine and proline in mung bean confer tolerance to cadmium stress. *Physiol Mol Biol Plants* 16:259–272.

Hossain, M. A., Hossain, M. D., Rohman, M. M., da Silva, J. A. T. and Fujita, M. 2012. Onion major compounds (flavonoids, organosulfurs) and highly expressed glutathione-related enzymes: Possible physiological interaction, gene cloning and abiotic stress response. In: Aguirre, C. B. and Jaramillo, L. M. (eds), *Onion Consumption and Health*, pp. 49–90. Nova Science Publishers, New York, NY, USA.

Hossain, M. A., Hossain, M. Z. and Fujita, M. 2009. Stress-induced changes of methylglyoxal level and glyoxalase I activity in pumpkin seedlings and cDNA cloning of glyoxalase I gene. *Australian J Crop Sci* 3:53–64.

Hu, S., Lau, K. W. K. and Wu, M. 2001. Cadmium sequestration in *Chlamydomonas reinhardtii*. *Plant Sci* 161:987–996. doi: 10.1016/S0168-9452(01)00501-5

Huang, H., Gao, H., Liu, B., Qi, T., Tong, J., Xiao, L. and Song, S. 2017. *Arabidopsis* MYB24 regulates jasmonate-mediated stamen development. *Front Plant Sci* 8:1525.

Hussain, D., Haydon, M. J., Wang, Y., Wong, E., Sherson, S. M., Young, J., Camakaris, J., Harper, J. F. and Cobbett, C. S. 2004. P-type ATPase heavy metal transporters with roles in essential zinc homeostasis in *Arabidopsis*. *Plant Cell* 16:1327–1339.

Ibrahim, H. A. and Abdellatif, Y. M. 2016. Effect of maltose and trehalose on growth, yield and some biochemical components of wheat plant under water stress. *Ann Agric Sci* 61:267–274.

Islam, M. M., Hoque, M. A., Okuma, E., Banu, M. N. A., Shimoishi, Y., Nakamura, Y. and Murata, Y. 2009. Exogenous proline and glycinebetaine increase antioxidant enzyme activities and confer tolerance to cadmium stress in cultured tobacco cells. *J Plant Physiol* 166:1587–1597.

Israr, M., Jewell, A., Kumar, D. and Sahi, S. V. 2011. Interactive effects of lead, copper, nickel and zinc on growth, metal uptake and antioxidative metabolism of *S. drummondii*. *J Hazard Mater* 186: 1520–1526.

Jabeen, N., Abbas, Z., Iqbal, M., Rizwan, M., Jabbar, A., Farid, M., Ali, S., Ibrahim, M. and Abbas, F. 2015. Glycinebetaine mediates chromium tolerance in mung bean through lowering of Cr uptake and improved antioxidant system. *Arch Agron Soil Sci* 62:648–662.

Jamil, A., Anwer, F. and Ashraf, M. 2005. Plants tolerance to biotic and abiotic stresses through modern genetic engineering techniques. In: Dris, R. (ed.), *Crops: Growth, Quality and Biotechnology*, pp. 1276–1299. WFL Publishers, Helsinki, Finland.

Jeon, J., Kim, N. Y., Kim, S., Kang, N. Y., Novak, O., Ku, S. J., et al. 2010. A subset of cytokinin two component signaling system plays a role in cold temperature stress response in *Arabidopsis*. *J Biol Chem* 285:23371–23386.

Jiang, W., Liu, D. and Hou, W. 2001. Hyperaccumulation of cadmium by roots, bulbs and shoots of garlic. *Bioresour Technol* 76:9–13.

Jiménez, A., Hernández, J. A., Pastori, G., del Río, L. A. and Sevilla, F. 1998. Role of the ascorbate-glutathione cycle of mitochondria and peroxisomes in the senescence of pea leaves. *Plant Physiol* 118:1327–1335. doi: 10.1104/pp.118.4.1327

John, R., Ahmad, P., Gadgil, K. and Sharma, S. 2009. Heavy metal toxicity: Effect on plant growth, biochemical parameters and metal accumulation by *Brassica juncea* L. *Int J Plant Prod* 3:65–76.

Juwarkar, A. S. and Shende, G. B. 1986. Interaction of Cd-Pb effect on growth yield and content of Cd, Pb in barley. *Ind J Environ Heal* 28:235–243.

Kacperska, A. 2004. Sensor types in signal transduction pathways in plant cells responding to abiotic stressors: Do they depend on stress intensity? *Physiol Plant* 122:159–168. doi: 10.1111/j.0031-9317.2004.00388.x

Kakkar, R. K. and Sawhney, V. K. 2002. Polyamine research in plants–a changing perspective. *Physiol Plant* 116:281–292.

Kaminek, M., Motika, V. and Vankova, R. 1997. Regulation of cytokinin content in plant cell. *Physiol Plant* 101:689–700.

Kampfenkel, K., Kushnir, S., Babiychuk, E., Inzé, D. and Van Montagu, M. 1995. Molecular characterization of a putative *Arabidopsis thaliana* copper transporter and its yeast homologue. *J Biol Chem* 270:28479–28486.

Kanechi, M., Hikosaka, Y. and Uno, Y. 2013. Application of sugarbeet pure and crude extracts containing glycinebetaine affects root growth, yield, and photosynthesis of tomato grown during summer. *Sci Hortic* 152:9–15.

Kenderešová, L., Stanová, A., Pavlovkin, J., Durišová, E., Nadubinská, M., Ciamporová, M., et al. 2012. Early Zn_2+ induced effects on membrane potential account for primary heavy metal susceptibility in tolerant and sensitive *Arabidopsis* species. *Ann Bot* 110:445–459.

Keramat, B., Kalantari, K. M. and Arvin, M. J. 2010. Effects of methyl jasmonate treatment on alleviation of cadmium damages in soybean. *J Plant Nutr* 33:1016–1025.

Kessler, A. and Baldwin, I. T. 2002. Plant responses to insect herbivory: The emerging molecular analysis. *Annu Rev Plant Biol* 53:299–328.

Keunen, E., Schellingen, K., Vangronsveld, J., Cuypers, A. 2016. Ethylene and metal stress: Small molecule, big impact. *Front. Plant Sci* 7:23.

Khan, N. A., Khan, M. I. R., Ferrante, A. and Poor, P. 2017. Ethylene: A key regulatory molecule in plants. *Front Plant Sci* 8:1782.

Khan, M. I. R. and Khan, N. A. 2014. Ethylene reverses photosynthetic inhibition by nickel and zinc in mustard through changes in PS II activity, photosynthetic nitrogen use efficiency, and antioxidant metabolism. *Protoplasma* 251:1007–1019.

Khan, M. I. R., Nazir, F., Asgher, M., Per, T. S. and Khan, N. A. 2015b. Selenium and sulfur influence ethylene formation and alleviate cadmium-induced oxidative stress by improving proline and glutathione production in wheat. *J Plant Physiol* 173:9–18.

Kikui, S., Sasaki, T., Maekawa, M., Miyao, A., Hirochika, H., Matsumoto, H., et al. 2005. Physiological and genetic analyses of aluminium tolerance in rice, focusing on root growth during germination. *J Inorg Biochem* 99:1837–1844. doi: 10.1016/j.jinorgbio.2005.06.031

Kim, D. Y., Bovet, L., Maeshima, M., Martinoia, E. and Lee, Y. 2007. The ABC transporter AtPDR8 is a cadmium extrusion pump conferring heavy metal resistance. *Plant J* 50:207–218.

Kim, H. J., Fonseca, J. M., Choi, J. H. and Kubota, C. 2007. Effect of methyl jasmonate on phenolic compounds and carotenoids of romaine lettuce (*Lactuca sativa* L.). *J Agric Food Chem* 55:10366–10372. doi: 10.1021/jf071927m

Kim, J. A., Agrawal, G. K., Rakwal, R., Han, K. S., Kim, K. N., Yun, Ch. H., Heu, S., Park, S. Y., Lee, Y. H. and Jwa, N. S. 2003. Molecular cloning and mRNA expression analysis of a novel rice (*Oryza sativa* L.) MAPK kinase kinase, OsEDR1, an ortholog of *Arabidopsis* AtEDR1, reveal its role in defense/stress signaling pathways and development. *Biochem Biophys Res Commun* 300:868–876.

Koch, K. E. 1996. Carbohydrate-modulated gene expression in plants. *Ann Rev Plant Physiol Plant Mol Biol* 47:509–540.

Koeduka, T., Matsui, K., Hasegawa, M., Akakabe, Y. and Kajiwara, T. 2005. Rice fatty acid alpha dioxygenase is induced by pathogen attack and heavy metal stress: Activation through jasmonate signaling. *J Plant Physiol* 162:912–920. doi: 10.1016/j.jplph.2004.11.003

Korenkov, V., Park, S. H., Cheng, N. H., Sreevidya, C., Lachmansingh, J., Morris, J., Hirschi, K. and Wagner, G. J. 2007. Enhanced Cd_2^+ selective root-tonoplast-transport in tobaccos expressing *Arabidopsis* cation exchangers. *Planta* 225:403–411.

Korkmaz, A., ¸Sirikçi, R., Kocaçınar, F., Děger, Ö. and Demirkırıan, A. R. 2012. Alleviation of salt-induced adverse effects in pepper seedlings by seed application of glycinebetaine. *Sci Hortic* 148:197–205.

Kováčik, J., Klejdus, B., Hedbavny, J. and Bačkor, M. 2010. Effect of copper and salicylic acid on phenolic metabolites and free amino acids in *Scenedesmus quadricauda* (Chlorophyceae). *Plant Sci* 178:307–311.

Krishnamurthy, A. and Rathinasabapathi, B. 2013. Auxin and its transport play a role in plant tolerance to arsenite-induced oxidative stress in *Arabidopsis thaliana. Plant Cell Environ* 36:1838–1849.

Kuznetsov, V., Radyukina, N. L. and Shevyakova, N. I. 2006. Polyamines and stress: Biological role, metabolism and regulation. *Russ J Plant Physiol* 53:583–604.

Lamb, C. and Dixon, R. 1997. The oxidative burst in plant disease resistance. *Annu Rev Plant Physiol Plant Mol Biol* 48:251–275.

Lanquar, V., Lelièvre, F., Bolte, S., Hamès, C., Alcon, C., Neumann, D., Vansuyt, G., Curie, C., Schröder, A., Krämer, U., Barbier-Brygoo, H. and Thomine, S. 2005. Mobilization of vacuolar iron by AtNRAMP3 and AtNRAMP4 is essential for seed germination on low iron. *EMBO J* 24:4041–4051.

Le Jean, M., Schikora, A., Mari, M., Briat, J. F. and Curie, C. 2005. A loss-of-function mutation in AtYSL1 reveals its role in iron and nicotinamine seed loading. *Plant J* 44:769–782.

Lee, S., Kim, Y. Y., Lee, Y. and An, G. 2007. Rice P1B-type heavy-metal ATPase, OsHMA9, is a metal efflux protein. *Plant Physiol* 145:831–842.

Lesko, K. and Simon-Sarkadi, L. 2002. Effect of cadmium stress on amino acid and polyamine content of wheat seedlings. *Period Polytech Ser Chem Eng* 46:65–71.

Li, D., Xu, X., Hu, X., Liu, Q., Wang, Z., Zhang, H., Wang, H., Wei, M., Wang, H., Liu, H. and Li, C. 2015. Genome-wide analysis and heavy metal-induced expression profiling of the HMA gene family in *Populus trichocarpa. Front Plant Sci* 6:1149. doi: 10.3389/fpls.2015.01149

Li, H. F., Gray, C., Mico, C., Zhao, F. J. and McGrath, S. P. 2009. Phytotoxicity and bioavailability of cobalt to plants in a range of soils. *Chemosphere* 75:979–986.

Li, S., Li, F., Wang, J., Zhang, W., Meng, Q., Chen, T. H. H., Murata, N. and Yang, X. 2011. Glycinebetaine enhances the tolerance of tomato plants to high temperature during germination of seeds and growth of seedlings. *Plant Cell Environ* 34:1931–1943.

Lin, Z., Zhong, S. and Grierson, D. 2009. Recent advances in ethylene research. *J Exp Bot* 60:3311–3336.

Liphadzi, M. S., Kirkham, M. B. and Paulsen, G. M. 2006. Auxin enhanced root growth for phytoremediation of sewage sludge amended soil. *Environ Technol* 27:695–704.

Lipiec, J. 2012. Crop responses to soil compaction. *Nordic Assoc Agric Sci Rep* 8:32.

Linger, P., Ostwald, A. and Haensler, J. 2005. *Cannabis sativa* L. growing on heavy metal contaminated soil: Growth, cadmium uptake and photosynthesis. *Biol Plant* 49:567–576.

Liu, D. H., Jiang, W. S. and Gao, X. Z. 2003. Effects of cadmium on root growth, cell division and nucleoli in root tip cells of garlic. *Biol Plant* 47:79–83.

Liu, D. H., Wang, M., Zou, J. H. and Jiang, W. S. 2006. Uptake and accumulation of cadmium and some nutrient ions by roots and shoots of maize (*Zea mays* L.). *Pak J Bot* 38:701–709.

Liu, J. H., Kitashiba, H., Wang, J., Ban, Y. and Moriguchi, T. 2007. Polyamines and their ability to provide environmental stress tolerance to plants. *Plant Biotechnol J* 24:117–126. doi: 10.5511/plantbiotechnology.24.117

Lomozik, L., Gasowska, A., Bregier-Jarzebowska, R. and Jastrzab, R. 2005. Coordination chemistry of polyamines and their interactions in ternary systems including metal ions, nucleosides and nucleotides. *Coord Chem Rev* 249:2335–2350.

López, M. L., Peralta-Videa, J. R., Benitez, T. and Gardea- Torresdey, J. L. 2005. Enhancement of lead uptake by alfalfa (*Medicago sativa*) using EDTA and a plant growth promoter. *Chemosphere* 61:595–598.

Lu, G., Coneva, V., Casaretto, J. A., Ying, S., Mahmood, K., Liu, F., Nambara, E., Bi, Y. M. and Rothstein, S. J. 2015. OsPIN5b modulates rice (*Oryza sativa*) plant architecture and yield by changing auxin homeostasis, transport and distribution. *Plant J* 83:913–925.

Lu, Y. P., Li, Z. S., Drozdowicz, Y. M., Hortensteiner, S., Martinoia, E. and Rea, P. A. 1998. AtMRP2, an *Arabidopsis* ATP binding cassette transporter able to transport glutathione S-conjugates and chlorophyll catabolites: Functional comparisons with AtMRP1. *Plant Cell* 10:267–282.

Lu, Y. P., Li, Z. S. and Rea, P. A. 1997. AtMRP1 gene of *Arabidopsis* encodes a glutathione S-conjugate pump: Isolation and functional definition of a plant ATP-binding cassette transporter gene. *Proc Natl Acad Sci USA* 94:8243–8248.

Lynch, J., and Brown, K. M. 1997. Ethylene and plant responses to nutritional stress. *Physiol Plant* 100:613–619. doi: 10.1111/j.1399-3054.1997.tb03067.x

Makela, P., Mantila, J., Hinkkanen, R., Pehu, E. and Peltnen-Sainio, P. 1996. Effect of foliar applications of glycinebetaine on stress tolerance, growth and yield of spring cereals and summer turnip rape in Finland. *J Agron Crop Sci* 176:223–234.

Maksymiec, W. 1997. Effect of copper on cellular processes in higher plants. *Photosynthetica* 34:321–342.

Maksymiec, W. and Krupa, Z. 2006. The effects of short-term exposition to Cd, excess Cu ions and jasmonate on oxidative stress appearing in *Arabidopsis thaliana*. *Environ Exp Bot* 57:187–194.

Maksymiec, W. and Krupa, Z. 2007. Effects of methyl jasmonate and excess copper on root and leaf growth. *Biol Plant* 51:322–326.

Maksymiec, W., Wianowska, D., Dawidowicz, A. L., Radkiewicz, S., Mardarowicz, M. and Krupa, Z. 2005. The level of jasmonic acid in *Arabidopsis thaliana* and *Phaseolus coccineus* plants under heavy metal stress. *J Plant Physiol* 162:1338–1346.

Martins, L. L., Mourato, M. P., Baptista, S., et al. 2014. Response to oxidative stress induced by cadmium and copper in tobacco plants (*Nicotiana tabacum*) engineered with the trehalose-6-phosphate synthase gene (AtTPS1). *Acta Physiol Plant* 36:755–765.

McLaughlin, M. J., Parker, D. R. and Clark, J. M. 1999. Metal and micronutrients-food safety issues. *Field Crops Res* 60:143–163. doi: 10.1016/S0378-4290(98)00137-3

Mehta, S. K. and Gaur, J. P. 1999. Heavy-metal-induced proline accumulation and its role in ameliorating metal toxicity in *Chlorella vulgaris*. *New Phytol* 143:253–259.

Menguer, P. K., Farthing, E., Peaston, K. A., Ricachenevsky, F. K., Fett, J. P. and Williams, L. E. 2013. Functional analysis of the rice vacuolar zinc transporter OsMTP1. *J Exp Bot* 64:2871–2883.

Minois, N. 2014. Molecular basis of the 'anti-aging' effect of spermidine and other natural polyamines—A mini-review. *Gerontology* 60:319–326.

Mira, H., Martinez, N. and Penarrubia, L. 2002. Expression of vegetative-storage-protein gene from *Arabidopsis* is regulated by copper, senescence and ozone. *Planta* 214:939–946.

Miransari, M. 2011. Hyperaccumulators, arbuscular mycorrhizal fungi and stress of heavy metal. *Biotechnol Adv* 29:645–653. doi: 10.1016/j.biotechadv.2011.04.006

Mishra, S. and Dubey, R. S. 2006. Heavy metal uptake and detoxificationmechanisms in plants. *Int J Agric Res* 1:122–141.

Mithofer, A., Schulze, B. and Boland, W. 2004. Biotic and heavy metal stress response in plants: Evidence for common signals. *FEBS Lett* 566:1–5.

Mittler, R. 2002. Oxidative stress, antioxidants, and stress tolerance. *Trends Plant Sci* 7:405–410.

Mizuno, T., Usui, K., Horie, K., Nosaka, S., Mizuno, N. and Obata, H. 2005. Cloning of three ZIP/NRAMP transporter genes from a Ni hyperaccumulator plant *Thlaspi japonicum* and their Ni 2þ-transport abilities. *Plant Physiol Biochem* 43:793–801.

Mockaitis, K. and Estelle, M. 2008. Auxin receptors and plant development: A new signaling paradigm. *Ann Rev Cell Develop Biol* 24:55–80.

Mohanpuria, P., Rana, N. K. and Yadav, S. K. 2007. Cadmium induced oxidative stress influence on glutathione metabolic genes of *Camellia sinensis* (L.) O. Kuntze. *Environ Toxicol* 22:368–374.

Monnet, F., Vaillant, N., Vernay, P., Coudret, A., Sallanon, H. and Hitmi, A. 2001. Relationship between PSII activity, CO_2 fixation, and Zn, Mn and Mg contents of *Lolium perenne* under zinc stress. *J Plant Physiol* 158:1137–1144.

Moreno, I., Norambuena, L., Maturana, D., Toro, M., Vergara, C., Orellana, A., et al. 2008. AtHMA1 is a thapsigargin-sensitive Ca_2+/heavy metal pump. *J Biol Chem* 283:9633–9641.

Mostofa, M. G., Seraj, Z. I. and Fujita, M. 2015. Interactive effects of nitric oxide and glutathione in mitigating copper toxicity of rice (*Oryza sativa* L.) seedlings. *Plant Signal Behav* 10:9915701–4.

Mudgal, V., Madaan, N. and Mudgal, A. 2010. Biochemical mechanisms of salt tolerance in plants: A review. *Int J Bot* 6:136–143.

Muller, B., Pantin, F., Génard, M., Turc, O., Freixes, S., Piques, M. and Gibon, Y. 2011. Water deficits uncouple growth from photosynthesis, increase C content, and modify the relationships between C and growth in sink organs. *J Exp Bot* 62:1715–1729.

Muthuchelian, K., Bertamini, M. and Nedunchezhian, N. 2001. Triacontanol can protect *Erythrina variegata* from cadmium toxicity. *J Plant Physiol* 158:1487–1490.

Mysliwa-Kurdziel, B., Prasad, M. N. V. and Stralka, K. 2004. Photosynthesis in heavy metal stress plants. In: Prasad, M. N. V. (ed.), *Heavy Metal Stress in Plants*, 3rd edn, pp. 146–181. Springer, Berlin.

Nahar, K., Hasanuzzaman, M., Alam, M. M., Rahman, A., Suzuki, T. and Fujita, M. 2016. Polyamine and nitric oxide crosstalk: Antagonistic effects on cadmium toxicity in mung bean plants through upregulating the metal detoxification, antioxidant defense and methylglyoxal detoxification systems. *Ecotoxicol Environ Saf* 126:245–255.

Nanjo, T., Fujita, M., Seki, M., Kato, T. and Tabata, S. 2003. Toxicity of free proline revealed in an *Arabidopsis* T-DNA-tagged mutant defcient in proline dehydrogenase. *Plant Cell Physiol* 44:541–548.

Navari-Izzo, F. 1998. Thylakoid-bound and stromal antioxidative enzymes in wheat treated with excess copper. *Physiol Plant* 104:630–638.

Nayyar, H. and Chander, S. 2004. Protective effects of polyamines against oxidative stress induced by water and cold stress in chickpea. *J Agron Crop Sci* 190:355–365.

Nikolić, N., Kojic, D., Pilipovic, A., et al. 2008. Responses of hybrid poplar to cadmium stress: Photosynthetic characteristics, cadmium and proline accumulation, and antioxidant enzyme activity. *Acta Biologica Cracoviensia Series Botanica* 50:95–103.

Noctor, G., Arisi, A. C., Jouanin, L. and Foyer, C. H. 1998. Manipulation of glutathione and amino acid biosynthesis in the chloroplast. *Plant Physiol* 118:471–482.

Noctor, G. and Foyer, C. H. 1998. Ascorbate and glutathione: Keeping active oxygen under control. *Ann Rev Plant Biol* 49:249–279.

Nuccio, M. L., Russell, B. L., Nolte, K. D., Rathinasabapathi, B., Gage, D. A. and Hanson, A. D. 1998. The endogenous choline supply limits glycine betaine synthesis in transgenic tobacco expressing choline monooxygenase. *Plant J* 16:487–496.

Orozco-Cárdenas, M. L., Narváez-Vásquez, J. and Ryan, C. A. 2001. Hydrogen peroxide acts as a second messenger for the induction of defense genes in tomato plants in response to wounding, systemin, and methyl jasmonate. *Plant Cell* 13:179–19.

Ortiz, D. F., Ruscitti, T., McCue, K. F. and Ow, D. W. 1995. Transport of metal-binding peptides by HMT1, a fission yeast ABC-type vacuolar membrane protein. *J Biol Chem* 270:4721–4728.

Ostrowski, M., Ciarkowska, A. and Jakubowska, A. 2016. The auxin conjugate indole-3-acetyl-aspartate affects responses to cadmium and salt stress in *Pisum sativum* L. *J Plant Physiol* 191:63–72.

Ouariti, O., Boussama, N., Zarrouk, M., Cherif, A. and Ghorbal, M. H. 1997. Cadmium and copper induced changes in tomato membrane lipids. *Phytochemistry* 45:1343–1350.

Ouzounidou, G., Giamparova, M., Moustakas, M. and Karataglis, S. 1995. Responses of maize (*Zea mays* L.) plants to copper stress. *Environ Exp Bot* 35:167–176.

Paivoke, H. 1983. The short term effect of zinc on growth anatomy and acid phosphate activity of pea seedlings. *Ann Bot* 20:307–309.

Paleg, L. G., Doughlas, T. J., Vandaal, A. and Keech, D. B. 1981. Proline and betaine protect enzymes against heat inactivation. *Aust J Plant Physiol* 8:107–114.

Peleg, Z. and Blumwald, E. 2011. Hormone balance and abiotic stress tolerance in crop plants. *Curr Opin Plant Biol* 14:290–295.

Pence, N. S., Larsen, P. B., Ebbs, S. D., Letham, D. L., Lasat, M. M., Garvin, D. F., Eide, D. and Kochian, L. V. 2000. The molecular physiology of heavy metal transport in the Zn/Cd hyperaccumulator *Thlaspi caerulescens*. *Proc Natl Acad Sci* 97:4956–4960.

Pandey, J. and Pandey, U. 2009. Accumulation of heavy metals in dietary vegetables and cultivated soil horizon in organic farming system in relation to atmospheric deposition in a seasonally dry tropical region of India. *Environ Monit Assess* 148:61–74.

Pandey, N. and Sharma, C. P. 2002. Effect of heavy metals Co_2^+, Ni_2^+, and Cd_2^+ on growth and metabolism of cabbage. *Plant Sci* 163:753–758.

Pandolfini, T., Gabbrielli, R. and Comparini, C. 1992. Nickel toxicity and peroxidase activity in seedlings of *Triticum aestivum* L. *Plant Cell Environ* 15:719–725.

Panuccio, M. R., Sorgona, A., Rizzo, M. and Cacco, G. 2009. Cadmium adsorption on vermiculite, zeolite and pumice: Batch experiment studies. *J Environ Manage* 90:364–374. doi: 10.1016/j.jenvman.2007.10.005

Park, E. J., Jeknic, Z. and Chen, T. H. H. 2006. Exogenous application of glycinebetaine increases chilling tolerance in tomato plants. *Plant Cell Physiol* 47:706–714.

Parker, D. R., Aguilera, J. J. and Thomason, D. N. 1992. Zinc-phosophate interactions in two cultivars of tomato has grown in chelator buffered nutrient solution. *Plant Soil* 143:163–177.

Petit-Paly, G., Franck, T., Brisson, L., Kevers, C., Chenieux, C. and Rideau, M. 1999. Cytokinin modulates catalase activity and coumarin accumulation in in vitro cultures of tobacco. *J Plant Physiol* 155:9–15.

Petö, A., Lehotai, N., Lozano-Juste, J., León, J., Tari, I., Erdei, L., et al. 2011. Involvement of nitric oxide and auxin in signal transduction of copper-induced morphological responses in *Arabidopsis* seedlings. *Ann Bot* 108:449–457. doi: 10.1093/aob/mcr176

Petrov, V., Hille, J., Mueller-Roeber, B. and Gechev, T. S. 2015. ROS-mediated abiotic stress-induced programmed cell death in plants. *Front Plant Sci* 6:69. doi: 10.3389/fpls.2015.00069

Pierik, R., Tholen, D., Poorter, H., Visser, E. J. and Voesenek, L. A. 2006. The Janus face of ethylene: Growth inhibition and stimulation. *Trends Plant Sci* 11:176–183.

Piotrowska, A., Bajguz, A., Godlewska-ylkiewicz, B., Czerpak, R. and Kaminska, M. 2009. Jasmonic acid as modulator of lead toxicity in aquatic plant *Wolffia arrhiza* (Lemnaceae). *Environ Exp Bot* 66:507–513. doi: 10.1016/j.envexpbot.2009.03.019

Piotrowska-Niczyporuk, A., Bajguz, A., Zambrzycka, E. and Godlewska-Żylkiewicz, B. 2012. Phytohormones as regulators of heavy metal biosorption and toxicity in green alga *Chlorella vulgaris* (Chlorophyceae). *Plant Physiol Biochem* 52:52–65.

Podlešáková, K., Ugena, L., Spíchal, L., Doležal, K. and De Diego, N. 2019. Phytohormones and polyamines regulate plant stress responses by altering GABA pathway. *New Biotechnol* 48:53–65.

Poonam, S., Kaur, H. and Geetika, S. 2013. Effect of jasmonic acid on photosynthetic pigments and stress markers in *Cajanus cajan* (L.) Millsp. Seedlings under copper stress. *Am J Plant Sci* 4:817–823.

Pospíšilová, J. 2003. Participitation of phytohormones in the stomatal regulation of gas exchange during water stress. *Biol Plant* 46:491–506.

Pourrut, B., Perchet, G., Silvestre, J., Cecchi, M., Guiresse, M. and Pinelli, E. 2008. Potential role of NADPH-oxidase in early steps of leadinduced oxidative burst in *Vicia faba* roots. *J Plant Physiol* 165:571–579.

Prakash, T. R., Swamy, P. M., Suguna, P. and Reddanna, P. 1990. Characterization and behaviour of 15-lipoxygenase during peanut cotyledonary senescence. *Biochem Biophys Res Commun* 172:462–470.

Prasad, M. N. V., Greger, M. and Landberg, T. 2001. *Acacia nilotica* L. bark removes toxic elements from solution: Corroboration from toxicity bioassay using *Salix viminalis* L. in hydroponic system. *Int J Phytoremed* 3:289–300.

Price, J., Laxmi, A., St. Martin, S. K. and Jang, J. C. 2004. Global transcription profiling reveals multiple sugar signal transduction mechanisms in *Arabidopsis*. *The Plant Cell* 16:2128–2150.

Qin, F., Kodaira, K. S., Maruyama, K., Mizoi, J., Phan Tran, L. S., Fujita, Y., et al. 2011. SPINDLY, a negative regulator of gibberellic acid signaling, is involved in the plant abiotic stress response. *Plant Physiol* 157:1900–1913.

Quartacci, M. F., Cosi, E. and Navari-Izzo, F. 2001. Lipids and NADPH-dependent superoxide production in plasma membrane vesicles from roots of wheat grown under copper deficiency or excess. *J Exp Bot* 52:77–84.

Rady, M. M., El-Yazal, M. A. S., Taie, H. A. A. and Ahmed, S. M. 2016. Response of wheat growth and productivity to exogenous polyamines under lead stress. *J Crop Sci Biotechnol* 19:363–371.

Rady, M. M. and Hemida, K. A. 2015. Modulation of cadmium toxicity and enhancing cadmium tolerance in wheat seedlings by exogenous application of polyamines. *Ecotoxicol Environ Saf* 119:178–185.

Rahman, A., Nahar, K., Hasanuzzaman, M. and Fujita, M. 2016. Manganese-induced cadmium stress tolerance in rice seedlings: Coordinated action of antioxidant defense, glyoxalase system and nutrient homeostasis. *Comptes Rendus Biologies* 339:462–474.

Rahoui, S., Chaoui, A., Ben, C., Rickauer, M., Gentzbittel, L. and El Ferjani, E. 2015. Effect of cadmium pollution on mobilization of embryo reserves in seedlings of six contrasted *Medicago truncatula* lines. *Phytochemistry* 111:98–106.

Rai, V., Vajpayee, P., Singh, S. N. and Mehrotra, S. 2004. Effect of chromium accumulation on photosynthetic pigments, oxidative stress defense system, nitrate reduction, proline level and eugenol content of *Ocimum tenuiflorum* L. *Plant Sci* 167:1159–1169.

Rascio, N. and Navari Izzo, F. 2011. Heavy metal hyperaccumulating plants: How and why do they do it? And what makes them so interesting? *Plant Sci* 180:169–181.

Rashotte, A. M., Mason, M. G., Hutchison, C. E., et al. 2006. A subset of *Arabidopsis* AP2 transcription factors mediates cytokinin responses in concert with a two-component pathway. *Proc Natl Acad Sci USA* 103:11081–11085.

Raza, M. A. S., Saleem, M. F., Jamil, M. and Khan, I. H. 2014. Impact of foliar applied glycinebetaine on growth and physiology of wheat (*Triticum aestivum* L.) under drought conditions. *Pak J Agric Sci* 51:327–334.

Reddy, A. M., Kumar, S. G., Jyotsnakumari, G., Thimmanayak, S. and Sudhakar, C. 2005. Lead induced changes in antioxidant metabolism of horsegram (*Macrotyloma uniflorum* (Lam.) Verdc.) and bengalgram (*Cicer arietinum* L.). *Chemosphere* 60:97–104.

Remans, T., Opdenakker, K., Smeets, K., Mathijsen, D., Vangronsveld, J. and Cuypers, A. 2010. Metal-specific and NADPH oxidase dependent changes in lipoxygenase and NADPH oxidase gene expression in *Arabidopsis thaliana* exposed to cadmium or excess copper. *Func Plant Biol* 37:532–544.

Rocchetta, I., Mazzuca, M., Conforti, V., Ruiz, L., Balzaretti, V. and Ríos deMolina, M. C. 2006. Effect of chromium on the fatty acid composition of two strains of *Euglena gracilis. Environ Pollut* 141:353–358.

Rodríguez-Serrano, M., Romero-Puertas, M. C., Zabalza, M. A., et al. 2006. Cadmium effect on oxidative metabolism of pea (*Pisum sativum* L.) roots. Imaging of reactive oxygen species and nitric oxide accumulation *in vivo. Plant Cell Environ* 29:1532–1544.

Romero-Puertas, M. C., Rodríguez-Serrano, M., Corpas, F. J., Gómez, M., del Río, L. A. and Sandalio, L. M. 2004. Cadmium induced sub cellular accumulation of O_2.– and H_2O_2 in pea leaves. *Plant Cell Environ* 27:1122–1134.

Ronzan, M., Piacentinia, D., Fattorinia, L., Della Roverea, F., Eicheb, E., Riemannc, M., Altamuraa, M. M. and Falasca, G. 2018. Cadmium and arsenic affect root development in *Oryza sativa* L. negatively interacting with auxin. *Environ Exp Bot* 151:64–75.

Roosens, N. H., Willems, G. and Saumitou-Laprade, P. 2008. Using *Arabidopsis* to explore zinc tolerance and hyperaccumulation. *Trends Plant Sci* 13:208–215.

Ros, R., Cooke, D. T., Martínez-Cortina, C. and Picazo, I. 1992. Nickel and cadmium-related changes in growth, plasma membrane lipid composition, ATPase hydrolytic activity and proton-pumping of rice (*Oryza sativa* L. cv. Bahia) shoots. *J Exp Bot* 43:1475–1481.

Rosa, M., Prado, C., Podazza, G., Interdonato, R., Gonzalez, J. A., Hilal, M. and Prado, F. E. 2009. Soluble sugars—Metabolism, sensing and abiotic stress: A complex network in the life of plants. *Plant Signal Behav* 4:388–393.

Rustgi, S., Joshi, A., Moss, H. and Riesz, P. 1977. ESR of spin trapped radicals in aqueous solutions of amino acids: Reactions of the hydroxyl radical. *Int J Radiat Biol Relat Stud Phys Chem Med* 31:415–440.

Saito, R., Yamamoto, H., Makino, A., Sugimoto, T. and Miyake, C. 2011. Methylglyoxal functions as Hill oxidant and stimulates the photoreductin of O_2 at photosystem I: A symptom of plant diabetes. *Plant Cell Environ* 34:1454–1464.

Sancenón, V., Puig, S., Mateu-Andrés, I., Dorcey, E., Thiele, D. J. and Peñarrubia, L. 2004. The *Arabidopsis* copper transporter COPT1 functions in root elongation and pollen development. *J Biol Chem* 279:15348–15355.

Sandmann, G. and Böger, P. 1980. Copper-mediated lipid peroxidation processes in photosynthetic membranes. *Plant Physiol* 66:797–800.

Sanjaya, Hsiao, P. Y., Su, R. C., Ko, S. S., Tong, C. G., Yang, R. Y. and Chan, M. T. 2008. Overexpression of *Arabidopsis thaliana* tryptophan synthase beta 1 (AtTSB1) in *Arabidopsis* and tomato confers tolerance to cadmium stress. *Plant Cell Environ* 31:1074–1085.

Sasaki, A., Yamaji, N., Yokosho, K. and Ma, J. F. 2012. Nramp5 is a major transporter responsible for manganese and cadmium uptake in rice. *Plant Cell* 24:2155–2167.

Schaaf, G., Schikora, A., Häberle, J., Vert, G., Ludewig, U., Briat, J. F., Curie, C. and von Wirén, N. 2005. A putative function for the *Arabidopsis* Fe-Phytosiderophore transporter homolog AtYSL2 in Fe and Zn homeostasis. *Plant Cell Physiol* 46:762–774.

Schaller, G. E. 2012. Ethylene and the regulation of plant development. *BMC Biology* 10:9.

Schat, H., Sharma, S. S. and Vooijs, R. 1997. Heavy metal induced accumulation of free proline in a metal tolerant and a non-tolerant ecotype of *Silene vulgaris*. *Physiol Plant* 101:477–482.

Schellingen, K., Van Der Straeten, D., Vandenbussche, F., Prinsen, E., Remans, T., Vangronsveld, J. and Cuypers, A. 2014. Cadmium-induced ethylene production and responses in *Arabidopsis thaliana* rely on ACS2 and ACS6 gene expression. *BMC Plant Biol* 14:214.

Sebastiani, L., Scebba, F. and Tognetti, R. 2004. Heavy metal accumulation and growth responses in poplar clones Eridano (*Populus deltoides* × *maximo wiczii*) and I-214 (*P.* × *euramericana*) exposed to industrial waste. *Environ Exp Bot* 52:79–88.

Sekhar, P. N., Amrutha, R. N., Sangam, S., Verma, D. P. and Kishor, P. B. 2007. Biochemical characterization, homology modeling and docking studies of ornithine deltaaminotransferase-- an important enzyme in proline biosynthesis of plants. *J Mol Graph Model* 26:709–719.

Serraj, R. and. Sinclair, T. R. 2002. Osmolyte accumulation: Can it really help increase crop yield under drought conditions? *Plant Cell Environ* 25:333–341.

Sgherri, C., Cosi, E. and Navari-Izzo, F. 2003. Phenols and antioxidative status of *Raphanus sativus* grown in copper excess. *Physiol Planta* 118:21–28.

Shahid, M., Khalid, S., Abbas, G., Shahid, N., Nadeem, M., Sabir, M., et al. 2015. Heavy metal stress and crop productivity. In: K. R. Hakeem (ed.), *Crop Production and Global Environmental Issues*, pp. 1–25. Springer International Publishing, Cham.

Shahid, M. A., Balal, R. M., Pervez, M. A., et al. 2014. Exogenous proline and proline-enriched *Lolium perenne* leaf extract protects against phytotoxic effects of nickel and salinity in *Pisumsativum* by altering polyamine metabolism in leaves. *Turkish J Bot* 38:914–926.

Shalaby, S. and Horwitz, B. A. 2015. Plant phenolic compounds and oxidative stress: Integrated signals in fungal–plant interactions. *Curr Genet* 61:347–357.

Shan, C. and Liang, Z. 2010. Jasmonic acid regulates ascorbate and glutathione metabolism in *Agropyron cristatum* leaves under water stress. *Plant Sci* 178:130–139.

Shanker, A. K. 2003. Physiological, biochemical and molecular aspects of chromium toxicity and tolerance in selected crops and tree species. *PhD Thesis*, Tamil Nadu Agricultural University, Coimbatore, India.

Sharma, P. and Dubey, R. S. 2007. Involvement of oxidative stress and role of antioxidative defense system in growing rice seedlings exposed to toxic levels of aluminium. *Plant Cell Reports* 26:2027–2038.

Sharma, S. S. and Dietz, K. J. 2009. The relationship between metal toxicity and cellular redox imbalance. *Trends Plant Sci* 14:43–50.

Shen, B., Jensen, R. C. and Bohnert, H. J. 1997. Increased resistance to oxidative stress in transgenic plants by targeting mannitol biosynthesis to chloroplasts. *Plant Physiol* 113:1177–1183.

Shevyakova, N. I., Ilina, E. N., Stetsenko, L. A. and Kuznetsov, V. I. V. 2010. Nickel accumulation in rape shoots (*Brassica napus* L.) increased by putrescine. *Int J Phytorem* 13:345–356.

Shevyakova, N. I., Shorina, M. V., Rakitin, V. Y. and Kuznetsov, V. V. 2006. Stress- dependent accumulation of spermidine and spermine in the halophyte *Mesembryanthemum crystalinum* under salinity conditions. *Russ J Plant Physiol* 53:739–745.

Shi, X. and Dalal, N. S. 1989. Chromium (V) and hydroxyl radical formation during the glutathione reductase-catalyzed reduction of chromium (VI). *Biochem Biophys Res* 163:627–634.

Shingu, Y., Kudo, T., Ohsato, S., Kimura, M., Ono, Y., Yamaguchi, I. and Hamamoto, H. 2005. Characterization of genes encoding metal tolerance proteins isolated from *Nicotiana glauca* and *Nicotiana tabacum*. *Biochem Biophys Res Comm* 331:675–680.

Siedlecka, A. and Baszynski, T. 1993. Inhibition of electron flow around photosystem I in chloroplasts of Cd-treated maize plants is due to Cd-induced iron deficiency. *Physiol Planta* 87:199–202.

Singh, I. and Shah, K. 2014. Exogenous application of methyl jasmonate lowers the effect of cadmium-induced oxidative injury in rice seedlings. *Phytochemistry* 108:57–66.

Singh, P. K. and Tewari, S. K. 2003. Cadmium toxicity induced changes in plant water relations and oxidative metabolism of *Brassica juncea* L. plants. *J Environ Biol* 24:107–117.

Singla-Pareek, S. L., Yadav, S. K., Pareek, A., Reddy, M. K. and Sopory, S. K. 2006. Transgenic tobacco overexpressing glyoxalase pathway enzymes grow and set viable seeds in zinc-spiked soils. *Plant Physiol* 140:613–623.

Sirhindi, G., Mir, M. A., Abd-Allah, E. F., Ahmad, P. and Gucel, S. 2016. Jasmonic acid modulates the physio-biochemical attributes, antioxidant enzyme activity, and gene expression in *Glycine max* under nickel toxicity. *Front Plant Sci* 7:951 doi: 10.3389/fpls.2016.00591

Siripornadulsil, S., Traina, S., Verma, D. P. and Sayre, R. T. 2002. Molecular mechanisms of proline-mediated tolerance to toxic heavy metals in transgenic microalgae. *Plant Cell* 14:2837–2847.

Somashekaraiah, B., Padmaja, K. and Prasad, A. 1992. Phytotoxicity of cadmium ions on germinating seedlings of mung bean (*Phaseolus mungo*): Involvement of lipid peroxides in chlorophyll degradation. *Physiol Plant* 85:85–89.

Song, W. Y., Yamaki, T., Yamaji, N., Ko, D., Jung, K. H., Fujii-Kashino, M., et al. 2014. A rice ABC transporter, OsABCC1, reduces arsenic accumulation in the grain. *PNAS* 111:15699–15704.

Spoel, S. H. and Dong, X. 2008. Making sense of hormone crosstalk during plant immune responses. *Cell Host Microb* 3:348–351. doi: 10.1016/j.chom.2008

Srivastava, S., Srivastava, A. K., Suprasanna, P. and D'Souza, S. E. 2013. Identification and profiling of arsenic stress-induced microRNAs in *Brassica juncea*. *J Exp Bot* 64:303–315.

Srivastava, A. K., Penna, S., Nguyen, D. V. and Tran, L. S. P. 2016. Multifaceted roles of aquaporins as molecular conduits in plant responses to abiotic stresses. *Crit Rev Biotechnol* 36:389–398.

Stadtman, E. R. 1993. Oxidation of free amino acids and amino acid residues in proteins by radiolysis and by metal-catalysed reactions. *Annu Rev Biochem* 62:797–821.

Stadtman, E. R. and Oliver, C. N. 1991. Metal-catalyzed oxidation of proteins. Physiological consequences. *J Biol Chem* 266:2005–2008.

Stiborova, M., Pitrichova, M. and Brezinova, A. 1987. Effect of heavy metal ions in growth and biochemical characteristic of photosynthesis of barley and maize seedlings. *Biol Plant* 29:453–467.

Stohs, S. J. and Bagchi, D. 1995. Oxidative mechanisms in the toxicity of metal ions. *Free Radic Biol Med* 18:321–336.

Stoyanova, S., Geuns, J., Hideg, E. and Van den Ende, W. 2011. The food additives inulin and stevioside counteract oxidative stress. *Int J Food Sci Nutr* 62:207–214.

Su, J. and Wu, R. 2004. Stress-inducible synthesis of proline in transgenic rice confers faster growth under stress conditions than that with constitutive synthesis. *Plant Sci* 166:941–948.

Subbarao, G. V., Wheeler, R. M., Levine, L. H. and Stutte, G. W. 2001. Glycine betaine accumulation, ionic and water relations of red-beet at contrasting levels of sodium supply. *J Plant Physiol* 158:767–776.

Sun, P., Tian, Q. Y., Chen, J. and Zhang, W. H. 2010a. Aluminium-induced inhibition of root elongation in *Arabidopsis* is mediated by ethylene and auxin. *J Exp Bot* 61:347–356.

Sun, X., Zhang, J., Zhang, H., et al. 2010b. The responses of *Arabidopsis thaliana* to cadmium exposure explored via metabolite profiling. *Chemosphere* 78:840–845. doi: 10.1016/j.chemosphere.2009.11.045

Sundaramoorthy, P., Chidambaram, A., Ganesh, K. S., Unnikannan, P. and Baskaran, L. 2010. Chromium stress in paddy: (i) nutrient status of paddy under chromium stress; (ii) phytoremediation of chromium by aquatic and terrestrial weeds. *C R Biol* 333:597–607.

Szechynska-Hebdam, M., Skrzypek, E., Dabrowska, G., Biesaga-Koscielniak, J., Filek, M. and Wedzony, M. 2007. The role of oxidative stress induced by growth regulators in the regeneration process of wheat. *Acta Physiol Plant* 29:327–337.

Taie, H. A., El-Yazal, M. A. S., Ahmed, S. M. and Rady, M. M. 2019. Polyamines modulate growth, antioxidant activity, and genomic DNA in heavy metal–stressed wheat plant. *Environ Sci Pollut Res* 26:22338–22350.

Talanova, V. V., Titov, A. F. and Boeva, N. P. 2001. Effect of increasing concentrations of heavy metals on the growth of barley and wheat seedlings. *Russian J Plant Physiol* 48:100–103.

Tan, Y. F., Toole, N., Taylor, N. L. and Harvey Millar, A. 2010. Divalent metal ions in plant mitochondria and their role in interactions with proteins and oxidative stress-induced damage to respiratory function. *Plant Physiol* 152:747–761.

Tang, C. F., Liu, Y. G., Zeng, G. M., Li, X., Xu. W. H., Li. C. F. and Yuan, X. Z. 2005. Effects of exogenous spermidine on antioxidant system responses of *Typha latifolia* L. under Cd_2+ stress. *J Integr Plant Biol* 47:428–434.

Tassi, E., Pouget, J., Petruzelli, G. and Barbafieri, M. 2008. The effects of exogenous plant growth regulators in the phytoextraction of heavy metals. *Chemosphere* 71:66–73.

Taulavuori, K., Prasad, M. N. V., Taulavuori, E. and Laine, K. 2005. Metal stress consequences on frost hardiness of plants at northern high latitudes: A review and hypothesis. *Environ Pollut* 135:209–220.

Thao, N. P., Khan, M. I. R., Thu, M. B. A., Hoang, X. L. T., Asgher, M., Khan, N. A., and Tran, L. S. P. T. 2015. Role of ethylene and its cross talk with other signaling molecules in plant responses to heavy metal stress. *Plant Physiol* 169:73–84.

Thao, N. P. and Tran, L. S. P. 2012. Potentials toward genetic engineering of drought-tolerant soybean. *Crit Rev Biotechnol* 32:349–362.

Thomas, J. C., Perron, M., LaRosa, P. C. and Smigocki, A. C. 2005. Cytokinin and the regulation of a tobacco metallothionein-like gene during copper stress. *Physiol Planta* 123:262–271.

Tseng, M. J., Liu, C. W. and Yiu, J. C. 2007. Enhanced tolerance to sulfur dioxide and salt stress of transgenic Chinese cabbage plants expressing both superoxide dismutase and catalase in chloroplasts. *Plant Physiol Biochem* 45:822–833. doi: 10.1016/j.plaphy.2007.07.011

Tsuchisaka, A., Yu, G., Jin, H., Alonso, J. M., Ecker, J. R., Zhang, X., et al. 2009. A combinatorial interplay among the 1-aminocyclopropane-1-carboxylate isoforms regulates ethylene biosynthesis in *Arabidopsis thaliana*. *Genetics* 183:979–1003. doi: 10.1534/genetics.109.107102

Tsukagoshi, H., Busch, W. and Benfey, P. N. 2010. Transcriptional regulation of ROS controls transition from proliferation to differentiation in the root. *Cell* 143:606–616.

Valluru, R. and Van den Ende, W. 2011. Myo-inositol and beyond – Emerging networks under stress. *Plant Science* 181:387–400.

Van de Poel, B. and Van Der Straeten, D. 2014. 1-aminocyclopropane-1-carboxylic acid (ACC) in plants: More than just the precursor of ethylene! *Front Plant Sci* 5:640. doi: 10.3389/fpls.2014.00640

Van den Ende, W. and Valluru, R. 2009. Sucrose, sucrosyl oligosaccharides, and oxidative stress: Scavenging and salvaging. *J Exp Bot* 60:9–18.

Villiers, F., Jourdain, A., Bastien, O., Leonhardt, N., Fujioka, S., Tichtincky, G., et al. 2012. Evidence for functional interaction between brassinosteroids and cadmium response in *Arabidopsis thaliana*. *J Exp Bot* 63:1185–1200. doi: 10.1093/jxb/err335

Vitti, A., Nuzzaci, M., Scopa, A., Tataranni, G., Remans, T., Vangronsveld, J., et al. 2013. Auxin and cytokinin metabolism and root morphological modifications in *Arabidopsis thaliana* seedlings infected with *Cucumber mosaic virus* (CMV) or exposed to cadmium. *Int J Mol Sci* 14:6889–6902. doi: 10.3390/ijms14046889

Vodnik, D., Gaberscik, A. and Gogala, N. 1999. Lead phytotoxicity in Norway spruce (*Picea abies* (L.) Karst.): The effect of Pb and zeatin-riboside on root respiratory potential. *Phyton* 39:155–159.

Walters, D. 2003. Resistance to plant pathogens: Possible roles for free polyamines and polyamine catabolism. *New Phytol* 159:109–115.

Wang, C., Lu, J., Zhang, S. H., Wang, P. F., Hou, J. and Qian, J. 2011. Effects of Pb stress on nutrient uptake and secondary metabolism in submerged macrophyte *Vallisneria natans*. *Ecotoxicol Environ Saf* 74:1297–1303. doi: 10.1016/j.ecoenv.2011.03.005

Wang, F., Zeng, B., Sun, Z. and Zhu, C. 2009. Relationship between proline and Hg_2+-induced oxidative stress in a tolerant rice mutant. *Arch Environ Contam Toxicol* 56:723–731.

Wang, H., Dai, B., Shu, X., Wang, H. and Ning, P. 2015. Effect of kinetin on physiological and biochemical properties of maize seedlings under arsenic stress. *Adv Mater Sci Engineer* 2015:1–7. Article ID 714646. doi: 10.1155/2015/714646

Wang, S. Y. 1999. Methyl jasmonate reduces water stress in strawberry. *J Plant Growth Regul* 18:127–134.

Wang, X., Shi, G., Xu, Q. and Hu, J. 2007. Exogenous polyamines enhance copper tolerance of *Nymphoides peltatum*. *J Plant Physiol* 164:1062–1070. doi: 10.1016/j.jplph.2006.06.003

Wang, X., Shi, G. X., Ma, G. Y., Xu, Q. S., Rafeek, K. and Hu, J. Z. 2004. Effects of exogenous spermidine on resistance of *Nymphoides peltatum* to Hg_2+ 918 stress. *J Plant Physiol Mol Biol* 30:69–74.

Wasternack, C. 2014. Action of jasmonates in plant stress responses and development-applied aspects. *Biotechnol Adv* 32:31–39.

Wasternack, C. and Parthier, B. 1997. Jasmonate-signalled plant gene expression. *Trends Biochem Sci* 2:302–307.

Weber, M., Harada, E., Vess, C., von Roepenack-Lahaye, E. and Clemens, S. 2004. Comparative microarray analysis of *Arabidopsis thaliana* and *Arabidopsis halleri* roots identifies nicotianamine synthase, a ZIP transporter and other genes as potential metal hyperaccumulation factors. *Plant J* 37:269–281.

Wei, D., Zhang, W., Wang, C., Meng, Q., Li, G., Chen, T. H. H. and Yang, X. 2017. Genetic engineering of the biosynthesis of glycinebetaine leads to alleviate salt-induced potassium efflux and enhances salt tolerance in tomato plants. *Plant Sci* 257:74–83.

Weinstein, L. H., Kaur-Sawhney, R., Venkat Rajam, M., Wettlaufer, S. H. and Galston, A. W. 1986. Cadmium-induced accumulation of putrescine in oat and bean leaves. *Plant Physiol* 82:641–645.

Wojas, S., Clemens, S., Hennig, J., Sklodowska, A. and Kopera, E. 2008. Overexpression of phytochelatin synthase in tobacco: Distinctive effects of AtPCS1 and CePCS genes on plant response to cadmium. *J Exp Bot* 59:2205–2219.

Wong, C. K. E., Jarvis, R. S., Sherson, S. M. and Cobbett, C. S. 2009. Functional analysis of the heavy metal binding domains of the Zn/Cd-transporting ATPase, HMA2, in *Arabidopsis thaliana. New Phytol* 181:79–88.

Wu, J. T., Chang, S. C. and Chen, K. S. 1995. Enhancement of intracellular proline level in cells of *Anacystis nidulans* (cyanobacteria) exposed to deleterious concentrations of copper. *J Phycol* 31:376–379.

Xia, X. J., Zhou, Y. H., Shi, K., Zhou, J., Foyer, C. H. and Yu, J. Q. 2015. Interplay between reactive oxygen species and hormones in the control of plant development and stress tolerance. *J Exp Bot* 66:2839–2856.

Xia, X. J., Zhou, Y. H., Ding, J., Shi, K., Asami, T., Chen, Z., et al. 2011. Induction of systemic stress tolerance by brassinosteroid in *Cucumis sativus. New Phytol* 191:706–720.

Xing, G. S., Zhou, G. K., Li, Z. X. and Cui, K. R. 2000. Studies of polyamine metabolism and β-*N* oxalyl-L-α, β- diamino-propionic acid accumulation in grass pea (*Lathyrus sativus*) under water stress. *Acta Bot Sin* 42:1039–1044.

Xu, J., Yin, H. and Li, X. 2009. Protective effects of proline against cadmium toxicity in micropropagated hyperaccumulator, *Solanum nigrum* L. *Plant Cell Rep* 28:325–333.

Xu, J. and Zhang, S. 2015. Ethylene biosynthesis and regulation in plants. In: Wen, C. K. (ed.), *Ethylene in Plants*, pp. 1–25. Springer, Dordrecht, The Netherlands.

Yadav, S. K., Singla-Pareek, S. L., Ray, M., Reddy, M. K. and Sopory, S. K. 2005. Methylglyoxal levels in plants under salinity stress are dependent on glyoxalase I and glutathione. *Biochem Biophys Res Commun* 337:61–67.

Yamaguchi, K., Takahashi, Y., Berberich, T., Imai, A., Miyazaki, A., Takahashi, T., et al. 2006. The polyamine spermine protects against high salt stress in *Arabidopsis thaliana. FEBS Lett* 580:6783–6788. doi: 10.1016/j.febslet.2006.10.078

Yan, Z. and Tam, N. F. Y. 2013. Effect of lead stress on anti-oxidative enzymes and stress related hormones in seedlings of *Excoecaria agallocha* Linn. *Plant Soil* 367:327–338.

Yancey, P. H. 1994. Compatible and counteracting solutes. In: Strange, K. (ed.), *Cellular and Molecular Physiology of Cell Volume Regulation*, pp. 81–109. CRC Press, Boca Raton, FL, USA.

Yao, W. Q., Lei, Y. K., Yang, P., Li, Q. S., Wang, L. L., He, B. Y., Xu, Z. M., Zhou, C. and Ye, H. J. 2018. Exogenous glycinebetaine promotes soil cadmium uptake by edible Amaranth grown during subtropical hot season. *Int J Environ Res Public Health* 15:1794. doi:10.3390/ijerph15091794

Ye, B., Muller, H. H., Zhang, J. and Gressel, J. 1997. Constitutively elevated levels of putrescine and putrescine generating enzymes correlated with oxidant stress resistance in *Conyza canadensis* and wheat. *Plant Physiol* 115:1443–1451.

Yiu, J. C., Juang, L. D., Fang, D. Y. T., Liu, C. W. and Wu, S. J. 2009. Exogenous putrescine reduces flooding-induced oxidative damage by increasing the antioxidant properties of Welsh onion. *Sci Hort* 120:306–314.

Yuan, H. M., Xu, H. H., Liu, W. C. and Lu, Y. T. 2013. Copper regulates primary root elongation through PIN1-mediated auxin redistribution. *Plant Cell Physiol* 54:766–778. doi: 10.1093/pcp/pct030

Zarei, S., Ehsanpour, A. A. and Abbaspour, J. 2012. The role of over expression of P5CS gene on proline, catalase, ascorbate peroxidase activity and lipid peroxidation of transgenic tobacco (*Nicotiana tabacum* L.) plant under in vitro drought stress. *J Cell Mol Res* 4:43–49.

Zelinová, V., Alemayehu, A., Bocová, B., Huttová, J. and Tamás, L. 2015. Cadmium-induced reactive oxygen species generation, changes in morphogenic responses and activity of some enzymes in barley root tip are regulated by auxin. *Biologia* 70:356–364.

Zengin, F. K. and Kirbag, S. 2007. Effects of copper on chlorophyll, proline, protein and abscisic acid level of sunflower (*Helianthus annuus* L.) seedlings. *J Environ Biol* 28:561–566.

Zhang, W. H. and Tyerman, S. D. 1999. Inhibition of water channels by $HgCl_2$ in intact wheat root cells. *Plant Physiol* 120:849–857.

Zhao, F. J., Lombi, E. and Mcgrath, S. P. 2003. Assessing the potential for zinc and cadmium phytoremediation with the hyper accumulator *Thlaspi. Plant Soil* 249:37–43.

Zhao, H., Ma, T., Wang, X., Deng, Y., Ma, H., Zhang, R. and Zhao, J. 2015. OsAUX1 controls lateral root initiation in rice (*Oryza sativa* L.). *Plant Cell Environ* 38:2208–2222.

Zhu, J. K. 2001. Cell signaling under salt, water and cold stresses. *Curr Opin Plant Biol* 4:401–406.

Zhu, X. F., Wang, Z. W., Dong, F., Lei, G. J., Shi, Y. Z., Li, G. X. and Zheng. S. J. 2013. Exogenous auxin alleviates cadmium toxicity in *Arabidopsis thaliana* by stimulating synthesis of hemicellulose 1 and increasing the cadmium fixation capacity of root cell walls. *J Hazard Mater* 263:398–403.

6 Role of Osmolytes in Regulation of Growth and Photosynthesis under Environmental Stresses

Rayees Ahmad Mir, Surendra Argal, and Mohammad Abass Ahanger

CONTENTS

6.1 INTRODUCTION

Environmental stresses, both abiotic and biotic, affect the survival and productivity of plants. Extreme environmental conditions hamper plant metabolism by initiating excess generation of toxic molecules like reactive oxygen species (ROS), resulting in reduced growth and yield performance (Ahanger et al. 2017a; Shao et al. 2008). ROS production occurs at major sites including the chloroplast, mitochondria, peroxisomes and apoplast (Mittler et al. 2004; Dietz et al. 2016), and among the key ROSs are included superoxide radical (O_2^-), hydroxyl radical (OH^-), hydroperoxyl radical (HO_2^-), hydrogen peroxide (H_2O_2), alkoxy radical (RO^-), peroxy radical ($ROO^{\cdot}$), singlet oxygen ($^1O_2$), etc. (Mittler et al. 2004; Ahanger et al. 2017a; Figure 6.1). ROS are highly reactive and toxic radicals affecting key cellular functions by damaging nucleic acids, lipids and proteins, thereby leading to metabolic dysfunction and programmed cell death (Apel and Hirt 2004, Mittler et al. 2004; Sharma et al. 2012; Ahmad et al. 2010; Ahanger et al. 2017a, 2018a).

Environmental stresses effect photosynthesis by influencing water and ion balance, electron transport in thylakoids and CO_2 supply by affecting stomatal functioning and the carbon reduction cycle (Allen and Ort, 2001; Khan et al. 2014; Khan et al. 2016; Ahmad et al. 2018). Both abiotic as well biotic stresses impart deleterious impacts on the photosynthetic performance of plants, resulting in significant growth reductions and yield loss (Wang et al. 2003; Strauss and Zangerl 2002; Maron and Kauffman 2006; Maron and Crone 2006, Mordecai 2011). The optimal growth and development of crop plants depends on the interaction of different intracellular organelles, and amongst them the chloroplast is the key site for photosynthesis wherein light and dark reactions of photosynthesis are carried out. The structural and functional stability of chloroplasts is

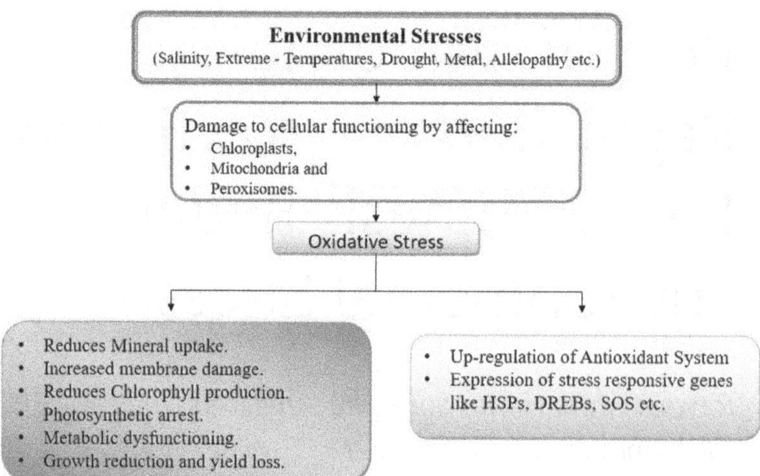

FIGURE 6.1 Environmental stresses and modulations in plant attributes.

highly sensitive to different environmental stresses, and key modulations are induced in response to stresses (Biswal et al. 2008, Saravanavel et al. 2011). It is well-established that environmental stresses affect the photosynthetic performance of the plant through stomatal as well as non-stomatal limitations (Rahnama et al. 2010; Khan et al. 2016).

Tolerance to stress-mediated growth alterations is a very complex process due to interactions occurring between stress factors and the different biological (physiological, biochemical and molecular) processes involved in the regulation of growth and development (Zhu 2002; Asharf and Harris 2004). Mechanisms initiated at the organelle, cellular or whole plant levels are manifestations of various key pathways driven by a set of regulatory mechanisms operating at transcriptional and translational levels (Ahanger et al. 2017b, c). Stress tolerance leads to the development of different defensive mechanisms in plants for withstanding adverse growth conditions. Among the key defence mechanisms contributing to stress tolerance are (a) accumulation of compatible solutes, (b) up-regulation of antioxidant system, and (c) synthesis of redox components and expression of stress-specific proteins like heat shock proteins (HSPs), dehydration responsive element binding proteins (DREBs), phytochelatins, etc. (Serraj and Sinclair 2002; Per et al. 2017; Ahanger et al. 2018b; Ahanger and Ahmad 2019; Figure 6.2). Compatible osmolytes including low molecular weight highly soluble compounds are usually non-toxic even at higher concentrations which leads to the protection of plants from stress-induced damage through osmotic adjustment, detoxification of ROS, maintenance of membrane integrity and enzymes/proteins stabilization (Ashraf and Foolad 2007; Per et al. 2017). Among the osmolytes are included proline, sucrose, polyols, trehalose, glycine betaine, K ion (inorganic osmolyte), etc. (Agarwal et al. 1999, 2009; Ashraf and Foolad 2007; Ahanger et al. 2014), which actively participate in the protection of plant structural and functional integrity. The present review addresses the recent developments in plant science research focussing on the physiological and biochemical roles of the osmolytes in alleviating stress-triggered alterations in photosynthesis and growth.

6.2 PHOTOSYNTHESIS AND ENVIRONMENTAL STRESSES

6.2.1 Drought/Water Stress and Photosynthesis

Water stress is one of the major environmental factors the world over, particularly in arid and semi-arid areas, limiting plant growth and yield significantly (Agarwal et al. 1999; Ahanger et al. 2014). Deficit water availability affects growth, enzyme activity, mineral uptake and assimilation (Pandey

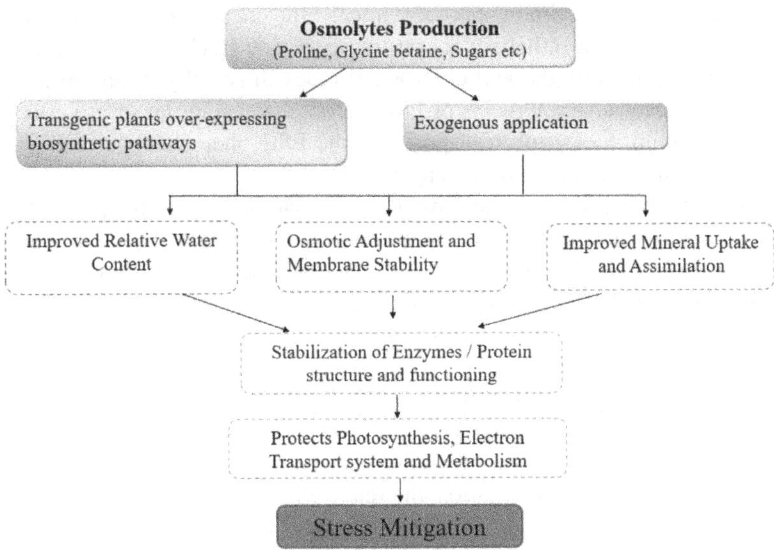

FIGURE 6.2 Osmolytes mediate stress mitigation through their active involvement in several key physiological and biochemical characteristics, leading to the protection of major plant functioning.

et al. 2004; Tomar and Agarwal 2013; Ahanger and Agarwal 2017a; Ahanger et al. 2017d), gene expression and photosynthesis (Zhang et al. 2018). Photosynthesis is the primary target of water deficit stress (Lawlor, 1995; Tiwari et al. 1998). Water stress induces closure of stomata thereby reducing the internal CO_2 concentration and hence affecting Calvin cycle functioning, and the consequent shortage of reducible coenzymes results in photo-inhibitory conditions (Horton et al. 1996). Under severe conditions stomatal conductance, internal CO_2 concentration and net assimilation decline (Flexas et al., 2000; Van Leeuwen et al., 2009). Under severe water stress, non-stomatal regulation of photosynthesis is observed including reduced photochemistry (Flexas et al. 2002a), activity of photosynthetic enzymes (Maroco et al., 2002) and mesophyll conductance (Flexas et al., 2002a). The closure of stomata during drought is either regulated through hydraulic signals (leaf water potential, cell turgor) or chemical signals (abscisic acid). During soil drying ABA synthesized in roots is transported to guard cells via the transpiration stream and induces stomatal closure after binding to the ABA receptor (Teiz and Zeiger 1998).

Experimentally, monitoring of the tissue water and chlorophyll contents has been used to study physiological changes, photosynthetic activity, stress severity and nutritional status for avoiding restrictions in crop growth and yield (Wu et al. 2008, Neto et al. 2017). For example, Goltsev et al. (2012) have reported that a 60% decline in RWC resulted in a parallel decrease in the ratio between the rate of excitation trapping in the photosystem II reaction centre and the rate of re-oxidation of reduced PSII acceptors. In addition a 20% decline in RWC suppresses electron transfer from reduced plastoquinone to the PSI reaction centre, however, re-oxidation of the photo-reduced primary quinone acceptor of PSII (Q_A^-) is inhibited below 15% RWC and the primary photochemical reactions in PSI and II are inactivated at less than 5% (Goltsev et al. 2012). Water shortage gradually decreases the PSII electron transport while improving capacity for non-photochemical quenching more slowly in anticipation of a large decrease in leaf relative water content (Zivcak et al. 2013). Water stress/drought reduces RWC and the synthesis of chlorophyll a and chlorophyll b, resulting in declined photochemical efficiency of PSII, photosynthetic performance index and stomatal conductance (Ghobadi et al. 2013), however, carotenoids and electrolyte leakage are increased (Mustafavi et al. 2016). Reduced water availability significantly reduces the net photosynthetic rate (P_n), stomatal conductance (G_s) and transpiration rate (E) but enhances immediate water use efficiency. In drought-stressed wheat, Wang et al. (2016) have reported a significant decline in the photochemical quenching coefficient (qP), maximum photochemical efficiency (F_v/F_m), quantum yield of PSII (Φ_{PSII}) and the apparent photosynthetic electron transport rate (ETR) inferred from chlorophyll *a* measurements. Drought reduces the leaf chlorophyll index drastically, resulting in a considerable decline in the fluorescence characteristics of PSII. In order to avoid the severe effects of drought, plants respond by declining electron transport and the PSII photochemical activity as has been reported in *Plectranthus scutellarioides* (Meng et al. 2016).

The impact of drought varies with the timing and severity as well as the plant species. Drought reduces stomatal conductance in C_3 plants more dramatically compared to C_4 species however leaf mortality has been seen to be slightly higher in C_4 than C_3 grasses (Taylor et al. 2011). A greater decline in stomatal conductance (*gs*) and related attributes of gas exchange than other physiological characters in shrubs and C_3 plants is more evident than in C_4 (Yan et al. 2016). C_4Pn is less sensitive than C_3Pn, which gives an advantage to C_4 plants in P_n. But their WUE decreased under drought conditions, demonstrating a great flexibility in C_4 WUE. For maintaining *Pn* under drought, plants sacrifice WUE significantly, exhibiting a great variation from herbs to trees. Moreover, *Gs* had a significant correlation with *Pn* and *Tr*, but insignificant with WUE, which is due to the effect of other factors like air flow, CO_2 concentration and relative humidity (Zhang et al. 2018).

6.2.2 Salinity and Photosynthesis

Soil salinity is a global threat and high soil salinity imparts physiological drought by preventing water uptake and inducing stomatal closure, thereby leading to restricted entry of CO_2 into the leaf (Nazar et al. 2011). Nearly 20% of the world's cultivated land and approximately half of irrigated land is affected by salinity (Zhu 2001). Salinity in soils and in irrigation water is one of the main abiotic factors in the whole agriculture world. Over the last 20 years, due to the increase in irrigation requirements in arid and semi-arid regions, the situation has worsened further (Munns and Gilliham 2015; Acosta-Motos et al. 2017a). Salinity stress results in stunted plant growth by reducing the leaf expansion and photosynthetic rate and restricting root growth, thereby resulting in a shift in the root-to-shoot allocation pattern (Rasmuson and Anderson 2002). It has been now well-established that salinity stress impairs electron transport in the chloroplast and mitochondria, leading to excess formation of ROS (Noctor et al. 2002) which ultimately inhibits photosynthesis. Salinity-induced oxidative damage imparts inhibitory effects on the key physiological and biochemical processes including photosynthesis, membrane functioning, enzyme activity, etc. (Mishra et al., 2006; Murillo-Amador et al., 2007; Ahanger and Agarwal 2017a). High salt concentrations may lead to the induction of axillary and adventitious buds on shoots, but the rate of photosynthesis, PSII efficiency and chlorophyll content decreases (Siler et al. 2007). *Arabidopsis* showed closure of stomata and inhibition of CO_2 fixation when exposed to sub-lethal salt (NaCl) concentrations, resulting in inhibition of electron transport via PSII, enhancement of cyclic electron flow around PSI and increased non-photochemical quenching of chlorophyll fluorescence (Stepien and Johnson 2009).

The morphological and nutritional alterations intensify with the increase in salt concentration in *Salvinia auriculata*, reducing the content of Ca^{2+}, K^+, Mg^{2+}, P and N ions (Gomes et al. 2011). Reduction in mineral ion uptake and the synthesis of photosynthetic pigments has been reported in various crop plants including rice (Ali et al. 2004), *Vicia faba* L. (Qados 2011), cotton (Saleh 2012), *Ociumum basilicm* L. (Heidari 2012), *Cajanus cajan* (Amuthavalli and Sivasankaramoorthy 2012), mustard (Fatma et al. 2014) and wheat (Ahanger and Agarwal 2017a). Salinity-triggered reduction in chlorophyll content is ascribed to membrane dysfunction (Gomes et al. 2011), loss of chloroplast membranes and alteration of the lamellae vesiculation (Ceccarelli et al. 2010), swelling of thylakoids and distraction of envelopes in mesophyll cell chloroplasts (Omoto et al. 2010), damage to chloroplasts by ROS (Ahanger et al. 2017a) and improved degradation (Bonales-Alatorre et al. 2013) leading to an impeded photosynthetic process by inducing oxidative stress (Acosta-Motos et al. 2017b).

High salinity causes alterations in both types of PSII heterogeneity, i.e. PSII antenna heterogeneity and PSII reducing side heterogeneity (Jajoo 2013). Salinity stress (NaCl) decreases the activity of

Rubisco (Aragao et al. 2005) and activation state, transcriptional level of *rbcL* gene, expression level of Rubisco large sub unit (LSU) (Hu et al. 2013) and Rubisco activase (Lee et al. 2014); however, it enhances the production of the chloroplastic fructose-bisphosphate aldolase enzyme along with carbohydrate-metabolizing pathways including the Kelvin cycle and glycolysis in canola seedlings with a significant decline in the amount of Cu/Zn SOD in the chloroplast resulting from an increase in OH^- radicals. The expression of photosynthesis proteins reduces due to the drop in osmotic pressure caused by water scarcity and stomata closure which also reduces the access to CO_2 (Alagoz et al. 2016). In salinity-stressed plants reduced chlorophyll content results from up-regulated chlorophyllase activity (Bertrand and Schoefs 1999).

Photosynthetic regulation under salinity stress is controlled by the transcription pattern of several genes involved in chlorophyll biosynthesis, the light signalling pathway and ROS scavenging. For example in rice, the chloroplast nucleoside diphosphate kinase 2 (NDPK2) encoded by *WSL12* plays an important role in the development of chloroplast and chlorophyll biosynthesis by regulating multiple gene expression levels of related genes and the salinity stress tolerance (Ye et al. 2016). In addition the proteins involved include the ones involved in ROS neutralization, thylakoid membrane organization, PSII activity, osmotic and ion homeostasis, ABA biosynthesis, gene expression and protein synthesis (see Suo et al. 2017).

6.2.3 Metal Stress and Photosynthesis

Toxic heavy metals enter into the environment through various processes like industrialization, excessive use of phosphate fertilizers, etc. and remain for long periods in the environment. They are incorporated in the main category of environmental pollutants, and their accumulation is potentially dangerous to almost all living beings (Benavides et al., 2005). Over the last few decades environmental contamination has witnessed a dramatic increase due to various anthropogenic activities (Gratao et al., 2005). Anthropogenic perturbations of natural biogeochemical cycles because of heavy industrialization, extensive mining, intensive agriculture and rapid urbanization have caused widespread and grave contamination of vital components of life on the planet; the accumulation of heavy metals is of paramount importance to emphasize for ecological, nutritional and environmental reasons (Nagajyoti et al. 2010; Ali et al. 2013).

The application of pesticides, chemical fertilizers, sewage sludge, dust, industrial waste and bad watering affects large areas of agricultural/cultivated land throughout the world (Bell et al., 2001; Passariello et al., 2002; Yang et al. 2005). Upon high heavy metal exposure, plants initially respond through the generation of ROS. ROS generation by different metals either directly through Haber–Weiss reactions or due to indirect consequences of metal toxicity (Mithofer et al., 2004) disrupts the electron transport chain (Qadir et al., 2004), metabolism of essential mineral elements, etc. (Dong et al., 2006; Asgher et al., 2014). Nevertheless, the effects of heavy metal pollution on plant development vary with the type of metal, soil characteristics and plant species, and the key alterations to physiological processes attributed to heavy metals include plant–water relationships, photosynthesis, cellular respiration and nitrogen metabolism (Romanowska et al., 2006; Khan et al. 2015; Ahmad et al. 2018). These effects have been correlated at cellular and molecular levels with oxidative damage causing enhanced lipid peroxidation, DNA damage and protein oxidation (Ahmad et al. 2010).

Like other stresses, heavy metals also influence photosynthesis significantly. In *Phaseolus vulgaris* L. heavy metals including Pb, Cu, Cd and Hg affected photosynthetic pigments and the overall photosynthesis with increasing heavy metal concentrations (Zengin and Munzuroglu 2005). High concentrations of heavy metals trigger loss of membrane selective permeability to cellular energy depletion and induce alterations in chloroplasts thereby affecting PSII, chlorophyll synthesis, water splitting and Calvin cycle enzymes (Linger et al. 2005; Basile et al. 2012). Excess copper inhibits the growth of the plant by impairing the cellular processes like photosynthetic electron transport (Yruela 2005), and increasing concentrations of copper have been reported to adversely reduce total chlorophyll and protein content in sunflower seedlings (Zengin and Kirbag 2007). Gene expression in response to metal stress influences the photosynthetic functioning of plants (Sudo et al. 2008). The addition of Cr or Ag shows a concentration-dependent inhibition in the activity of whole chain electron transport by reducing the level of the PSII catalysed reaction centre and the level of LHC (Babu et al. 2010).

Low concentrations of Cd have been observed to stimulate the growth of sorghum plants; however they imparted a significant reduction in growth by decreasing the chlorophyll content and improving the lipid peroxidation (Da-lin et al. (2011). Increased metal load decreases chlorophyll concentrations and the photosynthetic rate in *Phragmites australis* (Ayeni et al. 2012), thereby inducing large disturbances in ion uptake even at low concentrations and hence imparting metabolic disorders reflected in significant reductions in biomass, pigments, photosynthetic characteristics and plant growth (Lamhamdi et al. 2013; Bharwana et al. 2013). Higher Cd concentrations trigger the generation of ROS and lipid peroxidation, and decrease biomass and chlorophyll contents. This ROS accumulation mediates the inhibition of the PSII electron transport from OEC to Yz residue in D1 protein by replacing Ca^{2+} in the core of OEC. Furthermore increased electron transport efficiency from the QB to PSI acceptors has been concluded to be an important response of tall fescue against Cd toxicity (Huang et al. 2017). Cd and Zn restrict the photosynthetic electron transport leading to a significant decline in the efficiency of energy transformation in photosystem II (Paunov et al. 2018). Metal exposure affects the chloroplast structure and down-regulates the expression of photosynthesis-related genes (Song et al. 2014).

6.2.4 Heat/Temperature Stress and Photosynthesis

Changing temperatures also affect the growth and productivity of crops. The abrupt rise and fall in global temperatures have resulted in a considerable decline in crop growth and yield performance. Temperature is predicted to increase by about 1–3°C by the mid-21st century and about 2–5°C by the late 21st century (Eitzinger et al. 2010). This constantly rising ambient temperature is considered one of the most detrimental threats to crop production (Hatfield et al. 2011; Lobell et al. 2011; Wang et al. 2018). The chief component contributing to global temperature rise is the emission of greenhouse gases from various sources (Smith and Olesen 2010). High temperatures affect seed germination which is dependent on the intensity of stress and the plant species, affecting various physiological processes like photosynthesis, respiration, water relations and membrane stability. In addition, extreme temperatures alter the levels of hormones and primary and secondary metabolites, and increase ROS production (Wahid et al 2007). Temperature extremes reduce photosynthesis (Wahid et al. 2007; Khan et al. 2013) by disrupting the electron transport system and inactivating the oxygen-evolving enzymes of PSII, down-regulating the RuBP regeneration rate (Wise et al. 2004; Salvucci and Crafts-Brandner 2004). The inhibition of Rubisco enzyme activity mainly results from Rubisco activase inactivation which is highly sensitive to heat stress (Sharkey 2005; Salvucci and Crafts-Brandner 2004). High temperature stress affects physiology and biochemistry by affecting the transcriptional regulation of genes such as histones and DNA polymerases, and deregulating DNA methylation and activating transposons (Bita and Gerats 2013). The expression of genes in response to temperature extremes exhibits spatial and temporal variation; however genes related to the biosynthesis of storage compounds and cell growth are down-regulated (Mangelsen et al. 2011).

Exposure to high temperatures affects germination and seedling emergence and retards shoot and leaf development (Dash and Mohanty 2001). Deleterious effects of heat stress on seedling growth and leaf development are corelated with pigmentation sensitivity (Kariola et al. 2006) and PSII functioning (Khan et al. 2013). Different plant genotypes exhibit considerable variations in heat-stress tolerance (Ali et al. 1994; Dash and Mohanty 2001; Scheepens et al. 2018). High temperatures considerably reduce photosynthetic rate and stomatal conductance in *Glycine max* L. (Djanaguiraman et al. 2011) and sorghum (Yan et al. 2011). The decrease in photosynthesis is underlined by the structural and anatomical changes in the cell and organelles, mainly chloroplasts and mitochondria (Djanaguiraman et al. 2011). Temperature stress irreversibly affects various physiochemical processes and functions of the thylakoid membrane and chloroplast protein complexes including PSII (Brestic et al. 2012) and induces some reversible effects like increased photorespiration, and decreased activity of Rubisco and Rubisco activase (Mathur et al., 2014; Brestic et al. 2014, 2016). High temperatures also decrease pollen viability, pollen germination and seed set (Djanaguiraman et al. 2014).

Heat stress reduces photosynthetic performance and maximum quantum efficiency of PSII (Fv/Fm) reversibly when applied in the dark but causes significant de-epoxidation of the xanthophyll cycle pigments both under light and dark conditions (Buchner et al. 2015).The effect of temperature stress on photosynthesis and productivity varies with the plant species (C_3 and C_4) and the growth stage (Wang et al. 2016). During high temperatures the decreased Rubisco activation is associated with the biochemical limitation of the net CO_2 assimilation rate affecting the growth and biomass accumulation (Perdomo et al. 2017). The thickness of the leaf lamina, upper epidermis and palisade mesophyll increases, reflecting the higher assimilation of photosynthates (Yuan et al. 2017). Heat stress results in lipid peroxidation, inducing electrolyte leakage (Borriboon et al. 2018).

6.2.5 Allelopathy and Photosynthesis

Allelopathy may be any direct or indirect harmful or beneficial effect of one plant on other via the release of chemical compounds into the environment (Rice 1985). Allelochemicals in general are the secondary metabolites, containing different classes of chemical compounds like terpenoids, phenolics, organic cyanides and long chain fatty acids, etc. which influence several processes in the ecosystem by way of different mechanisms (Rizvi et al. 1992; Seigler 1996; Olofsdotter et al. 2002). Allelochemicals are released into the environment by way of foliar leaching, root exudation, residue decomposition, volatilization and debris incorporation (Inderjit and Keating 1999). Allelopathy is a natural phenomenon in which the production and release of the chemical compounds cause toxic effects in the receiving susceptible species by changing/modifying its molecular and biochemical processes against this toxic assault (Weir et al. 2004). Allelochemicals may affect growth and development by influencing the activity of different enzymes, cell division, cellular ultra-structure, membrane permeability and ion uptake, and have significant effects on photosynthesis and respiration as well (Gniazdowska and Bogatek 2005). Plants act together through release of chemical compounds/allelochemicals posing a negative impact on germination, growth, development and establishment of the receptor plant by altering various metabolic processes (Cruz-Ortega et al. 2007).

Patterson (1981) studied the effect of different phenolic compounds like caffeic acid, chlorogenic acid, t-cinnamic acid, p-coumaric acid, ferulic acid, gallic acid, p-hydroxybenzaldehyde, 5-sulfosalicylic acid, vanillic acid and vanillin at different concentrations and reported significant changes in the growth, photosynthesis, water relations and chlorophyll content of *Glycine max* L. Sorgoleone, a compound exudated by sorghum roots showing structural similarity with plastoquinone and acting as PSII inhibitor (Nimbal et al. 1996; Hejl and Koster 2004), alters photosynthetic activities such as the stomatal control over CO_2 supply, and light and dark reaction, leading to declined plant growth by affecting physiochemical processes including ion uptake and hydraulic conductivity (Blum et al. 1999). In addition sorgoleone has been suggested as an important inhibitor of p-hydroxyphenylpyruvate dioxygenase (HPPD) (reviewed in Dayan and Duke 2014), which is useful in the synthesis of α-tocopherol and plastoquinone. The inhibition of the HPPD enzyme results in decreased production of plastoquinone and affects the activity of phytoene desaturase (enzyme used in carotenoid synthesis), ultimately affecting photosynthesis (Meazza et al. 2002). Secondary metabolites including o-hydroxyphenyl acetic acid, ferulic acid and p-coumaric acid possess allelopathic properties while inhibiting the chlorophyll content of *Oryza sativa* cultivar TN67 (Yang et al. 2002). *Datura stramonium* leaf and seed extracts significantly reduced the chlorophyll contents of *Neonotonia wightii* and *Cenchrus ciliaris* (Elisante et al. 2013). Cellular pigments like chlorophyll a and carotenoids of *Phaeodacylum tricornutum* were considerably reduced by hydroquinone (HQ), and the ratio of carotenoids to chlorophyll 'a' showed increasing order under the extreme conditions caused by hydroquinone to resist environmental stresses (Yang et al. 2013). Allelochemicals may be helpful in controlling algal blooms; for example hydroquinone (HQ) has adverse effects on the growth, photosynthesis and other physiological activities of *Phaeodacylum tricornutum* (Yang et al. 2013). Under natural conditions plants influence each other via releasing bioactive or physiologically active compounds possessing inhibitory effects over other plants, and these compounds may be excreted from underground, above-ground plant organs or by the decomposition of plants parts. The treatment of aqueous extracts (15%) of *Mentha piperita* L. in *Helianthus annuus* L. showed stimulatory effects on the chlorophyll content and photochemical efficiency of PSII. With increasing concentration of peppermint extracts the photochemical and non-photochemical quenching and vitality index of PSII reduced (Skrzypek et al. 2015).

Various plastoquinone (PQ) analogue compounds competing with PQ for binding at the Q_B inhibit electron transfer from Q_A to Q_B (Wraight 1981). Johanningmeier et al. (1983) revealed that phenolic compounds, like dinoseb, inhibit the photosynthetic electron transport system in PSII particles isolated from spinach and *Chlamydomonas reinhardii* CW15 thylakoid. Later on, Roberts et al. (2003) reported the effects of 2,4,6-trinitrophenol (TNP) and other phenolic compounds like bromoxynil and dinoseb on PSII energetics. Phenolic compounds were bound to 90–95% of QB sites at saturating concentration reported in intact PSII. Moreover, allelochemicals adversely affect the stomatal conductance. Yu et al. (2003) conducted an allelopathic study on *Cucumis sativus* and revealed that allelopathic agents reduce leaf stomatal conductance, leaf transpiration and photosynthesis; however the exact mechanisms are largely unknown.

Benzoxaxolin-2-(*3H*)-one (BOA) is supposed to have toxic effects on physiology, including seed germination, growth and development (Chiapusio et al. 2004). BOA possesses allelopathic or phytotoxic potential and is mainly exuded by the members of poaceae like *Secale cereale*, *Zea mays* and *Triticum aestivum* (Batish et al. 2006) which not only affect the micro bodies but retard the growth of neighbouring plants also (Sanchez-Moreiras et al. 2011), reducing the net photosynthetic rate and quantum yield of PSII (Parizotto et al. 2017). Berberine, an allelochemical, enhances membrane permeability and induces oxidative damage in *M. aeruginosa* (Zhang et al. 2011) and represses the key photosynthesis genes such as psbD, psaA and psaB which ultimately induce disturbance in the algal photosynthetic process (Bi et al. 2012). Gao et al. (2018) studied in vivo four different phytotoxins, usnic acid (UA), salicylic acid, cinnamic acid and benzoic acid, which have multi-target negative effects on the photosynthetic process of *Chlamydomonas reinhardtii*, but the inhibition of PSII electron transport beyond QA at the acceptor side is the common original inhibiting site of photosynthesis in all cases. Furthermore UA decreases chlorophyll and carotenoid contents, also causing degradation of the D1/D2 proteins. SA destabilizes thylakoid membranes. Both CA and BA affect the photosynthetic process by reducing the efficiency of PSII electron transport. On the whole this decline in photosynthetic efficiency of plants under diverse environmental constraints might be responsible for the overall diminishment in plant growth and biomass production.

Analysis at the transcriptional level of the genome may provide evidence about the mode and action of allelochemicals and defence mechanisms against them. In soybean (*Glycine max*), the activity of genes was reported in response to environmental stresses (Wu et al. 2001; Hegab et al. 2016). Diallyl disulfide (DADS) is an important allelochemical found in *Allium sativum* L. which stimulates root growth in tomato. DADS treatment results in differential expression of genes (DEGs) involved in assimilatory sulphate reduction and glutathione metabolism, thereby increasing defensive activities in tomato (Cheng et al. 2016). Abdelmigid and Morsi (2017) have revealed DNA damage and variation in cysteine proteases at the transcriptional level by using a qualitative method of DNA fragmentation test (comet assay) in soybean under the influence of allelochemicals of litter produced by *Eucalyptus* trees.

6.3 ROLE OF OSMOLYTES UNDER STRESSFUL CONDITIONS

To avert the negative effects of different stresses plants have developed certain important adaptive mechanisms. One of the most important defensive mechanisms involved is the accumulation of compatible organic solutes (Serraj and Sinclair 2002; Ahanger et al. 2014). These solutes are low molecular weight, water-soluble organic compounds that are nontoxic at higher cellular concentrations and defend plants from unfavorable conditions through different ways like osmotic adjustment, protection of membranes, detoxification of ROS, enzyme/protein stabilization, etc. (Chen and Murata 2002; Hayat et al. 2012; Slama et al. 2015; Per et al. 2017; Annunziata et al. 2019). Yet their mechanism of action remains largely unknown regarding their contribution to stress tolerance. On the other hand, a few plants which have been genetically engineered express enzymes that catalyse the formation of various compatible solutes (Chen and Murata 2002; reviewed by Per et al. 2017). Used by cells and tissues under adverse environmental conditions to maintain cell volume, organic osmolytes include carbohydrates like sugars, polyols and derivatives, amino acids like glycine, proline and their derivatives and methylamines and methylsulfonium solutes including dimethylsulfonopropionate (Yancey 2005). They are synthesized at all developmental stages of the plant, and their biosynthesis is largely controlled by different environmental cues (Ahanger et al. 2014; Slama et al. 2015; Khan et al. 2014, 2015).

Glycine betaine (GB) is an important protective molecule synthesized by cells to maintain osmotic adjustment and is regarded as one of the most effective compatible solutes (Robinson and Jones 1986) involved in the stabilization of the oxygen-evolving complex of PSII and preventing the degradation of complex proteins, lipids and enzymes (Rajendrakumar et al. 1997; Sakamoto and Murata 2002). GB protects plants from Na interference by maintaining the osmotic equilibrium between the vacuole and cytoplasm (Subbarao et al. 2001), thereby improving plant tolerance to different environmental cues (Sakamoto and Murata 2002; Khan et al. 2009). The exogenous application and genetic engineering of GB metabolism improves plant tolerance to extreme conditions. Furthermore, under non-stress conditions, the accumulation of GB increases the yield potential (Chen and Murata 2008). GB accumulation imparts salt tolerance to transgenic cotton by protecting the cell membrane integrity and PSII activity, and GB-accumulating transgenic lines are of great agronomic value especially in soils ravaged by salinity (Zhang et al. 2009). It was noticed that the leaves of *Setaria italica* accumulate more GB under severe conditions than the root tissues (Ajithkumar and Panneerselvam 2013). The overaccumulation of GB in wheat (*Triticum aestivum* L.) lines resulting from the introduction of the betaine aldehyde dehydrogenase (BADH) gene enhances the salt tolerance by alleviating the impaired functions caused by salinity stress, particularly in photosynthetic apparatus and thylakoid membranes (Tian et al. 2017). Under different abiotic stresses GB accumulation has been associated with the signalling networks of most important phytohormones like abscisic acid, salicylic acid and ethylene (Xing and Rajashekar 2001; Jagendorf and Takabe 2001; Wang et al. 2010).

Proline is another important osmolyte involved in growth regulation under stress. Crop plants accumulating increased proline content exhibit greater stress tolerance (Agarwal et al. 2009; Ahanger et al. 2014; Ahanger and Agarwal 2017a, b). Exogenous proline application enhances leaf water potential in *Vicia faba* plants under salinity stress (Gadallah 1999) and protects plants against hazardous effects of stress (Hamilton and Heckathorn 2001). Compared with other osmolytes like GB, exogenous proline application was found effective in alleviating salinity-generated stress in tobacco cells (Ashraf and Foolad 2007). The application of proline exogenously also promotes K^+, Ca^+, P and N uptake in *Zea mays* plants under drought stress, resulting in improved growth and maintenance of the nutrient status (Ali et al. 2008). Proline alleviates the toxicity induced by cadmium in the growth of cultured tobacco Bright Yellow-2 (BY-2) cells (Islam et al. 2009). Furthermore, exogenous proline application overcomes the reduction in photosynthetic activity and leaf water relations in salt-stressed *Olea europaea* L. (Ben Ahmed et al. 2010). Compatible solutes like proline and trehalose play a protective role in the physiological responses of plants under unfavorable conditions, enabling them to better tolerate the ill effects of environmental stresses. Exogenously applied proline prevents salinity-induced reductions in growth and mineral nutrition by reducing the Na^+/K^+ ratio through improvement in the endogenous proline and transcript levels of pyrroline-5-carboxylatesynthetase (P5CS) and pyrroline-5-carboxylate reductase (P5CR), and a similar effect has been observed due to exogenous trehalose (Nounjan et al. 2012). In addition an exogenous osmoprotectant-mediated increase in osmolyte metabolism improves the growth recovery after stress release (Nounjan et al. 2012). Furthermore osmolyte application prevents oxidative damage by up-regulating the antioxidant system (Hasanuzzaman et al. 2014).

Proline accumulation in many plants under stress is a common response. Salinity stress enhances the proline accumulation in *Kosteletzkya virginica* seedlings and also up-regulates the expression of genes like *KvP5CS1*, *KvProT* and *KvOAT* depending on concentrations and time (Wang et al. 2015). As the key enzymes/genes for proline biosynthesis, the up-regulation of *KvP5CS1* played a more important role than *KvOAT* for proline accumulation under salt stress (Wang et al. 2015). Overexpression of the *SbPIP1* gene led to improved proline accumulation and synthesis of soluble sugars and played an important role in response to salt stress by reducing lipid peroxidation in halophytes and glycophytes, and such genes can be exploited to enhance the tolerance of important crops to salinity (Yu et al. 2015). Osmolytes are the major osmosis-regulating molecules in plants under extreme conditions and assist in quick adaptation to arid and semi-arid climatic conditions (Zhao et al. 2016). Salinity stress reduces photosynthetic pigments and total carbohydrates while increasing total soluble sugars, trehalose and proline contents in rice. Exogenous trehalose alleviates the harmful effects of salinity stress by enhancing the photosynthetic pigments and other physiological and biochemical parameters. A high solute concentration contributes to osmotic adjustment (Abdallah et al. 2016), leading to prevention of electrolyte leakage by bringing down the concentrations of ROS within normal ranges and hence avoiding the oxidative burst (Murmu et al. 2017). The accumulation of more osmolytes, like free sugars and proline, in *Pistacia vera* L. under salinity stress down-regulates sodium ion accumulation, lowering Na^+/K^+ in the shoots and maintaining nutrient contents, whereas a higher accumulation of toxic ions in the root prevents deleterious effects of ion toxicity in the leaf, thereby protecting the photosynthetic process (Rahneshan et al. 2018).

Djanaguiraman et al. (2005) have reported increased proline contents in sorghum and mung bean plants treated with leaf leachates of eucalyptus. In *Cassia occidentalis*, *Amaranthus viridis*, *Triticum aestivum*, *Pisum sativum* and *Cicer arietinum*, the alleviation of α-pinene-induced oxidative damage was found to correlate with enhanced accumulation of proline and antioxidant enzyme activity (Singh et al. 2006). Leaf leachates of *Cassia tora* induce proline accumulation in *Parthenium hysterophorus* leaves (Singh et al. 2006), and the generation of certain specific proteins against oxidative damage caused by allelopathic stress occurs (Mishra et al. 2006). Proline also acts as a metal chelator. Proline accumulation marks the regeneration of *Solanum nigrum* shoots from callus subjected to cadmium by protecting the plasma membrane through the reduction of ROS levels (Xu et al. 2009). Exogenous proline application enhanced its endogenous levels, which antagonized selenium toxicity by enhancing the growth of seedlings (Aggarwal et al. 2011).

The accumulation of osmolytes is regulated through molecular cascades. Environmental stresses primarily affect plasmalemma, leading to Ca^{2+} influx, cytoskeleton reorganization, up-regulation of mitogen-activated protein kinases (MAPK) and calcium-dependent protein kinases (CDPK), heat shock proteins (HSPs) and histidine kinase (HSK) (Sung et al. 2003). These signalling cascades ultimately lead to the generation of antioxidant molecules and compatible osmolytes for maintaining redox homeostasis, and cell water balance and osmotic adjustment (Bohnert et al. 2006). The stress-triggered accumulation of osmoprotectants regulates osmotic-dependent activities to protect cellular structures from extreme growth conditions by maintaining the cell–water balance, membrane stability and redox homeostasis (Farooq et al. 2008). The higher accumulation of carbohydrates like glucose and sucrose during heat stress has been considered as an important trait determining the tolerance potential of plants (Liu and Huang 2000), besides their antioxidant role in response to different stresses (Lang-Mladek et al. 2010). Sugar molecules regulate the signalling cascades at lower concentrations while carrying out ROS scavenging at higher concentrations (Sugio et al. 2009). A greater accumulation of osmolytes significantly regulates the plant growth under stressful conditions, integrating signalling events leading to the elicitation of efficient biochemical and physiological responses; however the actual/exact mechanisms are not known. The molecular mechanisms leading to the protection of the structural and functional capacity of the photosynthetic apparatus through osmolyte synthesis need an extensive research effort.

Sugars are produced by the photosynthesis mechanism in plants which regulates growth and development (Smeekens, 2000). Sugars like fructose, sucrose and trehalose act as osmoprotectants in maintaining osmotic adjustment, protecting membrane structure and functioning by scavenging ROS (Keunen et al. 2013; Singh et al. 2015). Amino acids are considered as precursors to and constituents of proteins and nucleic acids and affect the synthesis and activity of some enzymes, gene expression and redox-homeostasis (Rai 2002). Amino acid accumulation under salinity helps in maintaining K^+/Na^+ and transport across cells (Cuin and Shabala 2005). Amino acids provide resistance against various stresses by maintaining the cellular homeostasis, redox and energy status of plants (Szabados and Savoure 2010), and as osmolytes lead to the maintenance of ion transport, stomatal movements and heavy metal detoxification.

6.4 CONCLUSION

Environmental stresses significantly reduce photosynthetic performance through stomatal or non-stomatal limitations. Alterations in the electron transport chain through the overproduction of ROS and uneven modifications in the composition of the thylakoid membrane lead to the dysfunction of cellular components; however, the accumulation of osmolytes has a significant role in safeguarding the cellular structure and functioning by maintaining the intracellular levels of ROS, redox homeostasis and hence the photosynthetic efficiency (Figure 6.2). Osmolytes like proline and GB have a significant role in maintaining the osmotic adjustment under adverse stress conditions and enhancing the endogenous levels of osmolytes when applied exogenously, exhibiting variations with plant species and type as well as the severity of stress. Based on the physiological and biochemical reports presented in the present review it is imperative to stride towards the unravelling of the molecular and genetic regulation of osmolyte metabolism vis-à-vis growth regulation under stress conditions.

REFERENCES

Abdallah, M.S., Abdelgawad, Z.A. and El-Bassiouny, H.M.S. 2016. Alleviation of the adverse effects of salinity stress using trehalose in two rice varieties. *South African Journal of Botany* 103: 275–282.

Abdelmigid, H.M. and Morsi, M.M. 2017. Cytotoxic and molecular impacts of allelopathic effects of leaf residues of *Eucalyptus globulus* on soybean (*Glycine max*). *Journal* of Genetic Engineering *and* Biotechnology 15(2): 297–302.

Acosta-Motos, J.R., Hernandez, J.A., Alvarez, S., et al. 2017a. Long-term resistance mechanisms and irrigation critical threshold showed by *Eugenia myrtifolia* plants in response to saline reclaimed water and relief capacity. *Plant Physiology and Biochemistry* 111: 244–256.

Acosta-Motos, J., Ortuno, M., Bernal-Vicente, A., et al. 2017b. Plant responses to salt stress: Adaptive mechanisms. *Agronomy* 7(1): 18.

Agarwal, R.M., Pandey, R. and Gupta, S. 1999. Certain aspects of water stress induced changes and tolerance mechanisms in plants. *Journal of the Indian Botanical Society* 78: 255–269.

Agarwal, R.M., Tomar, N.S., Jatav, K.S., et al. 2009. Potassium induced changes in flowering plants. In: *Flower Retrospect and Prospect (Professsor Vishwambhar Puri Birth Centenary Volume)*. Ed. Y. Vimala. SR Scientific Publication, Delhi, pp. 158–186.

Aggarwal, M., Sharma, S., Kaur, N., et al. 2011. Exogenous proline application reduces phytotoxic effects of selenium by minimising oxidative stress and improves growth in bean (*Phaseolus vulgaris* L.) seedlings. *Biological Trace Element Research* 140(3): 354–367.

Ahanger, M.A. and Agarwal, R.M. 2017a. Potassium up-regulates antioxidant metabolism and alleviates growth inhibition under water and osmotic stress in wheat (*Triticum aestivum* L). *Protoplasma* 254(4): 1471–1486.

Ahanger, M.A. and Agarwal, R.M. 2017b. Salinity stress induced alterations in antioxidant metabolism and nitrogen assimilation in wheat (*Triticum aestivum* L) as influenced by potassium supplementation. *Plant Physiology and Biochemistry* 115: 449–460.

Ahanger, M.A. and Ahmad, P. 2019. Role of mineral nutrients in abiotic stress tolerance – revisiting the associated signalling mechanisms. In: *Plant Signalling Molecules*. Eds. M.I.R. Khan, P.S. Reddy, A. Ferrante and N.A. Khan. Elsevier Academic Press, USA, pp. 269–285.

Ahanger, M.A., Alyemeni, M.N., Wijaya, L., Alamri, S.A., Alam, P., Ashraf, M. and Ahmad, P. 2018a. Potential of exogenously sourced kinetin in protecting *Solanum lycopersicum* from NaCl-induced oxidative stress through up-regulation of the antioxidant system, ascorbate-glutathione cycle and glyoxalase system. *PLoS ONE* 13(9): e0202175. doi:10.1371/journal.pone.0202175.

Ahanger, M.A., Gul, F., Ahmad, P. and Akram, N.A. 2018b. Environmental stresses and metabolomics - deciphering the role of stress responsive metabolites. In: *Plant Metabolites and Regulation under Environmental Stress*. Eds. P. Ahmad, M.A. Ahanger, V.P. Singh, D.K. Tripathi, P. Alam and M.N. Alyemeni. Academic Press, pp. 53–67.

Ahanger, M.A., Tomar, N.S., Tittal, M., Argal, S. and Agarwal, R.M. 2017a. Plant growth under water/salt stress: ROS production; antioxidants and significance of added potassium under such conditions. *Physiology and Molecular Biology of Plants* 23(4): 731–744.

Ahanger, M.A., Akram, N.A., Ashraf, M., Alyemeni, M.N., Wijaya, L. and Ahmad, P. 2017b. Signal transduction and biotechnology in response to environmental stresses. *Biologia Plantarum* 61(3): 401–416.

Ahanger, M.A., Akram, N.A., Ashraf, M., Alyemni, M.N., Wijaya, L. and Ahmad, P. 2017c. Plant responses to environmental stresses – from gene to biotechnology. *Annals of Botany (AoB) Plants* 9(4): PLX025. doi:10.1093/aobpla/plx025.

Ahanger, M.A., Tittal, M., Mir, R.A. and Agarwal, R.M. 2017d. Alleviation of water and osmotic stress-induced changes in nitrogen metabolizing enzymes in *Triticum aestivum* L. cultivars by potassium. *Protoplasma* 254(5): 1953–1963.

Ahanger, M.A., Tyagi, S.R., Wani, M.R. and Ahmad, P. 2014. Drought tolerance: Roles of organic osmolytes, growth regulators and mineral nutrients. In: *Physiological Mechanisms and Adaptation Strategies in Plants under Changing Environment*, Volume Ist. Eds. P. Ahmad and M.R. Wani. Springer Science+Business Media, Inc., pp. 25–56.

Ahmad, P., Ahanger, M.A., Alyemeni, M.N., et al. 2018. Exogenous application of nitric oxide modulates osmolyte metabolism, antioxidants, enzymes of ascorbate-glutathione cycle and promotes growth under cadmium stress in tomato. *Protoplasma* 255(1): 79–93.

Ahmad, P., Jaleel, C.A., Salem, M.A., et al. 2010. Roles of enzymatic and non-enzymatic antioxidants in plants during abiotic stress. *Critical Reviews in Biotechnology* 30: 161–175.

Ajithkumar, I.P. and Panneerselvam, R. 2013. Osmolyte accumulation, photosynthetic pigment and growth of *Setaria italica* (L.) P. Beauv. under drought stress. *Asian Pacific Journal of Reproduction* 2(3): 220–224.

Alagoz, S.M., Toorchi, M. and Bandehagh, A. 2016. Canola seedling response to NaCl stress-a proteomic approach. *Notulae Botanicae Horti Agrobotanici Cluj-Napoca* 44(2): 361–366.

Ali, H., Khan, E. and Sajad, M.A. 2013. Phytoremediation of heavy metals—concepts and applications. *Chemosphere* 91(7): 869–881.

Ali, Q., Ashraf, M., Shahbaz, M. and Humera, H. 2008. Ameliorating effect of foliar applied proline on nutrient uptake in water stressed maize (*Zea mays* L.) plants. *Pakistan Journal of Botany* 40(1): 211–219.

Ali, Y., Aslam, Z., Ashraf, M.Y. and Tahir, G.R. 2004. Effect of salinity on chlorophyll concentration, leaf area, yield and yield components of rice genotypes grown under saline environment. *International Journal of Environmental Science and Technology* 1(3): 221–225.

Ali, Z.I., Mahalakshmi, V., Singh, M., et al. 1994. Variation in cardinal temperatures for germination among wheat (*Triticum aestivum*) genotypes. *Annals of Applied Biology* 125(2): 367–375.

Allen, D.J. and Ort, D.R. 2001. Impacts of chilling temperatures on photosynthesis in warm-climate plants. *Trends in Plant Science* 6(1): 36–42.

Amuthavalli, P. and Sivasankaramoorthy, S. 2012. Effect of salt stress on the growth and photosynthetic pigments of pigeon pea (*Cajanus cajan*). *Journal of Applied Pharmaceutical Science* 2(11): 131–133.

Apel, K. and Hirt, H. 2004. Reactive oxygen species: Metabolism, oxidative stress, and signal transduction. *Annual Review of Plant Biology* 55: 373–399.

Aragao, M.E.F.D., Guedes, M.M., Otoch, M.D.L.O., et al. 2005. Differential responses of ribulose-1, 5-bisphosphate carboxylase/oxygenase activities of two *Vigna unguiculata* cultivars to salt stress. *Brazilian Journal of Plant Physiology* 17(2): 207–212.

Asgher, M., Khan, N.A., Khan, M.I.R., et al. 2014. Ethylene production is associated with alleviation of cadmium-induced oxidative stress by sulfur in mustard types differing in ethylene sensitivity. *Ecotoxicology and Environmental Safety* 106: 54–61.

Ashraf, M.F.M.R. and Foolad, M. 2007. Roles of glycine betaine and proline in improving plant abiotic stress resistance. *Environmental and Experimental Botany* 59(2): 206–216.

Ashraf, M.P.J.C. and Harris, P.J.C. 2004. Potential biochemical indicators of salinity tolerance in plants. *Plant Science* 166(1): 3–16.

Ayeni, O., Ndakidemi, P., Snyman, R. and Odendaal, J. 2012. Assessment of metal concentrations, chlorophyll content and photosynthesis in phragmites australis along the Lower Diep River, CapeTown, South Africa. *Energy and Environment Research* 2(1): 128.

Babu, N.G., Sarma, P.A., Attitalla, I.H. and Murthy, S.D.S. 2010. Effect of selected heavy metal ions on the photosynthetic electron transport and energy transfer in the thylakoid membrane of the cyanobacterium, *Spirulina platensis. Acaemicd Journal of Plant Sciences* 3: 46–49.

Basile, A., Sorbo, S., Conte, B., et al. 2012. Toxicity, accumulation, and removal of heavy metals by three aquatic macrophytes. *International Journal of Phytoremediation* 14(4): 374–387.

Batish, D.R., Singh, H.P., Setia, N., Kaur, S. and Kohli, R.K. 2006. 2-Benzoxazolinone (BOA) induced oxidative stress, lipid peroxidation and changes in some antioxidant enzyme activities in mung bean (*Phaseolus aureus*). *Plant Physiology and Biochemistry* 44(11–12): 819–827.

Bell, F.G., Bullock, S.E.T., Halbich, T.F.J. and Lindsay, P. 2001. Environmental impacts associated with an abandoned mine in the Witbank Coalfield, South Africa. *International Journal of Coal Geology* 45(2–3): 195–216.

Ben Ahmed, C., Ben Rouina, B., Sensoy, S., Boukhriss, M. and Ben Abdullah, F. 2010. Exogenous proline effects on photosynthetic performance and antioxidant defense system of young olive tree. *Journal of Agricultural and Food Chemistry* 58(7): 4216–4222.

Benavides, M.P., Gallego, S.M. and Tomaro, M.L. 2005. Cadmium toxicity in plants. *Brazilian Journal of Plant Physiology* 17(1): 21–34.

Bertrand, M. and Schoefs, B. 1999. Photosynthetic pigment metabolism in plants during stress. In: *Handbook of Plant and Crop Stress*. Ed. M. Pessarkli. Marcel Dekker, New York, pp. 527–544.

Bharwana, S.A., Ali, S., Farooq, M.A., Iqbal, N., Abbas, F. and Ahmad, M.S.A. 2013. Alleviation of lead toxicity by silicon is related to elevated photosynthesis, antioxidant enzymes suppressed lead uptake and oxidative stress in cotton. *Journal of Bioremediation and Biodegradation* 4(4): 187.

Bi, X.D., Zhang, S.L., Zhang, B., et al. 2012. Effects of berberine on the photosynthetic pigments compositions and ultrastructure of Cyanobacterium *Microcystis aeruginosa*. In: *Advanced Materials Research,* volume 343. Ed. D. Wang. Trans Tech Publications, pp. 1117–1125.

Biswal, B., Raval, M.K., Biswal, U.C. and Joshi, P. 2008. Response of photosynthetic organelles to abiotic stress: Modulation by sulfur metabolism. In: *Sulfur Assimilation and Abiotic Stress in Plants*. Eds. N.A. Khan, S. Singh and S. Umar. Springer, Berlin, Heidelberg, pp. 167–191.

Bita, C.E. and Gerats, T. 2013. Plant tolerance to high temperature in a changing environment: Scientific fundamentals and production of heat stress-tolerant crops. *Frontiers in Plant Science* 4: 273.

Blum, A., Ribaut, J.M. and Poland, D. 1999. Towards standard assays of drought resistance in crop plants, Molecular approaches for the genetic improvement of cereals for stable production in water-limited environments. *International Workshop*, June 1999, CIMMYT, Mexico, pp. 29–35.

Bohnert, H.J., Gong, Q., Li, P. and Ma, S. 2006. Unraveling abiotic stress tolerance mechanisms–getting genomics going. *Current Opinion in Plant Biology* 9(2): 180–188.

Bonales-Alatorre, E., Pottosin, I., Shabala, L., et al. 2013. Differential activity of plasma and vacuolar membrane transporters contributes to genotypic differences in salinity tolerance in a Halophyte species, *Chenopodium quinoa. International Journal of Molecular Sciences* 14(5): 9267–9285. doi:10.3390/ijms14059267.

Borriboon, W., Lontom, W., Pongdontri, P., Theerakulpisut, P. and Dongsansuk, A. 2018. Effects of short-and long-term temperature on seed germination, oxidative stress and membrane stability of three rice cultivars (Dular, KDML105 and Riceberry). *Pertanika Journal of Tropical Agricultural Science* 41(1): 151–162.

Brestic, M., Zivcak, M., Kalaji, H.M., Carpentier, R. and Allakhverdiev, S.I. 2012. Photosystem II thermostability in situ: Environmentally induced acclimation and genotype-specific reactions in *Triticum aestivum* L. *Plant Physiology and Biochemistry* 57: 93–105.

Brestic, M., Zivcak, M., Kunderlikova, K. and Allakhverdiev, S.I. 2016. High temperature specifically affects the photoprotective responses of chlorophyll b-deficient wheat mutant lines. *Photosynthesis Research* 130(1–3): 251–266.

Brestic, M., Zivcak, M., Olsovska, K., Kalaji, H.M., Shao, H. and Hakeem, K.R. 2014. Heat signaling and stress responses in photosynthesis. In: *Plant Signaling: Understanding the Molecular Crosstalk*. Eds. K. Hakeem, R. Rehman and I. Tahir. Springer, New Delhi, pp. 241–256.

Buchner, O., Stoll, M., Karadar, M., Kranner, I. and Neuner, G. 2015. Application of heat stress in situ demonstrates a protective role of irradiation on photosynthetic performance in alpine plants. *Plant, Cell and Environment* 38(4): 812–826.

Ceccarelli, S., Grando, S., Maatougui, M., et al. 2010. Plant breeding and climate changes. *The Journal of Agricultural Science* 148(6): 627–637.

Chen, T.H. and Murata, N. 2008. Glycinebetaine: An effective protectant against abiotic stress in plants. *Trends in Plant Science* 13(9): 499–505.

Cheng, F., Cheng, Z.H. and Meng, H.W. 2016. Transcriptomic insights into the allelopathic effects of the garlic allelochemical diallyl disulfide on tomato roots. *Scientific Reports* 6: 38902.

Chiapusio, G., Pellissier, F. and Gallet, C. 2004. Uptake and translocation of phytochemical 2-benzoxazolinone (BOA) in radish seeds and seedlings. *Journal of Experimental Botany* 55(402): 1587–1592.

Cruz-Ortega, R., Lara-Núñez, A. and Anaya, A.L. 2007. Allelochemical stress can trigger oxidative damage in receptor plants: Mode of action of phytotoxicity. *Plant Signaling and Behavior* 2(4): 269–270.

Cuin, T.A. and Shabala, S. 2005. Exogenously supplied compatible solutes rapidly ameliorate NaCl-induced potassium efflux from barley roots. *Plant Cell Physiology* 46: 1924–1933.

Da-lin, L., Kai-qi, H., Jing-jing, M., Wei-wei, Q., Xiu-ping, W. and Shu-pan, Z. 2011. Effects of cadmium on the growth and physiological characteristics of sorghum plants. *African Journal of Biotechnology* 10(70): 15770–15776.

Dash, S. and Mohanty, N. 2001. Evaluation of assays for the analysis of thermo-tolerance and recovery potentials of seedlings of wheat (*Triticum aestivum* L.) cultivars. *Journal of Plant Physiology* 158(9): 1153–1165.

Dayan, F.E. and Duke, S.O. 2014. Natural compounds as next-generation herbicides. *Plant Physiology* 166(3): 1090–1105.

Dietz, K.J., Turkan, I. and Krieger-Liszkay, A. 2016. Redox-and reactive oxygen species-dependent signaling into and out of the photosynthesizing chloroplast. *Plant Physiology* 171(3): 1541–1550.

Djanaguiraman, M., Prasad, P.V., Boyle, D.L. and Schapaugh, W.T. 2011. High-temperature stress and soybean leaves: Leaf anatomy and photosynthesis. *Crop Science* 51(5): 2125–2131.

Djanaguiraman, M., Prasad, P.V., Murugan, M., Perumal, R. and Reddy, U.K. 2014. Physiological differences among sorghum (*Sorghum bicolor* L. Moench) genotypes under high temperature stress. *Environmental and Experimental Botany* 100: 43–54.

Djanaguiraman, M., Vaidyanathan, R., Annie Sheeba, J., Durga Devi, D. and Bangarusamy, U. 2005. Physiological responses of Eucalyptus globulus leaf leachate on seedling physiology of rice, sorghum and blackgram. *International Journal of Agriculture and Biology* 7: 35–38.

Dong, J., Wu, F. and Zhang, G. 2006. Influence of cadmium on antioxidant capacity and four microelement concentrations in tomato seedlings (*Lycopersicon esculentum*). *Chemosphere* 64(10): 1659–1666.

Eitzinger, J., Orlandini, S., Stefanski, R. and Naylor, R.E.L. 2010. Climate change and agriculture: Introductory editorial. *Journal of Agricultural Science, Cambridge* 148: 499–500.

Elisante, F., Tarimo, M.T. and Ndakidemi, P.A. 2013. Allelopathic effect of seed and leaf aqueous extracts of *Datura stramonium* on leaf chlorophyll content, shoot and root elongation of *Cenchrus ciliaris* and *Neonotonia wightii*. *American Journal of Plant Sciences* 4(12): 2332.

Farooq, M., Basra, S.M.A., Wahid, A., Cheema, Z.A., Cheema, M.A. and Khaliq, A. 2008. Physiological role of exogenously applied glycinebetaine to improve drought tolerance in fine grain aromatic rice (*Oryza sativa* L.). *Journal of Agronomy and Crop Science* 194(5): 325–333.

Fatma, M., Asgher, M., Masood, A. and Khan, N.A. 2014. Excess sulfur supplementation improves photosynthesis and growth in mustard under salt stress through increased production of glutathione. *Environmental and Experimental Botany* 107: 55–63.

Flexas, J., Bota, J., Escalona, J.M., Sampol, B. and Medrano, H. 2002a. Effects of drought on photosynthesis in grapevines under field conditions: An evaluation of stomatal and mesophyll limitations. *Functional Plant Biology* 29(4): 461–471.

Flexas, J., Briantais, J.M., Cerovic, Z., Medrano, H. and Moya, I. 2000. Steady-state and maximum chlorophyll fluorescence responses to water stress in grapevine leaves: A new remote sensing system. *Remote Sensing of Environment* 73(3): 283–297.

Gadallah, M.A.A. 1999. Effects of proline and glycinebetaine on *Vicia faba* responses to salt stress. *Biologia Plantarum* 42(2): 249–257.

Gao, Y., Liu, W., Wang, X., et al. 2018. Comparative phytotoxicity of usnic acid, salicylic acid, cinnamic acid and benzoic acid on photosynthetic apparatus of *Chlamydomonas reinhardtii*. *Plant Physiology and Biochemistry* 128: 1–12.

Ghobadi, M., Taherabadi, S., Ghobadi, M.E., Mohammadi, G.R. and Jalali-Honarmand, S. 2013. Antioxidant capacity, photosynthetic characteristics and water relations of sunflower (*Helianthus annuus* L.) cultivars in response to drought stress. *Industrial Crops and Products* 50: 29–38.

Gniazdowska, A. and Bogatek, R. 2005. Allelopathic interactions between plants. Multi site action of allelochemicals. *Acta Physiologiae Plantarum* 27(3): 395–407.

Goltsev, V., Zaharieva, I., Chernev, P., et al. 2012. Drought-induced modifications of photosynthetic electron transport in intact leaves: Analysis and use of neural networks as a tool for a rapid non-invasive estimation. *Biochimica et Biophysica Acta (BBA)-Bioenergetics* 1817(8): 1490–1498.

Gomes, M.A.D.C., Suzuki, M.S., Cunha, M.D. and Tullii, C.F. 2011. Effect of salt stress on nutrient concentration, photosynthetic pigments, proline and foliar morphology of *Salvinia auriculata* Aubl. *Acta Limnologica Brasiliensia* 23(2): 164–176.

Gratao, P.L., Polle, A., Lea, P.J. and Azevedo, R.A. 2005. Making the life of heavy metal-stressed plants a little easier. *Functional Plant Biology* 32(6): 481–494.

Hamilton, E.W. and Heckathorn, S.A. 2001. Mitochondrial adaptations to NaCl. Complex I is protected by anti-oxidants and small heat shock proteins, whereas complex II is protected by proline and betaine. *Plant Physiology* 126(3): 1266–1274.

Hasanuzzaman, M., Alam, M.M., Rahman, A., et al. 2014. Exogenous proline and glycine betaine mediated upregulation of antioxidant defense and glyoxalase systems provides better protection against salt-induced oxidative stress in two rice (*Oryza sativa* L.) varieties. *BioMed Research International* 757219. doi:10.1155/2014/757219.

Hatfield, J.L., Boote, K.J., Kimball, B.A., et al. 2011. Climate impacts on agriculture: Implications for crop production. *Agronomy Journal* 103(2): 351–370.

Hayat, S., Hayat, Q., Alyemeni, M.N., et al. 2012. Role of proline under changing environments: A review. *Plant Signal Behavior* 7(11): 1456–1466.

Hegab, M.M., Gabr, M.A., Al-Wakeel, S.A. and Hamed, B.A. 2016. Allelopathic potential of *Eucalyptus rostrata* leaf residue on some metabolic activities of *Zea mays* L. *University Journal of Plant Science* 4(2): 11–21.

Heidari, M. 2012. Effects of salinity stress on growth, chlorophyll content and osmotic components of two basil (*Ocimum basilicum* L.) genotypes. *African Journal of Biotechnology* 11(2): 379–384.

Hejl, A.M. and Koster, K.L. 2004. The allelochemical sorgoleone inhibits root H+-ATPase and water uptake. *Journal of Chemical Ecology* 30(11): 2181–2191.

Horton, P., Ruban, A.V. and Walters, R.G. 1996. Regulation of light harvesting in green plants. *Annual Review of Plant Biology* 47(1): 655–684.

Hu, T., Yi, H., Hu, L. and Fu, J. 2013. Stomatal and metabolic limitations to photosynthesis resulting from NaCl stress in perennial ryegrass genotypes differing in salt tolerance. *Journal of the American Society for Horticultural Science* 138(5): 350–357.

Huang, M., Zhu, H., Zhang, J., et al. 2017. Toxic effects of cadmium on tall fescue and different responses of the photosynthetic activities in the photosystem electron donor and acceptor sides. *Scientific Reports* 7(1): 14387.

Inderjit and Keating, K.I. 1999. Allelopathy: Principles, procedures, processes, and promises for biological control. *Advances in Agronomy* 67: 141–231.

Islam, M.M., Hoque, M.A., Okuma, E., et al. 2009. Exogenous proline and glycinebetaine increase antioxidant enzyme activities and confer tolerance to cadmium stress in cultured tobacco cells. *Journal of Plant Physiology* 166(15): 1587–1597.

Jagendorf, A.T. and Takabe, T. 2001. Inducers of glycinebetaine synthesis in barley. *Plant Physiology* 127(4): 1827–1835.

Jajoo, A. 2013. Changes in photosystem II in response to salt stress. In: *Ecophysiology and Responses of Plants under Salt Stress*. Eds. P. Ahmad, M.M. Azooz and M.N.V. Prasad. Springer, New York, pp. 149–168.

Johanningmeier, U., Neumann, E. and Oettmeier, W. 1983. Interaction of a phenolic inhibitor with photosystem II particles. *Journal of Bioenergetics and Biomembranes* 15(2): 43–66.

Kariola, T., Brader, G., Helenius, E., et al. 2006. Early responsive to dehydration 15, a negative regulator of abscisic acid responses in *Arabidopsis*. *Plant Physiology* 142(4): 1559–1573.

Keunen, E., Peshev, D., Vangronsveld, J., Ende, W. and Cuypers, A. 2013. Plant sugars are crucial players in the oxidative challenge during abiotic stress: Extending the traditional concept. *Plant Cell and Environment* 36: 1242–1255.

Khan, M.I.R., Asgher, M. and Khan, N.A. 2014. Alleviation of salt-induced photosynthesis and growth inhibition by salicylic acid involves glycine betaine and ethylene in mungbean (*Vigna radiata* L.). *Plant Physiology Biochemistry* 80: 67–74.

Khan, M.I.R., Iqbal, N., Masood, A., et al. 2013. Salicylic acid alleviates adverse effects of heat stress on photosynthesis through changes in proline production and ethylene formation. *Plant Signalling Behaviour* 8: 1–10. doi:10.4161/psb.26374.

Khan, M.I.R., Nazir, F., Asgher, M., et al. 2015. Selenium and sulfur influence ethylene formation and alleviate cadmium-induced oxidative stress by improving proline and glutathione production in wheat (*Triticum aestivum* L.). *Journal of Plant Physiology* 173: 9–18.

Khan, M.S., Yu, X., Kikuchi, A., Asahina, M. and Watanabe, K.N. 2009. Genetic engineering of glycine betaine biosynthesis to enhance abiotic stress tolerance in plants. *Plant Biotechnology* 26(1): 125–134.

Khan, N.A., Asgher, M., Per, T.S., et al. 2016. Ethylene potentiates sulfur-mediated reversal of cadmium inhibited photosynthetic responses in mustard. *Frontiers in Plant Science* 7: 1628. doi:10.3389/fpls.2016.01628.

Lamhamdi, M., El Galiou, O., Bakrim, A., et al. 2013. Effect of lead stress on mineral content and growth of wheat (*Triticum aestivum*) and spinach (*Spinacia oleracea*) seedlings. *Saudi Journal of Biological Sciences* 20(1): 29–36.

Lang-Mladek, C., Popova, O., Kiok, K., et al. 2010. Transgenerational inheritance and resetting of stress-induced loss of epigenetic gene silencing in *Arabidopsis*. *Molecular Plant* 3(3): 594–602.

Lawlor, D.W. 1995. The effects of water deficit on photosynthesis. In: *Environment and Plant Metabolism. Flexibility and Acclimation.* Ed. N. Smirnoff. BIOS Scientific Publisher, Oxford, pp. 129–160.

Lee, S.Y., Damodaran, P.N. and Roh, K.S. 2014. Influence of salicylic acid on rubisco and rubisco activase in tobacco plant grown under sodium chloride in vitro. *Saudi Journal of Biological Sciences* 21(5): 417–426.

Linger, P., Ostwald, A. and Haensler, J. 2005. *Cannabis sativa* L. growing on heavy metal contaminated soil: Growth, cadmium uptake and photosynthesis. *Biologia Plantarum* 49(4): 567–576.

Liu, X. and Huang, B. 2000. Carbohydrate accumulation in relation to heat stress tolerance in two creeping bentgrass cultivars. *Journal of the American Society for Horticultural Science* 125(4): 442–447.

Lobell, D.B., Schlenker, W. and Costa-Roberts, J. 2011. Climate trends and global crop production since 1980. *Science* 333(6042): 616–620.

Mangelsen, E., Kilian, J., Harter, K., Jansson, C., Wanke, D. and Sundberg, E. 2011. Transcriptome analysis of high-temperature stress in developing barley caryopses:early stress responses and effects on storage compound biosynthesis. *Molecular Plant* 4: 97–115.

Maroco, J.P., Rodrigues, M.L., Lopes, C. and Chaves, M.M. 2002. Limitations to leaf photosynthesis in field-grown grapevine under drought-metabolic and modelling approaches. *Functional Plant Biology* 29(4): 451–459.

Maron, J.L. and Crone, E. 2006. Herbivory: Effects on plant abundance, distribution and population growth. *Proceedings of the Royal Society B: Biological Sciences* 273(1601): 2575–2584.

Maron, J.L. and Kauffman, M.J. 2006. Habitat-specific impacts of multiple consumers on plant population dynamics. *Ecology* 87(1): 113–124.

Meazza, G., Scheffler, B.E., Tellez, M.R., et al. 2002. The inhibitory activity of natural products on plant p-hydroxyphenylpyruvate dioxygenase. *Phytochemistry* 60(3): 281–288.

Meng, L.L., Song, J.F., Wen, J., Zhang, J. and Wei, J.H. 2016. Effects of drought stress on fluorescence characteristics of photosystem II in leaves of *Plectranthus scutellarioides*. *Photosynthetica* 54: 414–421.

Mishra, A.N., Latowski, D. and Strzalka, K. 2006. The xanthophyll cycle activity in kidney bean and cabbage leaves under salinity stress. *Russian Journal of Plant Physiology* 53(1): 102–109.

Mittler, R., Vanderauwera, S., Gollery, M. and Van Breusegem, F. 2004. Reactive oxygen gene network of plants. *Trends in Plant Science* 9(10): 490–498.

Mordecai, E.A. 2011. Pathogen impacts on plant communities: Unifying theory, concepts, and empirical work. *Ecological Monographs* 81(3): 429–441.

Munns, R. and Gilliham, M. 2015. Salinity tolerance of crops–what is the cost? *New Phytologist* 208(3): 668–673.

Murillo-Amador, B., Yamada, S., Yamaguchi, T., et al. 2007. Influence of calcium silicate on growth, physiological parameters and mineral nutrition in two legume species under salt stress. *Journal of Agronomy and Crop Science* 193(6): 413–421.

Murmu, K., Murmu, S., Kundu, C.K. and Bera, P.S. 2017. Exogenous proline and glycine betaine in plants under stress tolerance. *International Journal of Current Microbiology and Applied Sciences* 6: 901–913.

Mustafavi, S.H., Shekari, F., Hatami-Maleki, H. and Nasiri, Y. 2016. Effect of water stress on some quantitative and qualitative traits of valerian (*Valeriana officinalis* L.) plants. *Bulletin of the University of Agricultural Sciences and Veterinary Medicine Cluj-Napoca. Horticulture* 73(1): 9–16.

Nagajyoti, P.C., Lee, K.D. and Sreekanth, T.V.M. 2010. Heavy metals, occurrence and toxicity for plants: A review. *Environmental Chemistry Letters* 8(3): 199–216.

Nazar, R., Iqbal, N., Syeed, S. and Khan, N.A. 2011. Salicylic acid alleviates decreases in photosynthesis under salt stress by enhancing nitrogen and sulfur assimilation and antioxidant metabolism differentially in two mungbean cultivars. *Journal of Plant Physiology* 168: 807–815.

Neto, A.J.S., Lopes, D.C., Pinto, F.A. and Zolnier, S. 2017. Vis/NIR spectroscopy and chemometrics for non-destructive estimation of water and chlorophyll status in sunflower leaves. *Biosystems Engineering* 155: 124–133.

Nimbal, C.I., Yerkes, C.N., Weston, L.A. and Weller, S.C. 1996. Herbicidal activity and site of action of the natural product sorgoleone. *Pesticide Biochemistry and Physiology* 54(1): 73–83.

Noctor, G., Veljovic-Jovanovic, S.O.N.J.A., Driscoll, S., Novitskaya, L. and Foyer, C.H. 2002. Drought and oxidative load in the leaves of C3 plants: A predominant role for photorespiration? *Annals of Botany* 89(7): 841–850.

Nounjan, N., Nghia, P.T. and Theerakulpisut, P. 2012. Exogenous proline and trehalose promote recovery of rice seedlings from salt-stress and differentially modulate antioxidant enzymes and expression of related genes. *Journal of Plant Physiology* 169(6): 596–604.

Omoto, E., Taniguchi, M. and Miyake, H. 2010. Effects of salinity stress on the structure of bundle sheath and mesophyll chloroplasts in NAD-Malic enzyme and PCK Type C4 plants. *Plant Production Science* 13(2): 169–176.

Olofsdotter, M., Jensen, L.B. and Courtois, B. 2002. Improving crop competitive ability using allelopathy – an example from rice. *Plant Breeding* 121: 1–9.

Pandey, R., Agarwal, R.M., Jeevaratnam, K., et al. 2004. Osmotic stress-induced alterations in rice (*Oryza sativa* L.) and recovery on stress release. *Plant Growth Regulation* 42: 79–87.

Parizotto, A.V., Marchiosi, R., Bubna, G.A., et al. 2017. Benzoxazolin-2-(3H)-one reduces photosynthetic activity and chlorophyll fluorescence in soybean. *Photosynthetica* 55(2): 386–390.

Passariello, B., Giuliano, V., Quaresima, S., et al. 2002. Evaluation of the environmental contamination at an abandoned mining site. *Microchemical Journal* 73(1–2): 245–250.

Patterson, D.T. 1981. Effects of allelopathic chemicals on growth and physiological responses of soybean (*Glycine max*). *Weed Science* 29: 53–59.

Paunov, M., Koleva, L., Vassilev, A., Vangronsveld, J. and Goltsev, V. 2018. Effects of different metals on photosynthesis: Cadmium and zinc affect chlorophyll fluorescence in durum wheat. *International Journal of Molecular Sciences* 19(3): 787.

Per, T.S., Khan, N.A., Reddy, P.S., et al. 2017. Approaches in modulating proline metabolism in plants for salt and drought stress tolerance: Phytohormones, mineral nutrients and transgenics. *Plant Physiology and Biochemistry* 115: 126–140.

Perdomo, J.A., Capo-Bauca, S., Carmo-Silva, E. and Galmés, J. 2017. Rubisco and rubisco activase play an important role in the biochemical limitations of photosynthesis in rice, wheat, and maize under high temperature and water deficit. *Frontiers in Plant Science* 8: 490.

Qadir, S., Qureshi, M.I., Javed, S. and Abdin, M.Z. 2004. Genotypic variation in phytoremediation potential of *Brassica juncea* cultivars exposed to Cd stress. *Plant Science* 167(5): 1171–1181.

Qados, A.M.A. 2011. Effect of salt stress on plant growth and metabolism of bean plant *Vicia faba* (L.). *Journal of the Saudi Society of Agricultural Sciences* 10(1): 7–15.

Rahnama, A., Poustini, K., Tavakkol-Afshari, R. and Tavakoli, A. 2010. Growth and stomatal responses of bread wheat genotypes in tolerance to salt stress. *International Journal of Biological, Biomolecular, Agricultural, Food and Biotechnological Engineering* 4(11): 787–792.

Rahneshan, Z., Nasibi, N. and Moghadam, A.A. 2018. Effects of salinity stress on some growth, physiological, biochemical parameters and nutrients in two pistachio (*Pistacia vera* L.) rootstocks. *Journal of Plant Interactions* 13(1): 73–82.

Rai, V.K. 2002. Role of amino acids in plant responses to stresses. *Biologia Plantarum* 45(4): 481–487.

Rasmuson, K.E. and Anderson, J.E. 2002. Salinity affects development, growth, and photosynthesis in cheatgrass. *Journal of Range Management* 55: 80–87.

Rice, E.L. 1985. Allelopathy—an overview. In: *Chemically Mediated Interactions between Plants and Other Organisms. Recent Advances in Phytochemistry*. Eds. G.A. Cooper-Driver, T. Swain and E.E. Conn. Springer, Boston, MA, pp. 81–105.

Rizvi, S.J.H., Haque, H., Singh, U.K. and Rizvi, V. 1992. A discipline called allelopathy. In: *Allelopathy: Basic and Applied Aspects*. Eds. S.J.H. Rizvi and H. Rizvi. Chapman & Hall, London, pp. 1–10.

Roberts, A.G., Gregor, W., Britt, R.D. and Kramer, D.M. 2003. Acceptor and donor-side interactions of phenolic inhibitors in photosystem II. *Biochimica et Biophysica Acta (BBA)-Bioenergetics* 1604(1): 23–32.

Robinson, S.P. and Jones, G.P. 1986. Accumulation of glycinebetaine in chloroplasts provides osmotic adjustment during salt stress. *Functional Plant Biology* 13(5): 659–668.

Romanowska, E., Wróblewska, B., Drozak, A. and Siedlecka, M. 2006. High light intensity protects photosynthetic apparatus of pea plants against exposure to lead. *Plant Physiology and Biochemistry* 44(5–6): 387–394.

Sakamoto, A. and Murata, N. 2002. The role of glycine betaine in the protection of plants from stress: Clues from transgenic plants. *Plant, Cell and Environment* 25(2): 163–171.

Saleh, B. 2012. Effect of salt stress on growth and chlorophyll content of some cultivated cotton varieties grown in Syria. *Communications in Soil Science and Plant Analysis* 43(15): 1976–1983.

Salvucci, M.E. and Crafts-Brandner, S.J. 2004. Relationship between the heat tolerance of photosynthesis and the thermal stability of Rubisco activase in plants from contrasting thermal environments. *Plant Physiology* 134(4): 1460–1470.

Sanchez-Moreiras, A.M., Martínez-Peñalver, A. and Reigosa, M.J. 2011. Early senescence induced by 2–3H-benzoxazolinone (BOA) in *Arabidopsis thaliana*. *Journal of Plant Physiology* 168(9): 863–870.

Saravanavel, R., Ranganathan, R. and Anantharaman, P. 2011. Effect of sodium chloride on photosynthetic pigments and photosynthetic characteristics of *Avicennia officinalis* seedlings. *Recent Research in Science and Technology* 3(4): 177–180.

Scheepens, J.F., Deng, Y. and Bossdorf, O. 2018. Phenotypic plasticity in response to temperature fluctuations is genetically variable, and relates to climatic variability of origin, in *Arabidopsis thaliana*. *AoB PLANTS* 10(4): 1–12. doi:10.1093/aobpla/ply043.

Seigler, D.S. 1996. Chemistry and mechanism of allelopathic interactions. *Agronomy Journal* 88: 876–885.

Serraj, R. and Sinclair, T.R. 2002. Osmolyte accumulation: Can it really help increase crop yield under drought conditions? *Plant, Cell and Environment* 25(2): 333–341.

Shao, H.B., Chu, L.Y., Jaleel, C.A. and Zhao, C.X. 2008. Water-deficit stress-induced anatomical changes in higher plants. *Comptes Rendus Biologies* 331(3): 215–225.

Sharkey, T.D. 2005. Effects of moderate heat stress on photosynthesis: Importance of thylakoid reactions, rubisco deactivation, reactive oxygen species, and thermotolerance provided by isoprene. *Plant, Cell and Environment* 28(3): 269–277.

Sharma, P., Jha, A.B., Dubey, R.S. and Pessarakli, M. 2012. Reactive oxygen species, oxidative damage, and antioxidative defense mechanism in plants under stressful conditions. *Journal of Botany* 2012: 217037.

Siler, B., Misic, D., Filipovic, B., Popovic, Z., Cvetic, T. and Mijovic, A. 2007. Effects of salinity on in vitro growth and photosynthesis of common centaury (*Centaurium erythraea* Rafn.). *Archives of Biological Sciences (Serbia)* 59(2): 129–134.

Singh, H.P., Batish, D.R., Kaur, S., Arora, K. and Kohli, R.K. 2006. Alfa pinene inhibits growth and induces oxidative stress in roots. *Annals of Botany* 98(6): 1261–1269.

Singh, M., Kumar, J., Singh, S., Singh, V., and Prasad, S. 2015. Roles of osmoprotectants in improving salinity and drought tolerance in plants: A review. *Reviews in Environmental Science and Bio/Technology* 14: 407–426.

Skrzypek, E., Repka, P., Stachurska-Swakoń, A., Barabasz-Krasny, B. and Możdżeń, K. 2015. Influence of extracts from peppermint (*Mentha* x *piperita* (L.) Hudson) on growth and activity of the PSII sunflower garden (*Helianthus annuus* L.). *Notulae Botanicae Horti AgrobotaniciCluj-Napoca* 43(2): 335–342.

Slama, I., Abdelly, C., Bouchereau, A., Flowers, T. and Savouré, A. 2015. Diversity, distribution and roles of osmoprotective compounds accumulated in halophytes under abiotic stress. *Annals of Botany* 115(3): 433–447.

Smith, P. and Olesen, J.E. 2010. Synergies between the mitigation of, and adaptation to, climate change in agriculture. *The Journal of Agricultural Science* 148(5): 543–552.

Song, Y., Chen, Q., Ci, D., Shao, X. and Zhang, D. 2014. Effects of high temperature on photosynthesis and related gene expression in poplar. *BMC Plant Biology* 14(1): 111.

Stepien, P. and Johnson, G.N. 2009. Contrasting responses of photosynthesis to salt stress in the glycophyte *Arabidopsis* and the halophyte *Thellungiella*: Role of the plastid terminal oxidase as an alternative electron sink. *Plant Physiology* 149(2): 1154–1165.

Strauss, S.Y. and Zangerl, A.R. 2002. Plant-insect interactions in terrestrial ecosystems. In: *Plant-Animal Interactions: An Evolutionary Approach*. Eds. C.M. Herrera and O. Pellmyr. Blackwell Science, Oxford, UK, pp. 77–106.

Subbarao, G.V., Wheeler, R.M., Levine, L.H. and Stutte, G.W. 2001. Glycine betaine accumulation, ionic and water relations of red-beet at contrasting levels of sodium supply. *Journal of Plant Physiology* 158(6): 767–776.

Sudo, E., Itouga, M., Yoshida-Hatanaka, K., Ono, Y. and Sakakibara, H. 2008. Gene expression and sensitivity in response to copper stress in rice leaves. *Journal of Experimental Botany*: 59(12): 3465–3474.

Sugio, A., Dreos, R., Aparicio, F. and Maule, A.J. 2009. The cytosolic protein response as a subcomponent of the wider heat shock response in *Arabidopsis*. *The Plant Cell* 21(2): 642–654.

Sung, D.Y., Kaplan, F., Lee, K.J. and Guy, C.L. 2003. Acquired tolerance to temperature extremes. *Trends in Plant Science* 8(4): 179–187.

Suo, J., Zhao, Q., David, L., Chen, S. and Dai, S. 2017. Salinity response in chloroplasts: Insights from gene characterization. *International Journal of Molecular Sciences* 18(5): 1011.

Szabados, L. and Savoure, A. 2010. Proline: A multifunctional amino acid. *Trends in Plant Science* 15(2): 89–97.

Taylor, S.H., Ripley, B.S., Woodward, F.I. and Osborne, C.P. 2011. Drought limitation of photosynthesis differs between C3 and C4 grass species in a comparative experiment. *Plant, Cell and Environment* 34(1): 65–75.

Teiz, L. and Zeiger, S.C.E. 1998. *Plant Physiology*, University of California, Los Angeles Sinauer Associates, Inc., Publisher.

Tian, F., Wang, W., Liang, C., Wang, X., Wang, G. and Wang, W. 2017. Overaccumulation of glycine betaine makes the function of the thylakoid membrane better in wheat under salt stress. *The Crop Journal* 5(1): 73–82.

Tiwari, H.S., Agarwal, R.M. and Bhatt, R.K. 1998. Photosynthesis, stomatal resistance and related characteristics, as influenced by potassium under normal water supply and water stress conditions in rice (*Oryza sativa* L.). *Indian Journal of Plant Physiology* 3: 314–316.

Tomar, N.S. and Agarwal, R.M. 2013. Influence of treatment of *Jatropha curcas* L. leachates and potassium on growth and phytochemical constituents of wheat (*Triticum aestivum* L.). *American Journal of Plant Sciences* 4: 1134–1150.

Van Leeuwen, C., Tregoat, O., Choné, X., Bois, B., Pernet, D. and Gaudillère, J.P. 2009. Vine water status is a key factor in grape ripening and vintage quality for red Bordeaux wine. How can it be assessed for vineyard management purposes? *Journal International des Sciences de la Vigne et du vin* 43: 121–134.

Wahid, A., Perveen, M., Gelani, S. and Basra, S.M. 2007. Pretreatment of seed with H_2O_2 improves salt tolerance of wheat seedlings by alleviation of oxidative damage and expression of stress proteins. *Journal of Plant Physiology* 164(3): 283–294.

Wang, D., Heckathorn, S.A., Mainali, K. and Tripathee, R. 2016. Timing effects of heat-stress on plant ecophysiological characteristics and growth. *Frontiers in Plant Science* 7: 1629.

Wang, H., Tang, X., Wang, H. and Shao, H.B. 2015. Proline accumulation and metabolism-related genes expression profiles in *Kosteletzkya virginica* seedlings under salt stress. *Frontiers in Plant Science* 6: 792.

Wang, L.J., Fan, L., Loescher, W., et al. 2010. Salicylic acid alleviates decreases in photosynthesis under heat stress and accelerates recovery in grapevine leaves. *BMC Plant Biology* 10(1): 34.

Wang, Q.L., Chen, J.H., He, N.Y. and Guo, F.Q. 2018. Metabolic reprogramming in chloroplasts under heat stress in plants. *International Journal of Molecular Sciences* 19(3): 849.

Wang, W., Vinocur, B. and Altman, A. 2003. Plant responses to drought, salinity and extreme temperatures: Towards genetic engineering for stress tolerance. *Planta* 218(1): 1–14.

Weir, T.L., Park, S.W. and Vivanco, J.M. 2004. Biochemical and physiological mechanisms mediated by allelochemicals. *Current Opinion in Plant Biology* 7(4): 472–479.

Wise, R.R., Olson, A.J., Schrader, S.M. and Sharkey, T.D. 2004. Electron transport is the functional limitation of photosynthesis in field-grown Pima cotton plants at high temperature. *Plant Cell and Environment* 27(6): 717–724.

Wraight, C.A. 1981. Oxidation-reduction physical chemistry of the acceptor quinone complex in bacterial photosynthetic reaction centers: Evidence for a new model of herbicide activity. *Israel Journal of Chemistry* 21(4): 348–354.

Wu, C., Niu, Z., Tang, Q. and Huang, W. 2008. Estimating chlorophyll content from hyperspectral vegetation indices: Modeling and validation. *Agricultural and Forest Meteorology* 148(8–9): 1230–1241.

Wu, X.L., He, C.Y., Wang, Y.J., et al. 2001. Construction and analysis of a genetic linkage map of soybean. *Yi chuan xue bao= Acta genetica Sinica* 28(11): 1051–1061.

Xing, W. and Rajashekar, C.B. 2001. Glycine betaine involvement in freezing tolerance and water stress in *Arabidopsis thaliana. Environmental and Experimental Botany* 46(1): 21–28.

Xu, J., Yin, H. and Li, X. 2009. Protective effects of proline against cadmium toxicity in micropropagated hyperaccumulator, *Solanum nigrum* L. *Plant Cell Reports* 28(2): 325–333.

Yan, K., Chen, P., Shao, H., Zhang, L. and Xu, G. 2011. Effects of short-term high temperature on photosynthesis and photosystem II performance in sorghum. *Journal of Agronomy and Crop Science* 197(5): 400–408.

Yan, W., Zhong, Y. and Shangguan, Z. 2016. A meta-analysis of leaf gas exchange and water status responses to drought. *Scientific Reports* 6: 209–217.

Yancey, P.H. 2005. Organic osmolytes as compatible, metabolic and counteracting cytoprotectants in high osmolarity and other stresses. *Journal of Experimental Biology* 208(15): 2819–2830.

Yang, C., Zhou, J., Liu, S., Fan, P., Wang, W. and Xia, C. 2013. Allelochemical induces growth and photosynthesis inhibition, oxidative damage in marine diatom *Phaeodactylum tricornutum. Journal of Experimental Marine Biology and Ecology* 444: 16–23.

Yang, C.M., Lee, C.N. and Chou, C.H. 2002. Effects of three allelopathic phenolics on chlorophyll accumulation of rice (*Oryza sativa*) seedlings: I. Inhibition of supply-orientation. *Botanical Bulletin of Academia Sinica* 43: 299–304.

Yang, X., Feng, Y., He, Z. and Stoffella, P.J. 2005. Molecular mechanisms of heavy metal hyperaccumulation and phytoremediation. *Journal of Trace Elements in Medicine and Biology* 18(4): 339–353.

Ye, W., Hu, S., Wu, L., et al. 2016. White stripe leaf 12 (WSL12), encoding a nucleoside diphosphate kinase 2 (OsNDPK2), regulates chloroplast development and abiotic stress response in rice (*Oryza sativa* L.). *Molecular Breeding* 36(5): 57.

Yruela, I. 2005. Copper in plants. *Brazilian Journal of Plant Physiology* 17(1): 145–156.

Yu, G.H., Zhang, X. and Ma, H.X. 2015. Changes in the physiological parameters of SbPIP1-transformed wheat plants under salt stress. *International Journal of Genomics* 2015(384356): 1–6.

Yu, J.Q., Ye, S.F., Zhang, M.F. and Hu, W.H. 2003. Effects of root exudates and aqueous root extracts of cucumber (*Cucumis sativus*) and allelochemicals, on photosynthesis and antioxidant enzymes in cucumber. *Biochemical Systematics and Ecology* 31(2): 129–139.

Yuan, L., Tang, L., Zhu, S., et al. 2017. Influence of heat stress on leaf morphology and nitrogen–carbohydrate metabolisms in two wucai (*Brassica campestris* L.) genotypes. *Acta Societatis Botanicorum Poloniae* 86(2): 3554.

Zengin, F.K. and Kirbag, S. 2007. Effects of copper on chlorophyll, proline, protein and abscisic acid level of sunflower (*Helianthus annuus* L.) seedlings. *Journal of Environmental Biology* 28(3): 561–566.

Zengin, F.K. and Munzuroglu, O. 2005. Effects of some heavy metals on content of chlorophyll, proline and some antioxidant chemicals in bean (*Phaseolus vulgaris* L.) seedlings. *Acta Biologica Cracoviensia Series Botanica* 47(2): 157–164.

Zhang, H., Dong, H., Li, W., Sun, Y., Chen, S. and Kong, X. 2009. Increased glycine betaine synthesis and salinity tolerance in AhCMO transgenic cotton lines. *Molecular Breeding* 23(2): 289–298.

Zhang, J., Jiang, H., Song, X., Jin, J. and Zhang, X. 2018. The responses of plant leaf CO2/H2O exchange and water use efficiency to drought: A meta-analysis. *Sustainability* 10(2): 551.

Zhang, S., Zhang, B., Dai, W. and Zhang, X. 2011. Oxidative damage and antioxidant responses in *Microcystis aeruginosa* exposed to the allelochemical berberine isolated from golden thread. *Journal of Plant Physiology* 168(7): 639–643.

Zhao, X., Tan, H. and Chen, G. 2016. Effect of organic osmolytes and ABA accumulated in twelve dominant desert plants of the Tengger Desert, China. *Research and Reviews: Journal of Botanical Sciences. Plant Biotechnology and its Applications* S3: 45–50.

Zhu, J.K. 2001. Plant salt tolerance. *Trends in Plant Science* 6(2): 66–71.

Zhu, J.K. 2002. Salt and drought stress signal transduction in plants. *Annual Review of Plant Biology* 53(1): 247–273.

Zivcak, M., Brestic, M., Balatova, Z., et al. 2013. Photosynthetic electron transport and specific photoprotective responses in wheat leaves under drought stress. *Photosynthesis Research* 117(1–3): 529–546. Information Classification: General

7 Role of Exogenous Salicylic Acid in Drought-Stress Adaptability in a Changing Environment

Cátia Brito, Lia-Tânia Dinis, José Moutinho-Pereira, and Carlos Correia

CONTENTS

7.1 INTRODUCTION

Among abiotic stresses, drought is perhaps the most responsible for decreased agricultural production worldwide (Wani et al. 2016), being highly exacerbated in the presence of other stressors. The water deficit has negative repercussions on water relations, nutrient uptake, carbon assimilation, canopy dimension, oxidative pathways, phenology and reproduction processes (Bacelar et al. 2006, 2007a, 2007b; Brito et al. 2018a, 2018b; Farooq et al. 2009; Petridis et al. 2012).

Concomitant with a rainfall decrease during summer, climate models predict a stronger inter- and intra-annual weather variability (IPCC 2013). These scenarios give even more prominence to the concept of drought adaptability, which integrates much more than drought resistance, with recovery capacity also playing a fundamental role in plant growth and survival (Chen et al. 2016). The recovery degree is species-dependent, and it is also influenced by the duration and intensity of previous drought, depending on compatible solute accumulation, phytohormone dynamics and carbon allocation between plant organs (Chaves, Flexas and Pinheiro 2009; De Diego et al. 2012, 2013).

Salicylic acid (SA) is a phytohormone increasingly recognized as an abiotic stress-tolerance enhancer, via SA-mediated control of major plant-metabolic processes (Khan et al. 2015); its exogenous application is an emerging tool in plant-drought tolerance (Kang et al. 2012; Nazar et al. 2015; Brito et al. 2018b, 2019). Nevertheless, the precise mechanisms supporting this tolerance are still under debate (Khan et al. 2015), and the influence of SA in recovery processes has been poorly studied.

In this context, the action mode by which SA induces drought tolerance and its influence on plant recovery capacity will be discussed. Moreover, the resulting effects on plants' physiological, biochemical, growth and yield responses will be elucidated.

7.2 SALICYLIC ACID CHARACTERIZATION AND ACTION ON ABIOTIC STRESS TOLERANCE

Salicylic acid ($C_7H_6O_3$) is an endogenously synthetized signaling molecule in plants (Khan et al. 2015). The word "salicylic" is derived from *Salix*, which is the Latin name for the willow tree (*Salix alba*). Salicin, the glucoside of salicylic alcohol, was first isolated in 1826 from willow bark, and a large amount of the substance was successfully isolated in 1828. Salicin was then converted into a sugar and an aromatic compound that, upon oxidation, becomes SA (Miura and Tada 2014). Chemically, SA belongs to a diverse group of plant phenolics (Kahn et al. 2015) that possess an aromatic ring with a hydroxyl group or its functional derivatives (Figure 7.1).

This molecule is known to regulate several physiological mechanisms in plants, such as flowering induction, hormonal status, photosynthesis, respiration, solutes transport and metabolism, antioxidant defense system, plant–water relations, minerals uptake and defense responses against infections and pathogen attacks (Abd El-Razek et al. 2013; Kumar 2014; Miura and Tada 2014; Khan et al. 2015; Nazar et al. 2015; Brito et al. 2018b, 2019). Moreover, increasing evidence suggests a key role played by SA in response to multiple abiotic stresses, both the natural accumulation in plants subjected to stressful conditions (Bandurska and Cieślak 2013; Choudhary and Agrawal 2014), and the induction of protective mechanisms by the exogenous application in plants subjected to major abiotic stresses, such as drought (Kang et al. 2012; Nazar et al. 2015), salinity (Fayez and Bazaid 2014; Khan, Asgher and Khan 2014), heat (Wang and Li 2006; Wang et al. 2010) and high light intensity (Zhao et al. 2011; Wang et al. 2014).

Nevertheless, the precise mechanisms by which SA induces plant tolerance against abiotic stresses remain little discussed, and more comprehensive investigations are needed in this direction (Khan et al. 2015). It is believed that abiotic stress tolerance induction is mainly related to the dual redox effect of SA, a first oxidative phase, characterized by a transient increase in ROS levels, that is followed by an increase in reducing power (Herrera-Vásquez, Salinas and Holuigue 2015). Some studies have reported that SA pre-treatment leads to an initial increase in H_2O_2 levels, possibly by the inhibition of some antioxidant enzymes, such as catalase (Gunes et al. 2007; Belkadhi et al. 2014; Hao et al. 2014). Then, the slight increase in ROS levels triggers defense responses, mediated by the increase in the antioxidant enzymatic system (e.g. catalase, ascorbate peroxidase, superoxide dismutase) or by the synthesis of antioxidant metabolites and protective compounds (e.g. glutathione, ascorbic acid, phenolic compounds, heat shock proteins (HSPs), osmolites) (Horváth, Szalai and Janda 2007; Wang et al. 2010; Misra and Misra 2012; Li et al. 2013; Khan, Asgher and Khan 2014; Herrera-Vásquez, Salinas and Holuigue 2015; Brito et al. 2018b, 2019). SA might also induce genes responsible for protective mechanisms (Jumali et al. 2011; Li et al. 2013), the production of signal transducers (e.g. ROS, protein kinases and ABA) (Sakhabutdinova et al. 2003; Horváth, Szalai and Janda 2007; Jesus

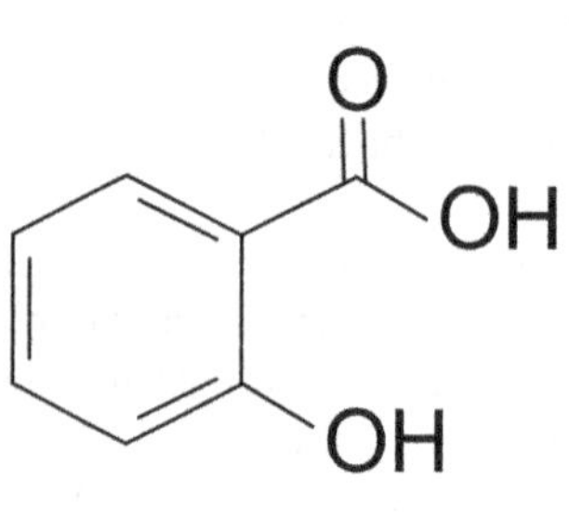

O
OH
OH

FIGURE 7.1 Salicylic acid chemical structure.

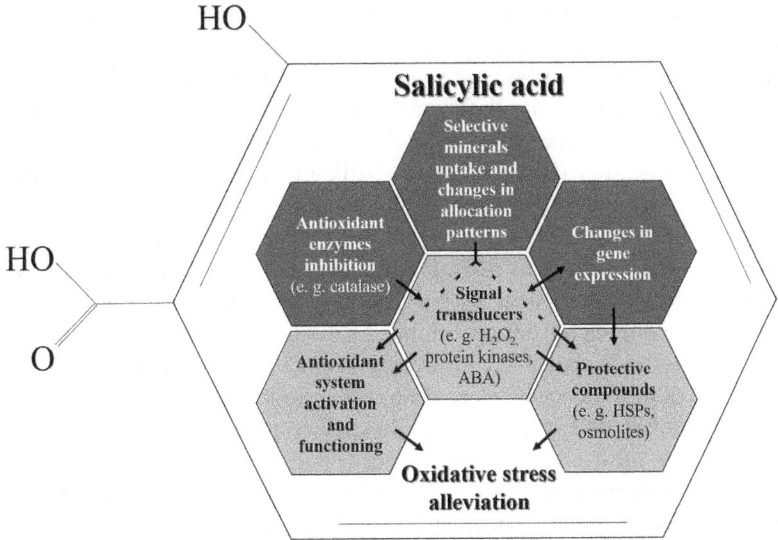

FIGURE 7.2 Schematization of salicylic acid action on the alleviation of oxidative stress.

et al. 2015) and the selective uptake of minerals (e.g. P, Fe, Mn and Zn) and changes in allocation patterns to potentiate antioxidant responses (unpublished results). A summary of the salicylic acid action mode in the alleviation of oxidative stress is presented in Figure 7.2. Nonetheless, lower concentrations of SA are generally more efficient than higher ones (Kang et al. 2012; Miura and Tada 2014). High concentrations can cause a high level of ROS and then oxidative stress that plants are unable to overcome (Horváth, Szalai and Janda 2007; Hara et al. 2012; Miura and Tada 2014).

The effects of exogenous SA on plant resistance to abiotic stress depend on a set of factors: (i) applied SA concentration (Kang et al. 2012; Agami et al. 2013; Jesus et al. 2015; Brito et al. 2018b), the dose for maximum stress tolerance ranging between 0.1 mM and 0.5 mM (Hara et al. 2012); (ii) method of SA administration, including pre-soaking, addition to the growth medium or foliar spray (Singh and Usha 2003; Kang et al. 2012); (iii) plant species and cultivars (Umebese, Olatimilehin and Ogunsusi 2009; Khalil et al. 2012); (iv) plant developmental stage (Umebese, Olatimilehin and Ogunsusi 2009; Ahmad, Murali and Marimathu 2014); (v) kind of stress (Sakhabutdinova et al. 2003; Fayez and Bazaid 2014); (vi) stress level (El-Tayeb 2005; Alam et al. 2013; Hashempour et al. 2014); (vii) system in which the study was carried out, from cell suspensions to the whole plant (Pál et al. 2013) and (viii) frequency of application (Shaaban, Abd El-Aal and Ahmed 2011). Still, an accurate SA application improves abiotic stress tolerance in plants by the modulation of several important aspects of plant function and structure, with relevant consequences on growth, yield and harvest quality (Brito et al. 2018b, 2018c, 2019).

7.3 EFFECT OF EXOGENOUS SALICYLIC ACID ON CROP DROUGHT ADAPTABILITY

7.3.1 Influence on Redox Status

Drought stress results in the increased generation of reactive oxygen species (ROS) due to energy accumulation, which increases the photooxidative effect (Waraich et al. 2011). Thus, ROS homeostasis is a convergence point to evaluate the plant stress status. The available evidence supports the theory that SA pre-treatment is related to the lower accumulation of ROS under water-deficit conditions (Alam et al. 2013; Nazar et al. 2015). Moreover, SA also improved the cell membrane integrity of droughted plants, a signal of lower oxidative stress (Agami et al. 2013; Alam et al. 2013; Fayez and Bazaid 2014; Jesus et al. 2015; Nazar et al. 2015). Additionally, there is also evidence that SA prevents photosynthetic pigment stress-induced degradation (Agami et al. 2013; Alam et al. 2013; Ahmad, Murali and Marimathu 2014; Fayez and Bazaid 2014; Brito et al. 2018b).

General responses to stress involve the signaling stress detection and, consequently, the increase in antioxidant responses (Farooq et al. 2009). Being the less reactive, H_2O_2 is the product of highly reactive species detoxification (Gill and Tuteja 2010; Pinto-Marijuan and Munne-Bosch 2014). Additionally, it is also a potent signaling molecule, due to its long half-life and the ability to cross cellular membranes (Petrov and Van Breusegem 2012). There is evidence that SA pre-treatment induces H_2O_2 production (Chen et al. 2007; Harfouche et al. 2008; Belkadhi et al. 2014), which in turn might induce antioxidant enzyme activity, decreasing cellular ROS levels (Arfan 2009). In fact, under drought conditions olive trees exhibited a distinct pattern of H_2O_2 and total ROS, with higher H_2O_2 accumulation in droughted SA-treated plants in comparison to SA-starved droughted controls, a reflection of the balance between each type of ROS (unpublished results). This response suggests that SA improves the capacity of drought stressed plants to scavenge and detoxify highly reactive oxygen species.

In general, strong evidence indicates that SA induces the enzymatic antioxidant system, including catalase, superoxide dismutase, ascorbate peroxidase, glutathione reductase, glutathione peroxidase, glutathione-S-transferase, polyphenol oxidase, monodehydroascorbate reductase and dehydroascorbate reductase activities (Singh and Usha 2003; Alam et al. 2013; Khan, Asgher and Khan 2014; Wang et al. 2014). SA was also found to increase the transcription of certain ascorbate-glutathione cycle-related genes, allowing the maintenance of higher concentrations of ascorbate and glutathione (Kang et al. 2013; Alam et al. 2013; Li et al. 2013; Khan, Asgher and Khan 2014), and the improvement of reduced-to-oxidized ascorbate and reduced-to-oxidized glutathione ratios (Wang and Li 2006; Alam et al. 2013). On the other hand, while the increase in the concentration of phenolic compounds in response to SA application was reported under drought conditions (Jesus et al. 2015), no influence and/or its decline seems to prevail (Agami et al. 2013; Fayez and Bazaid 2014), suggesting that other defense mechanisms might be activated. Meanwhile, the synthesis of stress proteins also plays a crucial role in drought-tolerance development (Farooq et al. 2009). The positive effect of SA on protein accumulation is commonly observed in stressed plants, either through developing mechanisms that prevent their degradation or by changing the protein patterns (Jalal, Moftah and Bafeel 2012; Kang et al. 2012; Kabiri, Nasibi and Farahbakhsh 2014; Brito et al. 2019). In fact, SA application prevented the drought-induced decline and/or induced protein overaccumulation in olive leaves, starting to stabilize during the recovery phase (unpublished results), and stimulated the expression of proteins associated with signal transduction, stress defense, photosynthesis, carbohydrate metabolism, protein metabolism and energy (Kang et al. 2012). SA was also found to be involved in the maintenance of total thiol levels in droughted olive trees (unpublished results), an interesting advantage as thiol groups are involved in the antioxidant defense system (Zagorchev et al. 2013).

The induction of osmolytes accumulation by SA application is also frequent (Khan et al. 2015), including soluble carbohydrates (Ahmad, Murali and Marimathu 2014; Jesus et al. 2015; Brito et al. 2018b), proline (Misra and Misra 2012; Ahmad, Murali and Marimathu 2014; Nazar et al. 2015) and glycinebetaine (Misra and Misra 2012; Khan, Asgher and Khan 2014). Some of these organic solutes can also protect cellular proteins, enzymes and cellular membranes and allow the metabolic machinery to continue functioning (Sanders and Arndt 2012).

7.3.2 Influence on Water Status and Hormonal Dynamics

The SA-induced osmolite accumulation, discussed above, certainly contributes to improving water status, since this mechanism decreases the osmotic potential, creating a soil–plant water gradient that enables the extraction of water from soil at water potential below the wilting point (Dichio et al. 2006). Indeed, an improvement in water status is commonly reported in droughted plants in response to SA application (Kang et al. 2012; Alam et al. 2013; Ahmad, Murali and Marimathu 2014; Jesus et al. 2015; Nazar et al. 2015; Brito et al. 2018b).

Salicylic acid also regulates several plant responses through signaling cross-talks with other phytohormones, a crucial mechanism for regulating plants' adaptation capacity in different environment conditions (Khan et al. 2015). For instance, the pretreatment with SA prevented salinity- and drought-induced decline in indole-3-acetic-acid (IAA) (Sakhabutdinova et al. 2003; Fahad and Bano 2012), while no significant effects of SA on IAA concentration were observed under normal conditions (Fahad and Bano 2012). Likewise, the immunohistochemical IAA signal in leaves was also found to be increased by SA application in olive trees under summer stress (Brito et al. 2019). Moreover, considering IAA function as a stress-signaling hormone in an early stage of stress (Jain and Khurana 2009; Sharma et al. 2015), more than preventing its degradation, SA may also induce its accumulation (Fahad and Bano 2012). The results of Brito and colleagues (2018b) suggest that SA might also influence IAA during drought recovery, as the signal intensity in SA-pre-treated leaves was higher than in the controls. On the other hand, SA seems to promote ABA accumulation under both drought and salinity conditions (Sakhabutdinova et al. 2003; Bandurska and Stroiński 2005; Szepesi et al. 2009; Jesus et al. 2015), while no significant effects of SA on ABA accumulation were recorded under normal conditions (Fahad and Bano 2012). Interestingly, even with a higher relative water content and stomatal conductance, SA-sprayed olive trees growing under field summer stress conditions displayed a leaf ABA signal intensity similar to control plants (Brito et al. 2019). On the other hand, in a different study carried out in pots, the persistence of the ABA immunohistochemical signal after a long period of drought recovery is less evident in SA-sprayed plants, which might reflect a "memory" of the worst water status presented by the non-sprayed plants during the drought stress period (Brito et al. 2018b). Salicylic acid was also found to restrict ethylene formation in drought and salinity conditions (Khan, Asgher and Khan 2014; Nazar et al. 2015).

The immunodetection of phytohormones has been proven to be an efficient tool to understand the hormone translocation in plants subjected to different stress situations and to clarify their role in regulating physiological responses (De Diego et al. 2013; Escandón et al. 2016; Jesus et al. 2015). For instance, the IAA signal seems to increase in the main vascular tissues of SA-sprayed leaves of olive trees under summer stress conditions, suggesting its transport (Brito et al. 2019). In the same experiment, SA also improved the ABA signal intensity in the main vascular tissues (Brito et al. 2019), signifying its transport and highlighting SA signaling in ABA accumulation (Jesus et al. 2015; Shakirova et al. 2003). Furthermore, the ABA signal in phloem was more intense in SA-sprayed leaves (Brito et al. 2019), which may be related to the involvement of ABA in photo-assimilate flow and distribution regulation (Peng, Lu and Zhang 2003).

7.3.3 Influence on Ionome

SA modulates mineral nutrient uptake and metabolism, improving its accumulation and changing its allocation among different plant organs, as demonstrated under drought and salinity conditions (El-Tayeb 2005; Gunes et al. 2007; Yildirim, Turan and Guvenc 2008; Nazar et al. 2015). In olive trees subjected to repeated cycles of drought and re-watering, SA changed the mineral nutrient stoichiometry in plant tissues and increased nutrient reserves (unpublished results). To illustrate, SA attenuated the drought-induced decline in nutrient root uptake efficiency, namely for Fe, Zn and Mn, and contributed to increasing the concentrations of P and S in stems, Fe and Mn in leaves and K, Mg, B, Fe, Mn and Zn in roots, despite the higher root biomass accumulation. As exogenous SA affected micronutrients to a greater extent than macronutrients, and because SA induced a change in allocation patterns, these responses mean that SA generates a selective behavior according to the plant needs, for instance of elements that act as antioxidant-enzyme cofactors (e.g. Fe, Mn and Zn) (unpublished results). Meanwhile, the higher K concentration observed in roots might be in order to improve osmotic adjustment, and the superior Mg might be related to protein synthesis and to its function as an enzymatic cofactor (Grusak 2001; Waraich et al. 2011), whereas the higher B and Fe, associated with the higher root biomass (unpublished results), claim their function in root cell elongation (Li et al. 2016; Hilo et al. 2017). On the other hand, the higher concentration of chlorophylls, as well a rise in the photosynthetic rate, observed in the SA-treated plants (Brito et al. 2018b) highlight that Fe is required for chlorophyll synthesis and photosynthesis (Rout and Sahoo 2015). Furthermore, the higher proline concentration in SA-treated mustard plants under drought was a result of increased N and S assimilation (Nazar et al. 2015).

7.3.4 Influence on Gas Exchange and Photosynthetic Capacity

It is common to find a positive influence of SA application on photosynthetic responses in drought-stressed plants. The first action seems to be related to the alleviation of stomatal limitations, by reducing the drought-induced stomatal closure, as described for tomato, mustard and olive plants (Hayat et al. 2008; Nazar et al. 2015; Brito et al. 2018b, 2019). In line with this, Rai, Sharma and Sharma (1986) described that SA reverses the stomatal closure induced by ABA and, as discussed earlier, prevents the salinity- and drought-induced decline in IAA (Fahad and Bano 2012; Sakhabutdinova et al. 2003), that in turn can stimulate stomatal opening (Peleg and Blumwald 2011) by impairing the ABA-inhibition response (Tanaka et al. 2006). The positive influence of SA in photosynthetic capacity was also related to higher carboxylation efficiency and increased protection of photosynthetic machinery in droughted plants, as shown by a lower intercellular-to-atmospheric CO_2 concentration ratio (C_i/C_a), in spite of higher stomatal conductance, and enhanced rubisco activity, PSII function, chlorophyll concentrations and chlorophyll a/b ratio or, at least, by changes in some of the reported variables (Ahmad, Murali and Marimathu 2014; Jesus et al. 2015; Nazar et al. 2015; Brito et al. 2018b, 2019). Moreover, the stomatal conductance and photosynthetic rate restauration after drought stress relief was improved by SA application in both young potted (Brito et al. 2018b) and field-established olive trees (Brito et al. 2019). In line with this, at the end of heat and high light stress imposition in wheat and grapevines, a faster recovery of the photosynthetic rate induced by SA, mainly through positive effects on PSII function, was also reported (Wang et al. 2010, 2014; Zhao et al. 2011).

Meanwhile, SA contributed to maintaining higher respiration rates in droughted olive trees, which suggests the use of consumption of carbohydrates for growth and/or defense mechanisms (Brito et al. 2018b), since this species is known to reduce its metabolism to conserve photosynthates to be used later on, under favorable conditions, for growth (Varone and Gratani 2015). SA was also found to induce a reduction of nighttime stomatal conductance (g_{night}) and transpiration (E_{night}) in droughted olive trees, along with higher concentrations of intercellular carbon dioxide ($C_{i\text{-}night}$) (unpublished results). This influence is very promising, contributing to ameliorate the plant–water balance, since the substantial water cost of g_{night} could represent a major problem for agronomic development (Resco de Dios et al. 2015), due to decreased daily water use efficiency, particularly under water-deficit conditions (Escalona et al. 2013; Medrano et al. 2015; Brito et al. 2018a). Furthermore, drought-stressed plants pretreated with SA even displayed a mass gain at the end of the night (unpublished results), the water losses possibly being compensated by the higher dew deposition in leaves (Brito et al. 2018a). Water compensation at nighttime can be of crucial importance in arid climates, where is quite difficult to reach high humidity during the drought season (Konrad et al. 2015).

7.3.5 Influence on Leaf Structure

There is evidence that SA affects the leaf and chloroplast structures of stressed plants, although no special consensus has been achieved. SA was found to increase the leaf thickness of droughted and salinity-stressed plants (Cárcamo et al. 2012; Agami et al. 2013), as well the number and diameter of vessels of droughted plants (Agami 2013). Conversely, the trichome layer thickness decreased with

SA foliar spray in plants under summer stress conditions (Brito et al. 2019). Meanwhile, stomatal density was found to be reduced and increased by SA application in plants under salinity (Cárcamo et al. 2012) and summer stress conditions (Brito et al. 2019), respectively. Moreover, SA was found to be involved in the chloroplast's centrifugal organization maintenance under salinity conditions (Cárcamo et al. 2012).

7.3.6 Influence on Growth, Yield and Harvest Quality

In line with the main SA-induced physiological and biochemical effects, plant growth and productivity are also modulated by SA. Indeed, several studies have reported an increase in plant growth and/or biomass accumulation under drought stress (Kang et al. 2012; Agami et al. 2013; Ahmad, Murali and Marimathu 2014; Fayez and Bazaid 2014; Nazar et al. 2015; Brito et al. 2018b). Interestingly, SA induces a higher allocation of dry matter into roots, in different species, such as tomato, wheat and olive tree, subjected to drought stress (Umebese, Olatimilehin and Ogunsusi 2009; Kang et al. 2012; Brito et al. 2018b) and in olive tree under salinity stress (Aliniaeifard, Hajilou and Tabatabaei 2016). This response is particularly important, not only due to carbohydrate storage, for posterior use in recovery processes, but also to improve water and mineral uptake from soil. In addition, SA was also found to improve whole-plant water-use efficiency in olive trees subjected to intermittent drought-recovery events (Brito et al. 2018b).

Research on SA influence on crop yield and harvest quality under drought conditions is very scarce. Still, in a study conducted in rainfed olive trees, a mean yield increase of 72% was reported over two consecutive growing seasons (Brito et al. 2018c). The influence on olives and olive oil phenolic compounds was more related to weather year-to-year variations, but the application of SA allowed higher production of these metabolites under mild summer conditions and the attenuation of their degradation due to frost events, while under severe summer conditions SA reduced the investment in secondary metabolites (Brito et al. 2018c).

Under normal growing conditions, SA is known to positively influence yield (Javaheri et al. 2012; Khalil et al. 2012; Abd El-Razek et al. 2013; Kazemi et al. 2013) and harvest quality, as in tomatoes where SA increased fruit skin diameter, the amounts of vitamin C and lycopene and the Brix index (Javaheri et al. 2012), in strawberry where SA enhanced total soluble solids, titratable acidity, phenolics and vitamin C (Kazemi et al. 2013), and in olives where SA increased fruit and pulp weight and oil contents (Khalil et al. 2012; Abd El-Razek et al. 2013).

7.4 CONCLUSIONS AND FUTURE DIRECTIONS

The implementation of short-term adaptation measures is fundamental to respond rapidly to the current and predicted adverse conditions. Moreover, plants' drought adaptability must be achieved by inducing the highest drought resistance and recovery capacities. In line with this, evidence has shown that the exogenous application of SA is a promising strategy to improve drought adaptability. Salicylic acid improves adaptative responses to drought by the direct and/or indirect modulation of several molecular, biochemical and physiological processes, including a better balance between ROS production and scavenging, reducing the oxidative stress. Meanwhile, the induced osmolyte accumulation and the optimization of the shoot-to-root ratio also contribute to maintain turgor and to a more favorable water status. An improved physiological performance was also achieved through signaling cross-talks with other phytohormones, selective mineral nutrient uptake, mainly of micronutrients and by changes in nutrient allocation patterns, according to the plant's needs. Moreover, by reducing both stomatal and non-stomatal limitations, SA contributes to higher photosynthetic capacity. The application of SA also contributes to a reduction in the investment in extra repair damages, while accelerating the restauration of the physiological functions, after stress relief. For all these reasons, SA attenuates the limitations imposed by drought in plant growth, yield and harvest quality. Still, the determination of the best time and frequency of applications will be necessary, and the optimum SA concentration for each species and environmental conditions is a prerequisite; once SA is applied beyond a certain range it might be detrimental. Moreover, studies of SA application in interaction with other agronomic practices, such as leguminous cover crops and deficit irrigation strategies, are needed in order to improve SA effectiveness under more severe stress conditions.

REFERENCES

Abd El-Razek, E., H. S. A. Hassan, and K. M. G. El-Din. 2013. Effect of foliar application with salicylic acid, benzyladenine and gibberellic acid on flowering, yield and fruit quality of olive trees (*Olea europaea* L.). *Middle East Journal of Scientific Research* 14: 1401–1406. doi:10.5829/idosi.mejsr.2013.14.11.3572.

Agami, R. 2013. Salicylic acid mitigates the adverse effect of water stress on lettuce (*Lactuca sativa* L.). *The Journal of Applied Sciences Research* 9: 5701–5711.

Ahmad, M. A., P. V. Murali, and G. Marimuthu. 2014. Impact of salicylic acid on growth, photosynthesis and compatible solute accumulation in *Allium cepa* L. subjected to drought stress. *International Journal of Agricultural and Food Science* 4: 22–30.

Alam, M., M. Hasanuzzaman, K. Nahar, and M. Fujita. 2013. Exogenous salicylic acid ameliorates short-term drought stress in mustard (*Brassica juncea* L.) seedlings by up-regulating the antioxidant defense and glyoxalase system. *Australian Journal of Crop Science* 7: 1053–1063.

Aliniaeifard, S., J. Hajilou, and S. J. Tabatabaei. 2016. Photosynthetic and growth responses of olive to proline and salicylic acid under salinity condition. *Notulae Botanicae Horti Agrobotanici Cluj-Napoca* 44, 579–585. doi:10.15835/nbha44210413.

Arfan, M. 2009. Exogenous application of salicylic acid through rooting medium modulates ion accumulation and antioxidant activity in spring wheat under salt stress. *International Journal of Agriculture and Biology* 11: 437–442.

Bacelar, E. A., J. M. Moutinho-Pereira, B. C. Gonçalves, H. F. Ferreira, and C. M. Correia. 2007a. Changes in growth, gas exchange, xylem hydraulic properties and water use efficiency of three olive cultivars under contrasting water availability regimes. *Environmental and Experimental Botany* 60: 183–192. doi:10.1016/j.envexpbot.2006.10.003.

Bacelar, E. A., D. L. Santos, J. M. Moutinho-Pereira, *et al.* 2007b. Physiological behaviour, oxidative damage and antioxidative protection of olive trees grown under different irrigation regimes. *Plant and Soil* 292, 1–12.

Bacelar, E. A., D. L. Santos, J. M. Moutinho-Pereira, *et al.* 2006. Immediate responses and adaptative strategies of three olive cultivars under contrasting water availability regimes: Changes on structure and chemical composition of foliage and oxidative damage. *Plant Science* 170: 596–605. doi:10.1016/j.plantsci.2005.10.014.

Bandurska, H., and A. Stroiński. 2005. The effect of salicylic acid on barley response to water deficit. *Acta Physiologiae Plantarum* 27: 379–386. doi:10.1007/s11738-005-0015-5.

Belkadhi, A., A. De Haro, P. Soengas, *et al.* 2014. Salicylic acid increases tolerance to oxidative stress induced by hydrogen peroxide accumulation in leaves of cadmium-exposed flax (*Linum usitatissimum* L.). *Journal of Plant Interactions* 9: 647–654. doi:10.1080/17429145.2014.890751.

Brito, C., L.-T. Dinis, H. Ferreira, J. Moutinho-Pereira, and C. Correia. 2018a. The role of nighttime water balance on *Olea europaea* plants subjected to contrasting water regimes. *Journal of Plant Physiology* 226: 56–63. doi:10.1016/j.jplph.2018.04.004.

Brito, C., L.-T. Dinis, M. Meijón, *et al.* 2018b. Salicylic acid modulates olive tree physiological and growth responses to drought and rewatering events in a dose dependent manner. *Journal of Plant Physiology* 230: 21–32. doi:10.1016/j.jplph.2018.08.004.

Brito, C., L.-T. Dinis, E. Silva, *et al.* 2018c. Kaolin and salicylic acid foliar application modulate yield, quality and phytochemical composition of olive pulp and oil from rainfed trees. *Scientia Horticulturae* 237: 176–183. doi:10.1016/j.scienta.2018.04.019.

Brito, C., L.-T. Dinis, A. Luzio, *et al.* 2019. Kaolin and salicylic acid alleviate summer stress in rainfed olive orchards by modulation of distinct physiological and biochemical responses. *Scientia Horticulturae* 246: 201–211. doi:10.1016/j.scienta.2018.10.059.

Cárcamo, H., R. Bustos, F. Fernández, and E. Bastías. 2012. Mitigating effect of salicylic acid in the anatomy of the leaf of *Zea mays* L. lluteño ecotype from the Lluta Valley (Arica-Chile) under NaCl stress. *Idesia* 30: 55–63. doi:10.4067/S0718-34292012000300007.

Chaves, M. M., J. Flexas, and C. Pinheiro. 2009. Photosynthesis under drought and salt stress: Regulation mechanisms from whole plant to cell. *Annals of Botany* 103: 551–560. doi:10.1093/aob/mcn125.

Chen, D., S. Wang, B. Cao, *et al.* 2016. Genotypic variation in growth and physiological response to drought stress and re-watering reveals the critical role of recovery in drought adaptation in *Maize* seedlings. *Frontiers in Plant Science* 6: 1–15. doi: 10.3389/fpls.2015.01241.

Chen, J., C. Zhu, L. P. Li, Z. Y. Sun, and X. B. Pan. 2007. Effects of exogenous salicylic acid on growth and H_2O_2-metabolizing enzymes in rice seedlings under lead stress. *Journal of Environmental Sciences* 19: 44–49. doi:10.1016/S1001-0742(07)60007-2.

Choudhary, K. K., and S. B. Agrawal. 2014. Cultivar specificity of tropical mung bean (*Vigna radiata* L.) to elevated ultraviolet-B: Changes in antioxidative defense system, nitrogen metabolism and accumulation of jasmonic and salicylic acids. *Environmental and Experimental Botany* 99: 122–132. doi:10.1016/j.envexpbot.2013.11.006.

Ciéslak, B. 2013. The interactive effect of water deficit and UV-B radiation on salicylic acid accumulation in barley roots and leaves. *Environmental and Experimental Botany* 94: 9–18. doi:10.1016/j.envexpbot.2012.03.001.

De Diego, N., F. Perez-Alfocea, E. Cantero, M. Lacuesta, and P. Moncalean. 2012. Physiological response to drought in radiata pine: Phytohormone implication at leaf level. *Tree Physiology* 32: 435–449. doi:10.1093/treephys/tps029.

De Diego, N., J. L. Rodriguez, I. C. Dodd, F. Perez-Alfocea, P. Moncalean, and M. Lacuesta. 2013. Immunolocalization of IAA and ABA in roots and needles of radiata pine (*Pinus radiata*) during drought and rewatering. *Tree Physiology* 33: 537–549. doi: 10.1093/treephys/tpt033.

Dichio, B., C. Xiloyannis, A. Sofo, and G. Montanaro. 2006. Osmotic regulation in leaves and roots of olive trees during a water deficit and rewatering. *Tree Physiology* 26: 179–185. doi:10.1093/treephys/26.2.179.

El-Tayeb, M. A. 2005. Response of barley grains to the interactive effect of salinity and salicylic acid. *Plant Growth Regulation* 45: 215–224. doi:10.1007/s10725-005-4928-1.

Escalona, J. M., S. Fuentes, M. Tomás, S. Martorell, J. Flexas, and H. Medrano. 2013. Responses of leaf night transpiration to drought stress in *Vitis vinifera* L. *Agricultural Water Management* 118: 50–58. doi:10.1016/j.agwat.2012.11.018.

Escandón, M., M. J. Cañal, J. Pascual, *et al.* 2016. Integrated physiological and hormonal profile of heat-induced thermotolerance in *Pinus radiata*. *Tree Physiology* 36: 63–77. doi:10.1093/treephys/tpv127.

Fahad, S., and A. Bano. 2012. Effect of salicylic acid on physiological and biochemical characterization of maize grown in saline area. *Pakistan Journal of Botany* 44: 1433–1438.

Farooq, M., A. Wahid, N. Kobayashi, D. Fujita, and S. M. A. Basra. 2009. Plant drought stress: Efects, mechanisms and management. *Agronomy for Sustainable Development* 29: 185–212. doi: 10.1051/agro:2008021.

Fayez, K. A., and S. A. Bazaid. 2014. Improving drought and salinity tolerance in barley by application of salicylic acid and potassium nitrate. *Journal of the Saudi Society of Agricultural Sciences* 13: 45–55. doi:10.1016/j.jssas.2013.01.001.

Gill, S. S., and N. Tuteja. 2010. Reactive oxygen species and antioxidant machinery in abiotic stress tolerance in crop plants. *Plant Physiology and Biochemestry* 48: 909–930. doi:10.1016/j.plaphy.2010.08.016.

Grusak, M. A. 2001. Plant macro- and micronutrient minerals. *Encyclopedia of Life Sciences* 1–5. doi:10.1038/npg.els.0001306.

Gunes, A., A. Inal, M. Alpaslan, F. Eraslan, E. G. Bagci, and N. Cicek. 2007. Salicylic acid induced changes on some physiological parameters symptomatic for oxidative stress and mineral nutrition in maize (*Zea mays* L.) grown under salinity. *Journal of Plant Physiology* 164: 728–736. doi:10.1016/j.jplph.2005.12.009.

Hao, W., H. Guo, J. Zhang, G. Hu, Y. Yao, and J. Dong. 2014. Hydrogen peroxide is involved in salicylic acid-elicited rosmarinic acid production in *Salvia miltiorrhiza* cell cultures. *The Scientific World Journal* 2014: 843764. doi:10.1155/2014/843764.

Hara, I., J. Furukawa, A. Sato, T. Mizoguchi, and K. Miura. 2012. Abiotic stress and role of salicylic acid in plants. In: *Abiotic Stress Responses in Plants - Metabolism, Productivity and Sustainability*, eds. P. Ahmad, and M. N. V. Prasad, pp. 235–252. Springer, New York.

Harfouche, A. L., E. Rugini, F. Mencarelli, R. Botondi, and R. Muleo. 2008. Salicylic acid induces H_2O_2 production and endochitinase gene expression but not ethylene biosynthesis in *Castanea sativa* in vitro model system. *Journal of Plant Physiology* 165: 734–744. doi:10.1016/j.jplph.2007.03.010.

Hashempour, A., M. Ghasemnezhad, R. Fotouhi Ghazvini, and M. M. Sohani. 2014. The physiological and biochemical responses to freezing stress of olive plants treated with salicylic acid. *Russian Journal of Plant Physiology* 61: 443–450. doi:10.1134/S1021443714040098.

Hayat, S., S. A. Hasan, Q. Fariduddin, and A. Ahmad. 2008. Growth of tomato (*Lycopersicon esculentum*) in response to salicylic acid under water stress. *Journal of Plant Interactions* 3: 297–304. doi:10.1080/17429140802320797.

Herrera-Vásquez, A., P. Salinas, and L. Holuigue. 2015. Salicylic acid and reactive oxygen species interplay in the transcriptional control of defense genes expression. *Frontiers in Plant Science* 6: 171. doi:10.3389/fpls.2015.00171.

Hilo, A., F. Shahinnia, U. Druege, *et al.* 2017. A specific role of iron in promoting meristematic cell division during adventitious root formation. *Journal of Experimental Botany* 68: 4233–4247. doi:10.1093/jxb/erx248.

Horváth, E., G. Szalai, and T. Janda. 2007. Induction of abiotic stress tolerance by salicylic acid signaling. *Journal of Plant Growth Regulation* 26: 290–300. doi:10.1007/s00344-007-9017-4.

IPCC. 2013. *Climate Change 2013: The Physical Science Basis. Contribution of Working Group I to the Fifth Assessment Report of the Intergovernmental Panel on Climate Change.* Cambridge University Press, New York.

Jain, M., and J. P. Khurana. 2009. Transcript profiling reveals diverse roles of auxin-responsive genes during reproductive development and abiotic stress in rice. *The FEBS Journal* 276: 3148–62. doi:10.1111/j.1742-4658.2009.07033.x.

Jalal, R. S., A. E. Moftah, and S. O. Bafeel. 2012. Effect of salicylic acid on soluble sugars, proline and protein patterns of shara (*Plectranthus tenuiflorus*) plants grown under water stress conditions. *International Research Journal of Agricultural Science and Soil Science* 2: 400–407.

Javaheri, M., K. Mashayekhi, A. R. Dadkhah, and T. F. Zaker. 2012. Effects of salicylic acid on yield and quality characters of tomato fruit (*Lycopersicum esculentum* Mill.). *International Journal of Agriculture and Crop Sciences* 4: 1184–187.

Jesus, C., M. Meijón, P. Monteiro, *et al.* 2015. Salicylic acid application modulates physiological and hormonal changes in *Eucalyptus globulus* under water deficit. *Environmental and Experimental Botaby* 118: 56–66. doi:10.1016/j.envexpbot.2015.06.004.

Jumali, S. S., I. M. Said, I. Ismail, and Z. Zainal. 2011. Genes induced by high concentration of salicylic acid in *'Mitragyna speciosa'*. *Australian Journal of Crop Science* 5: 296–303.

Kabiri, R., F. Nasibi, and H. Farahbakhsh. 2014. Effect of exogenous salicylic acid on some physiological parameters and alleviation of drought stress in *Nigella sativa* plant under hydroponic culture. *Plant Protection Science* 50: 43–51. doi:10.17221/56/2012-PPS.

Kang, G., G. Li, W. Xu, *et al.* 2012. Proteomics reveals the effects of salicylic acid on growth and tolerance to subsequent drought stress in wheat. *Journal of Proteome Research* 11: 6066–79. doi:10.1021/pr300728y.

Kang, G. Z., G. Z. Li, G. Q. Liu, *et al.* 2013. Exogenous salicylic acid enhances wheat drought tolerance by influence on the expression of genes related to ascorbate-glutathione cycle. *Biololia Plantarum* 57: 718–724. doi:10.1007/s10535-013-0335-z.

Kazemi, M. 2013. Foliar application of salicylic acid and calcium on yield, yield component and chemical properties of strawberry. *Bulletin of Environment, Pharmacology and Life Sciences* 2: 19–23.

Khalil, F., K. M. Qureshi, A. Khan, H. Fakharul, and N. Bibi. 2012. Effect of girdling and plant growth regulators on productivity in olive (*Olea europaea*). *Pakistan Journal of Agricultural Research* 25: 30–38.

Khan, M. I. R., M. Asgher, and N. A. Khan. 2014. Alleviation of salt-induced photosynthesis and growth inhibition by salicylic acid involves glycinebetaine and ethylene in mungbean (*Vigna radiata* L.). *Plant Physiology and Biochemestry* 80: 67–74. doi:10.1016/j.plaphy.2014.03.026.

Khan, M. I. R., M. Fatma, T. S. Per, N. A. Anjum, and N. A. Khan. 2015. Salicylic acid-induced abiotic stress tolerance and underlying mechanisms in plants. *Frontiers in Plant Science* 6: 462. doi:10.3389/fpls.2015.00462.

Konrad, W., J. Burkhardt, M. Ebner, and A. Roth-Nebelsick. 2015. Leaf pubescence as a possibility to increase water use efficiency by promoting condensation. *Ecohydrology* 8: 480–492. doi:10.1002/eco.1518.

Kumar, D. 2014. Salicylic acid signaling in disease resistance. *Plant Science* 228: 127–134. doi:10.1016/j.plantsci.2014.04.014.

Li, G., X. Peng, L. Wei, and G. Kang. 2013. Salicylic acid increases the contents of glutathione and ascorbate and temporally regulates the related gene expression in salt-stressed wheat seedlings. *Gene* 529: 21–325. doi: 10.1016/j.gene.2013.07.093.

Li, Q., Y. Liu, Z. Pan, S. Xie, and S.-A. Peng. 2016. Boron deficiency alters root growth and development and interacts with auxin metabolism by influencing the expression of auxin synthesis and transport genes. *Biotechnology and Biotechnological Equipment* 30: 661–668. doi:10.1080/13102818.2016.11 66985.

Medrano, H., M. Tomás, S. Martorell, *et al.* 2015. From leaf to whole-plant water use efficiency (WUE) in complex canopies: Limitations of leaf WUE as a selection target. *The Crop Journal* 3: 220–228. doi:10.1016/j.cj.2015.04.002.

Misra, N., and R. Misra. 2012. Salicylic acid changes plant growth parameters and proline metabolism in *Rauwolfia serpentina* leaves grown under salinity stress. *American-Eurasian Journal of Agricultural and Environmental Sciences* 12: 1601–1609.

Miura, K., and Y. Tada. 2014. Regulation of water, salinity, and cold stress responses by salicylic acid. *Frontiers in Plant Science* 5: 1–12. doi:10.3389/fpls.2014.00004.

Nazar, R., S. Umar, N. A. Khan, and O. Sareer. 2015. Salicylic acid supplementation improves photosynthesis and growth in mustard through changes in proline accumulation and ethylene formation under drought stress. *South African Journal of Botany* 98: 84–94. doi:10.1016/j.sajb.2015.02.005.

Pál, M., G. Szalai, V. Kovács, O. K. Gondor, and T. Janda. 2013. Salicylic acid-mediated abiotic stress tolerance. In: *Salicylic Acid: Plant Growth and Development*, eds. Hayat, S., A. Ahmad, and M. N. Alyemeni, pp. 183–248. Springer, New York.

Peleg, Z., and E. Blumwald. 2011. Hormone balance and abiotic stress tolerance in crop plants. *Current Opinion in Plant Biology* 14: 290–295. doi:10.1016/j.pbi.2011.02.001.

Peng, Y. B., Y. F. Lu, and D. P. Zhang. 2003. Abscisic acid activates ATPase in developing apple fruit especially in fruit phloem cells. *Plant, Cell and Environment* 26: 1329–1342. doi:10.1046/j.1365-3040.2003.01057.x.

Petridis, A., I. Therios, G. Samouris, S. Koundouras, and A. Giannakoula. 2012. Effect of water deficit on leaf phenolic composition, gas exchange, oxidative damage and antioxidant activity of four Greek olive (*Olea europaea* L.) cultivars. *Plant Physiololy and Biochemestry* 60: 1–11. doi:10.1016/j.plaphy.2012.07.014.

Petrov, V. D., and F. Van Breusegem. 2012. Hydrogen peroxide – a central hub for information flow in plant cells. *AoB Plants* pls014: 1–13. doi:10.1093/aobpla/pls014.

Pinto-Marijuan, M., and S. Munne-Bosch. 2014. Photo-oxidative stress markers as a measure of abiotic stress-induced leaf senescence: Advantages and limitations. *Journal of Experimental Botany* 65: 3845–3857. doi:10.1093/jxb/eru086.

Rai, V. K., S. S. Sharma, and S. Sharma. 1986. Reversal of ABA-induced stomatal closure by phenolic compounds. *Journal of Experimental Botany* 37: 129–134. doi:10.1093/jxb/37.1.129.

Resco de Dios, V., J. Roy, J. P. Ferrio, J. G. Alday, D. Landais, A. Milcu, and A. Gessle. 2015. Processes driving nocturnal transpiration and implications for estimating land evapotranspiration. *Scientific Reports* 5: 10975. doi:10.1038/srep10975.

Rout, G., and S. Sahoo. 2015. Role of iron in plant growth and metabolism. *Reviews in Agricultural Science* 3: 1–24 doi:10.7831/ras.3.1.

Sakhabutdinova, A. R., D. R. Fatkhutdinova, M. V. Bezrukova, and F. M. Shakirova. 2003. Salicylic acid prevents the damaging action of stress factors on wheat plants. *Bulgarian Journal of Plant Physiology* Special Issue: 314–319.

Sanders, G. J., and S. K. Arndt. 2012. Osmotic adjustment under drought conditions. In: *Plant Responses to Drought Stress - From Morphological to Molecular Features*, ed. R. Aroca, pp. 199–230. Springer, New York.

Shaaban, M. M., A. M. K. Abd El-Aal, and F. F. Ahmed. 2011. Insight into the effect of salicylic acid on apple trees growing under sandy saline soil. *Research Journal of Agriculture and Biological Sciences* 7: 150–156.

Shakirova, F. M., A. R. Sakhabutdinova, M. V. Bezrukova, R. A. Fatkhutdinova, and D. R. Fatkhutdinova. 2003. Changes in the hormonal status of wheat seedlings induced by salicylic acid and salinity. *Plant Science* 164: 317–322. doi:10.1016/S0168-9452(02)00415-6.

Sharma, E., R. Sharma, P. Borah, M. Jain, and J. P. Khurana. 2015. Emerging roles of auxin in abiotic stress responses. In: *Elucidation of Abiotic Stress Signaling in Plants: Functional Genomics Perspectives*, ed. G. K. Pandey, pp. 299–328. Springer, New York.

Singh, B., and K. Usha. 2003. Salicylic acid induced physiological and biochemical changes in wheat seedlings under water stress. *Plant Growth Regulation* 39: 137–141. doi:10.1023/A:1022556103536.

Szepesi, A., J. Csiszár, K. Gémes, *et al.* 2009. Salicylic acid improves acclimation to salt stress by stimulating abscisic aldehyde oxidase activity and abscisic acid accumulation, andincreases Na+ content in leaves without toxicity symptoms in *Solanum lycopersicum* L. *Journal of Plant Physiology* 166: 914–925. doi:10.1016/j.jplph.2008.11.012.

Tanaka, Y., T. Sano, M. Tamaoki, N. Nakajima, N. Kondo, and S. Hasezawa. 2006. Cytokinin and auxin inhibit abscisic acid-induced stomatal closure by enhancing ethylene production in *Arabidopsis*. *Journal of Experimental Botany* 57: 2259–2266. doi:10.1093/jxb/erj193.

Umebese, C., T. Olatimilehin, and T. Ogunsusi. 2009. Salicylic acid protects nitrate reductase activity, growth and proline in amaranth and tomato plants during water deficit. *American Journal of Agricultural and Biological Sciences* 4: 224–229. doi: 10.3844/ajabssp.2009.224.229.

Varone, L., and L. Gratani. 2015. Leaf respiration responsiveness to induced water stress in Mediterranean species. *Environmental and Experimental Botany* 109: 141–150. doi:10.1016/j.envexpbot.2014.07.018.

Wang, L.-J., L. Fan, W. Loescher, *et al.* 2010. Salicylic acid alleviates decreases in photosynthesis under heat stress and accelerates recovery in grapevine leaves. *BMC Plant Biology* 10: 34. doi:10.1186/1471-2229-10-34.

Wang, L.-J., and S.-H. Li. 2006. Salicylic acid-induced heat or cold tolerance in relation to Ca^{2+} homeostasis and antioxidant systems in young grape plants. *Plant Science* 170: 685–694. doi:10.1016/j.plantsci.2005.09.005.

Wang, Y., H. Zhang, P. Hou, *et al.* 2014. Foliar-applied salicylic acid alleviates heat and high light stress induced photoinhibition in wheat (*Triticum aestivum*) during the grain filling stage by modulating the psbA gene transcription and antioxidant defense. *Plant Growth Regulation* 73: 289–297. doi:10.1007/s10725-014-9889-9.

Wani, S. H., V. Kumar, V. Shriram, and S. K. Sah. 2016. Phytohormones and their metabolic engineering for abiotic stress tolerance in crop plants. *The Crop Journal* 4: 162–176. doi:10.1016/j.cj.2016.01.010.

Waraich, E., R. Ahmad, S. Ullah, M. Ashraf, and Ehsanullah. 2011. Role of mineral nutrition in alleviation of drought stress in plants. *Australian Journal of Crop Science* 5: 764–777.

Yildirim, E., M. Turan, and I. Guvenc. 2008. Effect of foliar salicylic acid applications on growth, chlorophyll, and mineral content of cucumber grown under salt stress. *Journal of Plant Nutrition* 31: 593–612. doi:10.1080/01904160801895118.

Zagorchev, L., C. E. Seal, I. Kranner, and M. Odjakova. 2013. A central role for thiols in plant tolerance to abiotic stress. *International Journal of Molecular Sciences* 14: 7405–7432. doi:10.3390/ijms14047405.

Zhao, H.-J., X.-J. Zhao, P.-F. Ma, *et al.* 2011. Effects of salicylic acid on protein kinase activity and chloroplast D1 protein degradation in wheat leaves subjected to heat and high light stress. *Acta Ecologica Sinica* 31: 259–263. doi:10.1016/j.chnaes.2011.06.006.

8 Role of Nitric Oxide in Plant Abiotic Stress Tolerance

Gábor Feigl, Árpád Molnár, Dóra Oláh, and Zsuzsanna Kolbert

CONTENTS

8.1 INTRODUCTION

8.1.1 NO Synthesis, Storage and Transport in Higher Plants

In land plants, the ability to produce and release NO was first demonstrated in the leaves of herbicide-treated soybean plants (Klepper 1979). Over the following 40 years, numerous studies revealed that NO is produced also in plants grown during normal conditions, and the active research has tried to explore the mechanism of NO synthesis. Presently, it is widely accepted that in higher plants NO can be synthetized by oxidative and reductive mechanisms involving enzymes and also non-enzymatic processes.

The oxidation of reduced nitrogen compounds is a possible way to release NO. For these reactions, reduced N compounds such as L-arginine, polyamines or hydroxylamine may be substrates. Using biochemical approaches L-arginine-dependent NO formation was detected in different plant species affected by abiotic stresses such as salt, elevated CO_2, iron deficiency and cadmium (Zhao et al. 2007, Du et al. 2015, Jin et al. 2011, Besson-Bard et al. 2009), although the gene for the mammalian-like nitric oxide synthase (NOS) enzyme has not been identified in land plants since. Therefore, Corpas and Barroso (2017) argue the hypothesis that L-arginine-dependent NOS-like activity in higher plants could be the result of cooperation between discrete proteins. In contrast to higher plants, in algae like *Ostreococcus tauri* and *Synechococcus* PCC 7335 NOS enzymes functionally and structurally similar to mammalian NOS have been characterized (Foresi et al. 2010, Correa-Aragunde et al. 2018). Furthermore, in the case of OtNOS its expression in *Arabidopsis thaliana* resulted in elevated NO production and also led to increased tolerance against salt and water stress (Foresi et al. 2015). Similarly, the expression of rat neuronal NOS improved the salt and drought tolerance of transgenic *Arabidopsis* (Shi et al. 2011). Also, polyamines (PAs) are good candidates for oxidative NO release; however, the mechanism is still unclear. Copper-amine oxidase1 (CuAO1) was found to be involved in PA-induced NO formation, and *cuao1-1* and *cuao1-2* mutants were shown to prevent PA-induced NO formation and insensitivity to abscisic acid (ABA) and osmotic stress, suggesting the involvement of PA-induced NO synthesis during osmotic stress (Wimalasekera et al. 2011). Moreover, exogenous hydroxylamine and salicylhydroxamate induced NO release in tobacco cell cultures, suggesting that these N compounds can be substrates for plant NO synthesis, although both the mechanism and the biological relevance of this reaction remain unexplained (Rümer et al. 2009).

It is clear that NO production is tightly connected to plant nitrate assimilation (Sanz-Luque et al. 2013) and reductive reactions lead to NO formation. The nitrate reductase (NR) enzyme was shown to directly reduce nitrite to NO; however, the NO-producing activity is 1% of its nitrate-reducing activity *in vitro* (Rockel et al. 2002). Moreover, a high nitrate concentration inhibits the NO-producing activity of NR; thus during normal circumstances the direct role of NR in plant NO synthesis may be moderate (Rockel et al. 2002). Despite these limitations, numerous data support the involvement of NR in NO production during stress situations like osmotic stress, cold stress, aluminium, copper or ozone exposure (Kolbert et al. 2010, Zhao et al. 2019, Wang et al. 2010, Sun et al. 2014, Hu et al. 2015a, Xu et al. 2012). In light of some recent evidence, the key process of NO synthesis can be realized with the indirect involvement of NR. The NR enzyme transfers electrons from NAD(P)H to the NO-forming nitrite reductase (NOFNiR) which catalyses the reduction of nitrite to NO *in vitro* and *in vivo* (Chamizo-Ampudia et al. 2016). This observation was made in *Chlamydomonas* but the authors argue that the NR-NOFNiR system can be an important NO source also in higher plants (Chamizo-Ampudia et al. 2017). However, the significance of the NR-NOFNiR system in NO synthesis during stress responses still has to be revealed. Besides NR, other putative enzymatic sources of NO using nitrite as a substrate have been suggested, e.g. peroxisomal enzyme xanthine oxidoreductase (XOR, Zhang et al. 1998) and root cell-specific nitrite:NO reductase (NiNOR, Stöhr et al. 2001). XOR is the ubiquitous molybdenum-containing enzyme found in peroxisomes and has been shown to reduce organic and inorganic nitrates and consequently produce NO (Godber et al. 2000). The enzyme NiNOR was found to be located in the plasma membrane of tobacco root cells and uses cytochrome *c* as an electron donor (Stöhr et al. 2001), but it has yet to be cloned and fully identified. Furthermore, the production of NO and ATP *via* cytochrome *c* oxidase and/or reductase and possibly by alternative oxidase was observed at the inner membrane of barley root mitochondria in hypoxic conditions (Stoimenova et al. 2007). Additional reductive processes of plant NO production without the involvement of enzymatic activities have been revealed, e.g. nitrite is reduced at acidic pH in the presence of ascorbate in the cell walls of barley aleurone layers, leading to the formation of NO (Bethke et al. 2004).

In addition to synthesis, removal also plays a role in the regulation of steady-state NO levels. Due to its relatively unstable character, NO can undergo numerous reactions. In the presence of molecular oxygen, it forms nitrite and nitrate, and it interacts with reactive oxygen species (ROS) superoxide to form peroxynitrite ($ONOO^-$) (Beckman et al. 1990). The overproduction of $ONOO^-$ in different organs of plants exposed to abiotic stresses (e.g. heavy metals) has been documented by numerous studies (Feigl et al. 2013, 2014, 2016, Corpas and Barroso 2014, Gzyl et al. 2016, Molnár 2018a, b, Kolbert et al. 2018). Moreover, NO may also react with proteins, like the non-symbiotic haemoglobins (nsHbs), and this interaction facilitates its oxidation to nitrate (Perazzolli et al. 2004, Hebelstrup et al. 2006). This interaction seems to be relevant during anoxia or hypoxia (Dordas et al. 2003, 2004, Perazzolli et al. 2005). Another possible mechanism for NO removal takes place with the involvement of the NR enzyme which can transfer electrons to the truncated haemoglobin THB1, catalysing the conversion of NO into nitrate by its dioxygenase activity (Sanz-Luque et al. 2015, Chamizo-Ampudia et al. 2017). Nitric oxide can initiate *S*-nitrosation reactions with thiol (SH)-containing proteins and peptides, resulting in the formation of low-molecular-weight *S*-nitrosothiols such as *S*-nitrosocysteine (CysNO) and *S*-nitrosoglutathione (GSNO) (Hogg, 2000; Foster et al. 2003) The *S*-nitrosothiols perform biological functions including NO liberation, trans-nitrosation and *S*-thiolation (Hogg, 2000; Stamler et al. 2001). The most abundant *S*-nitrosothiol is

GSNO which can non-enzymatically liberate NO or be reduced by the enzyme *S*-nitrosoglutathione reductase (GSNOR), yielding oxidized glutathione (GSSG) and NH_3 (Barroso et al. 2006; Corpas et al. 2008; Leterrier et al. 2011). In the leaf of cadmium-exposed pea plants, GSNO content significantly reduced; this was accompanied by a related reduction in GSNOR activity (Barroso et al. 2006). In the case of arsenate stress, *Arabidopsis thaliana* showed a significant reduction in GSNO content and elevation in GSNOR activity (Leterrier et al. 2012), similarly to sunflower plants exposed to mechanical stress or high temperatures (Chaki et al. 2010, 2011). These results indicate that the activity of GSNOR is influenced by abiotic stress stimuli and GSNO levels are regulated by GSNOR in stressed plants. Besides being an intracellular NO reservoir, GSNO may also be transported between cells, tissues and organs, implementing long-distance transport of the NO signal (Lindermayr 2018, Begara-Morales et al. 2018). The possible routes of NO synthesis and removal are depicted in Figure 8.1.

8.1.2 Perception and Transduction of NO Signal

The perception of the NO signal and its transduction is believed to be realized through posttranslational modifications (PTMs) such as *S*-nitrosation, tyrosine nitration and metal nitrosylation (Umbreen et al. 2018).

S-nitrosation is a reversible covalent chemical reaction affecting the cysteine thiol groups, and as a consequence, *S*-nitrosothiol (-SNO) is generated. This PTM is catalysed by, e.g. the higher oxides of NO or nitrosonium cation (NO^+), by metal–NO complexes, by low-molecular-weight *S*-nitrosothiols ((*S*-nitrosocysteine), CysNO) or by *S*-nitrosoglutathione (GSNO) (Lamotte et al. 2015). *S*-nitrosation may induce conformational changes of proteins, affecting their activities, subcellular localization and interactions or binding activities. In plant systems, Hu et al. (2015b) have provided the most comprehensive dataset of *S*-nitrosated proteins so far, since they identified 1,195

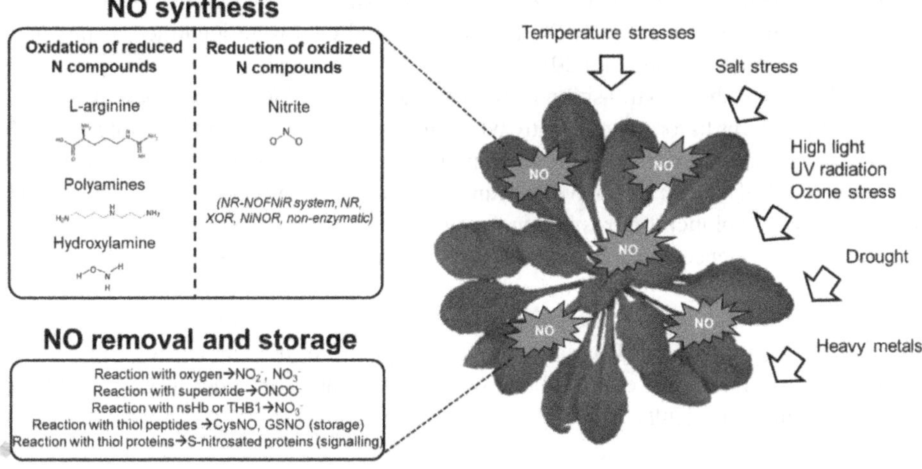

FIGURE 8.1 Routes of NO synthesis and removal in general and in abiotic-stressed plants. L-arginine, polyamines and hydroxylamine are possible substrates for oxidative NO synthesis. Reductive pathways use nitrite as a substrate and are mediated by enzymes like nitrate reductase-NO, forming the reductase system, xanthine oxidoreductase, nitrite:NO reductase. Additionally, the non-enzymatic reduction of nitrite may also lead to NO formation. Nitric oxide reacts with molecular oxygen or superoxide to form nitrite, nitrate or peroxynitrite, respectively. The interaction of NO with non-symbiotic haemoglobin (nsHb) or with truncated haemoglobin (THB1) leads to NO removal. Nitric oxide reacts with thiol peptides (cysteine or glutathione) and with thiol proteins, leading to the formation of *S*-nitrosated peptides and proteins. *S*-nitrosoglutathione (GSNO) serves as an NO reservoir and long-distance transport form, while *S*-nitrosation of proteins serves a signalling function.

endogenously *S*-nitrosated peptides in 926 proteins in the *Arabidopsis* proteome. Environmental stimuli like cold, salt, cadmium exposure or ozone induce modifications in the *S*-nitrosation of proteins like ribulose-1,5-bisphosphate carboxylase/oxygenase, ascorbate peroxidase (APX), catalase (CAT), guaiacol peroxidase, phytochelatins, dehydroascorbate reductase, isocitrate dehydrogenase, glutathione-*S*-transferase, phenylalanine ammonia lyase, etc. (Camejo et al. 2013, Abat and Deshwal 2009, Elviri et al. 2010, Vanzo et al. 2014, Ortega-Galisteo et al. 2012), as was recently reviewed by Fancy et al. (2017).

Additionally, NO is able to influence protein activity indirectly through the formation of ONOO- in a reaction with superoxide; therefore NO can be considered a direct scavenger of superoxide radical. Peroxynitrite formation leads to protein tyrosine nitration (PTN). PTN is an irreversible two-step posttranslational modification during which a nitro group ($-NO_2$) binds to the aromatic ring of tyrosine (Tyr) in the *ortho* position, resulting in the formation of 3-nitrotyrosine (Souza et al. 2008). In plant cells, PTN mostly inhibits the activity of the particular enzyme protein (reviewed by Kolbert et al. 2017); however, in animal systems, PTN-triggered activation or no change in activity also occurs (Yeo et al. 2015). Tyrosine nitration may either prevent or induce the tyrosine phosphorylation; thus despite its irreversible nature it can influence cell signalling after all (Souza et al. 2008). In an comprehensive study, 127 nitrated proteins were identified in wild-type *Arabidopsis thaliana* grown in control conditions (Lozano-Juste et al. 2011), indicating that plants possess a physiological nitroproteome (Kolbert et al. 2017). In the case of abiotic stress situations, protein nitration intensifies, as recently published in the cases of drought-stressed *Lotus japonicus* (Signorelli et al. 2018), salt-treated sunflower seedlings (Jain and Bathla 2018), cadmium-treated *Arabidopsis thaliana* (Liu et al. 2018) and selenium-exposed *Arabidopsis thaliana*, *Brassica juncea* (Molnár et al. 2018 ab) and *Astragalus membranaceus* (Kolbert et al. 2018). Through the above-mentioned PTMs (*S*-nitrosation and tyrosine nitration), NO is able to regulate ROS metabolism. NO has been shown to induce the *S*-nitrosation of NADPH oxidase and consequently inhibits its superoxide-producing activity (Yun et al. 2011). Remarkably, besides the direct scavenging of superoxide, NO also regulates its production. Other enzymes like glycolate oxidase, APX, superoxide dismutase (SOD), CAT, glutathione reductase (GR) and peroxidases (POD) were identified as targets of NO action (reviewed by Kolbert and Feigl 2018).

Nitric oxide can also bind to transition metal ions like iron (Fe^{2+} or Fe^{3+}), copper (Cu^{2+}) and zinc (Zn^{2+}) in metalloproteins to form metal–nitrosyl complexes (Russwurm and Koesling 2004). Metal-nitrosylation can effectively prevent peroxidation of the metal, thus making ROS formation impossible (Wink et al. 2001) which may have relevance in plants grown under stressful conditions. The biological significance of metal-nitrosylation, however, needs to be further analysed.

Collectively, the evidence points in the direction that, in plants exposed to environmental stresses, NO is perceived as a general stress signal and acts at the proteome level, influencing gene expression. The next part of the chapter discusses the recent evidence regarding the mitigating effect of exogenous NO during stresses such as salt, drought, temperature stress and heavy metals. The most relevant publications have been collected and organized in tables in order to give a better understanding of the numerous results.

8.2 ROLE OF NITRIC OXIDE IN PLANT ABIOTIC STRESS TOLERANCE

8.2.1 Salt Stress

Approximately 6% of the world's total land surface is subjected to salt stress, which is constantly increasing due to climate change caused by anthropogenic and natural activities (Munns and Tester, 2008; Setia et al. 2013). Salt stress affects plants at the germination, growth and developmental stages; at the physiological level it triggers osmotic and ionic stress, interrupting plant–water relations and disturbing cell division and expansion (Munns and Tester, 2008). In the presence of excess salt, plants are not able to take up water (osmotic stress), which leads to a decrease of stomatal conductance and photosynthetic activity. Moreover, ionic stress can be associated with nutrient imbalance as well. The alterations in the physiological processes due to salinity lead to the increased production of ROS, causing oxidative stress and leading to oxidative damage of cell membranes and other cellular components, such as proteins and DNA (Hasanuzzaman et al. 2013a, b, c).

NO is able to mitigate oxidative stress through different mechanisms; probably the most important way is the upregulation of the enzymatic and non-enzymatic antioxidant defence systems (Christou et al. 2014; Ahmad et al. 2016; Kong et al. 2016; da Silva et al. 2017). NO is also known to be a secondary antioxidant, counteracting the effect of produced ROS by either scavenging them or by the previously mentioned upregulation of antioxidant defence systems (Siddiqui et al. 2011; Hasanuzzaman et al. 2013a; Arora et al. 2016). Many recent studies discuss the effect of external NO on the antioxidant capacity of salt-stressed plants, and according to the results reviewed by Hasanuzzaman et al. (2018), NO treatment results in increased antioxidant activity and decreased oxidative damage, thus enhancing the salt-stress tolerance of the plants, as demonstrated in *Aegiceras corniculatum* (Chen et al. 2014), *Brassica juncea* (Fatma and Khan, 2014; Zhao et al. 2018), *Cicer arientium* (Sheokand et al. 2010), *Cucumis sativus* (Shi et al. 2007; Fan et al. 2007), *Fragaria X ananassa* (Kaya et al. 2018), *Gossypium hirsutum* (Liu et al. 2014), *Hordeum vulgare* (Li et al. 2008) and *Solanum lycopersicum* (Ahmad et al. 2018a). Also in tomato, sodium nitroprusside (SNP, NO donor) could slow down electron transport and inhibit the photochemical-quenching coefficient (Wu et al. 2010), while in *Oryza sativa* exogenous NO restored photosynthetic capacity (Uchida et al. 2002). In case of *Triticum aestivum*, SNP treatment restored the growth of salt-stressed seedlings (Zheng et al. 2009; Kausar et al. 2013). Moreover, in some cases stress-specific NO action was also demonstrated, e.g. sodium nitroprusside (SNP) decreased root-to-shoot Na^+ translocation in *Kosteletzkya virginica* (Guo et al. 2005), and SNP facilitated Na^+ compartmentalization in *Zea mays* (Zhang et al. 2006). Also in *Z. mays*, SNP treatment increased the nutrient content of plants suffering from salt stress (Çeli̇k and Eraslan, 2015), but alleviated oxidative stress was also reported (Klein et al. 2018). Literature data regarding exogenous NO effects on salt-stressed plants are presented in Table 8.1.

8.2.2 Water Stress

Drought is a major problem in agriculture with the changing climate, as it limits crop productivity. By definition, drought means a condition in which the moisture deficit is serious enough to negatively affect vegetation (White and Brinkmann, 1975), a state in which the available water cannot allow normal crop production. Drought has many effects on plants: water potential decrease causes stomatal closure and turgor loss (Yordanov et al. 2000), causing inhibited cell growth (Wu et al. 2008) and photosynthesis (Flexas and Medrano, 2002), among many other symptoms (Sidana et al. 2015). Successful mechanisms to survive drought stress include the regulation of stomatal closure, and coping with oxidative stress. The role of NO as a signal molecule is well-established; it activates antioxidant enzymes such as CAT, POD, APX and SOD under limited water conditions (Sidana et al. 2015), alleviating the effects of oxidative stress like chlorosis and apoptosis (Beligni and Lamattina, 1999). The exogenous application of NO through donors has been observed to alleviate drought-induced stress processes in numerous studies through the alleviation of oxidative stress and the decrease of photosynthesis, including in crops like *Brassica napus* (Akram et al. 2018), *Crambe abyssinica* (Batista et al. 2018), *Oryza sativa* (Farooq et al. 2009; Cao et al. 2018), *Triticum aestivum* (Xing et al. 2004; Boyarshinov and Asafova, 2011) and *Zea mays* (Hao et al. 2008; Majeed et al. 2018; Cheng et al. 2002) and in deciduous trees, like *Malus hupehensis* (Zhang et al. 2016) and *Populus przewalskii* (Lei et al. 2007). Moreover, Nω-nitro-L-arginine treatment of drought-stressed wheat roots enhanced their drought-tolerance by the upregulation of ABA synthesis (Zhao et al. 2001). Literature data regarding exogenous NO effects on drought-stressed plants are presented in Table 8.2.

TABLE 8.1
Effect of Exogenously Applied NO in Alleviation of Salt Stress

Species	Salt Stress	NO Treatment	Effect	Reference
Aegiceras corniculatum	350 mM NaCl	100 μM SNP	Decreased oxidative damage and peroxides content, increased glutathione and polyphenol levels.	Chen et al. (2014)
Brassica juncea	100 mM NaCl	50 or 100 μM SNP	Restored leaf area and dry matter production, enhanced photosynthesis.	Fatma and Khan (2014)
	200 mM NaCl	10 μM SNP	Counteracted growth inhibition, restored redox homeostasis.	Zhao et al. (2018)
Cicer arietinum	250 mM NaCl	0.2 or 1 mM SNP	Increased catalase (CAT), glutathione reductase (GR) and superoxide dismutase (SOD) activity; decreased lipid peroxidation and membrane damage.	Sheokand et al. (2010)
Cucumis sativus	100 mM NaCl	50 μM SNP	Enhanced antioxidant enzyme activities and decreased H_2O_2 accumulation in root mitochondria.	Shi et al. (2007)
	50 mM NaCl	10–400 μM SNP	Decreased membrane damage and lipid peroxidation, increased antioxidant capacity.	Fan et al. (2007)
Fragaria X ananassa	50 mM NaCl	0.1 mM SNP	Reduced electrolyte leakage, malondialdehyde (MDA) content and H_2O_2content. Increased antioxidant response.	Kaya et al. (2018)
Gossypium hirsutum	100 mM NaCl	0.1 mM SNP	Increased growth rate, photosynthesis and antioxidant enzyme activities. Alleviated inhibition of H^+-ATPase	Liu et al. (2014)
Hordeum vulgare	50 mM NaCl	50 μM SNP	Enhanced antioxidant capacity and increased ferritin accumulation.	Li et al. (2008)
Kosteletzkya virginica	100–400 mM NaCl	60 μM SNP	Decreased Na^+ translocation from the root to the shoot, preventing oxidative damage.	Guo et al. (2005)
Oryza sativa	100 mM NaCl	10 μM SNP	Restored growth inhibition, photosynthesis and antioxidant enzyme activity.	Uchida et al. (2002)
Solanum lycopersicum	100 mM NaCl	100 μM SNP	Slowed down electron transport rate and inhibited photochemical and non-photochemical quenching.	Wu et al. (2010)
	200 mM NaCl	50 μM SNAP (*S*-Nitroso-N-Acetyl-D,L-Penicillamine)	Improved SOD, CAT, ascorbate peroxidase (APX) and GR activities; elevated flavonoid, proline and glycine betaine synthesis.	Ahmad et al. (2018a)
Triticum aestivum	300 mM NaCl	0.1 mM SNP	Increased germination, seedling weight and respiration.	Zheng et al. (2009)
	150 mM NaCl	0.05, 0.1 or 0.15 mM SNP	Increased dry mass and overall plant size.	Kausar et al. (2013)
Zea mays	100 mM NaCl	10, 100 or 1000 μM SNP	Increased H^+-ATPase activity to facilitate Na^+ compartmentalization, leading to a high K^+/Na^+ ratio.	Zhang et al. (2006)
	40 mM NaCl	0.5 and 1 mM SNP	Increased N, Na, Cl, K, Ca, Mg, De, Zn and Mn content.	Çelik and Eraslan (2015)
	150 mM NaCl	10 μM DETA/NO (Diethylenetriamine/nitric oxide)	Reduced superoxide accumulation and lipid peroxidation, increased the activity of some SOD isoenzymes in leaves.	Klein et al. (2018)

TABLE 8.2
Effect of Exogenously Applied NO in Alleviation of Water Stress

Species	Drought Stress	NO Treatment	Effect	Reference
Brassica napus	Watering to 60% field capacity	0.02 mM SNP	Improved growth, chlorophyll *a*, glycinebetaine and total phenolics content, upregulated CAT activity and (elevated) total soluble protein content. Decreased relative membrane permeability, MDA, and POD activity.	Akram et al. (2018)
Crambe abyssinica	50% of the maximum water holding capacity	75 and 150 μM SNP	Maintenance of gas exchange and chlorophyll content, activation of antioxidant enzymes, increased intracellular NO and proline, decreased MDA and ROS levels.	Batista et al. (2018)
Malus hupehensis and *M. sieversii*	Watered to 40–45% of field moisture capacity	300 μM SNP	Alleviated ion leakage, MDA and soluble protein accumulation. Maintained water potential and content, increased antioxidant activity and photosynthesis.	Zhang et al. (2016)
Oryza sativa	Soil watered to 50% of field capacity	50, 100 or 150 μM SNP	Alleviated the effects of water deficit, maintained water potential, enhanced antioxidant response and photosynthetic activity.	Farooq et al. (2009)
	10% PEG-6000 (polyethylene glycol)	1 mM nitrate and/or 20 μM SNP	Alleviated oxidative damage (upregulation of antioxidant enzyme activities).	Cao et al. (2018)
Populus przewalskii	Soil watered to 25% of field capacity	0.2 mM SNP	External NO enhanced photosynthesis and the photochemical efficiency of PSII.	Lei et al. (2007)
Triticum aestivum	Drying until 20% of the root tips are lost	0.2 mM Nω-nitro-L-arginine	Enhanced drought tolerance by the upregulation of ABA synthesis.	Zhao et al. (2001)
	0.4 M mannitol	0.01–10 mM SNP	Improved photosynthetic rate and protein synthesis, maintained high relative water content and increased antioxidant enzyme activity – prevented excessive water loss.	Xing et al. 2004
	Drying from 5 min to 3 hours	50–500 μM SNP	Alleviated the effects of water deficit, elevated APX and CAT activities, decreased lipid peroxidation.	Boyarshinov and Asafova, 2011
Zea mays	Dehydration stress for 3 or 4 hours	0.2 mM SNP	Alleviated the effects of water deficit, increased SOD activity.	Hao et al. 2008
	"Withholding water to one set of pots"	100 μM SNP	Improved water status and chlorophyll content, alleviated oxidative stress (increased antioxidant enzyme activity).	Majeed et al. 2018
	–1.5 MPa PEG-6000 or dehydration	100 μM SIN or 100 μM SNP	Stimulated SOD activity and prevention of senescence.	Cheng et al. 2002

8.2.3 Low and High Temperatures

The relative tolerance to low temperatures is a fundamental factor of the different plant species' spread on the globe. The difference between freezing (exposure to sub-zero temperatures) and chilling (exposure to sub-optimal, but positive temperatures) is very important since, under temperate climates, most plants are able to tolerate chilling, but there is a big difference in their freezing tolerance (Baudouin and Jeandroz, 2015). Recent findings suggest that NO, among others, is an important component of signalling networks participating in low-temperature-stress responses. Numerous studies have found a correlation between exogenously applied NO and the enhanced chilling tolerance of plants, including monocotyledonous perennial *Anthurium andraeanum* (Liang et al. 2018), perennial herb *Chorispora bungeana* (Liu et al. 2010), monocotyledonous *Musa* spp. (Wang et al. 2013; Wu et al. 2014), important crop species like *Brassica campestris* (Fan et al. 2014), *Cucumis sativus* (Yang et al. 2011a), and *Triticum aestivum* (Esim et al. 2014) and woody plants, e.g. *Juglans regia* (Dong et al. 2018), *Mangifera indica* (Zaharah and Singh, 2011), *Prunus persica* (Zhu et al. 2010) and *Prunus salicina* (Singh et al. 2009b), mostly by the upregulation of antioxidant defence systems (Table 8.3).

High temperatures generate oxidative stress, lipid peroxidation, enzyme inactivation and disintegration of DNA in plants (Suzuki and Mittler, 2006); however, NO has the ability to protect them against heat stress by enhancing photosynthesis, as reported in *Chrysanthemum morifolium* (Yang et al. 2011b) and *Festuca arundinacea* (Chen et al. 2013); alleviating oxidative stress, like in *Oryza sativa* (Uchida et al. 2002) and *Phragmites australis* (Song et al. 2006); or both, as in *Phaseolus radiatus* (Yang et al. 2006) and *Solanum lycopersicum* (Siddiqui et al. 2017). In the important crop plants, *Triticum aestivum* (Karpets et al. 2015) and *Zea mays* (Li et al. 2013), external NO supplementation improved the survival rate and seedlings' resistance to heat damage. In the reduced endogenous NO mutant *Arabidopsis* line *nitric oxide associated1/resistant to inhibition by fosmidomycin1* (*noa1(rif1)*) heat tolerance was improved by NO donor pre-treatment, suggesting that calmodulin 3 is involved in NO signalling (Xuan et al. 2010), while NO donor treatment partially rescued the heat sensitivity of reduced-endogenous-H_2O_2-containing mutant lines (*atrbohB, atrbohD, atrbohB/D*) (Wang et al. 2014). Literature data regarding exogenous-NO effects on heat-stressed plants are presented in Table 8.4.

8.2.4 UV-B Exposure

Ultraviolet-B (UV-B, 280–315 nm) is a non-photosynthetically active radiation, but an abiotic factor that might lead to the increase of endogenous ROS and NO content in plants (A-H Mackerness et al. 2001; Zhang et al. 2003). Due to the depletion of the stratospheric ozone layer, high doses cause several negative effects, such as growth inhibition, decreased productivity, necrosis and oxidative stress (Frohnmeyer and Staiger, 2003). Though UV-B-mediated stress in plants occurs relatively rarely, the stress reaction depends on the duration and dose, immune status and reactive oxygen/nitrogen species homeostasis of the plant or the co-occurrence with other environmental factors (Yemets et al. 2015). Examples show that NO acts as an antioxidant and results in the detoxification of UV-B-induced ROS in numerous species, including the cyanobacteria *Spirulina platensis* (Xue et al. 2006, 2007, 2011) and green algae *Chlorella pyrenoidosa* (Chen et al. 2010), and important crop plants, such as *Glycine max* (Santa-Cruz et al. 2010), *Phaseolus vulgaris* (Shi et al. 2005), *Triticum aestivum* (Yang et al. 2013) and *Zea mays* (Zhang et al. 2003; Kim et al. 2010) (Table 8.5). Moreover, in the case of UV-B-stressed *Solanum tuberosum*, SNP application caused flavonoid, anthocyanin and shielding wax accumulation, all of which specifically protect plant tissues from UV-B damage (Beligni and Lamattina, 2001).

8.2.5 Heavy Metals

Due to natural or anthropogenic activities, the presence of heavy metals (HMs) in soils is a growing problem for agriculture and the environment as well. The mechanisms of HM toxicity in plants is not yet fully understood, but changes in the oxidative balance is a very important component (Laspina et al. 2005).

Exogenous NO is able to alleviate the negative effects of high concentrations of essential microelements, such as copper in *Arabidopsis thaliana* (Pető et al. 2013), *Lactuca sativa* (Shams et al. 2018), *Nicotiana tabacum* (Khairy et al. 2016) and *Oryza sativa* (Mostofa et al. 2014; Yu et al. 2005); zinc in *Carthamus tinctorius*, by reducing its translocation (Namdjoyan et al. 2017); and nickel in

TABLE 8.3
Effect of Exogenously Applied NO in Alleviation of Low Temperature-Induced Stress

Species	Chilling	NO Treatment	Effect	Reference
Anthurium andraeanum	15 days on 12/5°C (day/night)	0.2 mM SNP	Decreased MDA content and electrolyte leakage, improved photochemical efficiency, antioxidant enzyme activity (SOD, CAT, POD, APX) and antioxidant content (glutathione, ascorbic acid (AsA), phenolics), reduced accumulation of ROS.	Liang et al. (2018)
Brassica campestris ssp. *chinensis*	4 or 8 days at 5°C	100 μM SNP	Alleviated shoot growth inhibition. Enhanced antioxidant enzyme activities, chlorophyll and protein content. Reduced membrane permeability and MDA content.	Fan et al. (2014)
Chorispora bungeana	3 days at 0 or 4°C	0.1 mM SNP	Reduced ROS levels and increased APX, CAT, GR, POD and SOD activities. AsA and glutathione content increased.	Liu et al. (2010)
Cucumis sativus	15 days at 2°C	25 μL/L NO pre-treatment	Reduced increase in membrane permeability and lipid peroxidation, delayed increase in hydrogen peroxide and superoxide anion production. Increased activities of SOD, CAT, APX and POD.	Yang et al. (2011a)
Juglans regia	3 days at 4°C	100 μM SNP	Reduced electrolyte leakage and lipid peroxidation, improved photosynthetic efficiency. Increased levels of soluble sugar, proline, total phenol and glutathione.	Dong et al. (2018)
Mangifera indica	2 or 4 weeks' storage at 5°C	5, 10, 20 or 40 μL/L NO fumigation	Suppressed ethylene production and respiration rate. Delayed fruit colour development, softening, ripening.	Zaharah and Singh, (2011)
Musa spp.	Up to 20 days at 7°C	0.05 mM SNP	Lower chilling injury index, higher firmness, lower electrolyte leakage and MDA content. Higher (glutathione peroxidase) GPX, APX and GR activities and antioxidant capacity, reduced hydrogen peroxide and superoxide anion content. Accumulation of total phenolics and proline.	Wang et al. (2013)
	15 days at 7°C	60 μL/L NO gas for 3 h	Reduced electrolyte leakage and MDA content. Delayed increase in hydrogen peroxide and superoxide anion content. Higher SOD, CAT, POD and APX activity.	Wu et al. (2014)
Prunus persica	45 days at 5°C with intermittent warmings	15 μL/L NO gas	Extended post-harvest life with lower ethylene production. Side effects if intermittent warmings were counteracted.	Zhu et al. (2010)
Prunus salicina	5, 6 or 7 weeks' storage at 0°C	Fruit fumigation with 5, 10 or 20 μL/L NO gas	Suppression of respiration and ethylene production – delayed fruit ripening and alleviation of chilling injury symptoms.	Singh et al. (2009b)
Triticum aestivum	3 days at 5/2°C	0.1 mM SNP	Enhanced SOD, POX and CAT activity, decreased hydrogen peroxide, superoxide anion and MDA content.	Esim et al. (2014)

TABLE 8.4
Effect of Exogenously Applied NO in Alleviation of High Temperature-Induced Stress

Species	Heat Stress	NO Treatment	Effect	Reference
Arabidopsis thaliana	0–70 min at 45°C	20 μM SNP	Heat shock tolerance of reduced endogenous NO mutant line *noa1(rif1)* was improved by NO donor pre-treatment, suggesting that calmodulin 3 is involved in NO signalling.	Xuan et al. (2010)
	0–60 min at 45°C	20 μM SNP or SNAP	NO donor treatment partially rescued the heat sensitivity of reduced endogenous H_2O_2-containing mutant lines (*atrbohB, atrbohD, atrbohB/D*).	Wang et al. (2014)
Chrysanthemum morifolium	3, 6, 12 or 24 hours at 45°C	200 μM SNP	Enhanced photosynthesis, transpiration rate and stomatal conductance.	Yang et al. (2011b)
Festuca arundinacea	0–4 hours at 44°C	100 μM SNP	Improved functionality of PSII by the upregulated gene expression of PSII core proteins.	Chen et al. (2013)
Oryza sativa	5 hours at 50°C	1 μM SNP	Less severe heat damage after SNP pre-treatment and increased antioxidant enzyme activities.	Uchida et al. (2002)
Phaseolus radiatus	90 min at 45°C	0.15 mM SNP	Enhanced photochemical and antioxidant enzyme activity, membrane integrity.	Yang et al. (2006)
Phragmites australis	2 hours at 45°C	0.2 mM SNP or 0.2 mM SNAP	Lower increase in ion leakage and growth suppression. Increased activity of SOD, CAT and POX.	Song et al. (2006)
Solanum lycopersicum	4 h at 42°C	0.1 mM SNP	Combined with Ca reduced MDA and ROS levels, stimulated synthesis of compatible solutes and photosynthetic pigments. Enhanced antioxidant enzyme and Rubisco activities.	Siddiqui et al. (2017)
Triticum aestivum	1 min hardening at 42°C and 10 min shocking at 46°C	2 mM SNP	Increased seedling resistance to heat damage.	Karpets et al. (2015)
Zea mays	18 h at 48°C	50–400 μM SNP	Improved survival rate and decreased lipid peroxidation and survival rate.	Li Z. G. et al. (2015)

Eleusine coracana (Kotapati et al. 2017) and rice (Rizwan et al. 2018). External NO supplementation can also improve the physiological parameters of plants under an excess of toxic element aluminium in *Arachis hypogaea* (He et al. 2017; Dong et al. 2016), *Cassia tora* (Wang and Yang, 2005) and *Triticum aestivum* (Sun et al. 2014, 2015). Several studies report the positive effect of external NO on cadmium-stressed plants, like *Brassica juncea* (Verma et al. 2013; Per et al. 2017), *Cucumis sativus* (Yu et al. 2013), *Lolium perenne* (Chen et al. 2018), *Nicotiana tabacum* (Khairy et al. 2016), rice (Yang et al. 2016; Hsu and Kao, 2004; Xiong et al. 2009), *Solanum lycopersicum* (Ahmad et al. 2018b), *Typha angustifolia* (Zhao et al. 2016) and *Trifolium repens* (Liu et al. 2015). Moreover, exogenous NO was able to enhance chromium tolerance and translocation in *Festuca arundinacea* (Huang et al. 2018), and a number of studies report external NO-induced improvement of antioxidant systems in lead-stressed plants, including rice (Kaur et al. 2015), *Vigna radiata* (Singh et al. 2017b) and *Vigna unguiculata* (Sadeghipour, 2016). In *Lolium perenne*, NO is also able to decrease the root-to-shoot transport of Pb, besides the enhancement of the antioxidant response (Bai et al. 2015). Finally, NO supplementation can alleviate the negative effect of the metalloid arsenic by

TABLE 8.5
Effect of Exogenously Applied NO in Alleviation of UV-B-Radiation-Induced Stress

Species	UV Treatment	NO Treatment	Effect	Reference
Chlorella pyrenoidosa	4 hours	100–300 μM SNP	Restored activity of enzyme activities (GR, APX, nitrite/nitrate reductase, PM H^+-ATPase, glutamine synthetase, 5′-nucleotidase).	Chen et al. (2010)
Glycine max	100 min	800 μM SNP	SNP pre-treatment increased CAT and APX activities and lowered ROS content; chlorophyll degradation and ion leakage decreased.	Santa-Cruz et al. (2010)
Phaseolus vulgaris	nd	50 or 100 μM SNP	Partial alleviation of UV-B-induced damage: lower oxidative damage and chlorophyll loss, enhanced photosynthetic efficiency and antioxidant enzyme activities.	Shi et al. (2005)
Solanum tuberosum	nd	100 μM SNP	Higher flavonoid and anthocyanin content, increased shielding wax amount.	Beligni and Lamattina, (2001)
Spirulina platensis	nd	500 μM SNP	Alleviated decrease of chlorophyll, protein and biomass loss; increased proline and glutathione content.	Xue et al. (2006)
	6 hours	500 μM SNP	Less severe biomass and chlorophyll-*a* loss, increased SOD and CAT activity, decreased oxidative stress.	Xue et al. (2007)
	2, 4 or 6 hours	500 μM SNP	Partially restored nitrogenase activity.	Xue et al. (2011)
Triticum aestivum	8 hours for 8 days	100 μM SNP	Restored seedling biomass, leaf length, photochemical efficiency, ATPase activity, carotenoid and chlorophyll content.	Yang et al. (12013)
Zea mays	5 days	100 μM SNP	Increased protein content and reduced endo- and exoglucanase activity.	Zhang et al. (2003)
	3 days	0.1 mM SNP	Enhanced CAT and APX activity, decreased H_2O_2 and MDA content, alleviated growth inhibition and loss of photosynthetic efficiency.	Kim et al. (2010)

the upregulation of antioxidant responses in such species as rice (Singh et al. 2009a, 2016, 2017a), *Phaseolus vulgaris* (Talukdar, 2013), *Pistia stratoites* (Farnese et al. 2013), wheat (Hasanuzzaman and Fujita, 2013), *Vicia faba* (Mohamad et al. 2016) and *Vigna radiata* (Ismail, 2012) (Table 8.6).

8.3 CONCLUSIONS AND FUTURE PERSPECTIVES

Year after year, accumulating evidence proves that exogenously applied NO has diverse positive effects on plants suffering from abiotic stressors. NO treatment, in general, is responsible for several molecular adjustments, such as the induction of different stress-related response genes (antioxidants, ion channels, heat-shock proteins, metabolite synthesis genes) and the modification of

TABLE 8.6
Effect of Exogenously Applied NO in Alleviation of Heavy Metal-Induced Stress

Species	HM Treatment	NO Treatment	Effect	Reference
Arabidopsis thaliana	0, 5, 25 or 50 μM $CuSO_4$	10 μM SNP	Restored cell viability and decreased oxidative stress.	Pető et al. (2013)
Arachis hypogaea	100 μM $AlCl_3$	200 μM SNP	Alleviated Al-induced programmed cell death.	He et al. (2017)
	50, 100 or 200 μM $CdCl_2$	250 μM SNP	Enhanced antioxidant enzyme activities, photosynthesis; reduced root-to-shoot Cd translocation.	Dong et al. (2016)
Brassica juncea	5–200 μM $CdCl_2$	0.01–20 mM SNP	Low SNP treatment increased root growth, chlorophyll content and antioxidant enzyme activity, while reducing oxidative stress.	Verma et al. (2013)
	50 μM $CdCl_2$	100 μM SNP	Enhanced antioxidant response, plant growth and photosynthesis. Co-application of SNP and SO_4^{2-} resulted in better responses.	Per et al. (2017)
Carthamus tinctorius	500 μM $ZnSO_4$	100 μM SNP	Increased root and shoot dry weight, reduced Zn translocation and enhanced ascorbate-glutathione and glyoxalase system.	Namdjoyan et al. (2017)
Cassia tora	10 μM Al	0.4 mM SNP	Restored root elongation and inhibited Al accumulation in root tips. Alleviated oxidative stress.	Wang and Yang, (2005)
Cucumis sativus	100 μM $CdCl_2$	100 μM SNP	Induced antioxidant enzymes and alleviated chlorosis.	Yu et al. (2013)
Eleusine coracana	0.5 mM $NiCl_2$	0.2 mM SNP	Reduced Ni toxicity, restored growth, chlorophyll content and mineral concentration. Lowered oxidative stress.	Kotapati et al. (2017)
Festuca arundinacea	1, 5 and 10 mg/L Cr(VI)	100 μM SNP	Improved physiological and photosynthetical parameters; enhanced Cr tolerance and translocation.	Huang et al. (2018)
Lactuca sativa	100, 200 and 300 μM $CuSO_4$	50–300 μM SNP	Enhanced germination rate and seedling growth.	Shams et al. (2018)
Lolium perenne	100 and 150 μM $CdCl_2$	100 μM SNP	Improved plant growth, chlorophyll content, antioxidant capacity and Fe/Cu/Zn absorption. Lower Cd translocation.	Chen et al. (2018)
	500 μM $Pb(NO_3)$	50, 100 and 200 μM SNP	Enhanced antioxidant response and photosynthesis, decreased oxidative stress and Pb root-to-shoot transport.	Bai et al. (2015)
Nicotiana tabacum	0.02 mM $CdCl_2$	0.05 mM SNP	Enhanced chlorophyll and Rubisco content and Rubisco activase activity and plant growth.	Khairy et al. (2016)
	0.2 mM $CuSO_4$	0.05 mM SNP	Enhanced chlorophyll and Rubisco content and Rubisco activase activity and plant growth.	Khairy et al. (2016)
Oryza sativa	25 or 50 μM Na_2HAsO_4	50 μM SNP	Enhanced antioxidant enzyme activities, reduced oxidative stress.	Singh et al. (2009, 2016)
	25 μM $NaAsO_2$	30 μM SNP	Decreased arsenic accumulation and nitrate reductase activity, alleviated oxidative stress.	Singh et al. (2017a)

(Continued)

TABLE 8.6 (CONTINUED)
Effect of Exogenously Applied NO in Alleviation of Heavy Metal-Induced Stress

Species	HM Treatment	NO Treatment	Effect	Reference
	10 μM $CdCl_2$	30 μM SNAP	Increased antioxidant enzyme activity, plant growth, photosynthetic efficiency and phytochelatine content. Reduced ion leakage and hydrogen peroxide accumulation.	Yang et al. (2016)
	5 mM $CdCl_2$	100 μM SNP	Prevented oxidative stress and enhanced antioxidant responses.	Hsu and Kao, (2004)
	0.2 mM $CdCl_2$	0.1 mM SNP	Increased pectin and hemicellulose and decreased cellulose content of the roots. Increased Cd accumulation in the root cell walls, inhibited transport to the soluble fraction.	Xiong et al. (2009)
	50 or 200 μM $NiSO_4$	100 or 200 μM SNP	Improved seedling growth, photosynthesis and antioxidant capacity; reduced Ni-uptake and oxidative stress.	Rizwan et al. (2018)
	100 μM Cu	200 μM SNP and glutathione	Increased CAT, dehydroascorbate reductase (DHAR), GPX and GST activity while decreasing APX, GR, monodehydroascorbate reductase (MDHAR) and SOD activity. Lowered oxidative stress and Cu accumulation.	Mostofa et al. (2014)
	10 mM $CuSO_4$	100 μM SNP	Reduced toxicity and NH_4^+ accumulation.	Yu et al. (2005)
Phaseolus vulgaris	50 mM Na_2HAsO_4	100 μM SNP	Restored plant growth, reduced membrane damage and contained hydrogen peroxide content via antioxidant enzymatic reactions.	Talukdar, (2013)
Pistia stratoites	1.5 ml/L As	0.1 mg/L SNP	Upregulated antioxidant enzyme activity (APX, CAT, POX) and phytochelatin biosynthesis	Farnese et al. (2013)
Solanum lycopersicum	150 μM $CdCl_2$	100 μM SNP	Protected chlorophyll pigments, increased gas exchange and growth. Enhanced synthesis of glycine and betaine, increased antioxidant response.	Ahmad et al. (2018b)
Typha angustifolia	444.8 μM $CdCl_2$	100 μM SNP	Increased AsA content; alleviated activity of SOD, CAT and MDA and phytochelatine content.	Zhao et al. (2016)
Trifolium repens	100 μM $CdCl_2$	50, 100, 200 or 400 μM SNP	Lower concentrations of donor treatments alleviated growth inhibition and upregulated antioxidant enzymes. Mineral uptake was partly enhanced.	Liu et al. (2015)
Triticum aestivum	30 μM $AlCl_3$	250 μM SNP	Reduced root growth inhibition, callose deposition and oxidative stress; increased antioxidant response (enzymatic and non-enzymatic), upregulation of antioxidant enzyme gene expression (APX, GR, DHAR).	Sun et al. (2014, 2015)

(Continued)

TABLE 8.6 (CONTINUED)
Effect of Exogenously Applied NO in Alleviation of Heavy Metal-Induced Stress

Species	HM Treatment	NO Treatment	Effect	Reference
	0.25 and 0.5 mM Na_2HAsO_4	0.25 mM SNP	Increased antioxidant enzyme activities (CAT, GPX, GR, DHAR, MDHAR, Gly I and II); enhanced AsA, glutathione, proline, chlorophyll content and relative water content.	Hasanuzzaman and Fujita (2013)
	50 and 250 µM Pb	100 µM SNP	Reduced oxidative stress and lowered antioxidant enzyme activity.	Kaur et al. (2015)
Vicia faba	100, 200 and 400 µM Na_2HAsO_4	100 µM SNP	Alleviated growth inhibition; increased seed production, photosynthesis and mineral content. Decreased oxidative damage of the membranes.	Mohamed et al. (2016)
Vigna radiata	25, 50 and 100 µM Na_2HAsO_4	75 µM SNP	Improved seed germination and plant growth, decreased oxidative stress.	Ismail (2012)
	0.5, 1 or 2 g/kg Pb	250 µM SNP	Pre-treatment of seed decreased MDA content and enhanced antioxidant enzyme activities in the seedlings.	Singh et al. (2017b)
Vigna unguiculata	200 Pb mg/kg in soil	0, 0.5 and 1 mM SNP	Enhanced antioxidant response and photosynthesis; lower MDA accumulation.	Sadeghipour (2016)

already existing molecules by posttranslational modifications. These adjustments lead to the development of different tolerance mechanisms, like enhanced ionic or protein homeostasis, osmoprotection and ROS detoxification, which result in improved physiological homeostasis and the alleviation of abiotic-stress-induced growth reduction. Despite the high number of available publications on the topic, the vast majority of them focus on the physiological aspects of the relationship between abiotic stress and exogenous nitric oxide, and relatively little is known about the changes on the metabolomic, proteomic and genomic level in the background.

ACKNOWLEDGMENTS

This work was supported by the János Bolyai Research Scholarship of the Hungarian Academy of Sciences (Grant no. BO/00751/16/8) and by the National Research, Development and Innovation Office (Grant no. NKFI-8 PD 120962, NKFI-8 PD 131589, NKFI-6, K 120383 and NKFI-8 KH 129511) and by the EU-funded Hungarian grant EFOP-3.6.116-2016-00008. Zs. K. was supported by UNKP-18-4 New National Excellence Program of the Ministry of Human Capacities.

REFERENCES

A.-H.-Mackerness, S., John, C. F., Jordan, B., & Thomas, B. (2001). Early signaling components in ultraviolet-B responses: Distinct roles for different reactive oxygen species and nitric oxide. *FEBS Letters*, *489*(2–3):237–242.

Abat, J. K., & Deswal, R. (2009). Differential modulation of S-nitrosoproteome of *Brassica juncea* by low temperature: Change in S-nitrosylation of Rubisco is responsible for the inactivation of its carboxylase activity. *Proteomics*, *9*(18):4368–4380.

Ahmad, P., Abass Ahanger, M., Nasser Alyemeni, M., Wijaya, L., Alam, P., & Ashraf, M. (2018a). Mitigation of sodium chloride toxicity in *Solanum lycopersicum* L. by supplementation of jasmonic acid and nitric oxide. *Journal of Plant Interactions*, *13*(1):64–72.

Ahmad, P., Ahanger, M. A., Alyemeni, M. N., Wijaya, L., & Alam, P. (2018b). Exogenous application of nitric oxide modulates osmolyte metabolism, antioxidants, enzymes of ascorbate-glutathione cycle and promotes growth under cadmium stress in tomato. *Protoplasma*, *255*(1):79–93.

Ahmad, P., Latef, A. A. A., Hashem, C., Abd_Allah, E. F., Gucel, S., & Tran, L.-S. P. (2016). Nitric oxide mitigates salt stress by regulating levels of osmolytes and antioxidant enzymes in chickpea. *Frontiers in Plant Science*, *7*:347.

Akram, N. A., Iqbal, M., Muhammad, A., Ashraf, M., Al-Qurainy, F., & Shafiq, S. (2018). Aminolevulinic acid and nitric oxide regulate oxidative defense and secondary metabolisms in canola (*Brassica napus* L.) under drought stress. *Protoplasma*, *255*(1):163–174.

Arora, D., Jain, P., Singh, N., Kaur, H., & Bhatla, S. C. (2016). Mechanisms of nitric oxide crosstalk with reactive oxygen species scavenging enzymes during abiotic stress tolerance in plants. *Free Radical Research*, *50*:291–303.

Bai, X. Y., Dong, Y. J., Wang, Q. H., Xu, L. L., Kong, J., & Liu, S. (2015). Effects of lead and nitric oxide on photosynthesis, antioxidative ability, and mineral element content of perennial ryegrass. *Biologia Plantarum*, *59*(1):163–170.

Barroso, J. B., Corpas, F. J., Carreras, A., Rodríguez-Serrano, M., Esteban, F. J., Fernández-Ocana, A., et al. (2006). Localization of S-nitrosoglutathione and expression of S-nitrosoglutathione reductase in pea plants under cadmium stress. *Journal of Experimental Botany*, *57*(8):1785–1793.

Batista, P. F., Costa, A. C., Müller, C., de Oliveira Silva-Filho, R., da Silva, F. B., Merchant, A., et al. (2018). Nitric oxide mitigates the effect of water deficit in *Crambe abyssinica. Plant Physiology and Biochemistry*, *129*:310–322.

Baudouin, E., & Jeandroz, S. (2015). Nitric oxide as a mediator of cold stress response: A transcriptional point of view. In *Nitric Oxide Action in Abiotic Stress Responses in Plants*, eds. Khan, M., Mobin, M. Mohammad, F., & Corpas, F., pp. 129–139. Cham: Springer.

Beckmen, J. S., Beckman, T. W., Chen, J., Marshall, P. A., & Freeman, B. A. (1990). Apparent hydroxyl radical production by peroxynitrite: Implication for endothelial injury from nitric oxide and superoxide. *PNAS*, *87*:1620–1624.

Begara-Morales, J. C., Chaki, M., Valderrama, R., Sánchez-Calvo, B., Mata-Pérez, C., Padilla, M. N., et al. (2018). Nitric oxide buffering and conditional nitric oxide release in stress response. *Journal of Experimental Botany*, *69*(14):3425–3438.

Beligni, M. V., & Lamattina, L. (1999). Is nitric oxide toxic or protective? *Trends in Plant Science*, *4*:299–300.

Beligni, M. V., & Lamattina, L. (2001). Nitric oxide in plants: The history is just beginning. *Plant, Cell and Environment*, *24*(3):267–278.

Besson-Bard, A., Gravot, A., Richaud, P., Auroy, P., Duc, C., Gaymard, F., et al. (2009). Nitric oxide contributes to cadmium toxicity in *Arabidopsis* by promoting cadmium accumulation in roots and by upregulating genes related to iron uptake. *Plant Physiology*, *149*(3):1302–1315.

Bethke, P. C., Badger, M. R., & Jones, R. L. (2004). Apoplastic synthesis of nitric oxide by plant tissues. *The Plant Cell*, *16*(2):332–341.

Boyarshinov, A. V., & Asafova, E. V. (2011). Stress responses of wheat leaves to dehydration: Participation of endogenous NO and effect of sodium nitroprusside. *Russian Journal of Plant Physiology*, *58*(6):1034.

Camejo, D., del Carmen Romero-Puertas, M., Rodríguez-Serrano, M., Sandalio, L. M., Lázaro, J. J., Jiménez, A., & Sevilla, F. (2013). Salinity-induced changes in S-nitrosylation of pea mitochondrial proteins. *Journal of Proteomics*, *79*:87–99.

Cao, X., Zhu, C., Zhong, C., Zhang, J., Zhu, L., Wu, L., et al. (2018). Nitric oxide synthase-mediated early nitric oxide-burst alleviates drought-induced oxidative damage in ammonium supplied-rice roots. *BMC Plant Biology*, *19*(1):108.

Çelik, A., & Eraslan, F. (2015). Effects of exogenous nitric oxide on mineral nutrition and some physiological parameters of maize grown under salinity stress. *Ziraat Fakültesi Dergisi-Süleyman Demirel Üniversitesi*, *10*(1):55–64.

Chaki, M., Valderrama, R., Fernández-Ocana, A. M., Carreras, A., Gómez-Rodríguez, M. V., López-Jaramillo, J., et al. (2011). High temperature triggers the metabolism of S-nitrosothiols in sunflower mediating a process of nitrosative stress which provokes the inhibition of ferredoxin–NADP reductase by tyrosine nitration. *Plant, Cell and Environment*, *34*(11):1803–1818.

Chaki, M., Valderrama, R., Fernández-Ocaña, A. M., Carreras, A., Gómez-Rodríguez, M. V., Pedrajas, J. R., et al. (2010). Mechanical wounding induces a nitrosative stress by down-regulation of GSNO reductase and an increase in S-nitrosothiols in sunflower (*Helianthus annuus*) seedlings. *Journal of Experimental Botany*, *62*(6):1803–1813.

Chamizo-Ampudia, A., Sanz-Luque, E., Llamas, A., Galvan, A., & Fernandez, E. (2017). Nitrate reductase regulates plant nitric oxide homeostasis. *Trends in Plant Science*, *22*(2):163–174.

Chamizo-Ampudia, A., Sanz-Luque, E., Llamas, Á., Ocaña-Calahorro, F., Mariscal, V., Carreras, A., et al. (2016). A dual system formed by the ARC and NR molybdoenzymes mediates nitrite-dependent NO production in Chlamydomonas. *Plant, Cell and Environment*, *39*(10):2097–2107.

Chen, J., Xiao, Q., Wang, C., Wang, W. H., Wu, F. H., He, B. Y., et al. (2014). Nitric oxide alleviates oxidative stress caused by salt in leaves of a mangrove species, *Aegiceras corniculatum*. *Aquatic Botany*, *117*:41–47.

Chen, K., Chen, L., Fan, J., & Fu, J. (2013). Alleviation of heat damage to photosystem II by nitric oxide in tall fescue. *Photosynthesis Research*, *116*(1):21–31.

Chen, K., Song, L., Rao, B., Zhu, T., & Zhang, Y. T. (2010). Nitric oxide plays a role as second messenger in the ultraviolet-B irradiated green alga *Chlorella pyrenoidosa*. *Folia Microbiologica*, *55*(1):53–60.

Chen, W., Dong, Y., Hu, G., & Bai, X. (2018). Effects of exogenous nitric oxide on cadmium toxicity and antioxidative system in perennial ryegrass. *Journal of Soil Science and Plant Nutrition*, *18*(1):129–143.

Cheng, F. Y., Hsu, S. Y., & Kao, C. H. (2002). Nitric oxide counteracts the senescence of detached rice leaves induced by dehydration and polyethylene glycol but not by sorbitol. *Plant Growth Regulation*, *38*(3):265–272.

Christou, A., Manganaris, G. A., & Fotopoulos, V. (2014). Systemic mitigation of salt stress by hydrogen peroxide and sodium nitroprusside in strawberry plants via transcriptional regulation of enzymatic and non-enzymatic antioxidants. *Environmental and Experimental Botany*, *107*:46–54.

Corpas, F. J., & Barroso, J. B. (2014). Peroxynitrite (ONOO–) is endogenously produced in *Arabidopsis* peroxisomes and is overproduced under cadmium stress. *Annals of Botany*, *113*(1):87–96.

Corpas, F. J., & Barroso, J. B. (2017). Nitric oxide synthase-like activity in higher plants. *Nitric Oxide: Biology and Chemistry*, *68*:5.

Corpas, F. J., Chaki, M., Fernandez-Ocana, A., Valderrama, R., Palma, J. M., Carreras, A., et al. (2008). Metabolism of reactive nitrogen species in pea plants under abiotic stress conditions. *Plant and Cell Physiology*, *49*(11):1711–1722.

Correa-Aragunde, N., Foresi, N., Del Castello, F., & Lamattina, L. (2018). A singular nitric oxide synthase with a globin domain found in Synechococcus PCC 7335 mobilizes N from arginine to nitrate. *Scientific Reports*, *8*(1):12505.

da Silva, C. J., Fontes, E. P. B., & Modolo, L. V. (2017). Salinity-induced accumulation of endogenous H2S and NO is associated with modulation of the antioxidant and redox defense systems in *Nicotiana tabacum* L. cv. Havana. *Plant Science*, *256*:148–159.

Dong, N., Li, Y., Qi, J., Chen, Y., & Hao, Y. (2018). Nitric oxide synthase-dependent nitric oxide production enhances chilling tolerance of walnut shoots in vitro via involvement chlorophyll fluorescence and other physiological parameter levels. *Scientia Horticulturae*, *230*:68–77.

Dong, Y., Chen, W., Xu, L., Kong, J., Liu, S., & He, Z. (2016). Nitric oxide can induce tolerance to oxidative stress of peanut seedlings under cadmium toxicity. *Plant Growth Regulation*, *79*(1):19–28.

Dordas, C., Hasinoff, B. B., Igamberdiev, A. U., Manac'h, N., Rivoal, J., & Hill, R. D. (2003). Expression of a stress-induced hemoglobin affects NO levels produced by alfalfa root cultures under hypoxic stress. *The Plant Journal*, *35*(6):763–770.

Dordas, C., Hasinoff, B. B., Rivoal, J., & Hill, R. D. (2004). Class-1 hemoglobins, nitrate and NO levels in anoxic maize cell-suspension cultures. *Planta*, *219*(1):66–72.

Du, S., Zhang, R., Zhang, P., Liu, H., Yan, M., Chen, N., et al. (2015). Elevated CO2-induced production of nitric oxide (NO) by NO synthase differentially affects nitrate reductase activity in *Arabidopsis* plants under different nitrate supplies. *Journal of Experimental Botany*, *67*(3):893–904.

Elviri, L., Speroni, F., Careri, M., Mangia, A., di Toppi, L. S., & Zottini, M. (2010). Identification of in vivo nitrosylated phytochelatins in *Arabidopsis thaliana* cells by liquid chromatography-direct electrospray-linear ion trap-mass spectrometry. *Journal of Chromatography A*, *1217*(25):4120–4126.

Esim, N., Atici, O., & Mutlu, S. (2014). Effects of exogenous nitric oxide in wheat seedlings under chilling stress. *Toxicology and Industrial Health*, *30*(3):268–274.

Fan, H., Du, C., Xu, Y., & Wu, X. (2014). Exogenous nitric oxide improves chilling tolerance of Chinese cabbage seedlings by affecting antioxidant enzymes in leaves. *Horticulture, Environment, and Biotechnology*, *55*(3):159–165.

Fan, H., Guo, S., Jiao, Y., Zhang, R., & Li, J. (2007). Effects of exogenous nitric oxide on growth, active oxygen species metabolism, and photosynthetic characteristics in cucumber seedlings under NaCl stress. *Frontiers of Agriculture in China*, *1*(3):308–314.

Fancy, N. N., Bahlmann, A. K., & Loake, G. J. (2017). Nitric oxide function in plant abiotic stress. *Plant, Cell and Environment*, *40*(4):462–472.

Farnese, F. S., de Oliveira, J. A., Gusman, G. S., Leão, G. A., Ribeiro, C., Siman, L. I., & Cambraia, J. (2013). Plant responses to arsenic: The role of nitric oxide. *Water, Air, and Soil Pollution*, *224*(9):1660.

Farooq, M., Basra, S. M. A., Wahid, A., & Rehman, H. (2009). Exogenously applied nitric oxide enhances the drought tolerance in fine grain aromatic rice (*Oryza sativa* L.). *Journal of Agronomy and Crop Science*, *195*(4):254–261.

Fatma, M., & Khan, N. A. (2014). Nitric oxide protects photosynthetic capacity inhibition by salinity in Indian mustard. *Journal of Functional and Environmental Botany*, *4*:106–116.

Feigl, G., Kolbert, Z., Lehotai, N., Molnár, Á., Ördög, A., Bordé, Á., et al. (2016). Different zinc sensitivity of Brassica organs is accompanied by distinct responses in protein nitration level and pattern. *Ecotoxicology and Environmental Safety*, *125*:141–152.

Feigl, G., Kumar, D., Lehotai, N., Tugyi, N., Molnár, Á., Ördög, A., et al. (2013). Physiological and morphological responses of the root system of Indian mustard (*Brassica juncea* L. Czern.) and rapeseed (*Brassica napus* L.) to copper stress. *Ecotoxicology and Environmental Safety*, *94*:179–189.

Feigl, G., Lehotai, N., Molnár, Á., Ördög, A., Rodríguez-Ruiz, M., Palma, J. M., et al. (2014). Zinc induces distinct changes in the metabolism of reactive oxygen and nitrogen species (ROS and RNS) in the roots of two Brassica species with different sensitivity to zinc stress. *Annals of Botany*, *116*(4):613–625.

Flexas, J., & Medrano, H. (2002). Drought-inhibition of photosynthesis in C3 plants: Stomatal and non-stomatal limitations revisited. *Annals of Botany*, *89*:183–189.

Foresi, N., Correa-Aragunde, N., Parisi, G., Caló, G., Salerno, G., & Lamattina, L. (2010). Characterization of a nitric oxide synthase from the plant kingdom: NO generation from the green alga *Ostreococcus tauri* is light irradiance and growth phase dependent. *The Plant Cell*, *22*(11):3816–3830.

Foresi, N., Mayta, M. L., Lodeyro, A. F., Scuffi, D., Correa-Aragunde, N., García-Mata, C., et al. (2015). Expression of the tetrahydrofolate-dependent nitric oxide synthase from the green alga *Ostreococcus tauri* increases tolerance to abiotic stresses and influences stomatal development in *Arabidopsis*. *The Plant Journal*, *82*(5):806–821.

Foster, M. W., McMahon, T. J., & Stamler, J. S. (2003). S-nitrosylation in health and disease. *Trends in Molecular Medicine*, *9*(4):160–168.

Frohnmeyer, H., & Staiger, D. (2003). Ultraviolet-B radiation-mediated responses in plants. Balancing damage and protection. *Plant Physiology*, *133*(4):1420–1428.

Godber, B. L., Doel, J. J., Sapkota, G. P., Blake, D. R., Stevens, C. R., Eisenthal, R., & Harrison, R. (2000). Reduction of nitrite to nitric oxide catalyzed by xanthine oxidoreductase. *Journal of Biological Chemistry*, *275*(11):7757–7763.

Guo, Y., Tian, Z., Yan, D., Zhang, J., & Qin, P. (2005). Effects of nitric oxide on salt stress tolerance in *Kosteletzkya virginica*. *Life Science Journal*, *6*(1):67–75.

Gzyl, J., Izbiańska, K., Floryszak-Wieczorek, J., Jelonek, T., & Arasimowicz-Jelonek, M. (2016). Cadmium affects peroxynitrite generation and tyrosine nitration in seedling roots of soybean (*Glycine max* L.). *Environmental and Experimental Botany*, *131*:155–163.

Hao, G. P., Xing, Y., & Zhang, J. H. (2008). Role of nitric oxide dependence on nitric oxide synthase-like activity in the water stress signaling of maize seedling. *Journal of Integrative Plant Biology*, *50*(4):435–442.

Hasanuzzaman, M., & Fujita, M. (2013). Exogenous sodium nitroprusside alleviates arsenic-induced oxidative stress in wheat (*Triticum aestivum* L.) seedlings by enhancing antioxidant defense and glyoxalase system. *Ecotoxicology*, *22*(3):584–596.

Hasanuzzaman, M., Gill, S. S., & Fujita, M. (2013a). Physiological role of nitric oxide in plants grown under adverse environmental conditions. In *Plant Acclimation to Environmental Stress*, eds. Tuteja, N., & Singh Gill, S., pp. 269–322. New York: Springer.

Hasanuzzaman, M., Nahar, K., & Fujita, M. (2013b). Plant response to salt stress and role of exogenous protectants to mitigate salt-induced damages. In *Ecophysiology and Responses of Plants under Salt Stress*, eds. Ahmed, P., Azooz, M. M., & Prasad, M. N. V, pp. 25–87. New York: Springer.

Hasanuzzaman, M., Nahar, K., Fujita, M., Ahmad, P., Chandna, R., Prasad, M. N. V., & Ozturk, M. (2013c). Enhancing plant productivity under salt stress: Relevance of poly-omics. In *Salt Stress in Plants: Signaling, Omics and Adaptations*, eds. Ahmad, P., Azooz, M. M., & Prasad, M. N. V., pp. 113–156. New York: Springer.

Hasanuzzaman, M., Oku, H., Nahar, K., Bhuyan, M. B., Al Mahmud, J., Baluska, F., & Fujita, M. (2018). Nitric oxide-induced salt stress tolerance in plants: ROS metabolism, signaling, and molecular interactions. *Plant Biotechnology Reports*, *12*(2):77–92.

He, H., Huang, W., Oo, T. L., Gu, M., & He, L. F. (2017). Nitric oxide inhibits aluminum-induced programmed cell death in peanut (*Arachis hypoganea* L.) root tips. *Journal of Hazardous Materials*, *333*:285–292.

Hebelstrup, K. H., Hunt, P., Dennis, E., Jensen, S. B., & Jensen, E. Ø. (2006). Hemoglobin is essential for normal growth of *Arabidopsis* organs. *Physiologia Plantarum*, *127*(1):157–166.

Hogg, N. (2000). Biological chemistry and clinical potential of S-nitrosothiols. *Free Radical Biology and Medicine*, *28*(10):1478–1486.

Hsu, Y. T., & Kao, C. H. (2004). Cadmium toxicity is reduced by nitric oxide in rice leaves. *Plant Growth Regulation*, *42*(3):227–238.

Hu, J., Huang, X., Chen, L., Sun, X., Lu, C., Zhang, L., et al. (2015a). Site-specific nitrosoproteomic identification of endogenously S-nitrosylated proteins in *Arabidopsis*. *Plant Physiology*, *167*(4):1731–1746.

Hu, Y., You, J., & Liang, X. (2015b). Nitrate reductase-mediated nitric oxide production is involved in copper tolerance in shoots of hulless barley. *Plant Cell Reports*, *34*(3):367–379.

Huang, M., Ai, H., Xu, X., Chen, K., Niu, H., Zhu, H., et al. (2018). Nitric oxide alleviates toxicity of hexavalent chromium on tall fescue and improves performance of photosystem II. *Ecotoxicology and Environmental Safety*, *164*:32–40.

Ismail, G. S. M. (2012). Protective role of nitric oxide against arsenic-induced damages in germinating mung bean seeds. *Acta Physiologiae Plantarum*, *34*(4):1303–1311.

Jain, P., & Bhatla, S. C. (2018). Tyrosine nitration of cytosolic peroxidase is probably triggered as a long distance signaling response in sunflower seedling cotyledons subjected to salt stress. *PLoS one*, *13*(5):e0197132.

Jin, C. W., Du, S. T., Shamsi, I. H., Luo, B. F., & Lin, X. Y. (2011). NO synthase-generated NO acts downstream of auxin in regulating Fe-deficiency-induced root branching that enhances Fe-deficiency tolerance in tomato plants. *Journal of Experimental Botany*, *62*(11):3875–3884.

Karpets, Y. V., Kolupaev, Y. E., & Vayner, A. A. (2015). Functional interaction between nitric oxide and hydrogen peroxide during formation of wheat seedling induced heat resistance. *Russian Journal of Plant Physiology*, *62*(1):65–70.

Kaur, G., Singh, H. P., Batish, D. R., Mahajan, P., Kohli, R. K., & Rishi, V. (2015). Exogenous nitric oxide (NO) interferes with lead (Pb)-induced toxicity by detoxifying reactive oxygen species in hydroponically grown wheat (*Triticum aestivum*) roots. *PLoS One*, *10*(9):e0138713.

Kausar, F., Shahbaz, M., & Ashraf, M. (2013). Protective role of foliar-applied nitric oxide in *Triticum aestivum* under saline stress. *Turkish Journal of Botany*, *37*:1155–1165.

Kaya, C., Akram, N. A., & Ashraf, M. (2018). Influence of exogenously applied nitric oxide on strawberry (*Fragaria* × *ananassa*) plants grown under iron deficiency and/or saline stress. *Physiologia Plantarum*, *165*(2):247–263.

Khairy, A. I. H., Oh, M. J., Lee, S. M., & Roh, K. S. (2016). Nitric oxide overcomes Cd and Cu toxicity in in vitro-grown tobacco plants through increasing contents and activities of rubisco and rubisco activase. *Biochimie Open*, *2*:41–51.

Kim, T. Y., Jo, M. H., & Hong, J. H. (2010). Protective effect of nitric oxide against oxidative stress under UV-B radiation in maize leaves. *Journal of Environmental Science International*, *19*(12):1323–1334.

Klein, A., Hüsselmann, L., Keyster, M., & Ludidi, N. (2018). Exogenous nitric oxide limits salt-induced oxidative damage in maize by altering superoxide dismutase activity. *South African Journal of Botany*, *115*:44–49.

Klepper, L. (1979). Nitric oxide (NO) and nitrogen dioxide (NO2) emissions from herbicide-treated soybean plants. *Atmospheric Environment (1967)*, *13*(4):537–542.

Kolbert, Z., & Feigl, G. (2018). Cross-talk of reactive oxygen species and nitric oxide in various processes of plant development: Past and present. In *Reactive Oxygen Species in Plants: Boon or Bane - Revisiting the Role of ROS*, eds. Singh, V. P., Singh, S., Tripathi, D. K., Prasad, S. M., & Chauhan, D. K., pp. 261–289. Oxford: Wiley.

Kolbert, Z., Feigl, G., Bordé, Á., Molnár, Á., & Erdei, L. (2017). Protein tyrosine nitration in plants: Present knowledge, computational prediction and future perspectives. *Plant Physiology and Biochemistry*, *113*:56–63.

Kolbert, Z., Molnár, Á., Szőllősi, R., Feigl, G., Erdei, L., & Ördög, A. (2018). Nitro-oxidative stress correlates with Se tolerance of Astragalus species. *Plant and Cell Physiology*, *59*(9):1827–1843.

Kolbert, Z., Ortega, L., & Erdei, L. (2010). Involvement of nitrate reductase (NR) in osmotic stress-induced NO generation of *Arabidopsis thaliana* L. roots. *Journal of Plant Physiology*, *167*(1):77–80.

Kong, X., Wang, T., Li, W., Tang, W., Zhang, D., & Dong, H. (2016). Exogenous nitric oxide delays salt-induced leaf senescence in cotton (*Gossypium hirsutum* L.). *Acta Physiologiae Plantarum*, *38*:61.

Kotapati, K. V., Palaka, B. K., & Ampasala, D. R. (2017). Alleviation of nickel toxicity in finger millet (*Eleusine coracana* L.) germinating seedlings by exogenous application of salicylic acid and nitric oxide. *The Crop Journal*, *5*(3):240–250.

Lamotte, O., Bertoldo, J. B., Besson-Bard, A., Rosnoblet, C., Aimé, S., Hichami, S., et al. (2015). Protein S-nitrosylation: Specificity and identification strategies in plants. *Frontiers in Chemistry*, *2*:114.

Laspina, N. V., Groppa, M. D., Tomaro, M. L., & Benavides, M. P. (2005). Nitric oxide protects sunflower leaves against Cd-induced oxidative stress. *Plant Science*, *169*(2):323–330.

Lei, Y., Yin, C., & Li, C. (2007). Adaptive responses of *Populus przewalskii* to drought stress and SNP application. *Acta Physiologiae Plantarum*, *29*:519–526.

Leterrier, M., Airaki, M., Palma, J. M., Chaki, M., Barroso, J. B., & Corpas, F. J. (2012). Arsenic triggers the nitric oxide (NO) and S-nitrosoglutathione (GSNO) metabolism in *Arabidopsis*. *Environmental Pollution*, *166*:136–143.

Leterrier, M., Chaki, M., Airaki, M., Valderrama, R., Palma, J. M., Barroso, J. B., & Corpas, F. J. (2011). Function of S-nitrosoglutathione reductase (GSNOR) in plant development and under biotic/abiotic stress. *Plant Signaling and Behavior*, *6*(6):789–793.

Li, Q. Y., Niu, H. B., Yin, J., Wang, M. B., Shao, H. B., Deng, D. Z., et al. (2008). Protective role of exogenous nitric oxide against oxidative-stress induced by salt stress in barley (*Hordeum vulgare*). *Colloids and Surfaces B: Biointerfaces*, *65*(2):220–225.

Li, Z. G., Yang, S. Z., Long, W. B., Yang, G. X., & Shen, Z. Z. (2013). Hydrogen sulphide may be a novel downstream signal molecule in nitric oxide-induced heat tolerance of maize (*Zea mays* L.) seedlings. *Plant, Cell and Environment*, *36*(8):1564–1572.

Liang, L., Deng, Y., Sun, X., Jia, X., & Su, J. (2018). Exogenous nitric oxide pretreatment enhances chilling tolerance of Anthurium. *Journal of the American Society for Horticultural Science*, *143*(1):3–13.

Lindermayr, C. (2018). Crosstalk between reactive oxygen species and nitric oxide in plants: Key role of S-nitrosoglutathione reductase. *Free Radical Biology and Medicine*, *122*:110–115.

Liu, S., Dong, Y., Xu, L., & Kong, J. (2014). Effects of foliar applications of nitric oxide and salicylic acid on salt-induced changes in photosynthesis and antioxidative metabolism of cotton seedlings. *Plant Growth Regulation*, *73*(1):67–78.

Liu, S., Yang, R., Pan, Y., Ma, M., Pan, J., Zhao, Y., et al. (2015). Nitric oxide contributes to minerals absorption, proton pumps and hormone equilibrium under cadmium excess in *Trifolium repens* L. plants. *Ecotoxicology and Environmental Safety*, *119*:35–46.

Liu, S., Yang, R., Tripathi, D. K., Li, X., He, W., Wu, M., et al. (2018). The interplay between reactive oxygen and nitrogen species contributes in the regulatory mechanism of the nitro-oxidative stress induced by cadmium in *Arabidopsis*. *Journal of Hazardous Materials*, *344*:1007–1024.

Liu, Y., Jiang, H., Zhao, Z., & An, L. (2010). Nitric oxide synthase like activity-dependent nitric oxide production protects against chilling-induced oxidative damage in *Chorispora bungeana* suspension cultured cells. *Plant Physiology and Biochemistry*, *48*(12):936–944.

Lozano-Juste, J., Colom-Moreno, R., & León, J. (2011). In vivo protein tyrosine nitration in *Arabidopsis thaliana*. *Journal of Experimental Botany*, *62*(10):3501–3517.

Majeed, S., Nawaz, F., Naeem, M., & Ashraf, M. Y. (2018). Effect of exogenous nitric oxide on sulfur and nitrate assimilation pathway enzymes in maize (*Zea mays* L.) under drought stress. *Acta Physiologiae Plantarum*, *40*(12):206.

Mohamed, H. I., Latif, H. H., & Hanafy, R. S. (2016). Influence of nitric oxide application on some biochemical aspects, endogenous hormones, minerals and phenolic compounds of *Vicia faba* plant grown under arsenic. *Gesunde Pflanzen*, *68*(2):99–107.

Molnár, Á., Feigl, G., Trifán, V., Ördög, A., Szőllősi, R., Erdei, L., & Kolbert, Z. (2018a). The intensity of tyrosine nitration is associated with selenite and selenate toxicity in *Brassica juncea* L. *Ecotoxicology and Environmental Safety*, *147*:93–101.

Molnár, Á., Kolbert, Z., Kéri, K., Feigl, G., Ördög, A., Szőllősi, R., & Erdei, L. (2018b). Selenite-induced nitro-oxidative stress processes in *Arabidopsis thaliana* and *Brassica juncea*. *Ecotoxicology and Environmental Safety*, *148*:664–674.

Mostofa, M. G., Seraj, Z. I., & Fujita, M. (2014). Exogenous sodium nitroprusside and glutathione alleviate copper toxicity by reducing copper uptake and oxidative damage in rice (*Oryza sativa* L.) seedlings. *Protoplasma*, *251*(6):1373–1386.

Munns, R., & Tester, M. (2008). Mechanisms of salinity tolerance. *Annual Review of Plant Biology*, *59*:651–681.

Namdjoyan, S., Kermanian, H., Soorki, A. A., Tabatabaei, S. M., & Elyasi, N. (2017). Interactive effects of salicylic acid and nitric oxide in alleviating zinc toxicity of safflower (*Carthamus tinctorius* L.). *Ecotoxicology*, *26*(6):752–761.

Ortega-Galisteo, A. P., Rodríguez-Serrano, M., Pazmiño, D. M., Gupta, D. K., Sandalio, L. M., & Romero-Puertas, M. C. (2012). S-Nitrosylated proteins in pea (*Pisum sativum* L.) leaf peroxisomes: Changes under abiotic stress. *Journal of Experimental Botany*, *63*(5):2089–2103.

Per, T. S., Masood, A., & Khan, N. A. (2017). Nitric oxide improves S-assimilation and GLUTATHIONE production to prevent inhibitory effects of cadmium stress on photosynthesis in mustard (*Brassica juncea* L.). *Nitric Oxide*, *68*:111–124.

Perazzolli, M., Dominici, P., Romero-Puertas, M. C., Zago, E., Zeier, J., Sonoda, M. & Delledonne, M. (2004). *Arabidopsis* nonsymbiotic hemoglobin AHb1 modulates nitric oxide bioactivity. *Plant Cell*, *16*:2785–2794.

Perazzolli, M., Romero-Puertas, M. C., & Delledonne, M. (2005). Modulation of nitric oxide bioactivity by plant haemoglobins. *Journal of Experimental Botany*, *57*(3):479–488.

Pető, A., Lehotai, N., Feigl, G., Tugyi, N., Ördög, A., Gémes, K., et al. (2013). Nitric oxide contributes to copper tolerance by influencing ROS metabolism in *Arabidopsis*. *Plant Cell Reports*, *32*(12):1913–1923.

Rizwan, M., Mostofa, M. G., Ahmad, M. Z., Imtiaz, M., Mehmood, S., Adeel, M., et al. (2018). Nitric oxide induces rice tolerance to excessive nickel by regulating nickel uptake, reactive oxygen species detoxification and defense-related gene expression. *Chemosphere*, *191*:23–35.

Rockel, P., Strube, F., Rockel, A., Wildt, J., & Kaiser, W. M. (2002). Regulation of nitric oxide (NO) production by plant nitrate reductase in vivo and in vitro. *Journal of Experimental Botany*, *53*(366):103–110.

Rümer, S., Gupta Kapuganti, J., & Kaiser, W. M. (2009). Oxidation of hydroxylamines to NO by plant cells. *Plant Signaling and Behavior*, *4*(9):853–855.

Russwurm, M., & Koesling, D. (2004). NO activation of guanylyl cyclase. *The EMBO Journal*, *23*(22):4443–4450.

Sadeghipour, O. (2016). Pretreatment with nitric oxide reduces lead toxicity in cowpea (*Vigna unguiculata* [L.] walp.). *Archives of Biological Sciences*, *68*(1):165–175.

Santa-Cruz, D. M., Pacienza, N. A., Polizio, A. H., Balestrasse, K. B., Tomaro, M. L., & Yannarelli, G. G. (2010). Nitric oxide synthase-like dependent NO production enhances heme oxygenase up-regulation in ultraviolet-B-irradiated soybean plants. *Phytochemistry*, *71*(14–15):1700–1707.

Sanz-Luque, E., Ocaña-Calahorro, F., de Montaigu, A., Chamizo-Ampudia, A., Llamas, Á., Galván, A., & Fernández, E. (2015). THB 1, a truncated hemoglobin, modulates nitric oxide levels and nitrate reductase activity. *The Plant Journal*, *81*(3):467–479.

Sanz-Luque, E., Ocaña-Calahorro, F., Llamas, A., Galvan, A., & Fernandez, E. (2013). Nitric oxide controls nitrate and ammonium assimilation in *Chlamydomonas reinhardtii*. *Journal of Experimental Botany*, *64*(11):3373–3383.

Setia, R., Gottschalk, P., Smith, P., Marschner, P., Baldock, J., Setia, D., & Smith, J. (2013). Soil salinity decreases global soil organic carbon stocks. *Science of the Total Environment*, *465*:267–272.

Shams, M., Yildirim, E., Guleray, A. G. A. R., Ercisli, S., Dursun, A., Ekinci, M., & Raziye, K. U. L. (2018). Nitric oxide alleviates copper toxicity in germinating seed and seedling growth of *Lactuca sativa* L. *Notulae Botanicae Horti Agrobotanici Cluj-Napoca*, *46*(1):167–172.

Sheokand, S., Bhankar, V., & Sawhney, V. (2010). Ameliorative effect of exogenous nitric oxide on oxidative metabolism in NaCl treated chickpea plants. *Brazilian Journal of Plant Physiology*, *22*(2):81–90.

Shi, H. T., Li, R. J., Cai, W., Liu, W., Wang, C. L., & Lu, Y. T. (2011). Increasing nitric oxide content in *Arabidopsis thaliana* by expressing rat neuronal nitric oxide synthase resulted in enhanced stress tolerance. *Plant and Cell Physiology*, *53*(2):344–357.

Shi, Q., Ding, F., Wang, X., & Wei, M. (2007). Exogenous nitric oxide protects cucumber roots against oxidative stress induced by salt stress. *Plant Physiology and Biochemistry*, *45*(8):542–550.

Shi, S., Wang, G., Wang, Y., Zhang, L., & Zhang, L. (2005). Protective effect of nitric oxide against oxidative stress under ultraviolet-B radiation. *Nitric Oxide*, *13*(1):1–9.

Sidana, S., Bose, J., Shabala, L., & Shabala, S. (2015). Nitric oxide in drought stress signalling and tolerance in plants. In *Nitric Oxide Action in Abiotic Stress Responses in Plants*, eds. Khan, M., Mobin, M. Mohammad, F., & Corpas, F., pp. 95–114. Cham: Springer.

Siddiqui, M., Alamri, S. A., Mutahhar, Y. Y., Al-Khaishany, M. A., Al-Qutami, H. M., & Nasir Khan, M. A. (2017). Nitric oxide and calcium induced physiobiochemical changes in tomato (*Solanum lycopersicum*) plant under heat stress. Fresenius Environmental Bulletin, *26*:1663–1672.

Siddiqui, M. H., Al-Whaibi, M. H., & Basalah, M. O. (2011). Role of nitric oxide in tolerance of plants to abiotic stress. *Protoplasma*, *248*(3):447–455.

Signorelli, S., Corpas, F. J., Rodríguez-Ruiz, M., Valderrama, R., Barroso, J. B., Borsani, O., & Monza, J. (2018). Drought stress triggers the accumulation of NO and SNOs in cortical cells of *Lotus japonicus* L. roots and the nitration of proteins with relevant metabolic function. *Environmental and Experimental Botany, 161*:228–241.

Singh, A. P., Dixit, G., Kumar, A., Mishra, S., Kumar, N., Dixit, S., et al. (2017a). A protective role for nitric oxide and salicylic acid for arsenite phytotoxicity in rice (*Oryza sativa* L.). *Plant Physiology and Biochemistry, 115*:163–173.

Singh, H., Singh, A., Hussain, I., & Yadav, V. (2017b). Oxidative stress induced by lead in *Vigna radiata* L. seedling attenuated by exogenous nitric oxide. *Tropical Plant Research, 4*(2):225–234.

Singh, A. P., Dixit, G., Kumar, A., Mishra, S., Singh, P. K., Dwivedi, S., et al. (2016). Nitric oxide alleviated arsenic toxicity by modulation of antioxidants and thiol metabolism in rice (*Oryza sativa* L.). *Frontiers in Plant Science, 6*:1272.

Singh, H. P., Kaur, S., Batish, D. R., Sharma, V. P., Sharma, N., & Kohli, R. K. (2009a). Nitric oxide alleviates arsenic toxicity by reducing oxidative damage in the roots of *Oryza sativa* (rice). *Nitric Oxide, 20*(4):289–297.

Singh, S. P., Singh, Z., & Swinny, E. E. (2009b). Postharvest nitric oxide fumigation delays fruit ripening and alleviates chilling injury during cold storage of Japanese plums (*Prunus salicina* Lindell). *Postharvest Biology and Technology, 53*(3):101–108.

Song, L., Ding, W., Zhao, M., Sun, B., & Zhang, L. (2006). Nitric oxide protects against oxidative stress under heat stress in the calluses from two ecotypes of reed. *Plant Science, 171*(4):449–458.

Souza, J. M., Peluffo, G., & Radi, R. (2008). Protein tyrosine nitration—functional alteration or just a biomarker?. *Free Radical Biology and Medicine, 45*(4):357–366.

Stamler, J. S., Lamas, S., & Fang, F. C. (2001). Nitrosylation: The prototypic redox-based signaling mechanism. *Cell, 106*(6):675–683.

Stöhr, C., Strube, F., Marx, G., Ullrich, W. R., & Rockel, P. (2001). A plasma membrane-bound enzyme of tobacco roots catalyses the formation of nitric oxide from nitrite. *Planta, 212*(5–6):835–841.

Stoimenova, M., Igamberdiev, A. U., Gupta, K. J., & Hill, R. D. (2007). Nitrite-driven anaerobic ATP synthesis in barley and rice root mitochondria. *Planta, 226*(2):465–474.

Sun, C., Liu, L., Yu, Y., Liu, W., Lu, L., Jin, C., & Lin, X. (2015). Nitric oxide alleviates aluminum-induced oxidative damage through regulating the ascorbate-glutathione cycle in roots of wheat. *Journal of Integrative Plant Biology, 57*(6):550–561.

Sun, C., Lu, L., Liu, L., Liu, W., Yu, Y., Liu, X., et al. (2014). Nitrate reductase-mediated early nitric oxide burst alleviates oxidative damage induced by aluminum through enhancement of antioxidant defenses in roots of wheat (*Triticum aestivum*). *New Phytologist, 201*(4):1240–1250.

Suzuki, N., & Mittler, R. (2006). Reactive oxygen species and temperature stresses: A delicate balance between signaling and destruction. *Physiologia Plantarum, 126*(1):45–51.

Talukdar, D. (2013). Arsenic-induced oxidative stress in the common bean legume, *Phaseolus vulgaris* L. seedlings and its amelioration by exogenous nitric oxide. *Physiology and Molecular Biology of Plants, 19*(1):69–79.

Uchida, A., Jagendorf, A. T., Hibino, T., Takabe, T., & Takabe, T. (2002). Effects of hydrogen peroxide and nitric oxide on both salt and heat stress tolerance in rice. *Plant Science, 163*(3):515–523.

Umbreen, S., Lubega, J., Cui, B., Pan, Q., Jiang, J., & Loake, G. J. (2018). Specificity in nitric oxide signalling. *Journal of Experimental Botany, 69*(14):3439–3448.

Vanzo, E., Ghirardo, A., Merl-Pham, J., Lindermayr, C., Heller, W., Hauck, S. M., et al. (2014). S-nitroso-proteome in poplar leaves in response to acute ozone stress. *PLoS one, 9*(9):e106886.

Verma, K., Mehta, S. K., & Shekhawat, G. S. (2013). Nitric oxide (NO) counteracts cadmium induced cytotoxic processes mediated by reactive oxygen species (ROS) in *Brassica juncea*: Cross-talk between ROS, NO and antioxidant responses. *Biometals, 26*(2):255–269.

Wang, H. H., Huang, J. J., & Bi, Y. R. (2010). Nitrate reductase-dependent nitric oxide production is involved in aluminum tolerance in red kidney bean roots. *Plant Science, 179*(3):281–288.

Wang, L., Guo, Y., Jia, L., Chu, H., Zhou, S., Chen, K., et al. (2014). Hydrogen peroxide acts upstream of nitric oxide in the heat shock pathway in *Arabidopsis* seedlings. *Plant Physiology, 164*(4):2184–2196.

Wang, Y., Luo, Z., Du, R., Liu, Y., Ying, T., & Mao, L. (2013). Effect of nitric oxide on antioxidative response and proline metabolism in banana during cold storage. *Journal of Agricultural and Food Chemistry, 61*(37):8880–8887.

Wang, Y. S., & Yang, Z. M. (2005). Nitric oxide reduces aluminum toxicity by preventing oxidative stress in the roots of *Cassia tora* L. *Plant and Cell Physiology, 46*(12):1915–1923.

White, G. F., & Brinkmann, W. A. (1975). *Flood Hazard in the United States: A Research Assessment*, vol 6. Boulder: University of Colorado Boulder, Institute of Behavioral Science.

Wimalasekera, R., Villar, C., Begum, T., & Scherer, G. F. (2011). COPPER AMINE OXIDASE1 (CuAO1) of *Arabidopsis thaliana* contributes to abscisic acid-and polyamine-induced nitric oxide biosynthesis and abscisic acid signal transduction. *Molecular Plant*, *4*(4):663–678.

Wink, D. A., Miranda, K. M., Espey, M. G., Pluta, R. M., Hewett, S. J., Colton, C., et al. (2001). Mechanisms of the antioxidant effects of nitric oxide. *Antioxidants and Redox Signaling*, *3*(2):203–213.

Wu, B., Guo, Q., Li, Q., Ha, Y., Li, X., & Chen, W. (2014). Impact of postharvest nitric oxide treatment on antioxidant enzymes and related genes in banana fruit in response to chilling tolerance. *Postharvest Biology and Technology*, *92*:157–163.

Wu, Q. S., Xia, R. X., & Zou, Y. N. (2008). Improved soil structure and citrus growth after inoculation with three arbuscular mycorrhizal fungi under drought stress. *European Journal of Soil Biology*, *44*:122–128.

Wu, X. X., Ding, H. D., Chen, J. L., Zhang, H. J., & Zhu, W. M. (2010). Attenuation of salt-induced changes in photosynthesis by exogenous nitric oxide in tomato (*Lycopersicon esculentum* Mill. L.) seedlings. *African Journal of Biotechnology*, *9*:7837–7846.

Xing, H., Tan, L., An, L., Zhao, Z., Wang, S., & Zhang, C. (2004). Evidence for the involvement of nitric oxide and reactive oxygen species in osmotic stress tolerance of wheat seedlings: Inverse correlation between leaf abscisic acid accumulation and leaf water loss. *Plant Growth Regulation*, *42*(1):61–68.

Xiong, J., An, L., Lu, H., & Zhu, C. (2009). Exogenous nitric oxide enhances cadmium tolerance of rice by increasing pectin and hemicellulose contents in root cell wall. *Planta*, *230*(4):755–765.

Xu, M., Zhu, Y., Dong, J., Jin, H., Sun, L., Wang, Z., et al. (2012). Ozone induces flavonol production of Ginkgo biloba cells dependently on nitrate reductase-mediated nitric oxide signaling. *Environmental and Experimental Botany*, *75*:114–119.

Xuan, Y., Zhou, S., Wang, L., Cheng, Y., & Zhao, L. (2010). Nitric oxide functions as a signal and acts upstream of AtCaM3 in thermotolerance in *Arabidopsis* seedlings. *Plant Physiology*, *153*(4):1895–1906.

Xue, L., Li, S., Sheng, H., Feng, H., Xu, S., & An, L. (2007). Nitric oxide alleviates oxidative damage induced by enhanced ultraviolet-B radiation in cyanobacterium. *Current Microbiology*, *55*(4):294–301.

Xue, L., Li, S., Zhang, B., Shi, X., & Chang, S. (2011). Counteractive action of nitric oxide on the decrease of nitrogenase activity induced by enhanced ultraviolet-B radiation in cyanobacterium. *Current Microbiology*, *62*(4):1253–1259.

Xue, L. G., Li, S. W., Xu, S. J., An, L. Z., & Wang, X. L. (2006). Alleviative effects of nitric oxide on the biological damage of Spirulina platensis induced by enhanced ultraviolet-B. *Wei sheng wu xue bao= Acta Microbiologica Sinica*, *46*(4):561–564.

Yang, H., Wu, F., & Cheng, J. (2011a). Reduced chilling injury in cucumber by nitric oxide and the antioxidant response. *Food Chemistry*, *127*(3):1237–1242.

Yang, W., Sun, Y., Chen, S., Jiang, J., Chen, F., Fang, W., & Liu, Z. (2011b). The effect of exogenously applied nitric oxide on photosynthesis and antioxidant activity in heat stressed chrysanthemum. *Biologia Plantarum*, *55*(4):737.

Yang, J. D., Yun, J. Y., Zhang, T. H., & Zhao, H. L. (2006). Presoaking with nitric oxide donor SNP alleviates heat shock damages in mung bean leaf discs. *Botanical Studies*, *47*:129–136.

Yang, L., Han, R., & Sun, Y. (2013). Effects of exogenous nitric oxide on wheat exposed to enhanced ultraviolet-B radiation. *American Journal of Plant Sciences*, *4*(06):1285.

Yang, L., Ji, J., Harris-Shultz, K. R., Wang, H., Wang, H., Abd-Allah, E. F., et al. (2016). The dynamic changes of the plasma membrane proteins and the protective roles of nitric oxide in rice subjected to heavy metal cadmium stress. *Frontiers in Plant Science*, *7*:190.

Yemets, A. I., Krasylenko, Y. A., & Blume, Y. B. (2015). Nitric oxide and UV-B radiation. In *Nitric Oxide Action in Abiotic Stress Responses in Plants*, eds. Khan, M., Mobin, M. Mohammad, F., & Corpas, F., pp. 141–154. Cham: Springer.

Yeo, W. S., Kim, Y. J., Kabir, M. H., Kang, J. W., & Kim, K. P. (2015). Mass spectrometric analysis of protein tyrosine nitration in aging and neurodegenerative diseases. *Mass Spectrometry Reviews*, *34*(2):166–183.

Yordanov, I., Velikova, V., & Tsonev, T. (2000). Plant responses to drought, acclimation, and stress tolerance. *Photosynthetica, 38*:171–186.

Yu, C. C., Hung, K. T., & Kao, C. H. (2005). Nitric oxide reduces Cu toxicity and Cu-induced NH4+ accumulation in rice leaves. *Journal of Plant Physiology*, *162*(12):1319–1330.

Yu, L., Gao, R., Shi, Q., Wan, X., Wei, M., Yang, F., Pak, J. (2013). Exogenous application of sodium nitroprusside alleviated cadmium induced chlorosis, photosynthesis inhibition and oxidative stress in cucumber. *Pakistan Journal of Botany*, *45*(3):813–819.

Yun, B. W., Feechan, A., Yin, M., Saidi, N. B., Le Bihan, T., Yu, M., et al. (2011). S-nitrosylation of NADPH oxidase regulates cell death in plant immunity. *Nature*, *478*(7368):264.

Zaharah, S. S., & Singh, Z. (2011). Postharvest nitric oxide fumigation alleviates chilling injury, delays fruit ripening and maintains quality in cold-stored 'Kensington Pride' mango. *Postharvest Biology and Technology*, *60*(3):202–210.

Zhang, L., Li, X., Li, X., Wei, Z., Han, M., Zhang, L., & Li, B. (2016). Exogenous nitric oxide protects against drought-induced oxidativestress in Malus rootstocks. *Turkish Journal of Botany*, *40*(1):17–27.

Zhang, M., An, L., Feng, H., Chen, T., Chen, K., Liu, Y., et al. (2003). The cascade mechanisms of nitric oxide as a second messenger of ultraviolet B in inhibiting mesocotyl elongations. *Photochemistry and Photobiology*, *77*(2):219–225.

Zhang, Y., Wang, L., Liu, Y., Zhang, Q., Wei, Q., & Zhang, W. (2006). Nitric oxide enhances salt tolerance in maize seedlings through increasing activities of proton-pump and Na+/H+ antiport in the tonoplast. *Planta*, *224*(3):545–555.

Zhang, Z., Naughton, D., Winyard, P. G., Benjamin, N., Blake, D. R., & Symons, M. C. (1998). Generation of nitric oxide by a nitrite reductase activity of xanthine oxidase: A potential pathway for nitric oxide formation in the absence of nitric oxide synthase activity. *Biochemical and Biophysical Research Communications*, *249*(3):767–772.

Zhao, G., Zhao, Y., Yu, X., Kiprotich, F., Han, H., Guan, R., et al. (2018). Nitric oxide is required for melatonin-enhanced tolerance against salinity stress in rapeseed (*Brassica napus* L.) seedlings. *International Journal of Molecular Sciences*, *19*(7):1912.

Zhao, H., Jin, Q., Wang, Y., Chu, L., Li, X., & Xu, Y. (2016). Effects of nitric oxide on alleviating cadmium stress in *Typha angustifolia*. *Plant Growth Regulation*, *78*(2):243–251.

Zhao, M. G., Chen, L., Zhang, L. L., & Zhang, W. H. (2009). Nitric reductase-dependent nitric oxide production is involved in cold acclimation and freezing tolerance in *Arabidopsis*. *Plant Physiology*, *151*(2):755–767.

Zhao, M. G., Tian, Q. Y., & Zhang, W. H. (2007). Nitric oxide synthase-dependent nitric oxide production is associated with salt tolerance in *Arabidopsis*. *Plant Physiology*, *144*(1):206–217.

Zhao, Z., Chen, G., & Zhang, C. (2001). Interaction between reactive oxygen species and nitric oxide in drought-induced abscisic acid synthesis in root tips of wheat seedlings. *Functional Plant Biology*, *28*(10):1055–1061.

Zheng, C., Jiang, D., Liu, F., Dai, T., Liu, W., Jing, Q., & Cao, W. (2009). Exogenous nitric oxide improves seed germination in wheat against mitochondrial oxidative damage induced by high salinity. *Environmental and Experimental Botany*, *67*:222–227.

Zhu, L. Q., Zhou, J., & Zhu, S. H. (2010). Effect of a combination of nitric oxide treatment and intermittent warming on prevention of chilling injury of 'Feicheng' peach fruit during storage. *Food Chemistry*, *121*(1):165–170.

9 Role of Jasmonates in Plant Abiotic Stress Tolerance

Péter Poór, Zalán Czékus, and Attila Ördög

CONTENTS

9.1 INTRODUCTION

Abiotic stress factors such as high light, UV, heat, cold, drought, flooding, salt, and heavy metals impair cellular structures and disrupt the physiological functions of plants leading to growth perturbation, reduced fertility, premature senescence, and yield losses. Reducing these harmful environmental effects and thus preventing yield losses is particularly important in the face of climate change and a rapidly expanding world population which will require the production of more food and feed (Alexandratos and Bruinsma, 2012).

Plants respond to stressful environmental conditions at hormonal, transcriptional, biochemical, physiological, and morphological levels. Plant hormones and their signalling pathways integrate different environmental inputs and regulate plant responses to optimize their growth, survival, and reproduction (Cramer et al. 2011; Peleg and Blumwald, 2011). Traditionally, jasmonates (JAs), salicylic acid (SA), ethylene (ET), and abscisic acid (ABA) are the main stress phytohormones, whereas auxins (IAAs), gibberellins (GAs), cytokinins (CKs), and brassinosteroids (BRs) are associated with plant developmental processes. However, it is becoming increasingly evident that all plant hormones are in a close relationship and all of them regulate the stress and developmental responses of plants in a complex manner (Santino et al. 2013; Kazan, 2015). This review will briefly discuss recent studies that have revealed insights into the roles of JAs, including jasmonic acid and its derivatives, which play a role in plant defence reactions against various insect pests and pathogens (Wasternack and Hause, 2013), but can be also an important regulator of abiotic stress tolerance mechanisms (Goossens et al. 2016).

In this review, the physiological effects of several abiotic stress factors on plants and stress-induced changes in hormonal pathways will be summarized in model plants and crops to gain a better understanding of JAs-regulated processes at the physiological, biochemical, and molecular levels in plant stress responses. This knowledge can help to improve our ability to improve stress resistance in crops, facing future challenges in the changing environment.

9.2 METABOLISM AND SIGNALLING OF JASMONATES

The biosynthesis, perception, and action of JAs have already been extensively studied and well-documented (Goossens et al. 2016; Wasternack and Song, 2016). JAs are synthesized from α-linolenic acid (α-LeA/18:3) via the octadecanoid pathway in chloroplasts and peroxisomes. Firstly, α-LeA is produced by the coordinated actions of fatty acid desaturase (FAD) and phospholipase A1 (PLA) from the plastid membrane lipids. Then, α-LeA is converted sequentially to (13S)-hydroperoxyoctadecatrienoic acid (13-HPOT) by the action of 13-lipoxygenase (LOX), then to 12,13(S)-epoxyoctadecatrienoic acid (12,13-EOT) by the function of allene oxide synthase (AOS), and finally the oxidized intermediate is cyclized into (9S,13S)-12-oxo-phytodienoic acid (OPDA) by the activity of allene oxide cyclase (AOC). After being transported to the peroxisome, OPDA is reduced to 3-oxo-2-(cis-2′-pentenyl)-cyclopentane-1-octanoic acid (OPC-8:0) by OPDA reductase (OPR), which reaction consumes nicotinamide adenine dinucleotide phosphate (NADPH). OPC-8:0 is then activated to OPC-8:0 CoA by OPC-8:0 CoA ligase (OPCL) and converted to JA via β-oxidation catalysed by three different enzymes: acyl-CoA oxidase (ACX), multifunctional protein (MFP), and 3-ketoacyl-CoA thiolase (KAT), which reaction requires the participation of ATP and O_2. JA then is exported to the cytoplasm, where it may undergo amino acid conjugation with isoleucine (Ile) to form bioactive (+)-7-iso-JA-Ile by jasmonoyl-isoleucine synthetase (JAR1), or be metabolized to other inactive forms via hydroxylation, carboxylation, decarboxylation, methylation, esterification, sulphation, and glucosylation (Wasternack and Hause, 2013; Wasternack and Song, 2016; Wasternack and Strnad, 2018).

Bioactive JA, JA-Ile induces the interaction of the JA receptor CORONATINE INSENSITIVE1 (COI1) with the JA ZIM-domain (JAZ) family proteins, recruiting NOVEL INTERACTOR OF JAZ (NINJA; an adaptor protein) and TOPLESS (TPL; a co-repressor), leading to the ubiquitination and degradation of JAZ proteins via the 26S proteasome. Thus the downstream transcription factors (TFs) became de-repressed, allowing them to activate JA-responsive early genes and JA responses (Yan et al. 2007; Pauwels and Goossens, 2011; Wasternack and Hause, 2013; Wasternack and Song, 2016; Wasternack and Strnad, 2018).

JA signalling activators based on Wasternack and Song (2016) are the basic helix–loop–helix (bHLH) subgroup IIIe TFs: MYC2, MYC3, MYC4, MYC5; the TTG1/bHLH/MYB complex; R2R3-MYB TFs; AP2/ERF-domain TFs; EIN3/EIL1, the core TFs in the ethylene pathway; YABs modulating anthocyanin accumulation and chlorophyll degradation; NAC019, NAC055, and NAC072 regulating leaf senescence via upregulation of the chlorophyll catabolic genes; and DELLAs, the repressors in the GA pathway. JA signalling repressors are the bHLH subgroup IIId TFs and WRKY57 (Wasternack and Song, 2016). Several TFs mediated by JAs play a different regulation role under various abiotic stress conditions.

9.3 LIGHT STRESS

Although light is essential for plants, both as a source of energy for photosynthesis and as an environmental signal, excess light could have a damaging impact on photosynthetic efficiency by inducing photoinhibition and causing transcriptome changes, through the generation of reactive oxygen species (ROS), and inducing a wide-range of phytohormones like JAs (Takahashi and Badger, 2011; Demarsy et al. 2018). Light also regulates a whole range of developmental processes including germination, de-etiolation, stomatal development, circadian rhythm, and flowering (de Wit et al. 2016). At the same time, strong sunlight is associated with high levels of UV-B, which can damage DNA and other subcellular components, decrease the efficiency of photosynthesis, and destroy photosynthetic apparatus, inducing ROS production that also induces the production of phytohormones including SA, ET, and JA (Müller-Xing et al. 2014; Yin and Ulm, 2017). In contrast, low light and dark also induce new signalling and regulation pathways modulated by phytohormones, especially by JA and SA (Ballaré 2014).

Plants can use light as an informational signal, which is interpreted by photoreceptors. Photoreceptors usually absorb photons using a specific prosthetic chromophore and induce structural changes in the protein part of the receptor. Various photoreceptors are known in *Arabidopsis thaliana* model plants, namely the red/far-red light-sensing phytochromes (phyA–phyE), the blue light-sensing cryptochromes (cryl, cry2) and the phototropins (photl, phot2), the Zeitlupe family members (ZTL, FKF1, LKP2), and the UV-B receptor UVR8 (Demarsy et al. 2018). Based on analysis of *Arabidopsis thaliana* mutants, phyB is the primary photoreceptor for red light (R) perception in the regulation of the growth inhibition of the hypocotyl, while phyA senses far-red light (FR), inducing hypocotyl growth and shade avoidance. There is a close interaction between JAs, light signalling, and photomorphogenesis (Goossens et al. 2016). JA-Ile activates the JA pathway by binding to the COI1-JAZ co-receptor complex; besides this COI1 also modulates signal transduction in light-driven developmental processes, for instance in the shade-avoidance response. In high R/FR light, phyB suppresses the shade-avoidance response and enhances sensitivity to JA, promoting defence responses (Robson et al. 2010). In low R/FR light, phyB is inactivated, the suppression of shade responses is released, and sensitivity to JA is reduced (Moreno et al. 2009). The JA receptor COI1 is required for the FR-induced expression of the light-regulated genes and modulates flowering and shade responses, but exogenous JA suppresses this process. FR light increased the transcript levels of allene oxide synthase (*OsAOS1* and *OsAOS4*) in seedling shoots of rice (Haga and Iino, 2004), and the transcription of the JA biosynthesis gene ALLENE OXIDE CYCLASE1 (AOC1), the signalling genes JAZ1 and MYC2/JIN1/ZBF1, and the response gene VEGETATIVE STORAGE PROTEIN1 (VSP1) in seedlings of *Arabidopsis thaliana*, but this induction was attenuated in *coi1-16* mutants (Robson et al. 2010). In contrast, it was found that the mutants of phyA contained higher levels of the JA precursor OPDA (Robson et al. 2010), and the deficiency of phytochrome chromophore resulted in the overproduction of JA and activated COI1-dependent JA responses in *Arabidopsis thaliana* (Zhai et al. 2007). In addition, light-mediated phyA degradation is dependent on JA, based on the investigation of JA-deficient rice mutant, *hebiba*. In this mutant, the photodestruction of phyA is delayed but it was restored by exogenous application of JA (Riemann et al. 2009). Thus, JA plays important role in the inhibition of hypocotyl growth regulated by phyA and phyB. PhyA induces the expression of JA RESISTANT 1 (JAR1), which encodes the enzyme conjugating Ile to JA. JAR1/FAR-RED INSENSITIVE219 (FIN219) directly interacts with an E3-ligase, CONSTITUTIVE PHOTOMORPHOGENIC1 (COP1), a master regulator for TFs regulating hypocotyl growth under continuous FR light (Hsieh et al. 2000; Wang et al. 2011; Hsieh and Okamoto, 2014; Goossens et al. 2016). Based on the current model of Wasternack and Strnad (2018), phyA activation by FR light induced JA biosynthesis. JAs activate COI1 which promotes the degradation of JAZ1 and allows the MYC2 TF to induce gene expression. Under these conditions, JAR1/FIN219 also regulates the nuclear exclusion of COP1, allowing HY5 protein accumulation in the nucleus with the consequent inhibition of hypocotyl elongation.

JA signalling is also associated with blue light sensing (Svyatyna and Riemann, 2012). Riemann et al. (2008) observed that the mutation of JAR1 in rice resulted in longer coleoptiles compared to WT plants, indicating that *OsJar1* participates in the suppression of coleoptile elongation under blue light conditions. It was also observed in rice that blue light induced the expression of *OsAOS1*, the coding sequence of the key enzyme of the biosynthesis of JA, which was significantly higher in cryptochrome-overexpressing rice lines (*OsCRY1a* and *OsCRY1b*) (Hirose et al. 2006). MYC2 is an important regulator of JA and various light signals. It was found that the *jin1/myc2* mutant showed enhanced inhibition of hypocotyl elongation under blue light, suggesting that MYC2 is a negative regulator of blue light-mediated photomorphogenic growth (Yadav et al. 2005). A recent study found that overexpression of *FIN219* in WT plants resulted in a short-hypocotyl phenotype under blue light conditions, indicating that the FIN219 function is required for blue light signalling. Authors concluded that FIN219/JAR1 and CRY1 antagonize each other under blue light to modulate the photomorphogenic development of *Arabidopsis thaliana* seedlings (Chen et al. 2018a).

UV-B-mediated plant responses can be also dependent on JA signalling (Kazan and Manners, 2011; Vanhaelewyn et al. 2016). Solar UV-B was able to increase JA biosynthesis in *Nicotiana* species by enhancing the transcript levels of *LOX* and *AOS* (Izaguirre et al. 2003; Dinh et al. 2012), and promoted JA signalling in *Brassica oleracea* (Mewis et al. 2012). In addition, it was shown that UV-B induced JA accumulation in *Arabidopsis thaliana* (Mackerness et al. 1999) and in *Vigna radiata* (Choudhary and Agrawal 2014). Levels of phenolic compounds, which normally increase following UV-B exposure, decreased in the *jar-1 Arabidopsis thaliana* mutant, which is impaired in JA signalling (Caputo et al. 2006). It was also found that pre-treatment with MeJA counteracted the UV-B stress in *Hordeum vulgare* seedlings, where methyl JA (MeJA) increased the activity of superoxide dismutase (SOD), catalase (CAT), and peroxidase (POX) and elevated the proline content in leaves and roots, which can reduce the oxidative effects of UV-B stress (Fedina et al. 2009). Similarly, JA pre-treatment could moderate the harmful effects of UV-B on photosynthetic activity by increasing the maximal quantum efficiency (Fv/Fm) and the effective quantum efficiency (Φ_{PSII}), and the photosynthetic electron transport rate (ETR), as well as by decreasing nonphotochemical quenching (NPQ) in wheat seedlings (Liu et al. 2012). In other experiment with *Scutellaria baicalensis* it was found that JA pre-treatment could moderate the photosynthetic inhibition via the recovery of the chlorophyll content, the stomatal conductance, and the intercellular CO_2 concentration under UV-B exposure (Quan et al. 2018). UV-B failed to elicit an accumulation of JA in transgenic JA-deficient *Nicotiana attenuata* plants in which a lipoxygenase gene (*NaLOX3*) was silenced. Moreover, the UV-B-induced accumulation of JA and phenolic compounds was reduced in these plants, suggesting that the UV-mediated synthesis of these compounds requires JA biosynthesis. However, there are also JA-independent pathways in the elicitation of these phenolic compounds (Demkura et al. 2010). Moreover, this enhancing effect of UV-B on JA was not detected in *Arabidopsis* for other markers of the JA response (Demkura and Ballaré, 2012). Based on microarray data, UV-B radiation increased the expression of some JA-related genes, but this seemed highly dependent on the experimental conditions (Mazza and Ballaré, 2015).

Although light is essential for normal plant growth and development, high or excess light intensity can damage the photosynthetic apparatus and induce oxidative stress (Demarsy et al. 2018). Excess light can be sensed by photoreceptors such as phototropins and cryptochromes and relayed signals for chloroplast movement and changing the expression of several defence genes (Li et al. 2009). High light (PPFD = 1600 μmol quanta m^{-2} s^{-1}) elevated the gene expression of JA biosynthesis (e.g. *LOX3*, *AOS*, *OPR3*) and the protein amount of LOX-C and AOS both in WT and the *npq4* mutant (a key photoprotective protein involved in the process of feedback de-excitation or the qE component of NPQ) of *Arabidopsis thaliana* (Frenkel et al. 2009). The absorption of excess light energy in the chloroplasts results in the formation of singlet oxygen ($^1O_2$), triggering photooxidative stress and programmed cell death regulated by JA (Ramel et al. 2013a). Chlorophyll *b*-less *Arabidopsis thaliana* mutant (*chlorina1* [*ch1*]) is highly photosensitive due to a selective increase in the release of $^1O_2$ by photosystem II. JA biosynthesis was strongly induced in *ch1* mutants upon excess light and was repressed under acclimation conditions, suggesting the involvement of JA in $^1O_2$-induced cell death (Ramel et al. 2013b).

9.4 TEMPERATURE STRESS

Both low temperatures (including freezing) and heat stress can seriously affect plant growth and development. Plants have evolved sophisticated mechanisms involving altered molecular, biochemical, and physiological processes to tolerate temperature stresses. In the case of cold stress, the INDUCER OF CBF EXPRESSION 1 (ICE1) and ICE1-like bHLH TFs (e.g. ICE2) stimulate the transcription of C-REPEAT BINDING FACTORs (CBFs), which encode AP2/ERF family TFs. CBF TFs (CBF1, CBF2, and CBF3), by binding to the C-repeat (CRT)/dehydration-responsive elements, induce the expression of a large subset of COLD REGULATED (COR) genes, finally leading to an enhanced cold (freezing) stress tolerance in plants (Chinnusamy et al. 2007; Thomashow, 2010; Zhao et al. 2016a).

It is known that the exogenous application of JA significantly enhanced the freezing tolerance of *Arabidopsis thaliana* independently of cold acclimation (Aghdam and Bodbodak, 2013; Hu et al. 2013; 2017; Kazan, 2015; Per et al. 2018). In parallel, with the blocking of endogenous JA biosynthesis and signalling with the help of *lox2*, *aos*, and *jar1* mutant plants, hypersensitivity to freezing was observed. Moreover, cold induced the expression of *LOX1-4*, *AOS*, *AOC1-4*, and *JAR* JA biosynthesis and signalling genes, and JA elevated the transcript levels of the CBF/DREB1 signalling (Hu et al. 2013). Similarly, cold stress elevated endogenous JA content and the expression of *OsAOS1-2*, *OsOPR1*, *OsAOC*, and *OsLOX2* in rice (Du et al. 2013). It was also observed that cold-stimulated germination of dormant grains correlated with a transient increase in JA content and expression of JA biosynthesis genes (*TaAOS1-2*, *TaAOC1-3*) in wheat (Xu et al. 2016). The foliar application of JA confirmed that JA functioned downstream of ABA to activate the CBF pathway in the light quality-mediated cold tolerance of tomato (Wang et al. 2016). The exogenous application of MeJA significantly improved the quality of several tropical or subtropical fruits (Aghdam and Bodbodak, 2013), such as avocado (Meir et al. 1996), banana (Zhao et al. 2013; Ba et al. 2016), blueberry (Huang et al. 2015), garlic (Akan et al. 2019), lemon (Siboza et al. 2014), loquat fruit (Cao et al. 2009), mango (Gonzalez-Aguilar et al. 2000), orange (Rehman et al. 2018), guava fruit (Gonzalez-Aguilar et al. 2004), longkong fruit (Venkatachalam and Meenune, 2015), papaya (Gonzalez-Aguilar et al. 2003), peach (Meng et al. 2009; Jin et al. 2009; 2013), pepper (Fung et al. 2004; Shin et al. 2017), pineapple (Nilprapruck et al. 2008), pomegranate (Sayyari et al. 2011), and tomato (Ding et al. 2002; Zhang et al. 2012; Min et al. 2018) under cold storage (e.g. 2, 4 or 7°C). These studies indicate that the JA-induced cold tolerance of these fruits may also be associated with an increased accumulation of cryoprotective compounds, such as polyamine (Zhang et al. 2012), proline (Min et al. 2018), total phenols (Gonzalez-Aguilar et al. 2004; Jin et al. 2009; Meng et al. 2009), sugar (Gonzalez-Aguilar et al. 2004; Nilprapruck et al. 2008), and anthocyanin (Sayyari et al. 2011; Huang et al. 2015) or with increased antioxidant content, antioxidant enzyme activity, or gene expression such as SOD, CAT, and ascorbate peroxidase (APX) (Ding et al. 2002; Venkatachalam and Meenune, 2015; Cao et al. 2009; Jin et al. 2009; Sayyari et al. 2011; Siboza et al. 2014; Shin et al. 2017; Rehman et al. 2018). Moreover, the ratio of unsaturated/saturated fatty acids in MeJA-treated loquat fruit was also significantly higher than in control fruit (Cao et al. 2009). Furthermore, MYC2 in the activation of JA response is involved in the MeJA-induced chilling tolerance of banana fruit through physically interacting and likely functionally coordinating with ICE1. MeJA also induced the expression of ICE-CBF cold-responsive pathway genes including *MaCBF1*, *MaCBF2*, *MaCOR1*, *MaKIN2*, *MaRD2*, and *MaRD5* in banana fruit (Zhao et al. 2013). MYC2 is also an important component of MeJA-mediated chilling tolerance in tomato fruit (Min et al. 2018). The role of JAZ repressors as regulators of cold stress tolerance has also been shown. In banana, one lateral-organ boundaries domain (MaLBD5), which is induced by both cold and MeJA, physically interacted with MaJAZ1, a repressor of JA signalling, and transactivated the expression of *MaAOC2*. These results suggest that MaLBD5 may be partially associated with the biosynthesis of JA to mediate JA-induced cold tolerance of banana fruit (Ba et al. 2016). It was also revealed that JAZ1 and JAZ4 repressors physically interact with the ICE1 TF to inhibit its transcriptional activity. Consistent with these results, the overexpression of JAZ1 or JAZ4 represses the ICE–CBF signalling pathway and freezing stress responses in *Arabidopsis thaliana* (Hu et al. 2013).

Heat stress due to increased temperature is also a serious agricultural problem in many areas of the world. Heat stress affects membrane fluidity, metabolism, and cytoskeleton rearrangement, as well as resulting in the accumulation of unfolded proteins. Heat stress disturbs photosynthesis, respiration, water relations, and membrane stability, and also induces ROS production and modulates the hormone levels. In response to high temperatures, plants synthesize HEAT SHOCK PROTEINS (HSPs) that prevent denaturation and assist the refolding of damaged proteins (Wahid et al. 2007; Ruelland and Zachowski, 2010; Sharma and Laxmi, 2016). JA can be an important regulator of heat stress responses. Endogenous levels of JA, OPDA, and JA-Ile rose during heat stress in *Arabidopsis thaliana* exposed to 38°C for 8 h. Moreover, the exogenous application of low concentrations of

MeJA maintained cell viability in heat-stressed plants based on electrolyte leakage (EL) measurements. However, the authors did not find a close relationship between thermotolerance and MeJA-elicited HSP gene expression (Clarke et al. 2009). Zhang et al. (2015) observed that the mutation of the *Arabidopsis thaliana* co-chaperone SUPPRESSOR OF G2 ALLELE OF SKP1 (SGT1) impairs responses to JA. SGT1 is known to function as a cofactor of HSP90 in JA-COR (coronatine) responses and provides thermotolerance in plants. It was found that JA differentially regulates stress tolerance in a species- and organ-specific manner (Ozga et al. 2016). For example, the JA/COI1 signalling pathway is a key hub in regulating the stigma exsertion of tomato flowers induced by heat stress (Pan et al. 2019).

9.5 DROUGHT STRESS

Drought stress induces water loss and changes in the osmotic homeostasis, as well as the loss of turgidity in plant cells, which affects membrane tension perceived through changes in the activity of mechanosensitive ion channels, like calcium channels. The influx of calcium can be transduced through calcium-dependent kinases into activation of the NADPH oxidase RboH, generating apoplastic transient oxidative burst. At the organ level, the rapid closure of stomata will reduce additional loss of water. At the cellular level, the synthesis of antioxidants, osmotics, and late-embryogenesis abundant (LEA) proteins serves the tolerance mechanism regulated by phytohormones like JA (Peleg and Blumwald, 2011; Golldack et al. 2014; Riemann et al. 2015; Zhu, 2016).

Elevated JA content was measured in both the roots and shoots of etiolated *Zea mays* plantlets under desiccation stress induced by polyethylene glycol (PEG) application (Xin et al. 1997). Interestingly, the levels of JA-precursor OPDA and JA increased under water-deficit stress in the foliage and shoots of *Pinus pinaster* plants, but the responses of two populations of different provenances (Gredos and Bajo Tiétar) were different, suggesting a possible correlation with adaptations to diverse ecological conditions (Pedranzani et al. 2007). Water-deficit stress also induced ABA, JA, and proline accumulation but did not modify malondialdehyde (MDA) content in papaya seedlings (Mahouachi et al. 2012). Concentrations of endogenous ABA, JA, and free polyamines (putrescine, spermine, and spermidine), as well as polyamine oxidase (PAO) activity were elevated in the roots and leaves of tomato seedlings under PEG-induced drought stress (Zhang and Huang, 2013). JAs, particularly OPDA, were highly effective signalling molecules in the mediation of sunflower seedling responses to water stress (Fernández et al. 2012; Andrade et al. 2017). Savchenko et al. (2014) demonstrated that OPDA, as a drought-responsive regulator of stomatal closure, functioned most effectively together with ABA. Drought stress caused also an accumulation of JA, ABA, IAA, and SA in Kentucky bluegrass leaves (Krishnan and Merewitz, 2015a). Others found that OsbHLH148, a basic helix–loop–helix protein, interacts with OsJAZ proteins in the upstream signalling pathway of JA by forming an 'OsbHLH148-OsJAZ1-OsCOI1 signalling module' which eventually activates the expression of the *OsDREB1* gene, a marker of drought stress responses (Seo et al. 2011)

Results of Shan and Liang (2010) suggested that water stress-induced JA accumulation is an important signal that leads to the regulation of ascorbate (AsA) and glutathione (GSH) metabolism during drought stress tolerance in *Agropyron cristatum* leaves. Exogenous treatment with JA caused an increase in the content of ABA but not in that of proline and spermidine in the two barley genotypes (Bandurska et al. 2003). Pre-treatment with COR significantly increased the activity of SOD, CAT, APX, and gluthatione reductase (GR) in the leaves of water-stressed rice cultivars (Ai et al. 2008). Spraying drought-stressed seedlings with JA increased GR and glyoxalase (GLY) activities in *Brassica napus*, increased the monodehydroascorbate reductase (MDHAR) activity in *Brassica campestris*, and increased dehydroascorbate reductase (DHAR), GR, and GLY activities in *Brassica juncea*. JA improved the fresh weight, chlorophyll, and leaf relative water content (RWC) in all *Brassica* species (Alam et al. 2014). MeJA treatment mitigated the decline of the net photosynthetic rate (PN), stomatal conductance (gs), and water-use efficiency (WUE) induced by drought stress and enhanced the activities of SOD, APX, CAT, and reduced MDA content in wheat leaves under drought stress (Ma et al. 2014). MeJA application also improved the growth parameters as well as RWC, proline content, antioxidant activity, and essential oil percentage of summer savoury (*Satureja hortensis*) under drought stress condition (Miranshahi and Sayyari, 2016). Both MeJA and COR enhanced the growth and accumulation of dry matter in cauliflower seedlings during drought-stressed and rewatering conditions by increasing the accumulation of chlorophyll content, the net photosynthetic rate, the leaf relative water content and the endogenous ABAlevel , as well as activating the antioxidant enzymes (SOD, CAT, APX, and GR) and elevating the proline and soluble sugar content in the water-stressed leaves (Wu et al. 2012). COR also alleviated the effects of water deficiency stress on winter wheat seedlings (Li et al. 2010). Shan et al. (2015) found that nitric oxide (NO) participated in the regulation of the ascorbate-glutathione (AsA-GSH) cycle by exogenous JA in the leaves of wheat seedlings, and JA decreased MDA content and EL under drought stress. NO is also an important regulator of JA- and drought-induced stomatal closure (Suhita et al. 2003, 2004; Munemasa et al. 2007; Huang et al. 2009). JA induced a significant increase in the glycinebetaine levels in pear leaves when the plants were subjected to water stress (Gao et al. 2004). The beneficial role of exogenous MeJA treatment was confirmed by growth changes in parallel with the alteration in leaf gas exchange and chlorophyll contents of soybean plants under drought stress (Anjum et al. 2011). MeJA also plays role in the alleviation of water stress in soybean, regulating the content of saturated and unsaturated fatty acids, flavonoids, phenolic acid, and sugars in the shoots of the soybean genotypes (Mohamed and Latif, 2017).

There is close relationship between JA and other hormones in enhancing drought stress tolerance. Both exogenous JA and ABA ameliorate the adverse effects of drought stress through the accumulation of polyamines (putrescin, spermidine, spermine) and the production of trypsin inhibitor in soybean plants (Hassanein et al. 2009). For instance, a substantial interaction between ABA and JA was observed during their biosynthesis under water stress in *Arabidopsis*, where drought-associated stomatal closure is regulated mainly by ABA and weakly by JA, whereas JA plays a role in the formation of antioxidants regulating AsA and GSH metabolism (Brossa et al. 2011). JA content decreased in leaves but increased in the xylem sap in *Eucalyptus globulus* (Correia et al. 2014). A transient increase in JA content was measured after a few hours of drought stress in a citrus root, which returned to control levels 30 h after the onset of the stress conditions. This transient accumulation of JA was needed for ABA increase in the same conditions (de Ollas et al. 2013, 2015; de Ollas and Dodd, 2016), modifying stomatal responses under water-deficit stress (de Ollas et al. 2018). The putative crosstalk between JA and ABA was also investigated in tomato plants in response to drought, where ABA regulated the induction of the *OPR3* gene in roots (Muñoz-Espinoza et al. 2015). The results of Fugate et al. (2018) indicated that the exogenous application of MeJA delayed plant dehydration and protected the photosynthetic apparatus of sugar beet (*Beta vulgaris*) leaves from drought-induced impairment. Interestingly, exogenously applied MeJA improved the drought tolerance independently of the developmental stages of wheat plants (Anjum et al. 2016). MeJA priming also improved the chlorophyll content and photochemical efficiency under PEG stress in rice seedlings (Sheteiwy et al. 2018).

9.6 SALT STRESS

Salt stress is one of the most harmful environmental stresses in arid and semiarid regions, and can disrupt cellular structures and impair the physiological functions of plants, leading to growth perturbation, reduced fertility, premature senescence, and yield loss by inducing osmotic-, ionic-, and nitro-oxidative stress (Munns, 2002; Munns and Tester, 2008; Poór et al. 2015).

JA increased after NaCl treatments in many plant species like in *Artemisia sphaerocephala* and *Artemisia ordosiea* (Chen et al. 2018b), in *Citrus sinensis* and *Citrus macrophylla* root exudate (Vives-Peris et al. 2017), in *Cucumis sativus* leaves (Radhakrishnan and Lee, 2013), in *Glycine max* shoots (Hamayun et al. 2015; Liu et al. 2017), in *Oryza sativa* shoots (Hazman et al. 2015, 2016), in *Phaseolus vulgaris* shoots and roots (Farhangi-Abriz and Torabian, 2018), in Carrizo citrange

(*Poncirus trifoliata* L. Raf. × *Citrus sinensis* L. Osb.) leaves (Balfagón et al. 2018), in halophyte *Salvadora persica* leaves (Kumari and Parida, 2018), in *Sesuvium portulacastrum* shoots (Wali et al. 2016), and in *Solanum lycopersicum* seedlings (Andrade et al. 2005) and hairy root culture (Abdala et al. 2003). Salt stress elevated endogenous JA levels in the leaves of *Agrostis stolonifera* but decreased it in the roots of sensitive and tolerant genotypes of this plant species (Krishnan and Merewitz, 2015b). Severe salt stress (150 mM NaCl) induced a transient JA increase in the shoots of *Arabidopsis thaliana* between 15 and 60 min after treatment, and after 4 h in leaves. After 24 h, JA levels fell sharply in leaves and roots of the treated plants. The authors found similar changes in the levels of JA active conjugate and JA-Ile content (Prerostova et al. 2017). A similar transient increase in JA levels was observed in *Arabidopsis thaliana* upon a lower concentration of NaCl (Ding et al. 2016; Pilarska et al. 2016; Zarza et al. 2017) and in tomato leaves and roots (Ghanem et al. 2012). The concentrations of free JA and the JA-Ile conjugate significantly increased at the lowest salt concentration (50 mM), but then decreased at higher salt concentrations (200 mM) after 24 h in *Brassica rapa* seedlings (Pavlović et al. 2018). JA levels increased dramatically up to eight-fold in the *vte1* (lack α-tocopherol) *Arabidopsis thaliana* mutant after 72 h of 100 mM NaCl treatment (Cela et al. 2011; Ellouzi et al. 2013). Interestingly, Kurotani et al. (2015) observed that JAs have a negative effect on the viability of *Oryza sativa* leaves exposed to 600 mM NaCl.

150 mM NaCl induced the expression of *AtAOC3*, *AtJAZ1*, and *AtJAZ10* in *Arabidopsis thaliana* after 12 h (Pillai et al. 2018). The *AOC* gene from *Leymus mollis* was significantly expressed by 400 mM NaCl (Habora et al. 2013). JA-responsive JAZ genes were also upregulated by 150 mM NaCl in a COI1-dependent manner in the roots of *Arabidopsis thaliana* (Valenzuela et al. 2016). Liu et al. (2019) found that the overexpressing of *PnJAZ1* (jasmonate ZIM-domain gene, which encodes a nucleus-localized protein with conserved ZIM and Jas domains) showed increased tolerance to salt stress. The transcription of *EcJAZ* increased 4.2-fold due to salt stress in *Eleusine coracana* (Sen and Dutta, 2016). JAR1 and JAZ genes were upregulated during salt stress in the root tissue of *Gossypium arboreum* (Zhang et al. 2013) and *Gossypium hirsutum* leaf (Sun et al. 2017). The over-expression of *MdJAZ2* conferred an increased tolerance to salt and PEG stress during the seedling development of *Arabidopsis thaliana* (An et al. 2017). Wu et al. (2015) found that the suppression of *OsJAZ9* resulted in reduced salt tolerance in *Oryza sativa*. *JcMYB1* (from *Jatropha curcas*, *JcMYB1* is a member of the R2R3-MYB transcription factor subfamily) was upregulated by JA and ABA treatments and the overexpression of *JcMYB1* improved the salt and drought stress tolerance of transgenic tobacco (Li et al. 2015). A rice defence-related gene named 'jasmonic acid inducible pathogenesis-related class 10' (*JIOsPR10*) also showed upregulation under salt and drought stress conditions (Wu et al. 2016).

There is close interaction between JA and other phytohormones. Pre- or post-treatment with JA mostly further increased ABA content under salt stress in *Oryza sativa* seedlings (Seo et al. 2005), and ABA also increased due to JA in grapevine cell suspension culture (Ismail et al. 2012, 2014). ET signalling mutant *ein3-1* showed higher JA levels (Asensi-Fabado et al. 2012), but no differences in JA concentrations were found in the ABA biosynthesis *aba3* mutant upon 100 mM NaCl treatment (Asensi-Fabado and Munné-Bosch, 2011). Salt stress-induced *ETHYLENE RESPONSE FACTOR1* (*ERF1*) expression was also enhanced by ET and JA and suppressed by ABA treatment and in *abi1* and *abi2* knockout mutants, respectively (Cheng et al. 2013). ABA and SA significantly decreased but JA levels remained constant in old leaves of *Rosmarinus officinalis* exposed to 200 mM NaCl (Müller and Munné-Bosch, 2011). Halophyte plants (*Cakile maritima*, *Thellungiella salsuginea*) also displayed an enhanced accumulation of JA, ABA, and ET (Ellouzi et al. 2013; 2014). Similarly, the steady-state levels of JA and related compounds were higher in salt-tolerant tomato leaves than in salt-sensitive species, but JA levels in both tomato cultivars changed in response to salt stress (Pedranzani et al. 2003).

There are numerous studies pointing out the fact that the exogenous application of JAs shows increases in salt stress tolerance. The beneficial effect of JA was manifested under salt stress conditions by higher biomass, photosynthetic pigment and osmoprotectant levels, and decreased EL and lipid peroxidation in *Hibiscus esculentus* leaves (Azooz et al. 2015). A foliar spray of JA increased the lateral root growth, but reduced primary root and shoot growth as well as shoot dry weight upon 100 mM NaCl; however it did not alter the root water content under different levels of salt stress in *Brassica napus* (Farhangi-Abriz et al. 2019). A beneficial role of exogenous JA in the growth of *Arabidopsis thaliana* was also observed under 100 mM NaCl-induced senescence (Chen et al. 2017). Treatment with JA had a significant effect in enhancing protein quantity and quality under different levels of salinity in soybean (Farhangi-Abriz and Ghassemi-Golezani, 2016). JA-pre-treated salt-stressed barley accumulated lower content of Na^+ in the shoot tissue compared with only salt-stressed plants after several days of exposure to stress. In addition, pre-treatment with JA partially alleviated photosynthetic inhibition caused by salinity stress in barley. JA mediated salinity tolerance through arginine decarboxylase, ribulose 1,5-bisphosphate carboxylase/oxygenase (Rubisco) activase, and apoplastic invertase in this species (Walia et al. 2007). JA improved the Na^+ exclusion by decreasing the Na^+ uptake at the root surface of *Zea mays* (Shahzad et al. 2015). The net photosynthetic rate and chlorophyll contents, as well as the K^+/Na^+ ratio and salt gland density also changed beneficially after MeJA spraying in *Limonium bicolor* seedlings exposed to 300 mM NaCl (Yuan et al. 2019). The chlorophyll *a*/*b* ratio was also increased by JA under salt stress in *Oryza sativa*, and JA alleviated the decrease in RWC, MDA, and EL in salt-stressed plants (Baek et al. 2006). The leaf water potential, leaf photosynthetic rate, and maximum quantum yield of photosystem II (PSII) also remarkably recovered when 30 µM JA was applied 24 h after salt stress compared with the 40 mM NaCl-treated plants (Kang et al. 2005). JA can alleviate salt stress-induced oxidative stress. The JA-deficient tomato mutant *defenseless-1* (*def-1*) showed ROS-associated injury phenotype, which was associated with lower activity of both enzymatic antioxidants and non-enzymatic antioxidants upon 100 mM NaCl treatment (Abouelsaad and Renault, 2018).

9.7 HEAVY METAL STRESS

Heavy metal contamination has become a worldwide environmental concern with damaging impacts on the agriculture (Tchounwou et al. 2012). At the cellular level, an elevated quantity of heavy metals induces oxidative stress, which at high concentration leads to tissue damage and morpho-physiological changes in plants (Yadav 2010). The major signalling networks also work in metal stresses, including among others calcium, MAPK, and hormone signalling (Jalmi et al. 2018). Thus, JAs also influence plant defence responses to metal stress, such as aluminium (Al), arsenite (AsIII), arsenate (AsV), cadmium (Cd), chromium (Cr), copper (Cu), nickel (Ni), lead (Pb), and zinc (Zn) (Per et al. 2018).

Treatment with 50 µM Al promoted JA biosynthesis and signalling genes (LOX, 12-oxophytodienoate reductase, acyl-CoA oxidase, and jasmonate ZIM-domain proteins) after 48 h in Al-sensitive soybean lines (Huang et al. 2017). The application of 10 µM of MeJA simultaneously with 100 µM Al^{3+} increased Al-resistance in *Vaccinium corymbosum* by increasing phenolic compounds, while in the sensitive species it reduced oxidative damage through an increase in SOD activity (Ulloa-Inostroza et al. 2017).

Genes involved in the JA biosynthesis pathway were also upregulated after 20 and 80 µM of As(III) treatments in both the roots and shoots of rice plants. In the downstream signalling pathway, although the expression of *OsCOI1* was unaffected by As(III) stress, the expression of the eight *JAZ* genes was significantly increased (Yu et al. 2012). As treatment at 200 µM concentration was phytotoxic in *Brassica napus*; however, its combined application with MeJA resulted in a significant increase in leaf chlorophyll fluorescence and biomass production, and reduced MDA content compared with As-stressed plants. The application of MeJA minimized the oxidative stress in leaves via enhancing of the enzymatic activities and gene expression of antioxidant enzymes (SOD, APX, CAT, POD) and secondary metabolites (phenylalanine ammonia-lyase, polyphenol oxidase, cinnamyl alcohol dehydrogenase). The application of MeJA significantly reduced the As content in the leaves and roots of *Brassica napus* cultivars (Farooq et al. 2016). MeJA influenced the redox states of AsA and GSH, and the related enzymes involved in the AsA–GSH cycle in *Brassica napus* (Farooq et al. 2018a). Moreover, the upregulation of differential stress defence- and photosynthesis-related proteins was identified in *Brassica napus* leaves after 50 and 200 μM As treatment together with MeJA application, indicating an important role of MeJA in mitigating the adverse effects of As stress in *Brassica napus* (Farooq et al. 2018b).

Interestingly, in the first phase, after 7 h of exposure to Cu or Cd, a rapid increase in JA levels occurred, followed by a rapid decrease observed over 7 successive hours in *Arabidopsis thaliana* leaves (Maksymiec et al. 2005). 400 μM Cd decreased JA contents after 1 month in the roots of *Atriplex halimus* and *Suaeda fruticosa*, but an accumulation of JA and ABA was found in the leaves (Bankaji et al. 2014).

The exogenous application of JA reduced the Cd uptake in the leaves of *Brassica napus* (Ali et al. 2018), thereby reducing membrane damage and MDA content and increasing the essential nutrient uptake in this plant species (Ali et al. 2018), in *Capsicum frutescens* (Yan et al. 2013), and in *Kandelia obovata* (Chen et al. 2014). Furthermore, JA shields the chloroplast against the damaging effects of Cd, thereby increasing gas exchange and photosynthetic pigments in *Brassica napus* (Ali et al. 2018) and in *Glycine max* (Chaca et al. 2014), as well as the photosynthetic activity in *Brassica juncea* (Per et al. 2016). Moreover, JA modulates the antioxidant enzyme activity (APX, SOD, POD, and CAT) to strengthen the internal defence system in *Brassica napus* (Meng et al. 2009; Ali et al. 2018), in *Glycine max* (Noriega et al. 2012), in *Nicotiana tabacum* (Ma et al. 2017), in *Mentha arvensis* (Zaid and Mohammad 2018), in *Oryza sativa* (Singh and Shah 2014), in *Solanum lycopersicum* (Zhao et al. 2016b), in *Solanum nigrum* (Yan et al. 2015), and in *Vicia faba* leaves (Ahmad et al. 2017).

An enhancement of JA and shikimic acid (SHA) levels was observed in *Nicotiana langsdorffii* after 15-day-long 50 mg/kg Cr treatments (Scalabrin et al. 2016). Compared to Cr, Cu has an opposite effect. Both Cu and MeJA inhibited the root growth of *Phaseolus coccineus* and *Zea mays*. In addition, JA synthesis inhibitors (ibuprofen, salicylhydroxamic acid, and propyl gallate) partially reversed the inhibitory effect of Cu (Maksymiec and Krupa, 2007). Stoynova-Bakalova et al. (2009) found also that MeJA showed an inhibitory effect with the simultaneous addition of Cu in the growth and flavonol content of *Cucurbita pepo* cotyledon. Under Cu stress, MeJA did not modulate growth parameters such as leaf area, root growth, shoot and root fresh weight, and the shoot/root fresh weight ratio of *Phaseolus coccineus*, but the chlorophyll fluorescence parameter NPQ was increased by MeJA (Hanaka et al. 2015). In the same plant species, MeJA elevated MDA and anthocyanin levels in the leaves. However, MeJA treatments together with 50 μM Cu did not modify the proline and homoglutathione content. MeJA application after 5-day-Cu incubation caused an accumulation of malate and tartrate in roots (Hanaka et al. 2016). The activity of SOD and POD increased significantly in the presence of Cu after seed priming with JA in *Cajanus cajan*, and JA also promoted chlorophyll and carotenoid accumulation in the leaves (Poonam et al. 2013). Seed priming treatment with JA increased the protein, sugar, and vitamin contents, indicating the potential of JA in improving Cu stress tolerance in this plant species (Sirhindi et al. 2015).

The priming of soybean seeds with JA significantly improved the growth performance of soybean grown under excessive Ni. JA decreased Ni-induced membrane damage as evidenced by reduced levels of ROS by increasing the activity and gene expression of SOD, CAT, APX, POD, and MDA (Sirhindi et al. 2016), and decreasing lipoxygenase activity, and EL in Ni-stressed soybean plants (Mir et al. 2018). Exogenous JA treatment decreased ROS levels under Ni stress in *Alyssum inflatum* (Karimi and Ghasempour, 2019). The supplementation of plants with 1 nM JA restored the chlorophyll fluorescence, which was disturbed by 2 mM Ni in soybean (Sirhindi et al. 2016). Exogenous JA application regulated other hormone levels in the JA-treated plants. 0.5 μM JA stimulated the synthesis of active auxins and SA, contributing to enhanced mitotic activity in *Daphne jasminea* explants under Ni stress (Wiszniewska et al. 2018).

Time- and Pb concentration-dependent changes in the endogenous JA and SA levels were observed in *Excoecaria agallocha*. Increases of JA and SA levels were measured in the leaves and roots of *Excoecaria agallocha* at day 1 under low to moderate Pb, but they decreased under high Pb concentration, indicating that this species had fast responses and might have the ability to acclimate to low to moderate Pb levels (Yan and Tam, 2013). The exogenous application of JA had also concentration-dependent effects under Pb stress. Treatment with JA at the highest concentration (100 μM) resulted in the enhancement of heavy metal toxicity, leading to increasing in metal uptake and the formation of lipid peroxides as well as a decrease in fresh weight, chlorophyll *a*, carotenoid, monosaccharide, and soluble protein content in *Wolffia arrhiza*. In contrast, JA applied at 0.1 μM protected this plant species against Pb stress by inhibiting heavy metal accumulation, restoring plant growth and primary metabolite level. Moreover, JA at 0.1 μM activated the enzymatic antioxidants (CAT and APX) and non-enzymatic antioxidants (AsA, GSH) and, therefore, suppressed the oxidative destruction of cellular components induced by Pb (Piotrowska et al. 2009). The application of JA also modulated the AsA-GSH cycle in tomato plants under Pb stress. Treatment with JA enhanced the contents of osmolytes and metal chelating compounds and improved the photosynthetic efficiency in Pb-treated tomato leaves (Bali et al. 2018).

In contrast to Pb, JA (as well as its active metabolite jasmonate isoleucine) was suppressed in the presence of Zn nanoparticles in *Arabidopsis thaliana* (Vankova et al. 2017).

9.8 CONCLUSIONS AND FUTURE PERSPECTIVES

Despite the fact that the role of JA has been intensively studied in the past 20 years, there are many gaps in our knowledge of the signalling of JAs under various abiotic stresses. In this review we highlighted the current state of the physiological and molecular aspects of light-, temperature-, drought-, salt-, and heavy metal stress-modulated JA mechanisms in different plant species, genotypes, or organs, while the effect of exogenous JA application was also discussed (Figure 9.1). Moreover, the crosstalk between JA and other phytohormones under each abiotic stress was mentioned.

Based on the reviewed works we can conclude that all investigated environmental stresses elevated endogenous JA levels in various plant organs (leaf, stem, fruit, and root), but there can be significant differences in the hormone levels in various plant species and organs. Changes in JA levels in these organs could be crucial for the plants to survive severe environmental stresses. Fluctuation of the JA metabolism can be dependent on the dose and duration of stress treatments. Only a few authors investigated the impact of the different strengths of stresses on the selected plant species. Moreover, stress-induced JA accumulation shows distinct patterns, which can be altered by the stress in a time-dependent manner. Unfortunately, most of the authors determined JA levels only at one time-point after the stress exposure, although the increase may vary from some hours to a few days or weeks. The elevated JA can play roles in several signalling and metabolic processes mediated by other defence hormones in the stressed tissues, in which ABA, SA, and ET can be significant components. Numerous studies showed that the exogenous application of JAs, especially MeJA, increased the abiotic stress tolerance and improved the growth and development of various plant species under diverse stress conditions. Exogenous JA treatments decreased the sodium and heavy metal accumulation in several plant species. The beneficial effect of JA was manifested under environmental stress conditions by the alleviation of membrane damage based on EL measurements, by the reduction of lipid peroxidation based on MDA content, by enhancing the antioxidant capacity based on measurements of antioxidant enzyme activity (SOD, CAT, APX, POD) and antioxidant content (AsA and GSH), by increasing the photosynthetic efficiency based on chlorophyll fluorescence measurements, and by inducing the accumulation of defence-related compounds like polyamines, phenols, chelators, and osmoregulators.

In the future, the accurate description of the role of key enzymes in JA metabolism and the effects of JA mediated by other defence hormones would provide new insights into converging and diverging signalling pathways under severe environmental stress conditions. It would be interesting to further explore the role of JA with other defence hormones in plant responses to a combination of stress factors like drought and pathogen infection. Understanding the mechanism that can regulate

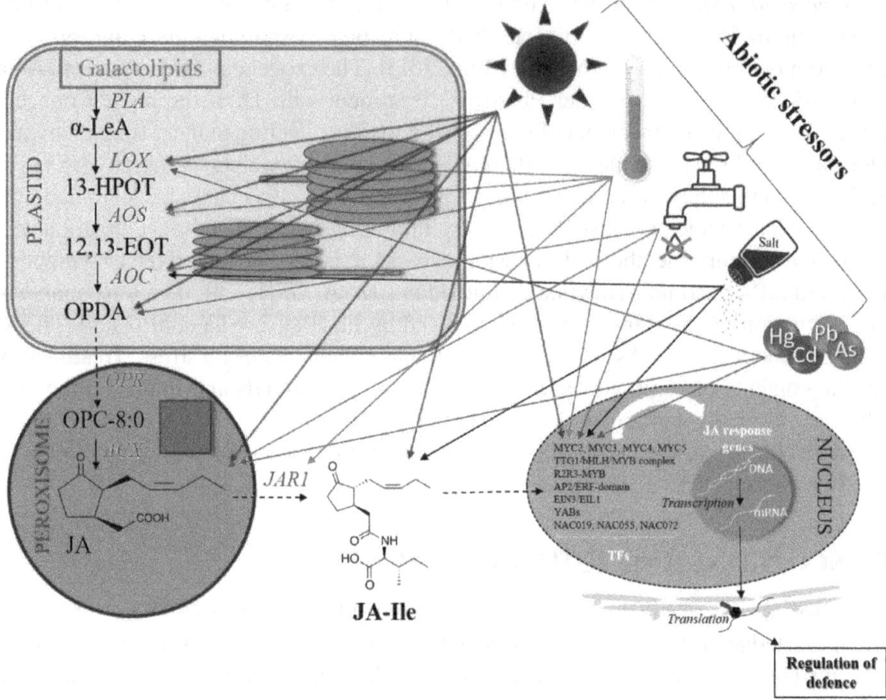

FIGURE 9.1 Effects of various abiotic stressors (light, temperature, water deficiency, salt, and heavy metals) on the synthesis of JA/JA-Ile from galactolipids. Arrows show the effects of the different stressors (see details in the text). Abbreviations for compounds: α-LeA, α-linolenic acid; 13-HPOT, (13S)-hydroperoxyoctadecatrienoic acid; OPDA, cis-(+)-12-oxophytodienoic acid; OPC-8 3,3-oxo-2-(2-pentenyl) cyclopentane-1-octanoic acid; JA, jasmonic acid; JA-Ile, JA conjugation with isoleucine. Abbreviations for enzymes/proteins: PLA, phospholipase A; LOX, 13-lipoxygenase; AOS, allene oxide synthase; AOC, allene oxide cyclase; OPR, OPDA reductase; ACX, acyl CoA-oxidase; JAR1, JA-amino acid synthetase. [Redrawn based on Wasternack and Song (2016).]

JA levels at the cellular, tissue, organ, or whole plant level is an important problem in current plant biology as well as in agriculture. Precise investigation of the effect of exogenous JA treatments can be used to increase the yield and abiotic stress tolerance in today's changing environment.

ACKNOWLEDGMENTS

We apologize to those colleagues whose work was not reviewed here. This work was supported by grants from the Hungarian Scientific Research Fund (OTKA FK 124871). P.P. supported by the János Bolyai Research Scholarship of the Hungarian Academy of Sciences (Grant no. BO/00743/19/8). No conflict of interest is declared.

REFERENCES

A.-H.-Mackerness, S., Surplus, S. L., Blake, P., et al. 1999. Ultraviolet-B-induced stress and changes in gene expression in *Arabidopsis thaliana*: Role of signalling pathways controlled by jasmonic acid, ethylene and reactive oxygen species. *Plant, Cell and Environment 22*(11):1413–1423.

Abdala, G., Miersch, O., Kramell, R., et al. 2003. Jasmonate and octadecanoid occurrence in tomato hairy roots. Endogenous level changes in response to NaCl. *Plant Growth Regulation 40*(1):21–27.

Abouelsaad, I., and Renault, S. 2018. Enhanced oxidative stress in the jasmonic acid-deficient tomato mutant *def-1* exposed to NaCl stress. *Journal of Plant Physiology 226*:136–144.

Aghdam, M. S., and Bodbodak, S. 2013. Physiological and biochemical mechanisms regulating chilling tolerance in fruits and vegetables under postharvest salicylates and jasmonates treatments. *Scientia Horticulturae 156*:73–85.

Ahmad, P., Alyemeni, M. N., Wijaya, L., Alam, P., Ahanger, M. A., and Alamri, S. A. 2017. Jasmonic acid alleviates negative impacts of cadmium stress by modifying osmolytes and antioxidants in faba bean (*Vicia faba* L.). *Archives of Agronomy and Soil Science 63*(13):1889–1899.

Akan, S., Gunes, N. T., and Yanmaz, R. 2019. Methyl jasmonate and low temperature can help for keeping some physicochemical quality parameters in garlic (*Allium sativum* L.) cloves. *Food Chemistry 270*:546–553.

Ai, L., Li, Z. H., Xie, Z. X., Tian, X. L., Eneji, A. E., and Duan, L. S. 2008. Coronatine alleviates polyethylene glycol-induced water stress in two rice (*Oryza sativa* L.) cultivars. *Journal of Agronomy and Crop Science 194*(5):360–368.

Alam, M. M., Nahar, K., Hasanuzzaman, M., and Fujita, M. 2014. Exogenous jasmonic acid modulates the physiology, antioxidant defense and glyoxalase systems in imparting drought stress tolerance in different *Brassica* species. *Plant Biotechnology Reports 8*(3):279–293.

Alexandratos, N., and Bruinsma, J. 2012. *World Agriculture Towards 2030/2050: The 2012 Revision*. ESA Working Paper, Vol. 12, No. 3. FAO, Rome. http://large.stanford.edu/courses/2014/ph240/yuan2/docs/ap106e.pdf

Ali, E., Hussain, N., Shamsi, I. H., Jabeen, Z., Siddiqui, M. H., and Jiang, L. X. 2018. Role of jasmonic acid in improving tolerance of rapeseed (*Brassica napus* L.) to Cd toxicity. *Journal of Zhejiang University-SCIENCE B 19*(2):130–146.

An, X. H., Hao, Y. J., Li, E. M., Xu, K., and Cheng, C. G. 2017. Functional identification of apple *MdJAZ2* in *Arabidopsis* with reduced JA-sensitivity and increased stress tolerance. *Plant Cell Reports 36*(2):255–265.

Andrade, A., Escalante, M., Vigliocco, A., del Carmen Tordable, M., and Alemano, S. 2017. Involvement of jasmonates in responses of sunflower (*Helianthus annuus*) seedlings to moderate water stress. *Plant Growth Regulation 83*(3):501–511.

Andrade, A., Vigliocco, A., Alemano, S., Miersch, O., Botella, M. A., and Abdala, G. 2005. Endogenous jasmonates and octadecanoids in hypersensitive tomato mutants during germination and seedling development in response to abiotic stress. *Seed Science Research 15*(4):309–318.

Anjum, S. A., Tanveer, M., Hussain, S., et al. 2016. Exogenously applied methyl jasmonate improves the drought tolerance in wheat imposed at early and late developmental stages. *Acta Physiologiae Plantarum 38*(1):25.

Anjum, S. A., Xie, X. Y., Farooq, M., et al. 2011. Effect of exogenous methyl jasmonate on growth, gas exchange and chlorophyll contents of soybean subjected to drought. *African Journal of Biotechnology 10*(47):9647–9656.

Asensi-Fabado, M. A., Cela, J., Müller, M., Arrom, L., Chang, C., and Munné-Bosch, S. 2012. Enhanced oxidative stress in the ethylene-insensitive (*ein3-1*) mutant of *Arabidopsis thaliana* exposed to salt stress. *Journal of Plant Physiology 169*(4):360–368.

Asensi-Fabado, M. A., and Munné-Bosch, S. 2011. The *aba3-1* mutant of *Arabidopsis thaliana* withstands moderate doses of salt stress by modulating leaf growth and salicylic acid levels. *Journal of Plant Growth Regulation 30*(4):456–466.

Azooz, M. M., Metwally, A., and Abou-Elhamd, M. F. 2015. Jasmonate-induced tolerance of Hassawi okra seedlings to salinity in brackish water. *Acta Physiologiae Plantarum 37*(4):77.

Ba, L. J., Kuang, J. F., Chen, J. Y., and Lu, W. J. 2016. MaJAZ1 attenuates the MaLBD5-mediated transcriptional activation of jasmonate biosynthesis gene *MaAOC2* in regulating cold tolerance of banana fruit. *Journal of Agricultural and Food Chemistry 64*(4):738–745.

Baek, M. H., Chung, B. Y., Kim, J. H., Wi, S. G., Kim, J. S., and Lee, I. J. 2006. Gamma radiation and hormone treatment as tools to reduce salt stress in rice (*Oryza sativa* L.). *Journal of Plant Biology 49*(3):257.

Balfagón, D., Zandalinas, S. I., and Gómez-Cadenas, A. 2018. High temperatures change the perspective: Integrating hormonal responses in citrus plants under co-occurring abiotic stress conditions. *Physiologia Plantarum 165*:183–197.

Bali, S., Kaur, P., Kohli, S. K., et al. 2018. Jasmonic acid induced changes in physio-biochemical attributes and ascorbate-glutathione pathway in *Lycopersicon esculentum* under lead stress at different growth stages. *Science of the Total Environment 645*:1344–1360.

Ballaré, C. L. 2014. Light regulation of plant defense. *Annual Review of Plant Biology 65*:335–363.

Bandurska, H., Stroiński, A., and Kubiś, J. 2003. The effect of jasmonic acid on the accumulation of ABA, proline and spermidine and its influence on membrane injury under water deficit in two barley genotypes. *Acta Physiologiae Plantarum 25*(3):279–285.

Bankaji, I., Sleimi, N., López-Climent, M. F., Perez-Clemente, R. M., and Gomez-Cadenas, A. 2014. Effects of combined abiotic stresses on growth, trace element accumulation, and phytohormone regulation in two halophytic species. *Journal of Plant Growth Regulation 33*(3):632–643.

Brossa, R., López-Carbonell, M., Jubany-Marí, T., and Alegre, L. 2011. Interplay between abscisic acid and jasmonic acid and its role in water-oxidative stress in wild-type, ABA-deficient, JA-deficient, and ascorbate-deficient *Arabidopsis* plants. *Journal of Plant Growth Regulation 30*(3):322–333.

Cao, S., Zheng, Y., Wang, K., Jin, P., and Rui, H. 2009. Methyl jasmonate reduces chilling injury and enhances antioxidant enzyme activity in postharvest loquat fruit. *Food Chemistry 115*(4):1458–1463.

Caputo, C., Rutitzky, M., and Ballaré, C. L. 2006. Solar ultraviolet-B radiation alters the attractiveness of *Arabidopsis* plants to diamondback moths (*Plutella xylostella* L.): Impacts on oviposition and involvement of the jasmonic acid pathway. *Oecologia 149*(1):81.

Cela, J., Chang, C., and Munné-Bosch, S. 2011. Accumulation of γ-rather than α-tocopherol alters ethylene signaling gene expression in the *vte4* mutant of *Arabidopsis thaliana. Plant and Cell Physiology 52*(8):1389–1400.

Chaca, M. P., Vigliocco, A., Reinoso, H., et al. 2014. Effects of cadmium stress on growth, anatomy and hormone contents in *Glycine max* (L.) Merr. *Acta Physiologiae Plantarum 36*(10):2815–2826.

Chen, H. J., Fu, T. Y., Yang, S. L., and Hsieh, H. L. 2018a. FIN219/JAR1 and cryptochromel antagonize each other to modulate photomorphogenesis under blue light in *Arabidopsis. PLoS Genetics 14*(3):e1007248.

Chen, X., Zhang, L., Miao, X., et al. 2018b. Effect of salt stress on fatty acid and α-tocopherol metabolism in two desert shrub species. *Planta 247*(2):499–511.

Chen, J., Yan, Z., and Li, X. 2014. Effect of methyl jasmonate on cadmium uptake and antioxidative capacity in *Kandelia obovata* seedlings under cadmium stress. *Ecotoxicology and Environmental Safety 104*:349–356.

Chen, Y., Wang, Y., Huang, J., et al. 2017. Salt and methyl jasmonate aggravate growth inhibition and senescence in *Arabidopsis* seedlings via the JA signaling pathway. *Plant Science 261*:1–9.

Cheng, M. C., Liao, P. M., Kuo, W. W., and Lin, T. P. 2013. The *Arabidopsis* ETHYLENE-RESPONSE-FACTOR1 regulates abiotic-stress-responsive gene expression by binding to different cis-acting elements in response to different stress signals. *Plant Physiology 162*:1566–1582.

Chinnusamy, V., Zhu, J., and Zhu, J. K. 2007. Cold stress regulation of gene expression in plants. *Trends in Plant Science 12*(10):444–451.

Choudhary, K. K., and Agrawal, S. B. 2014. Cultivar specificity of tropical mung bean (*Vigna radiata* L.) to elevated ultraviolet-B: Changes in antioxidative defense system, nitrogen metabolism and accumulation of jasmonic and salicylic acids. *Environmental and Experimental Botany 99*:122–132.

Clarke, S. M., Cristescu, S. M., Miersch, O., Harren, F. J., Wasternack, C., and Mur, L. A. 2009. Jasmonates act with salicylic acid to confer basal thermotolerance in *Arabidopsis thaliana. New Phytologist 182*(1):175–187.

Correia, B., Pintó-Marijuan, M., Castro, B. B., Brossa, R., López-Carbonell, M., Pinto, G. 2014. Hormonal dynamics during recovery from drought in two *Eucalyptus globulus* genotypes: From root to leaf. *Plant Physiology and Biochemistry 82*:151–160.

Cramer, G. R., Urano, K., Delrot, S., Pezzotti, M., and Shinozaki, K. 2011. Effects of abiotic stress on plants: A systems biology perspective. *BMC Plant Biology 11*(1):163.

de Ollas, C., Arbona, V., and GóMez-Cadenas, A. 2015. Jasmonoyl isoleucine accumulation is needed for abscisic acid build-up in roots of *Arabidopsis* under water stress conditions. *Plant, Cell and Environment 38*(10):2157–2170.

de Ollas, C., Arbona, V., Gómez-Cadenas, A., and Dodd, I. C. 2018. Attenuated accumulation of jasmonates modifies stomatal responses to water deficit. *Journal of Experimental Botany 69*(8):2103–2116.

de Ollas, C., and Dodd, I. C. 2016. Physiological impacts of ABA–JA interactions under water-limitation. *Plant Molecular Biology 91*(6):641–650.

de Ollas, C., Hernando, B., Arbona, V., and Gómez-Cadenas, A. 2013. Jasmonic acid transient accumulation is needed for abscisic acid increase in citrus roots under drought stress conditions. *Physiologia Plantarum 147*(3):296–306.

de Wit, M., Galvao, V. C., and Fankhauser, C. 2016. Light-mediated hormonal regulation of plant growth and development. *Annual Review of Plant Biology 67*:513–537.

Demarsy, E., Goldschmidt-Clermont, M., and Ulm, R. 2018. Coping with 'dark sides of the sun' through photoreceptor signaling. *Trends in Plant Science 23*(3):260–271.

Demkura, P. V., Abdala, G., Baldwin, I. T., and Ballaré, C. L. 2010. Jasmonate-dependent and-independent pathways mediate specific effects of solar ultraviolet B radiation on leaf phenolics and antiherbivore defense. *Plant Physiology 152*(2):1084–1095.

Demkura, P. V., and Ballaré, C. L. 2012. UVR8 mediates UV-B-induced *Arabidopsis* defense responses against *Botrytis cinerea* by controlling sinapate accumulation. *Molecular Plant 5*(3):642–652.

Ding, C. K., Wang, C., Gross, K. C., and Smith, D. L. 2002. Jasmonate and salicylate induce the expression of pathogenesis-related-protein genes and increase resistance to chilling injury in tomato fruit. *Planta 214*(6):895–901.

Ding, H., Lai, J., Wu, Q., et al. 2016. Jasmonate complements the function of *Arabidopsis* lipoxygenase3 in salinity stress response. *Plant Science 244*:1–7.

Dinh, S. N., Gális, I., and Baldwin, I. T. 2012. UVB radiation and HGL-DTGs provide durable resistance against mirid (*Tupiocoris notatus*) attack in field-grown *Nicotiana attenuata* plants. *Plant, Cell and Environment 36*:590–606.

Du, H., Liu, H., and Xiong, L. 2013. Endogenous auxin and jasmonic acid levels are differentially modulated by abiotic stresses in rice. *Frontiers in Plant Science 4*:397.

Ellouzi, H., Hamed, K. B., Cela, J., Müller, M., Abdelly, C., and Munné-Bosch, S. 2013. Increased sensitivity to salt stress in tocopherol-deficient *Arabidopsis* mutants growing in a hydroponic system. *Plant Signaling and Behavior 8*(2):e23136.

Ellouzi, H., Hamed, K. B., Hernández, I., et al. 2014. A comparative study of the early osmotic, ionic, redox and hormonal signaling response in leaves and roots of two halophytes and a glycophyte to salinity. *Planta 240*(6):1299–1317.

Farhangi-Abriz, S., Alaee, T., and Tavasolee, A. 2019. Salicylic acid but not jasmonic acid improved canola root response to salinity stress. *Rhizosphere 9*:69–71.

Farhangi-Abriz, S., and Ghassemi-Golezani, K. 2016. Improving amino acid composition of soybean under salt stress by salicylic acid and jasmonic acid. *Journal of Applied Botany and Food Quality 89*:243–248.

Farhangi-Abriz, S., and Torabian, S. 2018. Biochar increased plant growth-promoting hormones and helped to alleviates salt stress in common bean seedlings. *Journal of Plant Growth Regulation 37*(2):591–601.

Farooq, M. A., Gill, R. A., Islam, F., et al. 2016. Methyl jasmonate regulates antioxidant defense and suppresses arsenic uptake in *Brassica napus* L. *Frontiers in Plant Science 7*:468.

Farooq, M. A., Islam, F., Yang, C., et al. 2018a. Methyl jasmonate alleviates arsenic-induced oxidative damage and modulates the ascorbate–glutathione cycle in oilseed rape roots. *Plant Growth Regulation 84*(1):135–148.

Farooq, M. A., Zhang, K., Islam, F., et al. 2018b. Physiological and iTRAQ based quantitative proteomics analysis of methyl jasmonate induced tolerance in *Brassica napus* under arsenic stress. *Proteomics 18*:1–13 doi: 10.1002/pmic.201700290

Fedina, I., Nedeva, D., Georgieva, K., and Velitchkova, M. 2009. Methyl jasmonate counteract UV-B stress in barley seedlings. *Journal of Agronomy and Crop Science 195*(3):204–212.

Fernández, C., Alemano, S., Vigliocco, A., Andrade, A., and Abdala, G. 2012. Stress hormone levels associated with drought tolerance vs. sensitivity in sunflower (*Helianthus annuus* L.). In *Phytohormones and Abiotic Stress Tolerance in Plants*, eds. Khan, N. A., Nazar, R., Iqbal, N., and Anjum, N. A., 249–276. Springer, Berlin, Heidelberg.

Frenkel, M., Külheim, C., Jänkänpää, H. J., et al. 2009. Improper excess light energy dissipation in *Arabidopsis* results in a metabolic reprogramming. *BMC Plant Biology 9*(1):12.

Fugate, K. K., Lafta, A. M., Eide, J. D., et al. 2018. Methyl jasmonate alleviates drought stress in young sugar beet (*Beta vulgaris* L.) plants. *Journal of Agronomy and Crop Science 204*(6):566–576.

Fung, R. W., Wang, C. Y., Smith, D. L., Gross, K. C., and Tian, M. 2004. MeSA and MeJA increase steady-state transcript levels of alternative oxidase and resistance against chilling injury in sweet peppers (*Capsicum annuum* L.). *Plant Science 166*(3):711–719.

Gao, X. P., Wang, X. F., Lu, Y. F., et al. 2004. Jasmonic acid is involved in the water-stress-induced betaine accumulation in pear leaves. *Plant, Cell and Environment 27*(4):497–507.

Ghanem, M. E., Ghars, M. A., Frettinger, P., et al. 2012. Organ-dependent oxylipin signature in leaves and roots of salinized tomato plants (*Solanum lycopersicum*). *Journal of Plant Physiology 169*(11):1090–1101.

Golldack, D., Li, C., Mohan, H., and Probst, N. 2014. Tolerance to drought and salt stress in plants: Unraveling the signaling networks. *Frontiers in Plant Science 5*:151.

Gonzalez-Aguilar, G. A., Buta, J. G., and Wang, C. Y. 2003. Methyl jasmonate and modified atmosphere packaging (MAP) reduce decay and maintain postharvest quality of papaya 'Sunrise'. *Postharvest Biology and Technology 28*(3):361–370.

Gonzalez-Aguilar, G. A., Fortiz, J., Cruz, R., Baez, R., and Wang, C. Y. 2000. Methyl jasmonate reduces chilling injury and maintains postharvest quality of mango fruit. *Journal of Agricultural and Food Chemistry 48*(2):515–519.

Gonzalez-Aguilar, G. A., Tiznado-Hernandez, M. E., Zavaleta-Gatica, R., and Martınez-Téllez, M. A. 2004. Methyl jasmonate treatments reduce chilling injury and activate the defense response of guava fruits. *Biochemical and Biophysical Research Communications 313*(3):694–701.

Goossens, J., Fernández-Calvo, P., Schweizer, F., and Goossens, A. 2016. Jasmonates: Signal transduction components and their roles in environmental stress responses. *Plant Molecular Biology 91*(6):673–689.

Habora, M. E. E., Eltayeb, A. E., Oka, M., Tsujimoto, H., and Tanaka, K. 2013. Cloning of allene oxide cyclase gene from *Leymus mollis* and analysis of its expression in wheat–*Leymus* chromosome addition lines. *Breeding Science 63*(1):68–76.

Haga, K., and Iino, M. 2004. Phytochrome-mediated transcriptional up-regulation of ALLENE OXIDE SYNTHASE in rice seedlings. *Plant and Cell Physiology 45*(2):119–128.

Hamayun, M., Hussain, A., Khan, S. A., et al. 2015. Kinetin modulates physio-hormonal attributes and isoflavone contents of soybean grown under salinity stress. *Frontiers in Plant Science 6*:377.

Hanaka, A., Maksymiec, W., and Bednarek, W. 2015. The effect of methyl jasmonate on selected physiological parameters of copper-treated *Phaseolus coccineus* plants. *Plant Growth Regulation 77*(2):167–177.

Hanaka, A., Wójcik, M., Dresler, S., Mroczek-Zdyrska, M., and Maksymiec, W. 2016. Does methyl jasmonate modify the oxidative stress response in *Phaseolus coccineus* treated with Cu? *Ecotoxicology and Environmental Safety 124*:480–488.

Hassanein, R. A., Hassanein, A. A., El-Din, A. B., Salama, M., and Hashem, H. A. 2009. Role of jasmonic acid and abscisic acid treatments in alleviating the adverse effects of drought stress and regulating trypsin inhibitor production in soybean plant. *Australian Journal of Basic and Applied Sciences 3*:904–919.

Hazman, M., Hause, B., Eiche, E., Nick, P., and Riemann, M. 2015. Increased tolerance to salt stress in OPDA-deficient rice ALLENE OXIDE CYCLASE mutants is linked to an increased ROS-scavenging activity. *Journal of Experimental Botany 66*(11):3339–3352.

Hazman, M., Hause, B., Eiche, E., Riemann, M., and Nick, P. 2016. Different forms of osmotic stress evoke qualitatively different responses in rice. *Journal of Plant Physiology 202*:45–56.

Hirose, F., Shinomura, T., Tanabata, T., Shimada, H., and Takano, M. 2006. Involvement of rice cryptochromes in de-etiolation responses and flowering. *Plant and Cell Physiology 47*(7):915–925.

Hsieh, H. L., and Okamoto, H. 2014. Molecular interaction of jasmonate and phytochrome A signalling. *Journal of Experimental Botany 65*(11):2847–2857.

Hsieh, H. L., Okamoto, H., Wang, M., et al. 2000. *FIN219*, an auxin-regulated gene, defines a link between phytochrome A and the downstream regulator COP1 in light control of *Arabidopsis* development. *Genes and Development 14*(15):1958–1970.

Hu, Y., Jiang, L., Wang, F., and Yu, D. 2013. Jasmonate regulates the inducer of CBF expression–c-repeat binding factor/DRE binding factorl cascade and freezing tolerance in *Arabidopsis*. *The Plant Cell 25*:2907–2924.

Hu, Y., Jiang, Y., Han, X., Wang, H., Pan, J., and Yu, D. 2017. Jasmonate regulates leaf senescence and tolerance to cold stress: Crosstalk with other phytohormones. *Journal of Experimental Botany 68*(6):1361–1369.

Huang, A. X., She, X. P., Cao, B., Zhang, B., Mu, J., and Zhang, S. J. 2009. Nitric oxide, actin reorganization and vacuoles change are involved in PEG 6000-induced stomatal closure in *Vicia faba*. *Physiologia Plantarum 136*(1):45–56.

Huang, S. C., Chu, S. J., Guo, Y. M., et al. 2017. Novel mechanisms for organic acid-mediated aluminium tolerance in roots and leaves of two contrasting soybean genotypes. *AoB Plants 9*(6): plx064. doi: 10.1093/aobpla/plx064

Huang, X., Li, J., Shang, H., and Meng, X. 2015. Effect of methyl jasmonate on the anthocyanin content and antioxidant activity of blueberries during cold storage. *Journal of the Science of Food and Agriculture 95*(2):337–343.

Ismail, A., Riemann, M., and Nick, P. 2012. The jasmonate pathway mediates salt tolerance in grapevines. *Journal of Experimental Botany 63*(5):2127–2139.

Ismail, A., Seo, M., Takebayashi, Y., Kamiya, Y., Eiche, E., and Nick, P. 2014. Salt adaptation requires efficient fine-tuning of jasmonate signalling. *Protoplasma 251*(4):881–898.

Izaguirre, M. M., Scopel, A. L., Baldwin, I. T., and Ballaré, C. L. 2003. Convergent responses to stress. Solar ultraviolet-B radiation and *Manduca sexta* herbivory elicit overlapping transcriptional responses in field-grown plants of *Nicotiana longiflora*. *Plant Physiology 132*(4):1755–1767.

Jalmi, S. K., Bhagat, P. K., Verma, D., et al. 2018. Traversing the links between heavy metal stress and plant signaling. *Frontiers in Plant Science 9*:12.

Jin, P., Zheng, Y., Tang, S., Rui, H., and Wang, C. Y. 2009. A combination of hot air and methyl jasmonate vapor treatment alleviates chilling injury of peach fruit. *Postharvest Biology and Technology 52*(1):24–29.

Jin, P., Zhu, H., Wang, J., Chen, J., Wang, X., and Zheng, Y. 2013. Effect of methyl jasmonate on energy metabolism in peach fruit during chilling stress. *Journal of the Science of Food and Agriculture 93*(8):1827–1832.

Kang, D. J., Seo, Y. J., Lee, J. D., et al. 2005. Jasmonic acid differentially affects growth, ion uptake and abscisic acid concentration in salt-tolerant and salt-sensitive rice cultivars. *Journal of Agronomy and Crop Science 191*(4):273–282.

Karimi, N., and Ghasempour, H. R. 2019. Salicylic acid and jasmonic acid restrains nickel toxicity by ameliorating antioxidant defense system in shoots of metallicolous and non-metallicolous *Alyssum inflatum* Nâyr. populations. *Plant Physiology and Biochemistry 135*:450–459.

Kazan, K. 2015. Diverse roles of jasmonates and ethylene in abiotic stress tolerance. *Trends in Plant Science 20*(4):219–229.

Kazan, K., and Manners, J. M. 2011. The interplay between light and jasmonate signalling during defence and development. *Journal of Experimental Botany 62*(12):4087–4100.

Krishnan, S., and Merewitz, E. B. 2015a. Drought stress and trinexapac-ethyl modify phytohormone content within Kentucky bluegrass leaves. *Journal of Plant Growth Regulation 34*(1):1–12.

Krishnan, S., and Merewitz, E. B. 2015b. Phytohormone responses and cell viability during salinity stress in two creeping bentgrass cultivars differing in salt tolerance. *Journal of the American Society for Horticultural Science 140*(4):346–355.

Kumari, A., and Parida, A. K. 2018. Metabolomics and network analysis reveal the potential metabolites and biological pathways involved in salinity tolerance of the halophyte *Salvadora persica. Environmental and Experimental Botany 148*:85–99.

Kurotani, K. I., Hayashi, K., Hatanaka, S., et al. 2015. Elevated levels of CYP94 family gene expression alleviate the jasmonate response and enhance salt tolerance in rice. *Plant and Cell Physiology 56*(4):779–789.

Li, H. L., Guo, D., and Peng, S. Q. 2015. Molecular and functional characterization of the *JcMYB1*, encoding a putative R2R3-MYB transcription factor in *Jatropha curcas. Plant Growth Regulation 75*(1):45–53.

Li, X., Shen, X., Li, J., et al. 2010. Coronatine alleviates water deficiency stress on winter wheat seedlings. *Journal of Integrative Plant Biology 52*(7):616–625.

Li, Z., Wakao, S., Fischer, B. B., and Niyogi, K. K. 2009. Sensing and responding to excess light. *Annual Review of Plant Biology 60*:239–260.

Liu, H. R., Sun, G. W., Dong, L. J., et al. 2017. Physiological and molecular responses to drought and salinity in soybean. *Biologia Plantarum 61*(3):557–564.

Liu, S., Zhang, P., Li, C., and Xia, G. 2019. The moss jasmonate ZIM-domain protein PnJAZ1 confers salinity tolerance via crosstalk with the abscisic acid signalling pathway. *Plant Science 280*:1–11.

Liu, X., Chi, H., Yue, M., Zhang, X., Li, W., and Jia, E. 2012. The regulation of exogenous jasmonic acid on UV-B stress tolerance in wheat. *Journal of Plant Growth Regulation 31*(3):436–447.

Ma, C., Wang, Z. Q., Zhang, L. T., Sun, M. M., and Lin, T. B. 2014. Photosynthetic responses of wheat (*Triticum aestivum* L.) to combined effects of drought and exogenous methyl jasmonate. *Photosynthetica 52*(3):377–385.

Ma, Z., An, T., Zhu, X., et al. 2017. GR1-like gene expression in *Lycium chinense* was regulated by cadmium-induced endogenous jasmonic acids accumulation. *Plant Cell Reports 36*(9):1457–1476.

Mahouachi, J., Argamasilla, R., and Gómez-Cadenas, A. 2012. Influence of exogenous glycine betaine and abscisic acid on papaya in responses to water-deficit stress. *Journal of Plant Growth Regulation 31*(1):1–10.

Maksymiec, W., and Krupa, Z. 2007. Effects of methyl jasmonate and excess copper on root and leaf growth. *Biologia Plantarum 51*(2):322–326.

Maksymiec, W., Wianowska, D., Dawidowicz, A. L., Radkiewicz, S., Mardarowicz, M., and Krupa, Z. 2005. The level of jasmonic acid in *Arabidopsis thaliana* and *Phaseolus coccineus* plants under heavy metal stress. *Journal of Plant Physiology 162*(12):1338–1346.

Mazza, C. A., and Ballaré, C. L. 2015. Photoreceptors UVR8 and phytochrome B cooperate to optimize plant growth and defense in patchy canopies. *New Phytologist 207*(1):4–9.

Meir, S., Philosoph-Hadas, S., Lurie, S., et al. 1996. Reduction of chilling injury in stored avocado, grapefruit, and bell pepper by methyl jasmonate. *Canadian Journal of Botany 74*(6):870–874.

Meng, X., Han, J., Wang, Q., and Tian, S. 2009. Changes in physiology and quality of peach fruits treated by methyl jasmonate under low temperature stress. *Food Chemistry 114*(3):1028–1035.

Mewis, I., Schreiner, M., Nguyen, C. N., et al. 2012. UV-B irradiation changes specifically the secondary metabolite profile in broccoli sprouts: Induced signaling overlaps with defense response to biotic stressors. *Plant and Cell Physiology 53*(9):1546–1560.

Min, D., Li, F., Zhang, X., et al. 2018. *SlMYC2* involved in methyl jasmonate-induced tomato fruit chilling tolerance. *Journal of Agricultural and Food Chemistry 66*(12):3110–3117.

Mir, M. A., Sirhindi, G., Alyemeni, M. N., Alam, P., and Ahmad, P. 2018. Jasmonic acid improves growth performance of soybean under nickel toxicity by regulating nickel uptake, redox balance, and oxidative stress metabolism. *Journal of Plant Growth Regulation 37*(4):1195–1209.

Miranshahi, B., and Sayyari, M. 2016. Methyl jasmonate mitigates drought stress injuries and affects essential oil of summer savory. *Journal of Agricultural Science and Technology 18*(6):1635–1645.

Mohamed, H. I., and Latif, H. H. 2017. Improvement of drought tolerance of soybean plants by using methyl jasmonate. *Physiology and Molecular Biology of Plants 23*(3):545–556.

Moreno, J. E., Tao, Y., Chory, J., and Ballaré, C. L. 2009. Ecological modulation of plant defense via phytochrome control of jasmonate sensitivity. *Proceedings of the National Academy of Sciences 106*(12):4935–4940.

Müller, M., and Munné-Bosch, S. 2011. Rapid and sensitive hormonal profiling of complex plant samples by liquid chromatography coupled to electrospray ionization tandem mass spectrometry. *Plant Methods 7*(1):37.

Müller-Xing, R., Xing, Q., and Goodrich, J. 2014. Footprints of the sun: Memory of UV and light stress in plants. *Frontiers in Plant Science 5*:474.

Munemasa, S., Oda, K., Watanabe-Sugimoto, M., Nakamura, Y., Shimoishi, Y., and Murata, Y. 2007. The *coronatine-insensitive 1* mutation reveals the hormonal signaling interaction between abscisic acid and methyl jasmonate in *Arabidopsis* guard cells. Specific impairment of ion channel activation and second messenger production. *Plant Physiology 143*(3):1398–1407.

Munns, R. 2002. Comparative physiology of salt and water stress. *Plant, Cell and Environment 25*(2):239–250.

Munns, R., and Tester, M. 2008. Mechanisms of salinity tolerance. *Annual Review of Plant Biology 59*:651–681.

Muñoz-Espinoza, V. A., López-Climent, M. F., Casaretto, J. A., and Gómez-Cadenas, A. 2015. Water stress responses of tomato mutants impaired in hormone biosynthesis reveal abscisic acid, jasmonic acid and salicylic acid interactions. *Frontiers in Plant Science 6*:997.

Nilprapruck, P., Pradisthakarn, N., Authanithee, F., and Keebjan, P. 2008. Effect of exogenous methyl jasmonate on chilling injury and quality of pineapple (*Ananas comosus* L.) cv. Pattavia. *Science, Engineering and Health Studies 2*(2):33–42.

Noriega, G., Santa Cruz, D., Batlle, A., Tomaro, M., and Balestrasse, K. 2012. Heme oxygenase is involved in the protection exerted by jasmonic acid against cadmium stress in soybean roots. *Journal of Plant Growth Regulation 31*(1):79–89.

Ozga, J. A., Kaur, H., Savada, R. P., and Reinecke, D. M. 2016. Hormonal regulation of reproductive growth under normal and heat-stress conditions in legume and other model crop species. *Journal of Experimental Botany 68*(8):1885–1894.

Pan, C., Yang, D., Zhao, X., et al. 2019. Tomato stigma exsertion induced by high temperature is associated with the jasmonate signalling pathway. *Plant, Cell and Environment* 42:1205–1221.

Pauwels, L., and Goossens, A. 2011. The JAZ proteins: A crucial interface in the jasmonate signaling cascade. *The Plant Cell 23*(9):3089–3100.

Pavlović, I., Pěnčík, A., Novák, O., et al. 2018. Short-term salt stress in *Brassica rapa* seedlings causes alterations in auxin metabolism. *Plant Physiology and Biochemistry 125*:74–84.

Pedranzani, H., Racagni, G., Alemano, S., et al. 2003. Salt tolerant tomato plants show increased levels of jasmonic acid. *Plant Growth Regulation 41*(2):149–158.

Pedranzani, H., Sierra-de-Grado, R., Vigliocco, A., Miersch, O., and Abdala, G. 2007. Cold and water stresses produce changes in endogenous jasmonates in two populations of *Pinus pinaster* Ait. *Plant Growth Regulation 52*(2):111–116.

Peleg, Z., and Blumwald, E. 2011. Hormone balance and abiotic stress tolerance in crop plants. *Current Opinion in Plant Biology 14*(3):290–295.

Per, T. S., Khan, M. I. R., Anjum, N. A., Masood, A., Hussain, S. J., and Khan, N. A. 2018. Jasmonates in plants under abiotic stresses: Crosstalk with other phytohormones matters. *Environmental and Experimental Botany 145*:104–120.

Per, T. S., Khan, N. A., Masood, A., and Fatma, M. 2016. Methyl jasmonate alleviates cadmium-induced photosynthetic damages through increased S-assimilation and glutathione production in mustard. *Frontiers in Plant Science 7*:1933.

Pilarska, M., Wiciarz, M., Jajić, I., et al. 2016. A different pattern of production and scavenging of reactive oxygen species in halophytic *Eutrema salsugineum* (*Thellungiella salsuginea*) plants in comparison to *Arabidopsis thaliana* and its relation to salt stress signaling. *Frontiers in Plant Science 7*:1179.

Pillai, S. E., Kumar, C., Patel, H. K., and Sonti, R. V. 2018. Overexpression of a cell wall damage induced transcription factor, OsWRKY42, leads to enhanced callose deposition and tolerance to salt stress but does not enhance tolerance to bacterial infection. *BMC Plant Biology 18*(1):177.

Piotrowska, A., Bajguz, A., Godlewska-Żyłkiewicz, B., Czerpak, R., and Kamińska, M. 2009. Jasmonic acid as modulator of lead toxicity in aquatic plant *Wolffia arrhiza* (*Lemnaceae*). *Environmental and Experimental Botany 66*(3):507–513.

Poonam, S., Kaur, H., and Geetika, S. 2013. Effect of jasmonic acid on photosynthetic pigments and stress markers in *Cajanus cajan* (L.) Millsp. seedlings under copper stress. *American Journal of Plant Sciences 4*(04):817.

Poór, P., Laskay, G., and Tari, I. 2015. Role of nitric oxide in salt stress-induced programmed cell death and defense mechanisms. In *Nitric Oxide Action in Abiotic Stress Responses in Plants*, eds. Khan, M. N., Mobin, M., Firoz, M., and Corpas, F., 193–219. Springer, Cham.

Prerostova, S., Dobrev, P. I., Gaudinova, A., et al. 2017. Hormonal dynamics during salt stress responses of salt-sensitive *Arabidopsis thaliana* and salt-tolerant *Thellungiella salsuginea*. *Plant Science 264*:188–198.

Quan, J., Song, S., Abdulrashid, K., Chai, Y., Yue, M., and Liu, X. 2018. Separate and combined response to UV-B radiation and jasmonic acid on photosynthesis and growth characteristics of *Scutellaria baicalensis*. *International Journal of Molecular Sciences 19*(4):1194. doi: 10.3390/ijms19041194

Radhakrishnan, R., and Lee, I. J. 2013. Regulation of salicylic acid, jasmonic acid and fatty acids in cucumber (Cucumis sativus L.) by spermidine promotes plant growth against salt stress. *Acta Physiologiae Plantarum 35*(12):3315–3322.

Ramel, F., Ksas, B., and Havaux, M. 2013a. Jasmonate: A decision maker between cell death and acclimation in the response of plants to singlet oxygen. *Plant Signaling and Behavior 8*(12):e26655. doi: 10.4161/psb.26655

Ramel, F., Ksas, B., Akkari, E., et al. 2013b. Light-induced acclimation of the *Arabidopsis chlorina1* mutant to singlet oxygen. *The Plant Cell 25*(4):1445–1462.

Rehman, M., Singh, Z., and Khurshid, T. 2018. Methyl jasmonate alleviates chilling injury and regulates fruit quality in 'Midknight' Valencia orange. *Postharvest Biology and Technology 141*:58–62.

Riemann, M., Bouyer, D., Hisada, A., et al. 2009. Phytochrome A requires jasmonate for photodestruction. *Planta 229*(5):1035–1045.

Riemann, M., Dhakarey, R., Hazman, M., Miro, B., Kohli, A., and Nick, P. 2015. Exploring jasmonates in the hormonal network of drought and salinity responses. *Frontiers in Plant Science 6*:1077.

Riemann, M., Riemann, M., and Takano, M. 2008. Rice *JASMONATE RESISTANT 1* is involved in phytochrome and jasmonate signalling. *Plant, Cell and Environment 31*(6):783–792.

Robson, F., Okamoto, H., Patrick, E., et al. 2010. Jasmonate and phytochrome A signaling in *Arabidopsis* wound and shade responses are integrated through JAZ1 stability. *The Plant Cell 22*(4):1143–1160.

Ruelland, E., and Zachowski, A. 2010. How plants sense temperature. *Environmental and Experimental Botany* 69(3):225–232.

Santino, A., Taurino, M., De Domenico, S., et al. 2013. Jasmonate signaling in plant development and defense response to multiple (a) biotic stresses. *Plant Cell Reports 32*(7):1085–1098.

Savchenko, T., Kolla, V., Wang, C. Q., et al. 2014. Functional convergence of oxylipin and ABA pathways controls stomatal closure in response to drought. *Plant Physiology 164*(3):1151–1160.

Sayyari, M., Babalar, M., Kalantari, S., et al. 2011. Vapour treatments with methyl salicylate or methyl jasmonate alleviated chilling injury and enhanced antioxidant potential during postharvest storage of pomegranates. *Food Chemistry 124*(3):964–970.

Scalabrin, E., Radaelli, M., and Capodaglio, G. 2016. Simultaneous determination of shikimic acid, salicylic acid and jasmonic acid in wild and transgenic *Nicotiana langsdorffii* plants exposed to abiotic stresses. *Plant Physiology and Biochemistry 103*:53–60.

Sen, S., and Dutta, S. K. 2016. Cloning, characterization, and subcellular localization of a novel JAZ repressor from *Eleusine coracana*. *Biologia Plantarum 60*(4):715–723.

Seo, H. S., Kim, S. K., Jang, S. W., Choo, Y. S., Sohn, E. Y., and Lee, I. J. 2005. Effect of jasmonic acid on endogenous gibberellins and abscisic acid in rice under NaCl stress. *Biologia Plantarum 49*(3):447–450.

Seo, J. S., Joo, J., Kim, M. J., et al. 2011. OsbHLH148, a basic helix-loop-helix protein, interacts with OsJAZ proteins in a jasmonate signaling pathway leading to drought tolerance in rice. *The Plant Journal 65*(6):907–921.

Shahzad, A. N., Pitann, B., Ali, H., Qayyum, M. F., Fatima, A., and Bakhat, H. F. 2015. Maize genotypes differing in salt resistance vary in jasmonic acid accumulation during the first phase of salt stress. *Journal of Agronomy and Crop Science 201*(6):443–451.

Shan, C., and Liang, Z. 2010. Jasmonic acid regulates ascorbate and glutathione metabolism in *Agropyron cristatum* leaves under water stress. *Plant Science 178*(2):130–139.

Shan, C., Zhou, Y., and Liu, M. 2015. Nitric oxide participates in the regulation of the ascorbate-glutathione cycle by exogenous jasmonic acid in the leaves of wheat seedlings under drought stress. *Protoplasma 252*(5):1397–1405.

Sharma, M., and Laxmi, A. 2016. Jasmonates: Emerging players in controlling temperature stress tolerance. *Frontiers in Plant Science 6*:1129.

Sheteiwy, M. S., Gong, D., Gao, Y., Pan, R., Hu, J., and Guan, Y. 2018. Priming with methyl jasmonate alleviates polyethylene glycol-induced osmotic stress in rice seeds by regulating the seed metabolic profile. *Environmental and Experimental Botany 153*:236–248.

Shin, S. Y., Park, M. H., Choi, J. W., and Kim, J. G. 2017. Gene network underlying the response of harvested pepper to chilling stress. *Journal of Plant Physiology 219*:112–122.

Siboza, X. I., Bertling, I., and Odindo, A. O. 2014. Salicylic acid and methyl jasmonate improve chilling tolerance in cold-stored lemon fruit (*Citrus limon*). *Journal of Plant Physiology 171*(18):1722–1731.

Singh, I., and Shah, K. 2014. Exogenous application of methyl jasmonate lowers the effect of cadmium-induced oxidative injury in rice seedlings. *Phytochemistry 108*:57–66.

Sirhindi, G., Mir, M. A., Abd-Allah, E. F., Ahmad, P., and Gucel, S. 2016. Jasmonic acid modulates the physio-biochemical attributes, antioxidant enzyme activity, and gene expression in *Glycine max* under nickel toxicity. *Frontiers in Plant Science 7*:591.

Sirhindi, G., Sharma, P., Singh, A., Kaur, H., and Mir, M. 2015. Alteration in photosynthetic pigments, osmolytes and antioxidants in imparting copper stress tolerance by exogenous jasmonic acid treatment in *Cajanus cajan. International Journal of Plant Physiology and Biochemistry 7*(3):30–39.

Stoynova-Bakalova, E., Nikolova, M. I. L. E. N. A., and Maksymiec, W. 2009. Effects of Cu2+, cytokinins and jasmonate on content of two flavonols identified in zucchini cotyledons. *Acta Biological Cracoviensia Series Botanica 51*(2):77–83.

Suhita, D., Kolla, V. A., Vavasseur, A., and Raghavendra, A. S. 2003. Different signaling pathways involved during the suppression of stomatal opening by methyl jasmonate or abscisic acid. *Plant Science 164*(4):481–488.

Suhita, D., Raghavendra, A. S., Kwak, J. M., and Vavasseur, A. 2004. Cytoplasmic alkalization precedes reactive oxygen species production during methyl jasmonate-and abscisic acid-induced stomatal closure. *Plant Physiology 134*(4):1536–1545.

Sun, H., Chen, L., Li, J., et al. 2017. The JASMONATE ZIM-domain gene family mediates JA signaling and stress response in cotton. *Plant and Cell Physiology 58*(12):2139–2154.

Svyatyna, K., and Riemann, M. 2012. Light-dependent regulation of the jasmonate pathway. *Protoplasma 249*(2):137–145.

Takahashi, S., and Badger, M. R. 2011. Photoprotection in plants: A new light on photosystem II damage. *Trends in Plant Science 16*(1):53–60.

Tchounwou, P. B., Yedjou, C. G., Patlolla, A. K., and Sutton, D. J. 2012. Heavy metal toxicity and the environment. In *Molecular, Clinical and Environmental Toxicology*, ed. Luch, A., 133–164. Springer, Basel.

Thomashow, M. F. 2010. Molecular basis of plant cold acclimation: Insights gained from studying the CBF cold response pathway. *Plant Physiology 154*(2):571–577.

Ulloa-Inostroza, E. M., Alberdi, M., Meriño-Gergichevich, C., and Reyes-Díaz, M. 2017. Low doses of exogenous methyl jasmonate applied simultaneously with toxic aluminum improve the antioxidant performance of *Vaccinium corymbosum. Plant and Soil 412*(1–2):81–96.

Valenzuela, C. E., Acevedo-Acevedo, O., Miranda, G. S., et al. 2016. Salt stress response triggers activation of the jasmonate signaling pathway leading to inhibition of cell elongation in *Arabidopsis* primary root. *Journal of Experimental Botany 67*(14):4209–4220.

Vanhaelewyn, L., Prinsen, E., Van Der Straeten, D., and Vandenbussche, F. 2016. Hormone-controlled UV-B responses in plants. *Journal of Experimental Botany 67*(15):4469–4482.

Vankova, R., Landa, P., Podlipna, R., et al. 2017. ZnO nanoparticle effects on hormonal pools in *Arabidopsis thaliana. Science of the Total Environment 593*:535–542.

Venkatachalam, K., and Meenune, M. 2015. Effect of methyl jasmonate on physiological and biochemical quality changes of longkong fruit under low temperature storage. *Fruits 70*(2):69–75.

Vives-Peris, V., Gómez-Cadenas, A., and Pérez-Clemente, R. M. 2017. Citrus plants exude proline and phytohormones under abiotic stress conditions. *Plant Cell Reports 36*(12):1971–1984.

Wahid, A., Gelani, S., Ashraf, M., and Foolad, M. R. 2007. Heat tolerance in plants: An overview. *Environmental and Experimental Botany 61*(3):199–223.

Wali, M., Gunsè, B., Llugany, M., et al. 2016. High salinity helps the halophyte *Sesuvium portulacastrum* in defense against Cd toxicity by maintaining redox balance and photosynthesis. *Planta 244*(2):333–346.

Walia, H., Wilson, C., Condamine, P., Liu, X., Ismail, A. M., and Close, T. J. 2007. Large-scale expression profiling and physiological characterization of jasmonic acid-mediated adaptation of barley to salinity stress. *Plant, Cell and Environment 30*(4):410–421.

Wang, F., Guo, Z., Li, H., et al. 2016. Phytochrome A and B function antagonistically to regulate cold tolerance via abscisic acid-dependent jasmonate signaling. *Plant Physiology 170*(1):459–471.

Wang, J. G., Chen, C. H., Chien, C. T., and Hsieh, H. L. 2011. FAR-RED INSENSITIVE 219 modulates CONSTITUTIVE PHOTOMORPHOGENIC 1 activity via physical interaction to regulate hypocotyl elongation in *Arabidopsis*. *Plant Physiology 156*(2):631–646.

Wasternack, C., and Hause, B. 2013. Jasmonates: Biosynthesis, perception, signal transduction and action in plant stress response, growth and development. An update to the 2007 review in Annals of Botany. *Annals of Botany 111*(6):1021–1058.

Wasternack, C., and Song, S. 2016. Jasmonates: Biosynthesis, metabolism, and signaling by proteins activating and repressing transcription. *Journal of Experimental Botany 68*(6):1303–1321.

Wasternack, C., and Strnad, M. 2018. Jasmonates: News on occurrence, biosynthesis, metabolism and action of an ancient group of signaling compounds. *International Journal of Molecular Sciences 19*(9):2539.

Wiszniewska, A., Muszyńska, E., Hanus-Fajerska, E., Dziurka, K., and Dziurka, M. 2018. Evaluation of the protective role of exogenous growth regulators against Ni toxicity in woody shrub *Daphne jasminea*. *Planta 248*(6):1365–1381.

Wu, H., Wu, X., Li, Z., Duan, L., and Zhang, M. 2012. Physiological evaluation of drought stress tolerance and recovery in cauliflower (*Brassica oleracea* L.) seedlings treated with methyl jasmonate and coronatine. *Journal of Plant Growth Regulation 31*(1):113–123.

Wu, H., Ye, H., Yao, R., Zhang, T., and Xiong, L. 2015. OsJAZ9 acts as a transcriptional regulator in jasmonate signaling and modulates salt stress tolerance in rice. *Plant Science 232*:1–12.

Wu, J., Kim, S. G., Kang, K. Y., et al. 2016. Overexpression of a pathogenesis-related protein 10 enhances biotic and abiotic stress tolerance in rice. *The Plant Pathology Journal 32*(6):552.

Xin, Z. Y., Zhou, X., and Pilet, P. E. 1997. Level changes of jasmonic, abscisic, and indole-3yl-acetic acids in maize under desiccation stress. *Journal of Plant Physiology 151*(1):120–124.

Xu, Q., Truong, T. T., Barrero, J. M., Jacobsen, J. V., Hocart, C. H., and Gubler, F. 2016. A role for jasmonates in the release of dormancy by cold stratification in wheat. *Journal of Experimental Botany 67*(11):3497–3508.

Yadav, S. K. 2010. Heavy metals toxicity in plants: An overview on the role of glutathione and phytochelatins in heavy metal stress tolerance of plants. *South African Journal of Botany 76*(2):167–179.

Yadav, V., Mallappa, C., Gangappa, S. N., Bhatia, S., and Chattopadhyay, S. 2005. A basic helix-loop-helix transcription factor in *Arabidopsis*, MYC2, acts as a repressor of blue light-mediated photomorphogenic growth. *The Plant Cell 17*(7):1953–1966.

Yan, Y., Stolz, S., Chételat, A., et al. 2007. A downstream mediator in the growth repression limb of the jasmonate pathway. *The Plant Cell 19*(8):2470–2483.

Yan, Z., Chen, J., and Li, X. 2013. Methyl jasmonate as modulator of Cd toxicity in *Capsicum frutescens* var. *fasciculatum* seedlings. *Ecotoxicology and Environmental Safety 98*:203–209.

Yan, Z., and Tam, N. F. Y. 2013. Effects of lead stress on anti-oxidative enzymes and stress-related hormones in seedlings of *Excoecaria agallocha* Linn. *Plant and Soil 367*(1–2):327–338.

Yan, Z., Zhang, W., Chen, J., and Li, X. 2015. Methyl jasmonate alleviates cadmium toxicity in *Solanum nigrum* by regulating metal uptake and antioxidative capacity. *Biologia Plantarum 59*(2):373–381.

Yin, R., and Ulm, R. 2017. How plants cope with UV-B: From perception to response. *Current Opinion in Plant Biology 37*:42–48.

Yu, L. J., Luo, Y. F., Liao, B., et al. 2012. Comparative transcriptome analysis of transporters, phytohormone and lipid metabolism pathways in response to arsenic stress in rice (*Oryza sativa*). *New Phytologist 195*(1):97–112.

Yuan, F., Liang, X., Li, Y., Yin, S., and Wang, B. 2019. Methyl jasmonate improves tolerance to high salt stress in the recretohalophyte *Limonium bicolor*. *Functional Plant Biology 46*(1):82–92.

Zaid, A., and Mohammad, F. 2018. Methyl jasmonate and nitrogen interact to alleviate cadmium stress in mentha arvensis by regulating physio-biochemical damages and ROS detoxification. *Journal of Plant Growth Regulation 37*(4):1331–1348.

Zarza, X., Atanasov, K. E., Marco, F., et al. 2017. *Polyamine oxidase* 5 loss-of-function mutations in *Arabidopsis thaliana* trigger metabolic and transcriptional reprogramming and promote salt stress tolerance. *Plant, Cell and Environment 40*(4):527–542.

Zhai, Q., Li, C. B., Zheng, W., et al. 2007. Phytochrome chromophore deficiency leads to overproduction of jasmonic acid and elevated expression of jasmonate-responsive genes in *Arabidopsis. Plant and Cell Physiology 48*(7):1061–1071.

Zhang, C., and Huang, Z. 2013. Effects of endogenous abscisic acid, jasmonic acid, polyamines, and polyamine oxidase activity in tomato seedlings under drought stress. *Scientia Horticulturae 159*:172–177.

Zhang, X., Sheng, J., Li, F., Meng, D., and Shen, L. 2012. Methyl jasmonate alters arginine catabolism and improves postharvest chilling tolerance in cherry tomato fruit. *Postharvest Biology and Technology 64*(1):160–167.

Zhang, X., Yao, D., Wang, Q., et al. 2013. mRNA-seq analysis of the *Gossypium arboreum* transcriptome reveals tissue selective signaling in response to water stress during seedling stage. *PloS One 8*(1):e54762. doi: 10.1371/journal.pone.0054762

Zhang, X. C., Millet, Y. A., Cheng, Z., Bush, J., and Ausubel, F. M. 2015. Jasmonate signalling in *Arabidopsis* involves SGT1b–HSP70–HSP90 chaperone complexes. *Nature Plants 1*(5):15049.

Zhao, C., Zhang, Z., Xie, S., Si, T., Li, Y., and Zhu, J. K. 2016a. Mutational evidence for the critical role of *CBF* genes in cold acclimation in *Arabidopsis. Plant Physiology 171*(4):2744–2759.

Zhao, S., Ma, Q., Xu, X., Li, G., and Hao, L. 2016b. Tomato jasmonic acid-deficient mutant *spr2* seedling response to cadmium stress. *Journal of Plant Growth Regulation 35*(3):603–610.

Zhao, M. L., Wang, J. N., Shan, W., et al. 2013. Induction of jasmonate signalling regulators MaMYC2s and their physical interactions with *MaICE1* in methyl jasmonate-induced chilling tolerance in banana fruit. *Plant, Cell and Environment 36*(1):30–51.

Zhu, J. K. 2016. Abiotic stress signaling and responses in plants. *Cell 167*(2):313–324.

10 Emerging Trends of Proline Metabolism in Abiotic Stress Management

Sukhmeen Kaur Kohli, Vandana Gautum, Shagun Bali, Parminder Kaur, Anket Sharma, Palak Bakshi, Bilal Ahmad Mir, and Renu Bhardwaj

CONTENTS

10.1 INTRODUCTION

Plants are continuously exposed to varying climatic conditions. The harsh environmental conditions retard the growth and development of plants depending upon the degree and severity of stress (Roychoudhary et al. 2015). Plants exposed to unfavorable environmental conditions acclimatize by adopting certain strategies including the sequestration of various metabolites (Murmu et al. 2017). Among these secondary plant products, compatible solutes are low molecular weight, non-toxic, organic compounds with high solubility. They bestow protection against abiotic stresses by regulating the cellular osmotic potential, lowering reactive oxygen species (ROS) generation, maintenance of membrane permeability and protein/enzyme stabilization (Ashraf and Fooland 2007). These solutes include sucrose, proline, trehalose, polyols and quaternary ammonium compounds (QACs) such as proline betaine, glycine betaine, piperolate betaine and alanine betaine, etc. (Hayat et al. 2012; Roychoudhury and Chakraborty 2013).

Proline is a proteinogenic amino acid with conformational rigidity and is a pre-requisite for primary metabolism. In recent years, studies have been carried out to understand proline metabolism that have enhanced our knowledge of the molecular alterations involved in proline synthesis and catabolism during abiotic stress conditions (Iqbal et al. 2014; Roychoudhary and Chakraborty 2013).

The physiological role of proline in response to various abiotic factors (heat, chilling, drought, salinity) has been experimentally explored, by reviewing the effects of its exogenous application (Rady and Mohamed 2018; Wu et al. 2017; Agami et al. 2016; Zourari et al. 2016; Kibria et al. 2016; Singh et al. 2015; Al-Hassan et al. 2015). Besides acting as a compatible solute, proline also acts as a protein hydrotype. It controls the NADP/NADPH ratio by increasing the cytoplasmic acidosis. Proline has been extensively reported for its potential role in osmotic adjustment in plant cells. More recently the stress regulatory properties of proline have been affirmed by a plethora of researchers. In cytoplasm, proline is a molecular chaperone and protects the structure of proteins and maintains the cytosol pH and modulates the cellular-redox balance (Hayat et al. 2012). The regulation of mitochondrial oxidative phosphorylation and synthesis of adenosine triphosphate (ATP) in the absence of stressed conditions are also regulated by proline, which provides required reducing agents (Kunji et al. 2016). Keeping in view the imperative participation of proline in abiotic stress acclimatization and providing protection, this book chapter describes proline biosynthesis, toxicity and its role as a stress manager. Moreover, approaches for proline metabolism engineering have been critically examined and reviewed.

10.2 PROLINE SYNTHESIS

Proline synthesis has been reported to take place through two pathways (Figure 10.1). The main pathway for proline synthesis is glutamate cascade. It works in three steps. In the first step, the pyrroline-5-carboxylate synthetase (P5CS) enzyme reduces the glutamate to glutamate-semialdehyde (GSA). P5CS is the rate-limiting enzyme for the proline synthesis. The reduced nicotinamide adenine dinucleotide phosphate (NADPH) and ATP are consumed in the first step. In the second step, glutamate-semialdehyde is converted into pyrroline-5-carboxylate (P5C), and in the third and final step, pyrroline-5-carboxylate reductase reduces P5C intermediate using NADPH to synthesize proline (Fu et al. 2017; Hayat et al. 2012; Szepesi and Szőllősi 2018). Another pathway for the synthesis of proline is the ornithine pathway where the ornithine-delta-aminotransferase (δ-OAT) enzyme transforms ornithine to glutamic semialdehyde (GSA). Further, GSA is immediately changed to P5C, which is transformed to proline by *P5CR* (Roosens et al. 1998). The ornithine pathway makes use of arginine to generate P5C and glutamate in the mitochondria. Mitochondrial P5C can be recycled to proline in the cytosol by *P5CR* (Fu et al. 2017).

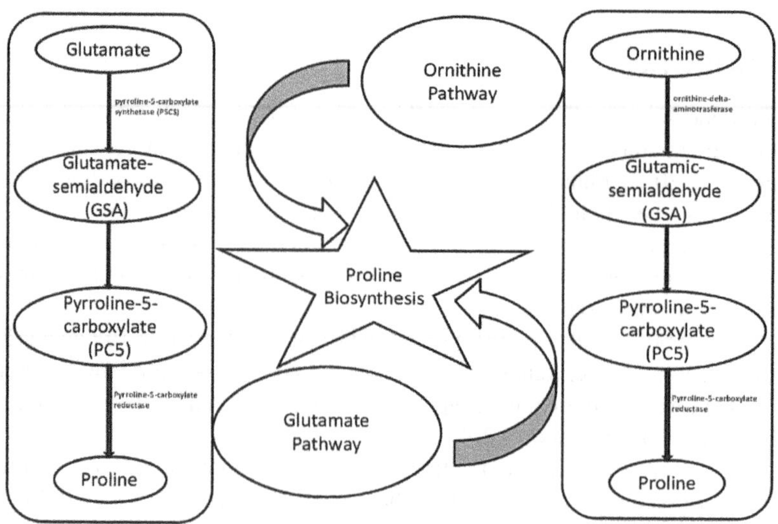

FIGURE 10.1 Schematic diagram of proline biosynthesis.

Funck et al. (2008), reported a normal accumulation of proline in the delta-aminotransferase mutants of *Arabidopsis* and observed that plants could not mobilize nitrogen from arginine or ornithine. They further suggested that the chief function of δ-OAT is to maintain arginine degradation rather than regulation of proline biosynthesis. Also, the δ-OAT is more abundant in mitochondria which leads to the straight consumption of the δ-OAT-produced P5C by *P5CR*, due to the localization of the latter in the cytosol or plastids (Funck et al. 2008). Glutamate seems to be the main precursor to stress-induced proline accumulation in plants, as the ornithine pathway mostly makes possible nitrogen recycling from arginine to glutamate.

10.2.1 Proline Metabolism

Proline metabolism includes its synthesis and degradation. The synthesis takes place by the glutamate and ornithine pathways. The degradation involves the oxidation of proline to pyrroline-5-carboxylate (P5C). The proline dehydrogenase (ProDH) and P5C dehydrogenase (P5CDH) enzymes act one after another to oxidize proline (Figure 10.2). The flavin adenine dinucleotide (FAD) acts as an electron acceptor and results in the oxidation of proline into pyrroline-5-carboxylate, and gets reduced to FADH which is the reduced acceptor. The reaction is catalyzed by two functional isoforms of an enzyme namely, proline dehydrogenase 1 (ProDH1) and proline dehydrogenase 2 (ProDH2). Proline dehydrogenase 1 is extensively articulated in plants and is supposed to be the principal isoform (Funck et al. 2010). The expression of ProDH2 is considerably lower than that of ProDH1 (Szabados and Savoure 2010). Just like the proline biosynthesis, the first step of proline degradation is the rate-limiting step and ProDH is the rate-limiting enzyme. The

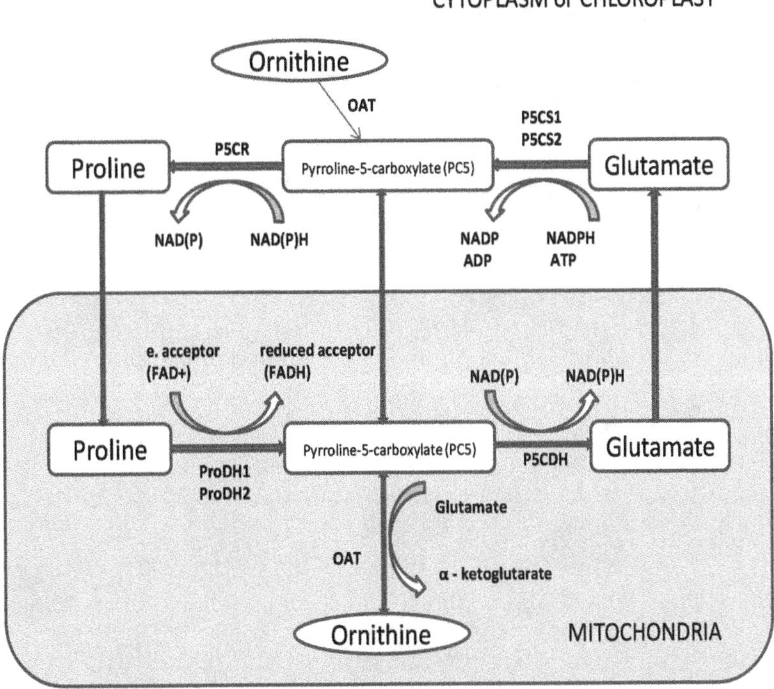

FIGURE 10.2 Proline metabolism pathways. (Abbreviations: OAT: ornithine-δ-aminotransferase; P5CS: pyrroline-5-carboxylate synthetase; P5CR: pyrroline-5-carboxylate reductase; P5CDH: pyrroline-5-carboxylate dehydrogenase; ProDH: proline dehydrogenase; NADP: nicotinamide adenine dinucleotide phosphate; NADPH: reduced nicotinamide adenine dinucleotide phosphate; ATP: adenosine tri-phosphate; ADP: adenosine di-phosphate; FAD: flavin adenine dinucleotide; FADH: reduced flavin adenine dinucleotide.)

pyrroline-5-carboxylate is further converted to glutamate by the action of the pyrroline-5-carboxylate dehydrogenase (P5CDH) enzyme which also reduces the nicotinamide adenine dinucleotide phosphate (NADP) to NADPH. The reaction takes place inside the mitochondria as both of these enzymes (ProDH and P5CDH) are present in the matrix side of the internal membrane of mitochondria (Per et al. 2017; Verslues and Sharma 2010). The P5CDH is programmed by one gene which has been recognized in *Nicotiana tabacum* and *Arabidopsis* (Kishore et al. 2005). ProDH is an oxygen-reliant flavo-protein (Elthon and Stewart 1982) that is also known as proline oxidase (POX) and is encoded by two genes. The catabolism of proline takes place in the mitochondria, which suggests that it might contribute carbon to the TCA cycle.

10.2.2 Molecular Fluxes in Proline Metabolism under Stress

Studies on proline metabolism in plants have been carried out in the past few years, but the molecular alterations involved in proline metabolism are not much explored. Proline anabolism is induced and its catabolism is suppressed during dehydration, but rehydration stimulates the inverse regulation. Various transcription factors involved in the regulation of proline metabolism are shown in the Figure 10.3. In plants, proline anabolism is modulated by two genes, i.e. *P5CS1* and *P5CS2*. The *P5CS1* gene encodes for the stress-specific *P5CS* isoform identified in *Arabidopsis* plants, and the *P5CS2* gene functions as a housekeeping gene and is actively present in the meristematic tissues and cell cultures (Verslues and Sharma 2010).

P5CS2 was recognized as one of the targets of CONSTANS (CO) which acts as a transcriptional activator for stimulating flowering under long day conditions (Samach et al. 2000). It can also be induced by ROS, salicylic acid and virulent bacteria which further stimulate a hypersensitive response. P5CS1 is triggered by salt-induced osmotic stress and is also stimulated by the abscisic acid (ABA) insensitive 1 controlled pathway and stress-related responses (Iqbal et al. 2014). The

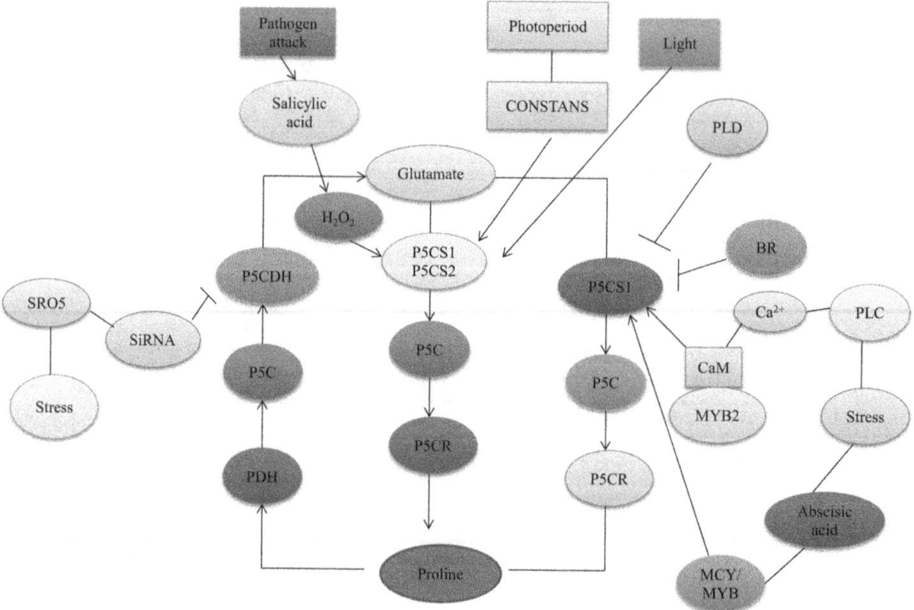

FIGURE 10.3 Various transcription factors involved in the regulation of proline metabolism (modified after Szabados and Savoure [2010]). (Crosstalk of various hormones and transcription factors involved in the regulation of proline metabolism under diverse environmental factors (stress, light, photoperiod). These environmental factors upregulate or inhibit the various transcription factors associated with the activation of proline biosynthesis. Arrows indicate up-regulation and lines ending in bars indicate inhibition.)

activation of p5cs1 and accumulation of proline is stimulated by light and suppressed by brassinosteroids. Phospholipase D (PLD) acts as negative regulator of proline under non-stressful conditions, while phospholipase C (PLC) acts as a positive regulator. PLC, in crosstalk with calcium signaling, activates p5cs transcription and enhances the accumulation of proline in plants under non-stressful conditions. Calcium signals may be regulated by CaM4 calmodulin and interplay with the myb2 transcription factor and enhance the transcript level of p5cs1 (Iqbal et al. 2014). ABA-dependent and -independent signal transduction pathways may affect proline accumulation. Calcium signaling also plays an important role in the accumulation of proline by mediating the ABA-dependent pathway (Roychoudhury and Chakarborty 2013; Parre et al. 2007).

Transcription factors that regulate abiotic stress responses by the ABA-dependent pathway are associated with MYC/MYB families. The over-expression of GmMYB76 in transgenic *Glycine max* lines showed higher transcript levels of responsive transcription factors such as dehydration29B (rd29B), rd1, dehydration-responsive element binding protein2A (DREB2A), P5CS, early dehydration inducible 10 (erd 10) and cold-regulated 78/responsive to dehydration29A (cor78/rd29A) (Liao et al. 2008; Agarwal and Jha 2010). On the other hand, the transcript levels of cor6.6, cor78/rd29A, cor15a and rd29B declined in GmMYB92 transgenic lines; however rd17, DREB2A and P5CS showed higher transcript levels (Liao et al. 2008). It has been revealed that alterations in the transcript levels of stress-responsive genes depend upon the nature of host plants (Roychoudhury and Chakarborty 2013). The exogenous application of phytohormones such as abscisic acid, indol-3-butyric acid and kinetin stimulates proline accumulation (Egamberdieva et al. 2017; Planchet et al. 2014). Cytokinin reduced the concentration of AtP5CS2 mRNA in *Arabidopsis* (Hu et al. 1992). The upregulation of proline biosynthesis is induced by osmotic and light stress (dark adaptation), and the catabolism of proline is stimulated in dark and non-stressful conditions and is directed by PDH and P5CDH (Abraham et al. 2003). The transcription of PDH is induced by rehydration and proline is suppressed by dehydration; consequently proline degradation is suppressed during abiotic stress. The transcription of pdh1 is inhibited during daylight and triggered in darkness; thus, light has a divergent influence on the transcription of pscs1 and pdh1. Promoter studies of pdh1 suggested that proline, hypo-osmolarity and the responsive element (PRE) motif sequence ACTCAT are important for the induction of the pdh gene. A transcription factor such as basic leucine zipper protein (bZIP) comprising AtbZIP-2, AtbZIP-11, AtbZIP-44 and AtbZIP-53 binds to the PRE motif (Lehmann et al. 2010).

In *Arabidopsis* plants the P5CDH transcripts were observed to be expressed in the absence of proline, although their expression was elevated with enhanced proline accumulation (Deuschle et al. 2001). A short analysis revealed a similarity with the PRE motif on the promoter of p5cdh in cereals and *Arabidopsis*. In *Linum usitatissimum*, fis1 encodes p5cdh which is induced during wounding and virulent pathogen attack. Overlapping of the natural antisense region of 3′UTR of P5CDH and salt, stimulated similar RCD ONE 5 (SRO5) gene, producing small interfering RNA such as 24-nt and 21-nt SiRNA endogenously in *Arabidopsis* which leads to the splitting of P5CDH RNA and reduction of its expression during stress (Mitchell et al. 2006; Ayliffe et al. 2002). Gene silencing at the transcriptional level can be used as a potential tool for the control of the activity of P5CDH.

10.3 PROLINE TOXICITY

Apart from the beneficial role of the exogenous application of proline, it causes toxicity when applied or accumulated at higher concentrations. Heuer (2003) reported that treatment with proline (1 and 10 mM) caused growth inhibitory effects in *Lycopersicon esculentum* under salt stress and a reduction in growth and accumulation of Na^+ and Cl^{-1} was observed. A study conducted by Roy et al. (1993) demonstrated that the application of proline at higher concentrations (40 mM, 50 mM) caused a decline in growth and the K^+/Na^+ ratio in *Oryza sativa*. It has been elucidated that treatment with proline at a low concentration (1 mM) stimulated *in-vitro* organogenesis of hypocotyl explants of *Arabidopsis* whereas a high concentration (10 mM) inhibits the growth (Hare et al. 2003).

An experiment conducted by Rodriguez and Heyser (1988) demonstrated that the application of proline (10 mM) inhibited the growth of suspension cultures of *Distichlis spicata*. Proline application significantly decreased the root growth of *Oryza sativa* (Chen and Kao 1995). Mutant plants were hypersensitive to proline treatment while wild-type plants showed normal growth. Plants accumulate proline under stress conditions, leading to degradation after its release. Δ^1-pyrroline-5-carboxylate dehydrogenase (P5CDH) is associated with proline degradation. A study conducted by Deuschle et al. (2005) suggested that proline toxicity is regulated by GSA (Glu semialdehyde)/P5C (Δ^1-pyrroline-5-carboxylate) accumulation. It was further suggested that proline degradation was not detectable in p5cdh mutants but did not cause alterations in the growth phenotype under normal conditions, demonstrating that proline degradation is not important for vegetative growth in plants. The exogenous application of proline enhanced ROS generation, programmed cell death, callose deposition and DNA laddering. These mutants showed hypersensitivity towards proline (Deuschle et al. 2005). Rajendrakumar et al. (1997) demonstrated that higher concentrations of proline destabilize the DNA helix, reduce the DNA melting point, enhance susceptibility towards S1 nuclease and also elevate insensitivity to DNAase1.

10.4 PROLINE AS STRESS MANAGER

Under abiotic stress conditions, the plant synthesizes enormous amounts of various compatible solutes or osmoprotectants including proline. Proline plays a defensive role in protecting plants under abiotic stress conditions by buffering the concentration of ROS, regulating the cellular redox-potential and maintaining the cellular membrane integrity and structure of various enzymes and proteins (Liang et al. 2013). It also acts as a signaling molecule, and its transportation serves as a metabolic signal (Zhang and Becker 2015; Verslues and Sharma 2010). It supplies energy for growth by the induction of the gene expression of PDH and P5CDH which donate electrons for the respiratory chain (Liang et al. 2013). Proline regulates mitochondrial respiration by protecting the complex II of the mitochondrial electron transport chain during salt stress. Therefore, the breakdown of proline molecules in the mitochondria serves as an important regulator of cellular ROS balance and can change various other regulating procedures (Szabados and Savouré 2010).

It was also reported that proline treatment scavenges ROS in fungi and yeast by inhibiting PCD (Chen and Deckman 2005). Proline also protects human cells from oxidative damage and the peroxidation of membranes occurring as a result of heavy metal toxicity and cancer (Krishnan et al. 2008). The major role of proline is the enhancement of the expression of stress-responsive genes in the promoter region (e.g. ACTCAT, *PRE*) under salt stress (Chinnusamy et al. 2005). Various functions played by proline in stress management are summarized in Figure 10.4.

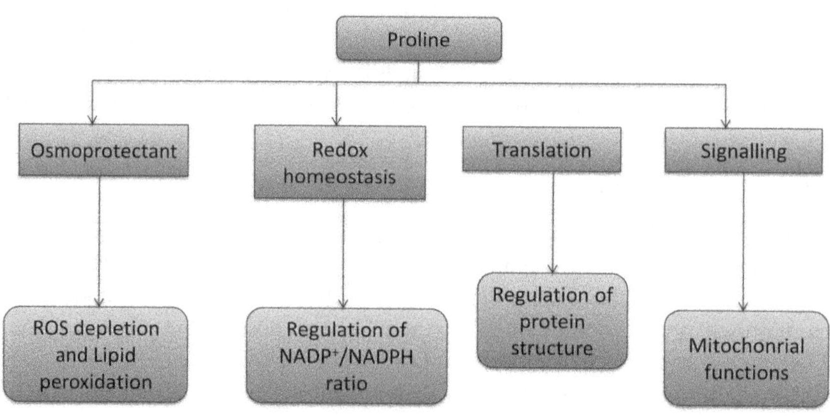

FIGURE 10.4 Role of proline in stress management.

10.4.1 Proline and Growth Parameters

Proline has been found to play a regulatory function in the growth and development of plants by acting as a signaling molecule. During abiotic stress, proline acts as an osmoprotectant, and its exogenous application improves plant growth under stress (Csonka and Hanson 1991). The exogenous application of proline at low concentrations results in the amelioration of the effects of damage caused by salt stress in *Oryza sativa* (Roy et al. 1993). Being an osmoprotectant, proline is considered a solute with defensive properties; however, under different conditions, the presence of exogenous proline can be injurious to plants and can impair plant growth and cell division. OsProT, a gene family in *Oryza sativa* plants which specifically transports proline, has been identified by employing the southern blot technique (Igarashi et al. 2000). A tissue-specific over-expression of HvProT mRNA in the root caps of *Hordeum vulgare* plants exposed to salt stress (Ueda et al. 2001). Similar over-expression of this mRNA in *Arabidopsis* plants' root tip led to the enhancement of proline content and subsequently resulted in replenished growth under normal conditions (Ueda et al. 2008; Per et al. 2017). Exogenous application of proline has been observed to cause deleterious effects on growth, specifically of the roots of *Petunia* spp. and *Thellungiella halophila* plants (Yamada et al. 2005). Similarly, Hare et al. (2003) reported that the deletion of p5cs mutants results in fatal impacts on embryo development and the plant shows abnormal growth patterns. Hence, it was concluded that proline has an imperative role in the growth and development of plants.

Proline is also an important amino acid and is an essential element of proline-rich proteins (PRP) which could be an additional way of regulating plant growth. It has been reported that sickle (Sic), a hydroxyproline-rich protein, plays a vital role in the stress management and development of *Arabidopsis* plants (Zhan et al. 2012). The reduction in the expression of the *sic* gene by miRNA-regulated post-transcriptional changes results in the production of plants with deferred growth characteristics, reduced activity and delayed flowering and maturity which shows the importance of proline in the normal development of the plant (Roychoudhury et al. 2015). Various reports on the role of proline in modulating plant growth and development have been tabulated in Table 10.1.

10.4.2 Proline and Plant–Water Relation

Abiotic stress in close proximity to plants directly affects the plant water status. As stress results in the restricted uptake of water, the hampered ascent of sap and stomatal opening and closing leads to an imbalance in the content of various solutes and decreased leaf water potential (Roychoudhury and Chakraborty 2013). Plants also have low osmotic potential in soil and fail to regulate the normal metabolic activities and turgidity of cells. At the same stage, plants also absorb enormous amounts of Na^+ and Cl^-. High uptake of Na^+ is the main reason for ion-related harm and results in a series of events such as the malfunctioning of protein synthesis and enzyme activity (Tester and Davenport 2003; Maathuis 2013). The regulation of osmotic adjustments is considered an important aspect that regulates plant tolerance under water deficit conditions and helps in the protection of plant cell organelles such as the chloroplast and cell membrane (Martinez et al. 2004).

To survive under high levels of salt or water deficit conditions, plants modify their metabolism. They reduce the osmotic potential of cells by storing various inorganic ions or compatible solutes to maintain the required turgor pressure. Proline is one of the major osmoregulators that is accumulated in plants to regulate low potential. Its accretion in plant cells results in extra absorption of water under extreme conditions (water deficit or high salt), thus counteracting the water shortage conditions (Fahad et al. 2017; de Oliveira et al. 2013; Cha-um and Kirdmanee 2009; Mousa and Abdel-Aziz 2008). In plants facing osmotic imbalance, the main route of proline generation is from glutamate via D1-pyrroline-5-carboxylate (P5C) through two consecutive reductions. These reactions are catalyzed by the rate-limiting enzymes P5C synthetase (P5CS) and P5C reductase (*P5CR*) (Iqbal et al. 2015; Khan et al. 2015). It was reported by Miller et al. (2005) that increased expression of the *P5CS* gene along with down-regulation of dehydrogenase genes (*PDH*) improves proline

TABLE 10.1
Role of Proline in Modulation of Growth and Plant–Water Relations in Plants under Various Abiotic Stresses

Plant	Stress	Proline Concentration	Effect on Growth and Plant–Water Relation	Reference
Triticum aestivum	Salinity	10 mM	Shoot length, shoot fresh weight and dry weight were slightly replenished.	Rady and Mohamed (2018)
Triticum aestivum	Drought	6 mM	Seedling growth in terms of shoot height, fresh and dry weight of root, relative water content were enhanced.	Bekka et al. (2018)
Triticum aestivum	Heavy metal (copper)	80 mM	The fresh weights of roots and shoots were enhanced.	Noreen et al. (2018)
Onobrychivi-ciaefolia scop.	Salinity	2.5 mmol/L	Growth in terms of fresh weight of root and shoot was improved.	Wu et al. (2017)
Zea mays	Short-term drought	1, 10, 20 mM	Leaf water potential and membrane damage was improved.	Demiralay et al. (2017)
Chenopodium quinoa	Drought	12.5, 25 mM	Enhanced root and shoot lengths, fresh weight and dry weight of roots and shoots.	Elewa et al. (2017)
Ocimum basilicum L.	Water (100% and 60% evapotrans-piration)	1 Mm	Reduced water stress and improved growth.	Agami et al. (2016)
Chilli	Salinity	0.4, 0.6, 0.8, 1.0, 1.2 Mm	Increase in root length, shoot length, fresh and dry weight.	Butt et al. (2016)
Phoenix dactylifera L.	Heavy metal (cadmium)	20 Mm	Improves mineral nutrition and growth.	Zouari et al. (2016)
Triticum aestivum	Drought	25, 50 mM	Plant growth, grain and straw yield were replenished.	Kibria et al. (2016)
Oryza sativa	Salinity	25, 50 mM	Plant growth and grain yield was enhanced.	Bhushan et al. (2016)
Olea europaea	Salinity	15 mM	Shoot and root fresh weight and dry weight were enhanced.	Aliniaeifard et al. (2016)
Zea mays	Salinity	25 mM	Increased root growth was recorded. Plants' dry weight and grain yield were also elevated.	Alam et al. (2016)
Solanum melongena	Heavy metal (arsenic)	25 mM	Increase in plant growth.	Singh et al. (2015)
Oryza sativa	Salinity	5 mM	Leaf relative water content level was elevated.	Hasanuzzaman et al. (2014)
Saccharum officinarum L.	Salinity	20 mM	Membrane integrity in terms of electrolyte leakage was improved.	Medeiros et al. (2015)
Cordia myxa	Salinity	0.75, 150 mg/L	Improved growth characteristic viz. stem height and diameter, number of leaves and leaf area, water content and dry material.	Al Mayahi and Fayadh (2015)
Phaseolus vulgaris	Salinity	5 mM	Growth attributes were improved.	Abdellhamid et al. (2013)

(Continued)

TABLE 10.1 (CONTINUED)
Role of Proline in Modulation of Growth and Plant–Water Relations in Plants under Various Abiotic Stresses

Plant	Stress	Proline Concentration	Effect on Growth and Plant–Water Relation	Reference
Triticum aestivum	Salinity	50, 100 mM	Root fresh weight, shoot fresh weight, root length, shoot length and plant dry weight were elevated.	Talat et al. (2013)
Solanum lycopersicum	Salinity	10 mg/L	Enhancement in leaf area, growth (root and shoot length) and fruit yield.	Kahlaoui et al. (2013)
Lepidium sativum	Drought	1, 5, 10 mM	Shoot length, number leaves, root length, fresh and dry weights were elevated.	Khalil and El-Noemani (2012)
Oryza sativa	Salinity	10 mM	Increased the fresh and dry weights.	Nounjan et al. (2012)
Oryza sativa	Salinity	10 mmol/L	Fresh weight, dry weight were elevated.	Nounjan and Theerakulpisut (2012)
Oryza sativa	Salinity	1, 5, 10 Mm	Increased cellular activity and improved growth.	Deivania et al. (2011)
Olea europaea L. cv. Chemlali	Salinity	25, 50 mM	Increased leaf water content.	Ben Ahmed et al. (2010)
Leymus chinensis	Alkaline stress and salinity	50, 100, 200, 500 mg/L	Improved levels of relative water content were observed.	Sun and Hong (2010)
Phaseolus vulgaris L.	Heavy metal (selenium)	50 µM	Improved growth.	Aggarwal et al. (2010)
Zea mays and *Vicia faba*	Salinity	100 ppm	Dry mass, water content, leaf area were enhanced.	El-Samad et al. (2011)
Sorghum bicolor L.)	Salinity	50, 100 mM	Proline alleviated the adverse effects of salt stress by improving plant growth.	Nawaz et al. (2010)
Nicotiana tabaccum	Heavy metal (cadmium)	1, 10 mM	Cell growth, fresh and dry weight were improved.	Islam et al. (2009)
Brassica napus L.	Salinity	1, 5 mM	Growth improvement was observed.	Athar and Ashraf (2009)
Zea mays	Drought	30, 60 mM	Improved growth attributes viz. shoot and fresh weight and dry weight.	Ali et al. (2007)

biosynthesis (Miller et al. 2005). Over-expression of the mutated gene (*VaP5CS129A*) of *Vigna aconitifolia*, responsible for high proline content in transgenic tobacco, showed enhanced tolerance to heat, salt and drought stress (Cvikrová et al. 2012; Gubis et al. 2007; Hong et al. 2000). The application of proline exogenously under high salt conditions leads to a significant enhancement in the water status of leaves in *Vicia faba* (Ben Ahmed et al. 2010). It was reported that even low doses of exogenous proline application were more effective than foliar application of ABA in enhancing the turgidity of leaves and stomatal functions (Hayat et al. 2013). *Eurya emarginata* plants given a treatment of 10 mM proline and 200 mM NaCl for 35 days showed an instant reduction of P5CS; however, PDH activity remained unchanged. The exogenous proline compensated for the decrease in endogenous synthetic proline and significantly affected the salinity tolerance of *Eurya emarginata*, through diverse protective effects on water relations, ionic and osmotic adjustment and antioxidant defense (Zheng et al. 2015). Table 10.1 lists various reports showing the role of proline in modulating plant–water relations in plants under various abiotic stresses.

10.4.3 Proline, Redox Balance, and Antioxidative Defense Responses

ROS are incessantly synthesized in plants as a result of several metabolic pathways (Ahanger et al. 2017a; Gill and Tuteja 2010). Proline participates greatly in protecting plants against stressful conditions. They induce the formation of xylem elements, lignin in the cell wall and various other developmental mechanisms (Das and Roychoudhury 2014). Abiotic stresses cause toxicity to plants through oxidative stress caused by the production of harmful ROS, ultimately impairing the growth of plants (Gill and Tuteja 2010; Sharma et al. 2017; Kohli et al. 2018). The imbalance in the production and scavenging of ROS (i.e. reduction in scavenging) leads to toxic symptoms in plants cells in response to oxidative stress (Kaur and Asthir 2015; Ahanger et al. 2017a). The generation of ROS above the optimal limits of plants results in damage to biomolecules (nucleic acid, proteins and lipids), disruption of the chlorophyll structure and the uncontrolled efflux of potassium ions from the cells (Chen and Dickman 2005; Ahanger and Agarwal 2017).

Proline is recognized as one of the most formidable ROS scavengers (Chen and Dickman 2005). The reduction of ROS is also favored by the deactivation of superoxide anion ($^1O_2$) by proline (Matysik et al. 2002). Proline also helps in the reduction of oxidative stress by controlling the accumulation of H_2O_2 and hydroxyl radicals in plants, mainly by converting the harmful forms of these ROS into non-toxic hydroxyproline derivatives (Kaul et al. 2008). It is also believed that the proline-proline cycle plays a major role in the scavenging of hydroxyl radicals. Due to H ions sequestration, proline captures the hydroxyl radicals which leads to the formation of P5C, and finally, enzymes like *P5CR*/NADPH help in recycling P5C back to proline (Signorelli et al. 2013). Moreover, proline-mediated regulation of the enzymes of the antioxidative system is also responsible for the control of ROS accumulation in plant cells under stress conditions (Kaur and Asthir 2015). Proline helps in the detoxification of singlet oxygen free radicals under cadmium toxicity by enhancing the activity of the SOD enzyme (Xu et al. 2009). Drought-induced oxidative stress is regulated by the endogenous accumulation of proline followed by a decrease in the lipid peroxidation, suggesting a possible role of endogenous proline (DeCampos et al. 2011).

Proline also regulates the excessive accumulation of malondialdehyde in the stressed cells, by the formation of stable H-bound water molecules around the protein molecules, protecting the protein ultrastructure from oxidative damage (Chutipaijit et al. 2009). The antioxidative enzymes like ascorbate peroxidase, glutathione reductase, monodehydroascorbate reductase (MDHAR) and dehydroascorbate reductase (DHAR) are the primary enzymes of the ascorbate-glutathione (ASH-GSH) cycle (Hoque et al. 2007). The enhanced activities of enzymes like SOD, CAT and POD after the application of proline under stress conditions also aid the hydroxyl radical scavenging, thereby regulating the oxidative stress (Hayat et al. 2012; Ashraf and Foolad 2007). Exogenous proline application acts as a vital ROS scavenger (Kaul et al. 2008). Similarly, in tobacco cells exposed to salinity stress, proline application was shown to enhance the activities of POD, SOD, CAT, glutathione-s-transferase and detoxifying enzyme viz. methyl glyoxal, subsequently resulting in regulation of the redox status (Hoque et al. 2008; Islam et al. 2009). Table 10.2 lists various reports showing the role of proline in modulating the redox balance and antioxidative defense responses. The contents of other oxidative stress markers like H_2O_2 and electrolyte leakage are also controlled by the exogenous application of osmoprotectants (Zouari et al. 2016), and this may be due to the fact that proline either has a direct antioxidant function responsible for ROS scavenging, or it modulates the redox balance of plant cells in an indirect way by modulating the antioxidative defense system (Ashraf and Foolad 2007). Furthermore, proline also improves the DPPH radical scavenging efficiency of plant cells by modulating the biosynthesis of secondary plant metabolites like phenolics and flavonoids (Zouari et al. 2018).

In a p5cs1 *Arabidopsis* mutant (hypersensitive to salt), the activities of the antioxidative enzymes of the GSH cycle were lowered, resulting in the hyper-accumulation of H_2O_2 and the elevation of MDA levels and chlorophyll damage (Szekely et al. 2008). Comparatively, a relatively untreated rose plant was shown to sustain elevation in Mn-dependent SOD activity in response to proline

TABLE 10.2
Role of Proline in Modulation of Redox Homeostasis and Antioxidative Defense in Plants under Various Abiotic Stresses

Plant	Stress	Proline Concentration	Effect on Redox Balance and Antioxidative Defense Response	Reference
Triticum aestivum	Salinity	10 mM	Content of ascorbic acid and glutathione was enhanced. Activities of SOD, CAT, acerbate peroxidase (APOX) and glutathione peroxidase (GPOX) were elevated.	Rady and Mohamed (2018)
Triticum aestivum	Drought	6 mM	Electrolyte leakage and MDA level were lowered. Protein level was enhanced. CAT, GPOX and APOX activity were modulated.	Bekka et al. (2018)
Triticum aestivum	Heavy metal (copper)	80 mM	Enzyme activity viz. CAT, POD and SOD were elevated. Modulation of total soluble proteins was recorded.	Noreen et al. (2018)
Onobrychivi-ciaefolia scop.	Salinity	2.5 mmol/L	Levels of MDA were reduced significantly, consequently lowering lipid peroxidation.	Wu et al. (2017)
Zea mays	Short-term drought	1, 10, 20 mM	MDA levels were reduced.	Demiralay et al. (2017)
Chenopodium quinoa	Drought	12.5, 25 mM	Free radical scavenging activity was elevated. Levels of total proteins were also increased.	Elewa et al. (2017)
Triticum aestivum	Drought	25, 50 mM	CAT, GPOX, APOX, POD enzyme activities were significantly enhanced.	Kibria et al. (2016)
Oryza sativa	Salinity	25, 50 mM	GPOX enzyme activity was elevated.	Bhushan et al. (2016)
Olea europaea	Salinity	15 mM	Increase in POD enzyme activity was observed.	Aliniaeifard et al. (2016)
Chili	Salinity	0.4, 0.6, 0.8, 1.2 mM	SOD and CAT activities were enhanced.	Butt et al. (2016)
Oryza sativa	Salinity	5 mM	ASH, GAH, APOX, MDHAR, DHAR, GR, GPOX and CAT activities were increased.	Hasanuzzaman et al. (2014)
Saccharum officinarum L.	Salinity	20 mM	POD, CAT and APOX enzyme activities were elevated.	Medeiros et al. (2015)
Lens culinaris	Drought	15 mM	GST and glyoxalase I activity was enhanced and levels of GSH were also elevated. H_2O_2 content was lowered.	Molla et al. (2014)
Phaseolus vulgaris	Salinity	5 mM	Activities of antioxidant enzymes including SOD, CAT and POD were elevated. Content of ascorbic acid was also elevated.	Abdellhamid et al. (2013)
Oryza sativa	Salinity	10 mmol/L	SOD, POD, CAT, APOX activities were enhanced. H_2O_2 levels were significantly lowered.	Nounjan and Theerakulpisut (2012)
Olea europaea L. cv. Chemlali	Salinity	25, 50 mM	Increase in SOD, CAT, APOX and PPO activities.	Ben Ahmed et al. (2010)
Leymus chinensis	Alkaline stress and Salinity	50, 100, 200, 500 mg/L	CAT activity was significantly enhanced. APOX activity was slightly elevated. MDA levels were lowered.	Sun and Hong (2010)
Nicotiana tabaccum	Heavy metal (cadmium)	1, 10 mM	CAT, POD and SOD activity was increased. Level of MDA was lowered.	Islam et al. (2009)

application. This subsequently resulted in a two-fold decline in superoxide anion radical levels at the petal senescence stage (Kumar et al. 2010). In plants, the MAPK cascade and ROS generation and scavenging network interplay to regulate the antioxidative defense signaling (Pitzschke et al. 2009; Sinha et al. 2011). The MAPK signaling cascade is comprised of three kinases, i.e. MAPK kinase kinase (MEKK), MAPK kinases (MKKs) and MAPKs (MKPs) (Sinha et al. 2011). The elevation in the H_2O_2 content under stress conditions activates the ROS scavenging enzymes which further stimulates the phosphorylation cycle involving *MKK* 4/5 and *MKK* 3/6 kinases (Kovtun et al. 2000; Zhang and Becker 2015). Hyper-osmotic stress was reported to induce comparatively similar alteration patterns in *MPK20* and *PRODH* expression in *Arabidopsis* plants under stress (Moutsafa et al. 2008). Furthermore, a recent genomic analysis of cotton plants with elevated contents of H_2O_2 showed over-expression of *MPK20* (Zhang et al. 2014). It might be plausible that ROS production in response to altered proline metabolism induced *MPK20* expression enhancement. The MAPK of *Zea mays* plants viz. *ZmMKK4* was over-expressed in tobacco and *Arabidopsis* plants in response to elevated P5CS activity (Kong et al. 2011a), which resulted in an increase in proline levels and tolerance to hyper-osmotic stress (Kong et al. 2011b). Moreover, the over-expression of *ZmMKK4* led to enhancement in the activity of peroxidases and lowered the ROS content (Kong et al. 2011 a,b)

10.4.4 Proline and Photosynthetic Efficacy

Abiotic stress conditions like heavy metals, salinity, temperature, drought and pesticide result in the reduction of the photosynthetic efficacy of plants (Kohli et al. 2017; Sharma et al. 2016; Zouari et al. 2016; Reddy et al. 2015; Molinari et al. 2007). However, proline is known to recover the process of photosynthetic efficacy which ultimately reduces the phytotoxicity caused by abiotic factors (Hayat et al. 2012). It is believed that the protective role of proline during adverse environmental cues is due to its potential to stabilize the membrane proteins and mitochondrial electron transport chain (ETC) complex II and regulate the key photosynthetic enzymes like Rubisco (Oukarroum et al. 2012; Kaushal et al. 2011; Holmström et al. 2000; Hare et al. 1998). Proline also regulates the movements of stomata with the upper and lower surfaces responding according to the proline concentration (i.e. they have a different response to each proline concentration). Additionally, abaxial stomata possess more resistance as compared to adaxial stomata under proline treatment (Rajagopal 1981).

Proline is also known to protect photosystem II (PS II) by protecting the thylakoid membranes from oxidative stress (Szabados and Savouré 2010). Moreover, it is also known that NADPH is required during the process of endogenous synthesis of proline, and it leads to $NADP^+$ generation which plays an important role in the photosynthetic electron transport chain (ETC) (Hare and Cress 1997). Proline also protects the photosynthetic apparatus from osmotic stresses by enhancing the production of photosynthetic pigments in the chloroplasts. The generation of $NADP^+$ is controlled by *P5CS1* genes which ultimately play a key role in plant development (Kruger and von Schaewen 2003). During salinity stress, the improvement in PS II efficiency and ETC of leaves after proline treatment is due to its protective effect on the chloroplast structure (Messedi et al. 2016). The study of Saradhi and Mohanty (1993) also support this fact of proline-mediated protection of the chloroplast in a saline environment. Moreover, the proline-mediated improvement in net CO_2 assimilation results in balancing the chloroplast redox-homeostasis, which ultimately improves the photosynthetic efficacy of plants (Messedi et al. 2016).

In sugarcane (genetically modified with over-producing efficiency of proline), it was observed that transgenic lines were the least affected by water stress as indicated by less damage to the photosynthetic machinery accompanied by enhanced efficiency of PS II (Molinari et al. 2007). The damage caused to the thylakoid membrane due to the generation of free radicals was noticed to reduce after proline application (Sivakumar et al. 2000). Improved photosynthetic performance of plants after proline application under stressful conditions is also anticipated due to the proline-mediated biosynthesis of photosynthetic pigments (Agami 2014; Khan et al. 2010). Moreover, proline also regulates the utilization efficiency of CO_2 and controls water loss through transpiration, which ultimately helps in the recovery of photosynthesis under saline conditions (Raven 2002). Furthermore, proline also regulates cellular turgor, leading to improvement in the conductance of stomata, which results in better CO_2 assimilation (Kamran et al. 2009). Regulation of potassium content by proline in guard cells is also one of the possible reasons behind the improved stomatal conductance (Ashraf and Foolad 2007). Table 10.3 lists various reports showing the role of proline in modulating the photosynthetic efficacy of plants under various stresses.

10.4.5 Proline as Metal Chelater

Another strategy by which proline combats abiotic stresses is chelation with toxic metal ions. It was reported by Liang et al. (2013) that high levels of proline are uncommon in metal-tolerant plant species. The participation of proline in metal chelation was suggested by Sharma et al. (1998). They reported proline to combat stress by protecting the important redox enzymes from zinc- and cadmium-induced inhibition by the formation of proline–metal complexes. A copper–proline complex was identified in *Armeria maritima*, a copper-tolerant plant (Farago and Mullen 1979). Another observation, made by Irtelli et al. (2004), reported proline along with histidine and nicotinamide which act as major copper chelators in *Brassica curinata* exposed to low and high levels of copper. Proline forms intercellular complexes with toxic metal ions by enhancing synthesis of other metal-chelating substances including organic acid, phytochelatins, polysaccharides and metallothionine (Emamverdian et al. 2015).

10.4.6 Proline and Other Osmolytes

Osmolytes are small organic molecules which a play protective role in combating various environmental stress conditions. They are also known as "osmoprotectants" or "compatible solutes". Other than proline, glycine betaine (GB), sugars and polyols are biologically imperative osmolytes (Ahanger et al. 2014). Proline and GB are quaternary amines, and are the most compatible solutes synthesized during abiotic stress conditions in plants (Verbruggen and Hermans 2008). Adverse stress conditions often disturb cells' redox homeostasis and various other mechanisms, in response to alteration in the oxidative balance in plant cells. The accumulation of various osmolytes under salinity and drought stress provides cell tolerance to stressed plants without interrupting other cellular functions (Nxele et al. 2017; Patade et al. 2011). The primary defense response to maintain osmotic pressure in the cell is to accumulate the osmoprotectants (Veeranagamalliaiah et al. 2007). Osmotic adjustments in cells are directly dependent upon the concentration of solutes and osmotic pressure to absorb water from the surroundings. In order to maintain the pressure, these compounds restore water in various cellular reactions (Giri 2011). Thus, the functional properties of various osmolytes play a vital role in maintaining the turgor pressure of the cells and mitigating drought stress symptoms. Proline is considered a source of carbon and nitrogen and plays an effective role in ameliorating stress conditions (Jain et al. 2001). Moreover, Hayat et al. (2012) suggested that proline acts as an osmolyte, scavenges ROS by increasing antioxidative defense activity and stabilizes biomolecule structures. Many researchers also revealed the accumulation of proline during water deficits (Choudhary et al. 2005), salinity (Yoshiba et al. 1995), heavy metals (Schat et al. 1997) and oxidative stress (Yang et al. 2009).

A study conducted by Al Hassan et al. (2015) focuses on the positive correlation between proline accumulation and salt stress along with the significant role of proline in cellular osmotic adjustment during salt stress. In tomato, it is generally accepted that proline is the only major osmolyte and glycine betaine is not able to accumulate in natural conditions. However, the exogenous application of glycine betaine to tomato led to an enhancement in its stress resistance potential against drought (Rezaei et al. 2012) and salinity (Chen et al. 2009). In many salt-tolerant plants, GB is reported to be accumulated in high quantities in the chloroplasts and plastids. Its level of accumulation is

TABLE 10.3
Role of Proline in Modulation of Photosynthetic Efficacy of Plants under Various Abiotic Stresses

Plant	Stress	Proline Concentration	Effect on Photosynthetic Efficacy	Reference
Triticum aestivum	Drought	6 mM	Total chlorophyll content was improved.	Bekka et al. (2018)
Triticum aestivum	Heavy metal (copper)	80 mM	Chlorophyll (total), Chla and Chl b levels were enhanced.	Noreen et al. (2018)
Zea mays	Short-term drought	1, 10, 20 mM	Chlorophyll content, gas exchange attributes in terms of net photosynthetic rate, transpiration rate, stomatal conductance, sub-stomatal CO_2 levels were improved.	Demiralay et al. (2017)
Chenopodium quinoa	Drought	12.5, 25 mM	Chlorophyll, Chl a, Chl b and carotenoid content was enhanced.	Elewa et al. (2017)
Triticum aestivum	Drought	25, 50 mM	Increase in chlorophyll content was observed.	Kibria et al. (2016)
Olea europaea	Salinity	15 mM	Photosynthetic rate, transpiration rate, stomatal conductance and chlorophyll levels were enhanced.	Aliniaeifard et al. (2016)
Chili	Salinity	0.4, 0.6, 0.8, 1.2 mM	Photosynthetic rate and transpiration rate were enhanced.	Butt et al. (2016)
Trigonella foenumgraec-um L. cv. Giza 30	Drought	25 mM	Increase in levels of photosynthetic pigments.	Elhamid et al. (2016)
Cordia myxa	Salinity	0.75, 150 mg/L	Endogenous chlorophyll and calcium levels were enhanced.	Al Mayahi and Fayadh, (2015)
Oryza sativa	Salinity	5 mM	Chlorophyll levels replenished.	Hasanuzzaman et al. (2014)
Pisum sativum	Heavy metal and salinity	60 mM	Chlorophyll content and gas exchange attributes (intercellular CO_2 concentration (Ci), net photosynthesis rate (Pn) and stomatal conductance) were improved.	Shahid et al. (2014)
Phaseolus vulgaris	Salinity	5 mM	Elevation in carotenoid content was recorded.	Abdellhamid et al. (2013)
Triticum aestivum	Salinity	50 and 100 mM	Chlorophyll a and b, carotenoid levels were increased. Sub-stomatal CO_2 and water use efficiency was not significantly affected. Net CO_2 assimilation rate and transpiration rate were improved.	Talat et al. (2013)
Cicer arietinum	Heavy metal (cadmium)	20 mM	Enhancement in photosynthetic attributes (activity of carbonic anhydrase) and yield characteristics.	Hayat et al. (2013)
Capsicum annuum	Salinity	1, 5, 10 mM	Salicylic acid and proline application led to improvement in photosynthetic attributes.	Jasim (2012)
Oryza sativa	Salinity	10 mmol/L	Enhanced activity of antioxidative enzymes resulted in improved photosynthetic attributes.	Nounjan and Theerakulpisut (2012)
Cucumis melo L.	Salinity	0.2 mM	Net photosynthetic rate and actual efficiency of PS II of leaves were enhanced.	Yan et al. (2011)

(Continued)

TABLE 10.3 (CONTINUED)
Role of Proline in Modulation of Photosynthetic Efficacy of Plants under Various Abiotic Stresses

Plant	Stress	Proline Concentration	Effect on Photosynthetic Efficacy	Reference
Olea europaea L. cv. Chemlali	Salinity	25, 50 mM	Photosynthetic activity and leaf chlorophyll levels increased.	Ben Ahmed et al. (2010)
Zea mays	Drought	30, 60 mM	Total chlorophyll and gas exchange parameters including photosynthetic rate and stomatal conductance were improved.	Ali et al. (2007)

correlated with the stress. During stress conditions, GB maintains the quaternary structures of enzymes and proteins, and membrane integrity (Chen et al. 2009). ROS produced during stress conditions are not directly scavenged by GB (Nawaz and Ashraf 2010). A recent study conducted by Chakraborty and Sairam (2017) reported increased amounts of GB and trehalose in *B. juncea* cultivars under salt stress. Biosynthetic genes of GB and trehalose, i.e. *BADH* and *T6PS*, showed up-regulation in *B. juncea* cultivars. Wani et al. (2013) reported that genetically engineered crop is capable of producing more GB under salt stress, elevating resistance against salt stress and balancing osmoregulation and water status in the leaves of plants (Wani et al. 2013).

Gamma aminobutyric acid (GABA), a non-protein amino acid, also plays an effective role in enhancing plant tolerance to stress (Giri 2011). It plays a potent role in osmotic balance and pH regulation as well as preventing the accumulation of ROS (Barbosa et al. 2010). Exogenous application of GABA under salt stress alters the gene expression of various mechanisms in *Caragana intermedia* (Shi et al. 2010). A study conducted by Al Hassan et al. (2015) demonstrated that, during salt stress in *Solanum lycopersicum*, the endogenous glycine betaine level increases (50–60 µmol/g) with an increase in proline content. Similarly in drought stress, an increase in glycine betaine content was observed with increased proline content. Total soluble sugars act as osmoprotectants in plants, and their accumulation in plants plays a significant role in osmotic adjustment (Kaur and Athir 2015). Abiotic stress conditions like drought increase monosaccharide levels which is considered an early response while a later increase in fructan has been reported as a delayed response (Kerepesi and Galiba 2000). Sugars play an effective role in seed development and maturation. Another investigation suggested that during drought, salinity and temperature stress soluble sugar levels increased, while under irradiation (high) and heavy metal toxicity, a low sugar level was reported (Gill et al. 2001). Trehalose, a disaccharide easily detectable in resurrection plants, has the unique property of protecting biological molecules from desiccation by its water re-absorption ability (Delorge et al. 2014). Raffinose family oligosaccharides have also been involved in ameliorating the effects of various stresses like temperature (low and high), dehydration, etc. (El Sayed et al. 2014). Sugars are well-known signal transducers apart from their osmoprotective characteristic during various stress conditions (Radomiljac et al. 2013). The exogenous application of 1 mM proline caused a significant increase in soluble sugars during water stress conditions in *Ocimum basilicum* L. (Agami et al. 2016). A similar observation was made by Elewa et al. (2017) in *Chenopodium quinoa* plants exposed to drought stress.

10.5 ENGINEERING PLANTS FOR PROLINE BIOSYNTHESIS

There is little knowledge about the post-translational modification of enzymes of the proline metabolic pathway. Schertl et al. (2014) suggested a redox-sensitive modification in ProDH1. Various enzymes of proline metabolism like P5CS and ProDH undergo transcriptional and post-transcriptional

TABLE 10.4
Plants Engineered for Proline Biosynthesis

S. No.	Source	Host	Gene	Reference
1	*Vigna aconitifolia*	*Sorghum bicolor*	P5CSF129A	Reddy et al. (2015)
2	*Solanum torvum*	*Glycine max*	StP5CS	Zhang et al. (2015)
3	*Vigna aconitifolia*	*Cajanus cajan*	P5CSF129A	Surekha et al. (2014)
4	*Not known*	*Olea europaea*	P5CS	Behelgardy et al. (2012)
5	*Arabidopsis*	*Nicotiana tabacum*	P5CS	Rastgar et al. (2011)
6	*Vigna aconitifolia*	*Cicer arietinum*	P5CS	Ghanti et al. (2011)
7	*Vigna aconitifolia*	*Oryza sativa*	P5CS	Karthikeyan et al. (2011)
8	*Vigna aconitifolia*	*Nicotiana tabacum*	P5CSF129A	Pospisilova et al. (2011)
9	*Vigna aconitifolia*	*Swingle citrumelo*	P5CSF129A	De Campos et al. (2011)
10	*Vigna aconitifolia*	*Oryza sativa*	P5CS	Kumar et al. (2010)

regulation. P5CDH undergoes post-transcriptional regulation while *P5Cr* is redox-sensitive (Giberti et al. 2014). Many researchers pointed towards a genetic engineering approach in order to modify the biosynthetic pathway of various osmolytes and stress-tolerant genes (Wani et al. 2013). The *P5CR* gene of *A. thaliana* was expressed in soybean in order to protect the production of soybean from frequent drought conditions in South Africa (de Ronde et al. 2004). Various other biosynthesis mechanisms can be modulated in order to raise the stress tolerance of species. A study documented by Tan et al. (2013) showed that expression of the *Medicago falcate myo-Inositol phosphate synthase* gene in transgenic tobacco plants leads to an increase in sugar content, resulting in increased resistance against stress conditions. In various studies the trehalose biosynthetic pathway is a choice target system in order to raise genetically engineered stress-tolerant plants (Delorge et al. 2014). *E. coli* genes (*OtsA* and *OtsB*) were used initially for drought-tolerant transgenic plants (Pilon-Smits et al. 1998), but currently *E. coli* genes with trehalose phosphate synthase/phosphatase (*TPSP*) have been used for rice transgenic plants (Garg et al. 2002). *TPSP* predicts that crops will be genetically modified and commercialized in the near future accordingly. The studies showing the source and host of the genes used in order to form transgenic plants with proline biosynthetic genes have been compiled in Table 10.4.

10.6 PROLINE VS. PHYTOHORMONES

Plant hormones act as messengers that work together with the proline metabolic pathway to enhance plant tolerance to adverse conditions (Iqbal et al. 2014; Ahanger et al. 2018a). A study conducted by Verslues and Bray (2005) showed the role of ABA in proline accumulation by regulating P5CS and *P5CR* expression. ABA stimulates proline biosynthesis by promoting P5CS1 expression under salt stress in *Arabidopsis* (Abraham et al. 2003). SA also induces proline metabolism and upregulates stress tolerance (Nazar et al. 2015). Another investigation by Yusuf et al. (2017) suggested that 24-epibrassinolide (EBL) application enhanced endogenous proline levels in aluminum- and salt-stressed plants. A recent study done by Xiong et al. (2018) reported that during salt stress, 5-aminolevulinic acid (ALA) application enhances the P5CS expression as compared to non-treated plants. Proline biosynthesis and accumulation also depend upon nitrogen status (Sanchez et al. 2001) and nitrogen metabolism (Ahanger and Agarwal 2017; Ahanger et al. 2017b). The increased accumulation of proline due to kinetin (Ahanger et al. 2018b) and jasmonic acid (Ahmad et al. 2018) has been reported to improve salinity tolerance in tomato. Nitrogen supply and proline accumulation showed a positive correlation in sugar beet under stressed conditions (Monreal et al. 2007).

10.7 CONCLUSION AND FUTURE PROSPECTS

Proline is well-acknowledged to modulate physiological development in plants under abiotic stresses. An understanding of the mechanism of proline biosynthesis and degradation in response to abiotic stress is significant for plants. Studying various physiochemical attributes including growth, plant–water relations, redox balance, antioxidative defense mechanism, photosynthetic efficacy and other metabolites altered in response to the exogenous application of proline confirmed its imperative role in the protection of plants from alterations in abiotic factors. In plants, proline accumulation is regulated by multiple genes and responds to a variety of conditions. Recognition of the various promoters related to proline synthesis and metabolism and their genetically engineered promoters has elevated our understanding of proline metabolism. Engineering approaches targeting promoters and their related genes have great potential in the near future. With the recent development of bioinformatic tools for the analysis of complete genome sequence and 3D structures, various biosynthetic genes viz. P5CS and *P5CR* have been found to be highly conserved in their action in bacteria as well as eukaryotes. Further, exhaustive insight is currently required to elucidate proline signaling cascades, metabolism and transcription factors that are influenced by exogenous proline application and have a significant role in developing more tolerant plants.

REFERENCES

Abdelhamid, M. T., M. M. Rady, A. S. Osman, and M. A. Abdalla. 2013. Exogenous application of proline alleviates salt-induced oxidative stress in *Phaseolus vulgaris* L. plants. *The Journal of Horticultural Science and Biotechnology* 88(4):439–446.

Abraham, E., G. Rigo, G. Szekely, R. Nagy, C. Koncz, and L. Szabados. 2003. Light-dependent induction of proline biosynthesis by abscisic acid and salt stress is inhibited by brassinosteroid in *Arabidopsis. Plant Molecular Biology* 51(3):363–372.

Agami, R. A. 2014. Applications of ascorbic acid or proline increase resistance to salt stress in barley seedlings. *Biologia Plantarum* 58(2):341–347.

Agami, R. A., R. A. Medani, I. A. Abd El-Mola, and R. S. Taha. 2016. Exogenous application with plant growth promoting rhizobacteria (PGPR) or proline induces stress tolerance in basil plants (*Ocimum basilicum* L.) exposed to water stress. *International Journal of Environment and Agricultural Research* 2(5):78.

Agarwal, P. K., and B. Jha. 2010. Transcription factors in plants and ABA dependent and independent abiotic stress signalling. *Biologia Plantarum* 54(2):201–12.

Aggarwal, M., S. Sharma, N. Kaur, et al. 2011. Exogenous proline application reduces phytotoxic effects of selenium by minimising oxidative stress and improves growth in bean (*Phaseolus vulgaris* L.) seedlings. *Biological Trace Element Research* 140(3):354–367.

Ahanger, M. A., and R. M. Agarwal. 2017. Salinity stress induced alterations in antioxidant metabolism and nitrogen assimilation in wheat (*Triticum aestivum* L) as influenced by potassium supplementation. *Plant Physiology and Biochemistry* 115:449–460.

Ahanger, M. A., M. Ashraf, A. Bajguz, and P. Ahmad. 2018a. Brassinosteroids regulate growth in plants under stressful environments and crosstalk with other potential phytohormones. *Journal of Plant Growth Regulation* 37(4):1007–1024.

Ahanger, M. A., M. N. Alyemeni, L. Wijaya, S. A. Alamri, P. Alam, M. Ashraf, and P. Ahmad. 2018b. Potential of exogenously sourced kinetin in protecting *Solanum lycopersicum* from NaCl-induced oxidative stress through up-regulation of the antioxidant system, ascorbate-glutathione cycle and glyoxalase system. *PLoS ONE* 13(9):e0202175. doi: 10.1371/journal. pone.0202175

Ahanger, M. A., N. S. Tomar, M. Tittal, S. Argal, and R. M. Agarwal. 2017a. Plant growth under water/salt stress: ROS production; antioxidants and significance of added potassium under such conditions. *Physiology and Molecular Biology of Plants* 23(4):731–744.

Ahanger, M. A., M. Tittal, R. A. Mir, and R. M. Agarwal. 2017b. Alleviation of water and osmotic stress-induced changes in nitrogen metabolizing enzymes in *Triticum aestivum* L. cultivars by potassium. *Protoplasma* 254(5):1953–1963.

Ahanger, M. A., S. R. Tyagi, M. R. Wani, and P. Ahmad. 2014. Drought tolerance: Roles of organic osmolytes, growth regulators and mineral nutrients. In: P. Ahmad and M. R. Wani (eds) *Physiological Mechanisms and Adaptation Strategies in Plants under Changing Environment*, pp. 25–56. Springer, Science+Business media, inc.

Ahmad, P., M. A. Ahanger, M. N. Alyemeni, L. Wijaya, P. Alam, and M. Ashraf. 2018. Mitigation of sodium chloride toxicity in *Solanum lycopersicum* L. by supplementation of jasmonic acid and nitric oxide. *Journal of Plant Interactions* 13(1):64–72.

Al Hassan, M., M. M. M. Futertes, F. J. R. Sanchez, O. Vicente, and M. Boscaiu. 2015. Effects of salt and water stress on plant growth and on accumulation of osmolytes and antioxidant compounds in cherry tomato. *Notulae Botanicae Horti Agrobotanici Cluj-Napoca* 43(1):1–11.

Al Mayahi, M. Z., and M. H. Fayadh. 2015. Effect of exogenous proline application on salinity tolerance of *Cordia myxa* L. seedlings. Effect on vegetative and physiological characteristics. *Journal of Natural Sciences Research* 5(24):118–125.

Alam, R., D. K. Das, M. R. Islam, Y. Murata, and M. A. Hoque. 2016. Exogenous proline enhances nutrient uptake and confers tolerance to salt stress in maize (*Zea mays* L.). *Progressive Agriculture* 27(4):409–417.

Ali, Q., M. Ashraf, and H. U. R. Athar. 2007. Exogenously applied proline at different growth stages enhances growth of two maize cultivars grown under water deficit conditions. *Pakistan Journal of Botany* 39(4):1133–1144.

Aliniaeifard, S., J. Hajilou, and S. J. Tabatabaei. 2016. Photosynthetic and growth responses of olive to proline and salicylic acid under salinity condition. *Notulae Botanicae Horti Agrobotanici Cluj-Napoca* 44(2):579–585.

Ashraf, M., and M. R. Foolad. 2007. Role of glycine betaine and proline in improving plant abiotic stress resistance. *Environmental and Experimental Botany* 59:206–216.

Athar, H. R., and M. Ashraf. 2009. Strategies for crop improvement against salinity and drought stress: An overview. In: M. Ashraf, M. Ozturk, and H. Athar (eds) *Salinity and* Water *Stress*, pp. 1–16. Springer, Dordrecht.

Ayliffe, M. A., J. K. Roberts, H. J. Mitchell, et al. 2002. A plant gene up-regulated at rust infection sites. *Plant Physiology* 129(1):169–180.

Barbosa, J. M., N. K. Singh, J. H. Cherry, and R. D. Locy. 2010. Nitrate uptake and utilization is modulated by exogenous γ-aminobutyric acid in *Arabidopsis thaliana* seedlings. *Plant Physiology and Biochemistry* 48(6):443–450.

Bekka, S., O. Abrous-Belbachir, and R. Djebbar. 2018. Effects of exogenous proline on the physiological characteristics of *Triticum aestivum* L. and *Lens culinaris* Medik. under drought stress. *Acta Agriculturae Slovenica* 111(2), 477–491.

Behelgardy, M. F., N. Motamed, and F. R. Jazii. 2012. Expression of the P5CS gene in transgenic versus non-transgenic olive (*Olea europaea*) under salinity stress. *World Applied Sciences Journal* 18(4):580–583.

Ben Ahmed, C., B. Ben Rouina, S. Sensoy, M. Boukhriss, and F. Ben Abdullah. 2010. Exogenous proline effects on photosynthetic performance and antioxidant defense system of young olive tree. *Journal of Agricultural and Food Chemistry* 58(7):4216–4222.

Bhusan, D., D. K. Das, M. Hossain, Y. Murata, and M. A. Hoque. 2016. Improvement of salt tolerance in rice (*Oryza sativa* L.) by increasing antioxidant defense systems using exogenous application of proline. *Australian Journal of Crop Science* 10(1):50.

Butt, M., C. M. Ayyub, M. Amjad, and R. Ahmad. 2016. Proline application enhances growth of chilli by improving physiological and biochemical attributes under salt stress. *Pakistan Journal of Agricultural Sciences* 53(1):43–49.

Chakraborty, K., and R. K. Sairam. 2017. Induced-expression of osmolyte biosynthesis pathway genes improves salt and oxidative stress tolerance in *Brassica* species. *Indian Journal of Experimental Biology* 55:711–721.

Cha-um, S., and C. Kirdmanee. 2009. Proline accumulation, photosynthetic abilities and growth characters of sugarcane (*Saccharum officinarum* L.) plantlets in response to ıso-osmotic salt and water-deficit stress. *Agricultural Sciences in China* 8:51–58.

Chen, C., and M. B. Dickman. 2005. Proline suppresses apoptosis in the fungal pathogen *Colletotrichum trifolii. Proceedings of the National Academy of Sciences* 102(9):3459–3464.

Chen, S., N. Gollop, and B. Heuer. 2009. Proteomic analysis of salt-stressed tomato (*Solanum lycopersicum*) seedlings: Effect of genotype and exogenous application of glycinebetaine. *Journal of Experimental Botany* 60(7):2005–2019.

Chen, S. L., and C. H. Kao. 1995. Cd induced changes in proline level and peroxidase activity in roots of rice seedlings. *Plant Growth Regulation* 17(1):67–71.

Chinnusamy, V., A. Jagendorf, and J. K. Zhu. 2005. Understanding and improving salt tolerance in plants. *Crop Science* 45(2):437–448.

Choudhary, N. L., R. K. Sairam, and A. Tyagi. 2005. Expression of delta1-pyrroline-5-carboxylate synthetase gene during drought in rice (*Oryza sativa* L.). *Indian Journal of Biochemistry and Biophysics* 42:366–370.

Chutipaijit, S., S. Cha-Um, and K. Sompornpailin. 2009. Differential accumulations of proline and flavonoids in indica rice varieties against salinity. *Pakistan Journal of Botany* 41(5):2497–2506.

Csonka, L. N., and A. D. Hanson. 1991. Prokaryotic osmoregulation: Genetics and physiology. *Annual Reviews in Microbiology* 45(1):569–606.

Cvikrová, M., L. Gemperlová, J. Dobrá, et al. 2012. Effect of heat stress on polyamine metabolism in proline-over-producing tobacco plants. *Plant Science* 182:49–58.

Das, K., and A. Roychoudhury. 2014. Reactive oxygen species (ROS) and response of antioxidants as ROS-scavengers during environmental stress in plants. *Frontiers in Environmental Science* 2:53.

De Campos, M. K. F., K. de Carvalho, F. S. de Souza, et al. 2011. Drought tolerance and antioxidant enzymatic activity in transgenic 'Swingle'citrumelo plants over-accumulating proline. *Environmental and Experimental Botany* 72(2):242–250.

de Oliveira, A. B., N. L. Alencar, and E. Gomes-Filho. 2013. Comparison between the water and salt stress effects on plant growth and development. In: S. Akıncı (ed.) *Responses of Organisms to Water Stress.* IntechOpen.

de Ronde, J. A., W. A. Cress, G. H. J. Krüger, R. J. Strasser, and J. Van Staden. 2004. Photosynthetic response of transgenic soybean plants, containing an *Arabidopsis* P5CR gene, during heat and drought stress. *Journal of Plant Physiology* 161(11):1211–1224.

Deivanai, S., R. Xavier, V. Vinod, K. Timalata, and O. F. Lim. 2011. Role of exogenous proline in ameliorating salt stress at early stage in two rice cultivars. *Journal of Stress Physiology and Biochemistry* 7(4):157–174.

Delorge, I., M. Janiak, S. Carpentier, and P. Van Dijck. 2014. Fine tuning of trehalose biosynthesis and hydrolysis as novel tools for the generation of abiotic stress tolerant plants. *Frontiers in Plant Science* 5:147.

Demiralay, M., C. Altuntaş, A. Sezgin, R. Terzi, and A. Kadioğlu. 2017. Application of proline to root medium is more effective for amelioration of photosynthetic damages as compared to foliar spraying or seed soaking in maize seedlings under short-term drought. *Turkish Journal of Biology* 41(4):649–660.

Deuschle, K., D. Funck, G. Forlani, H. Stransky, A. Biehl, D. Leister, E. van der Graaff, R. Kunze, and W. B. Frommer. 2005. The role of Δ1-pyrroline-5-carboxylate dehydrogenase in proline degradation. *The Plant Cell* 16(12):3413–3425.

Deuschle, K., D. Funck, H. Hellmann, K. Däschner, S. Binder, and W. B. Frommer. 2001. A nuclear gene encoding mitochondrial Δ1-pyrroline-5-carboxylate dehydrogenase and its potential role in protection from proline toxicity. *The Plant Journal* 27(4):345–356.

Egamberdieva, D., S. J. Wirth, A. A. Alqarawi, E. F. Abd_Allah, and A. Hashem. 2017. Phytohormones and beneficial microbes: Essential components for plants to balance stress and fitness. *Frontiers in Microbiology* 8:2104.

Elewa, T. A., M. S. Sadak, and A. M. Saad. 2017. Proline treatment improves physiological responses in quinoa plants under drought stress. *Bioscience Research* 14(1):21–33.

Elhamid, E. M. A., M. S. Sadak, and M. M. Tawfik. 2016. Physiological response of Fenugreek plant to the application of proline under different water regimes. *Research Journal of Pharmaceutical, Biological and Chemical Sciences* 7(3):580–586.

El-Samad, H. A., M. A. K. Shaddad, and N. Barakat. 2011. Improvement of plants salt tolerance by exogenous application of amino acids. *Journal of Medicinal Plants Research* 5(24):5692–5699.

ElSayed, A. I., M. S. Rafudeen, and D. Golldack. 2014. Physiological aspects of raffinose family oligosaccharides in plants: Protection against abiotic stress. *Plant Biology* 16(1):1–8.

Elthon, T. E., and C. R. Stewart. 1982. Proline oxidation in corn mitochondria: Involvement of NAD, relationship to ornithine metabolism, and sidedness of the inner membrane *Plant Physiology* 70(2):567–572.

Emamverdian, A., Y. Ding, F. Mokhberdoran, and Y. Xie. 2015. Heavy metal stress and some mechanisms of plant defense response. *The Scientific World Journal* 756120:1–18.

Fahad, S., A. A. Bajwa, U. Nazir, S. A. Anjum, A. Farooq, A. Zohaib, S. Sadia, W. Nasim, S. Adkins, S. Saud, and M. Z. Ihsan. 2017. Crop production under drought and heat stress: Plant responses and management options. *Frontiers in Plant Science* 8:1147.

Farago, M. E., and W. A. Mullen. 1979. Plants which accumulate metals. Part IV. A possible copper-proline complex from the roots of *Armeria maritima. Inorganica Chimica Acta* 32:L93–L94.

Fu, Y., H. Ma, S. Chen, T. Gu, and J. Gong. 2017. Control of proline accumulation under drought via a novel pathway comprising the histone methylase CAU1 and the transcription factor ANAC055. *Journal of Experimental Botany* 69(3):579–588.

Funck, D., S. Eckard, and G. Muller. 2010. Non-redundant functions of two proline dehydrogenase isoforms in *Arabidopsis. BMC Plant Biology* 10:70.

Funck, D., B. Stadelhofer, and W. Koch. 2008. Ornithine-α- aminotransferase is essential for arginine catabolism but not for proline biosynthesis. *BMC Plant Biology* 8:40.

Garg, A. K., J. K. Kim, and T. G. Owens, et al. 2002. Trehalose accumulation in rice plants confers high tolerance levels to different abiotic stresses. *Proceedings of the National Academy of Sciences* 99(25):15898–15903.

Ghanti, S. K. K., K. G. Sujata, B. V. Kumar, et al. 2011. Heterologous expression of P5CS gene in chickpea enhances salt tolerance without affecting yield. *Biologia Plantarum* 55(4):634.

Giberti, S., D. Funck, and G. Forlani. 2014. Delta1-Pyrroline-5-carboxylate reductase from *Arabidopsis thaliana*: Stimulation or inhibition by chloride ions and feedback regulation by proline depend on whether NADPH or NADH acts as co-substrate. *New Phytologist* 202:911–919.

Gill, P. K., A. D. Sharma. P. Singh, and S. S. Bhullar. 2001. Effect of various abiotic stresses on the growth, soluble sugars and water relations of *sorghum* seedlings grown in light and darkness. *Bulgarian Journal of Plant Physiology* 27(1–2):72–84.

Gill, S. S., and N. Tuteja. 2010. Reactive oxygen species and antioxidant machinery in abiotic stress tolerance in crop plants. *Plant Physiology and Biochemistry* 48(12):909–930.

Giri, J. 2011. Glycinebetaine and abiotic stress tolerance in plants. *Plant Signaling and Behavior* 6(11):1746–1751.

Gubiš, J., R. Vaňková, V. Červená, et al. 2007. Transformed tobacco plants with increased tolerance to drought. *South African Journal of Botany* 73(4):505–511.

Hare, P. D., and W. A. Cress. 1997. Metabolic implications of stress-induced proline accumulation in plants. *Plant Growth Regulation* 21(2):79–102.

Hare, P. D., W. A. Cress, and J. Van Staden. 1998. Dissecting the roles of osmolyte accumulation during stress. *Plant, Cell and Environment* 21(6):535–553.

Hare, P. D., W. A., Cress, and J. Van Staden. 2003. A regulatory role for proline metabolism in stimulating *Arabidopsis thaliana* seed germination. *Plant growth Regulation* 39(1):41–50.

Hasanuzzaman, M., M. Alam, A. Rahman, M. Hasanuzzaman, K. Nahar, and M. Fujita. 2014. Exogenous proline and glycine betaine mediated upregulation of antioxidant defense and glyoxalase systems provides better protection against salt-induced oxidative stress in two rice (*Oryza sativa* L.) varieties. *BioMed Research International* 757219:1–17.

Hayat, S., Q. Hayat, M. N. Alyemeni, A. S. Wani, J. Pichtel, and A. Ahmad. 2012. Role of proline under changing environments: A review. *Plant Signaling and Behavior* 7(11):1456–1466.

Hayat, S., Q. Hayat, M. N. Yemeni, and A. Ahmad. 2013. Proline enhances antioxidative enzyme activity, photosynthesis and yield of *Cicer arietinum* L. exposed to cadmium stress. *Acta Botanica Croatica* 72(2):323–335.

Heuer, B. 2003. Influence of exogenous application of proline and glycinebetaine on growth of salt-stressed tomato plants. *Plant Science* 165:693–639.

Holmström, K. O., S. Somersalo, A. Mandal, T. E. Palva, and B. Welin. 2000. Improved tolerance to salinity and low temperature in transgenic tobacco producing glycine betaine. *Journal of Experimental Botany* 51(343):177–185.

Hong, Z., K. Lakkineni, Z. Zhang, and D. P. S. Verma. 2000. Removal of feedback inhibition of Δ1-pyrroline-5-carboxylate synthetase results in increased proline accumulation and protection of plants from osmotic stress. *Plant Physiology* 122(4):1129–1136.

Hoque, M. A., M. N. Banu, Y. Nakamura, Y. Shimoishi, and Y. Murata. 2008. Proline and glycinebetaine enhance antioxidant defense and methylglyoxal detoxification systems and reduce NaCl-induced damage in cultured tobacco cells. *Journal of Plant Physiology* 165(8):813–824.

Hoque, M. A., M. N. A. Banu, E. Okuma, et al. 2007. Exogenous proline and glycinebetaine increase NaCl-induced ascorbate–glutathione cycle enzyme activities, and proline improves salt tolerance more than glycinebetaine in tobacco Bright Yellow-2 suspension-cultured cells. *Journal of Plant Physiology* 164(11):1457–1468.

Hu, C. A. A., A. J. Delauney, and D. P. S. Verma. 1992. A bifunctional enzyme (Dl-pyrroline-5-carboxylate synthetase) catalyzes the first two steps in proline biosynthesis in plants. *Proceedings of the National Academy of Sciences of the United States of America* 89:9354–9358.

Igarashi, Y., Y. Yoshiba, T. Takeshita, S. Nomura, J. Otomo, K. Yamaguchi-Shinozaki, and K. Shinozaki. 2000. Molecular cloning and characterization of a cDNA encoding proline transporter in rice. *Plant and Cell Physiology* 41(6):750–756.

Iqbal, N., S. Umar, and N. A. Khan. 2015. Nitrogen availability regulates proline and ethylene production and alleviates salinity stress in mustard (*Brassica juncea*). *Journal of Plant Physiology* 178:84–91.

Iqbal, N., S. Umar, N. A. Khan, and M. I. R. Khan. 2014. A new perspective of phytohormones in salinity tolerance: Regulation of proline metabolism. *Environmental and Experimental Botany* 100:34e42.

Irtelli, B., M. Quartacci, and F. Navari. 2004. Chelate-assisted phytoextraction of Cd by *Brassica juncea*: The role of phenolic acids in metal tolerance. *Acta Physiologiae Plantarum* 26:239–239.

Islam, M. M., M. A. Hoque, E. Okuma, M. N. Banu, Y. Shimoishi, Y. Nakamura, and Y. Murata. 2009. Exogenous proline and glycinebetaine increase antioxidant enzyme activities and confer tolerance to cadmium stress in cultured tobacco cells. *Journal of Plant Physiology* 166(15):1587–97.

Jain, M., G. Mathur, S. Koul, and N. Sarin. 2001. Ameliorative effects of proline on salt stress-induced lipid peroxidation in cell lines of groundnut (*Arachis hypogaea* L.). *Plant Cell Reports* 20(5):463–468.

Jasim, A. H. 2012. Effect of salt stress, application of salicylic acid and proline on seedlings growth of sweet pepper (*Capsicum annum* L.). *Euphrates Journal of Plant Cell ReportsAgriculture Science* 4(3):191–205.

Kahlaoui, B., M. Hachicha, J. Teixeira, E. Misle, F. Fidalgo, and B. Hanchi. 2013. Response of two tomato cultivars to field-applied proline and salt stress. *Journal of Stress Physiology and Biochemistry* 9(3):357–365.

Kamran, M., M. Shahbaz, M. Ashraf, and N. A. Akram. 2009. Alleviation of drought-induced adverse effects in spring wheat (*Triticum aestivum* L.) using proline as a pre-sowing seed treatment. *Pakistan Journal of Botany* 41(2):621–632.

Karthikeyan, A., S. K. Pandian, and M. Ramesh. 2011. Transgenic indica rice cv. ADT 43 expressing a D1-pyrroline-5-carboxylate synthetase (P5CS) gene from *Vigna aconitifolia* demonstrates salt tolerance. *Plant Cell, Tissue and Organ Culture* 107(3):383–395.

Kaul, S., S. S. Sharma, and I. K. Mehta. 2008. Free radical scavenging potential of L-proline: Evidence from in vitro assays. *Amino Acids* 34(2):315–320.

Kaur, G., and B. Asthir. 2015. Proline: A key player in plant abiotic stress tolerance. *Biologia Plantarum* 59(4):609–619.

Kaushal, N., K. Gupta, K. Bhandhari, S. Kumar, P. Thakur, and H. Nayyar. 2011. Proline induces heat tolerance in chickpea (*Cicer arietinum* L.) plants by protecting vital enzymes of carbon and antioxidative metabolism. *Physiology and Molecular Biology of Plants* 17(3):203.

Kerepesi, I., and G. Galiba. 2000. Osmotic and salt stress-induced alteration in soluble carbohydrate content in wheat seedlings. *Crop Science* 40(2):482–487.

Khan, A., I. Iram, S. Amin, N. Humera, A. Farooq, and M. Ibrahim. 2010. Alleviation of adverse effects of salt stress in brassica (*Brassica campestris*) by pre-sowing seed treatment with ascorbic acid. *American-Eurasian Journal of Agricultural and Environmental Science* 7(5):557–560.

Khan, M. I. R., F. Nazir, M. Asgher, T. S. Per, and N. A. Khan. 2015. Selenium and sulfur influence ethylene formation and alleviate cadmium-induced oxidative stress by improving proline and glutathione production in wheat. *Journal of Plant Physiology* 173:9–18.

Khalil, S. E., and A. A. El-Noemani. 2012. Effect of irrigation intervals and exogenous proline application in improving tolerance of garden cress plant (*Lepidium sativum* L.) to water stress. *Journal of Applied Sciences Research* 8(1):157–167.

Kibria, M. G., F. Kaniz, M. A. Matin, and M. A., Hoque. 2016. Mitigating water stress in wheat (BARI Gom-26) by exogenous application of proline. *Fundamental and Applied Agriculture* 1(3):118–123.

Kishore, K., S. Sangam, R. N. Amrutha, et al. 2005. Regulation of proline biosynthesis, degradation, uptake and transport in higher plants. *Current Science* 88(3):424–438.

Kohli, S. K., N. Handa, S. Bali, et al. 2018. Modulation of antioxidative defense expression and osmolyte content by co-application of 24-epibrassinolide and salicylic acid in Pb exposed Indian mustard plants. *Ecotoxicology and Environmental Safety* 147:382–393.

Kohli, S. K., N. Handa, A. Sharma, V. Kumar, P. Kaur, and R. Bhardwaj. 2017. Synergistic effect of 24-epibrassinolide and salicylic acid on photosynthetic efficiency and gene expression in *Brassica juncea* L. under Pb stress. *Turkish Journal of Biology* 41(6):943–953.

Kong, X., J. Pan, M. Zhang, X. I. Xing, Y. A. Zhou, Y. Liu, D. Li, and D. Li. 2011a. ZmMKK4, a novel group C mitogen-activated protein kinase kinase in maize (*Zea mays*), confers salt and cold tolerance in transgenic *Arabidopsis*. *Plant, Cell and Environment* 34(8):1291–303.

Kong, X., L. Sun, Y. Zhou, M. Zhang, Y. Liu, J. Pan, and D. Li. 2011b. ZmMKK4 regulates osmotic stress through reactive oxygen species scavenging in transgenic tobacco. *Plant Cell Reports* 30(11):2097.

Kovtun, Y., W. L. Chiu, G. Tena, and J. Sheen. 2000. Functional analysis of oxidative stress-activated mitogen-activated protein kinase cascade in plants. *Proceedings of the National Academy of Sciences* 97(6):2940–2945.

Krishnan, N., M. B. Dickman, and D. F. Becker. 2008. Proline modulates the intracellular redox environment and protects mammalian cells against oxidative stress. *Free Radical Biology and Medicine* 44(4):671–681.

Kruger, N. J., and A. von Schaewen. 2003. The oxidative pentose phosphate pathway: Structure and organisation. *Current Opinion in Plant Biology* 6(3):236–246.

Kumar, V., V. Shriram, P. K. Kishor, N. Jawali, and M. G. Shitole. 2010. Enhanced proline accumulation and salt stress tolerance of transgenic indica rice by overexpressingP5CSF129A gene. *Plant Biotechnology Reports* 4(1):37–48.

Kunji, E. R., A. Aleksandrova, M. S. King, H. Majd, V. L. Ashton, E. Cerson, R. Springett, M. Kibalchenko, S. Tavoulari, P. G. Crichton, and J. J. Ruprecht. 2016. The transport mechanism of the mitochondrial ADP/ATP carrier. *Biochimica et Biophysica Acta (BBA)-Molecular Cell Research* 1863(10):2379–2393.

Lehmann, S., D. Funck, L. Szabados, and D. Rentsch. 2010. Proline metabolism and transport in plant development. *Amino Acids* 39(4):949–962.

Liang, X., L. Zhang, S. K. Natarajan, and D. F. Becker. 2013. Proline mechanisms of stress survival. *Antioxidants and Redox Signaling* 19(9):998–1011.

Liao, Y., H. F. Zou, H. W. Wang, W. K. Zhang, B. Ma, J. S. Zhang, and S. Y. Chen. 2008. Soybean GmMYB76, GmMYB92, and GmMYB177 genes confer stress tolerance in transgenic *Arabidopsis* plants. *Cell Research* 18(10):1047.

Maathuis, F. J. 2013. Sodium in plants: Perception, signalling, and regulation of sodium fluxes. *Journal of Experimental Botany* 65(3):849–858.

Martinez, J. P., S. Lutts, A. Schanck, M. Bajji, and J. M. Kinet. 2004. Is osmotic adjustment required for water stress resistance in the Mediterranean shrub *Atriplex halimus* L. *Plant Physiology* 161:1041–1051.

Matysik, J., B. B. Alia, and P. Mohanty. 2002. Molecular mechanisms of quenching of reactive oxygen species by proline under stress in plants. *Current Science* 82(5):525–532.

Medeiros, L., M. Jaislanny, M. Medeiros De A. Silva, et al. 2015. Effect of exogenous proline in two sugarcane genotypes grown in vitro under salt stress. *Acta Biológica Colombiana* 20(2):57–63.

Messedi, D., F. Farhani, K. B. Hamed, N. Trabelsi, R. Ksouri, and C. A. Habib-ur-Rehman Athar. 2016. Highlighting the mechanisms by which proline can confer tolerance to salt stress in cakilemaritima. *Pakistan Journal of Botany* 48(2):417–427.

Miller, G., H. Stein, A. Honig, Y. Kapulnik, and A. Zilberstein. 2005. Responsive modes of *Medicago sativa* proline dehydrogenase genes during salt stress and recovery dictate free proline accumulation. *Planta* 222(1):70–79.

Mitchell, H. J., M. A. Ayliffe, K. Y. Rashid, and A. J. Pryor. 2006. A rust-inducible gene from flax (fis1) is involved in proline catabolism. *Planta* 223(2):213–222.

Molinari, H. B. C., C. J. Marur, E. Daros, et al. 2007. Evaluation of the stress-inducible production of proline in transgenic sugarcane (*Saccharum* spp.): Osmotic adjustment, chlorophyll fluorescence and oxidative stress. *Physiologia Plantarum* 130(2):218–229.

Molla, M. R., M. R. Ali, M. Hasanuzzaman, M. H. Al-Mamun, A. Ahmed, M. A. N. Nazim-ud-Dowla, and M. M. Rohman. 2014. Exogenous proline and betaine-induced upregulation of glutathione transferase and glyoxalase I in lentil (*Lens culinaris*) under drought stress. *Notulae Botanicae Horti Agrobotanici Cluj-Napoca* 42(1):73–80.

Monreal, J. A., E. T. Jimenez, E. Remesal, R. Morillo-Velarde, S. Garcia-Maurino, and C. Echevarria. 2007. Proline content of sugar beet storage roots: Response to water deficit and nitrogen fertilization at field conditions. *Environmental and Experimental Botany* 60(2):257–267.

Mousa, H. R., and S. M. Abdel-Aziz. 2008. Comparative response of drought tolerant and drought sensitive maize genotypes to water stress. *Australian Journal of Crop Science* 1:31–36.

Moustafa, K., D. Lefebvre-De Vos, A. S. Leprince, A. Savoure, and C. Laurie 'Re. 2008. Analysis of the *Arabidopsis* mitogen-activated protein kinase families: Organ specificity and transcriptional regulation upon water stresses. *Scholarly Research Exchange* 143656:1–12.

Murmu, K., S. Murmu, C. K. Kundu, and P. S. Bera. 2017. Exogenous proline and glycine betaine in plants under stress tolerance. *International Journal of Current Microbiology Applied Sciences* 6(9):901–913.

Nawaz, K., and M. Ashraf. 2010. Exogenous application of glycinebetaine modulates activities of antioxidants in maize plants subjected to salt stress. *Journal of Agronomy and Crop Science* 196(1):28–37.

Nawaz, K., A. Talat, K. Hussain, and A. Majeed. 2010. Induction of salt tolerance in two cultivars of sorghum (*Sorghum bicolor* L.) by exogenous application of proline at seedling stage. *World Applied Sciences Journal* 10(1):93–99.

Nazar, R., S. Umar, N. A. Khan, and O. Sareer. 2015. Salicylic acid supplementation improve photosynthesis and growth in mustard through changes in proline accumulation and ethylene formation under drought stress. *South African Journal of Botany* 98:84–94.

Noreen, S., M. S. Akhter, T. Yaamin, and M. Arfan. 2018. The ameliorative effects of exogenously applied proline on physiological and biochemical parameters of wheat (*Triticum aestivum* L.) crop under copper stress condition. *Journal of Plant Interactions* 13(1):221–230.

Nounjan, N., P. T. Nghia, and P. Theerakulpisut. 2012. Exogenous proline and trehalose promote recovery of rice seedlings from salt-stress and differentially modulate antioxidant enzymes and expression of related genes. *Journal of Plant Physiology* 169(6):596–604.

Nounjan, N., and P. Theerakulpisut. 2012. Effects of exogenous proline and trehalose on physiological responses in rice seedlings during salt-stress and after recovery. *Plant, Soil and Environment* 58(7):309–315.

Nxele, X., A. Klein, and B. K. Ndimba. 2017. Drought and salinity stress alters ROS accumulation, water retention, and osmolyte content in sorghum plants. *South African Journal of Botany* 108:261–266.

Oukarroum, A., S. El Madidi, and R. J. Strasser. 2012. Exogenous glycine betaine and proline play a protective role in heat-stressed barley leaves (*Hordeum vulgare* L.): A chlorophyll a fluorescence study. *Plant Biosystems-An International Journal Dealing with all Aspects of Plant Biology* 146(4):1037–1043.

Parre, E., M. A. Ghars, A. S. Leprince, L. Thiery, D. Lefebvre, M. Bordenave, L. Richard, C. Mazars, C. Abdelly, and A. Savouré. 2007. Calcium signaling via phospholipase C is essential for proline accumulation upon ionic but not nonionic hyperosmotic stresses in *Arabidopsis*. *Plant Physiology* 144(1):503–512.

Patade, V. Y., S. Bhargava, and P. Suprasanna. 2011. Salt and drought tolerance of sugarcane under iso-osmotic salt and water stress: Growth, osmolytes accumulation, and antioxidant defense. *Journal of Plant Interactions* 6(4):275–282.

Per, T. S., N. A. Khan, P. S. Reddy, A. Masood, M. Hasanuzzaman, M. I. Khan, and N. A. Anjum. 2017. Approaches in modulating proline metabolism in plants for salt and drought stress tolerance: Phytohormones, mineral nutrients and transgenics. *Plant Physiology and Biochemistry* 115:126–140.

Pilon-Smits, E. A., N. Terry, T. Sears, et al. 1998. Trehalose-producing transgenic tobacco plants show improved growth performance under drought stress. *Journal of Plant Physiology* 152(4–5):525–532.

Pitzschke, A., A. Djamei, F. Bitton, and H. Hirt. 2009. A major role of the MEKK1–MKK1/2–MPK4 pathway in ROS signalling. *Molecular Plant* 2(1):120–127.

Planchet, E., I. Verdu, J. Delahaie, C. Cukier, C. Girard, M. C. Morère-Le Paven, and A. M. Limami. 2014. Abscisic acid-induced nitric oxide and proline accumulation in independent pathways under water-deficit stress during seedling establishment in *Medicago truncatula*. *Journal of Experimental Botany* 65(8):2161–2170.

Pospisilova, J., D. Haisel, and R. Vankova. 2011. Responses of transgenic tobacco plants with increased proline content to drought and/or heat stress. *American Journal of Plant Science* 2:318e324.

Radomiljac, J., J. Whelan, and M. van der Merwe. 2013. Coordinating metabolite changes with our perception of plant abiotic stress responses: Emerging views revealed by integrative—omic analyses. *Metabolites* 3(3):761–786.

Rady, M. M., and G. F. Mohamed. 2018. Improving salt tolerance in *Triticum aestivum* (L.) plants irrigated with saline water by exogenously applied proline or potassium. *Plants and Agriculture Research* 8(2):193–199.

Rajagopal, V. 1981. The influence of exogenous proline on the stomatal resistance in *Vicia faba*. *Physiologia Plantarum* 52(2):292–296.

Rajendrakumar, C. S. V., T. Suryanarayana, and A. R. Reddy. 1997. DNA helix destabilization by proline and betaine: Possible role in the salinity tolerance process. *FEBS Letters* 410:201–205.

Rastgar, J. F., A. Yamchi, M. Hajirezaei, A. R. Abbasi, and A. A. Karkhane. 2011. Growth assessments of *Nicotiana tabaccum* cv. Xanthitrans formed with *Arabidopsis thaliana* P5CS under salt stress. *African Journal of Biotechnology* 10:8539e8552.

Raven, J. A. 2002. Selection pressures on stomatal evolution. *New Phytologist* 153(3):371–386.

Reddy, P. S., G. Jogeswar, G. K. Rasineni, et al. 2015. Proline over-accumulation alleviates salt stress and protects photosynthetic and antioxidant enzyme activities in transgenic sorghum [*Sorghum bicolor* (L.) Moench]. *Plant Physiology and Biochemistry* 94:104–113.

Rezaei, M. A., J. Jokar, M. Ghorbanli, B. Kaviani, and A. Kharabian-Masouleh. 2012. Morpho physiological improving effects of exogenous glycine betaine on tomato (*Lycopersicum esculentum* Mill.) cv. PS under drought stress conditions. *Plant Omics* 5:79–86.

Rodriguez, M. M., and J. W. Heyser. 1988. Growth inhibition by exogenous proline and its metabolism in saltgrass (*Distichlis spicata*) suspension cultures. *Plant Cell Reports* 7(5):305–308.

Roosens, N. H., T. T. Thu., H. M. Iskandar, and M. Jacobs. 1998. Isolation of the ornithine-d amino transferase cDNA and effect of salt stress on its expression in *Arabidopsis thaliana*. *Plant Physiology* 117:263–271.

Roy, D., N. Basu, A. Bhunia, and S. Banerjee. 1993. Counteraction of exogenous L-proline with NaCl in salt-sensitive cultivar of rice. *Biologia Plantarum* 35:69–72.

Roychoudhury, A., A. Banerjee, and V. Lahiri. 2015. Metabolic and molecular-genetic regulation of proline signaling and its cross-talk with major effectors mediates abiotic stress tolerance in plants. *Turkish Journal of Botany* 39(6):887–910.

Roychoudhury, A., and M. Chakraborty. 2013. Biochemical and molecular basis of varietal difference in plant salt tolerance. *Annual Review and Research in Biology* 3(4):422–454.

Samach, A., H. Onouchi, S. E. Gold, et al. 2000. Distinct roles of CONSTANS target genes in reproductive development of *Arabidopsis*. *Science* 288(5471):1613–1616.

Sanchez, E., L. R. Lopez-Lefebre, P. C. García, R. M. Rivero, J. M. Ruiz, and L. Romero. 2001. Proline metabolism in response to highest nitrogen dosages in green bean plants (*Phaseolus vulgaris* L. cv. Strike). *Journal of Plant Physiology* 158:593e598.

Saradhi, P. P., and P. Mohanty. 1993. Proline in relation to free radical production in seedlings of *Brassica juncea* raised under sodium chloride stress. In: N. J. Barrow (ed.) *Plant Nutrition—From Genetic Engineering to Field Practice*, pp, 731–734. Springer, Dordrecht.

Schat, H., S. S. Sharma, and R. Vooijs. 1997. Heavy metal-induced accumulation of free proline in a metal-tolerant and a non-tolerant ecotype of *Silene vulgaris*. *Physiologia Plantarum* 101:477–482.

Schertl, P., C. Cabassa, K. Saadallah, M. Bordenave, A. Savoure, and H. P. Braun. 2014. Biochemical characterization of proline dehydrogenase in *Arabidopsis* mitochondria. *The FEBBS Journal* 281(12):2794–2804.

Shahid, M. A., R. M. Balal, M. A. Pervez, et al. 2014. Exogenous proline and proline-enriched *Lolium perenne* leaf extract protects against phytotoxic effects of nickel and salinity in *Pisum sativum* by altering polyamine metabolism in leaves. *Turkish Journal of Botany* 38:914–926.

Sharma, A., V. Kumar, R. Singh, A. K. Thukral, and R. Bhardwaj. 2016. Effect of seed pre-soaking with 24-epibrassinolide on growth and photosynthetic parameters of *Brassica juncea* L. in imidacloprid soil. *Ecotoxicology and Environmental Safety* 133:195–201.

Sharma, A., S. Thakur, V. Kumar, A. K. Kesavan, A. K. Thukral, and R. Bhardwaj. 2017. 24-epibrassinolide stimulates imidacloprid detoxification by modulating the gene expression of *Brassica juncea* L. *BMC Plant Biology* 17(1):56.

Sharma, S. S., H. Schat, and R. Vooijs. 1998. In vitro alleviation of heavy metal-induced enzyme inhibition by proline. *Phytochemistry* 49(6):1531–1535.

Shi, S. Q., Z. Shi, Z. P. Jiang, et al. 2010. Effects of exogenous GABA on gene expression of *Caragana intermedia* roots under NaCl stress: Regulatory roles for H_2O_2 and ethylene production. *Plant, Cell and Environment* 33(2):149–162.

Signorelli, S., E. L. Coitiño, O. Borsani, and J. Monza. 2013. Molecular mechanisms for the reaction between •OH radicals and proline: Insights on the role as reactive oxygen species scavenger in plant stress. *The Journal of Physical Chemistry B* 118(1):37–47.

Singh, M., V. P. Singh, G. Dubey, and S. M. Prasad. 2015. Exogenous proline application ameliorates toxic effects of arsenate in *Solanum melongena* L. seedlings. *Ecotoxicology and Environmental Safety* 117:164–173.

Sinha, K., M. Jaggi, B. Raghuram, and N. Tuteja. 2011. Mitogen-activated protein kinase signaling in plants under abiotic stress. *Plant Signaling and Behavior* 6(2):196–203.

Sivakumar, P., P. Sharmila, and P. P. Saradhi. 2000. Proline alleviates salt-stress-induced enhancement in ribulose-1, 5-bisphosphate oxygenase activity. *Biochemical and Biophysical Research Communications* 279(2):512–515.

Sun, Y. L., and S. K. Hong. 2010. Exogenous proline mitigates the detrimental effects of saline and alkaline stresses in *Leymus chinensis* (Trin.). *Journal of Plant Biotechnology* 37(4):529–538.

Surekha, C., K. N. Kumari, L. V. Aruna, G. Suneetha, A. Arundhati, and P. K. Kishor. 2014. Expression of the *Vigna aconitifolia* P5CSF129A gene in transgenic pigeonpea enhances proline accumulation and salt tolerance. *Plant Cell, Tissue and Organ Culture* 116:27e36.

Szabados, L., and A. Savoure. 2010. Proline: A multifunctional amino acid. *Trends in Plant Science* 15(2):89–97.

Székely, G., E. Ábrahám, Á. Cséplő, G. Rigó, L. Zsigmond, J. Csiszár, F. Ayaydin, N. Strizhov, J. Jásik, E. Schmelzer, and C. Koncz. 2008. Duplicated P5CS genes of *Arabidopsis* play distinct roles in stress regulation and developmental control of proline biosynthesis. *The Plant Journal* 53(1):11–28.

Szepesi, Á., and R. Szőllősi. 2018. Mechanism of proline biosynthesis and role of proline metabolism enzymes under environmental stress in plants. In: P. Ahmad, M. A. Ahanger, V. P. Singh, D. K. Tripathi, P. Alam, and M. N. Alyemeni (eds) *Plant Metabolites and Regulation under Environmental Stress*, pp. 337–353. Academic Press.

Talat, A., K. Nawaz, K. Hussian, et al. 2013. Foliar application of proline for salt tolerance of two wheat (*Triticum aestivum* L.) cultivars. *World Applied Science Journal* 22(4):547–554.

Tan, J., C. Wang, B. Xiang, R. Han, and Z. Guo. 2013. Hydrogen peroxide and nitric oxide mediated cold- and dehydration-induced myo-inositol phosphate synthase that confers multiple resistances to abiotic stresses. *Plant, Cell and Environment* 36(2):288–299.

Tester, M., and R. Davenport. 2003. Na+ tolerance and Na+ transport in higher plants. *Annals of Botany* 91(5):503–527.

Ueda, A., W. Shi, K. Sanmiya, M. Shono, and T. Takabe. 2001. Functional analysis of salt-inducible proline transporter of barley roots. *Plant and Cell Physiology* 42(11):1282–1289.

Ueda, A., W. Shi, T. Shimada, H. Miyake, and T. Takabe. 2008. Altered expression of barley proline transporter causes different growth responses in *Arabidopsis. Planta* 227(2):277–286.

Veeranagamallaiah, G., P. Chandraobulreddy, G. Jyothsnakumari, and C. Sudhakar. 2007. Glutamine synthetase expression and pyrroline-5-carboxylate reductase activity influence proline accumulation in two cultivars of foxtail millet (*Setaria italica* L.) with differential salt sensitivity. *Environmental and Experimental Botany* 60:239–244.

Verbruggen, N., and C. Hermans. 2008. Proline accumulation in plants: A review. *Amino Acids* 35(4):753–759.

Verslues, P. E., and S. Sharma. 2010. Proline metabolism and its implications for plant-environment interaction. *The Arabidopsis Book/American Society of Plant Biologists* 8:e0140.

Wani, A. S., A. Ahmad, S. Hayat, and Q. Fariduddin. 2013. Salt-induced modulation in growth, photosynthesis and antioxidant system in two varieties of *Brassica juncea. Saudi Journal of Biological Sciences* 20(2):183–193.

Wu, G. Q., R. J. Feng, S. J. Li, and Y. Y. Du. 2017. Exogenous application of proline alleviates salt-induced toxicity in sainfoin seedlings. *The Journal of Animal and Plant Sciences* 27(1):246–251.

Xiong, J. L., H. C. Wang, X. Y. Tan, C. L. Zhang, and M. S. Naeem. 2018. 5-aminolevulinic acid improves salt tolerance mediated by regulation of tetrapyrrole and proline metabolism in *Brassica napus* L. seedlings under NaCl stress. *Plant Physiology and Biochemistry* 124:88–99.

Xu, J., H. Yin, and X. Li. 2009. Protective effects of proline against cadmium toxicity in micropropagated hyperaccumulator, *Solanum nigrum* L. *Plant Cell Reports* 28(2):325–333.

Yamada, M., H. Morishita, K. Urano, et al. 2005. Effects of free proline accumulation in petunias under drought stress. *Journal of Experimental Botany* 56(417):1975–1981.

Yan, Z., S. Guo, S. Shu, J. Sun, and T. Tezuka. 2011. Effects of proline on photosynthesis, root reactive oxygen species (ROS) metabolism in two melon cultivars (*Cucumis melo* L.) under NaCl stress. *African Journal of Biotechnology* 10(80):18381–18390.

Yang, S. L., S. S. Lan, and M. Gong. 2009. Hydrogen peroxide-induced proline and metabolic pathway of its accumulation in maize seedlings. *Journal of Plant Physiology* 166(15):1694–1699.

Yoshiba, Y., T. Kiyosue, T. Katagiri, et al. 1995. Correlation between the induction of a gene for delta 1-pyrroline-5-carboxylate synthetase and the accumulation of proline in *Arabidopsis thaliana* under osmotic stress. *The Plant Journal* 7:751–760.

Yusuf, M., Q. Fariduddin, T. A. Khan, and S. Hayat. 2017. Epibrassinolide reverses the stress generated by combination of excess aluminum and salt in two wheat cultivars through altered proline metabolism and antioxidants. *South African Journal of Botany* 112:391–398.

Zhan, X., B. Wang, H. Li, R. Liu, R. K. Kalia, J. K. Zhu, and V. Chinnusamy. 2012. *Arabidopsis* proline-rich protein important for development and abiotic stress tolerance is involved in microRNA biogenesis. *Proceedings of the National Academy of Sciences* 109(44):18198–18203.

Zhang, L., and D. Becker. 2015. Connecting proline metabolism and signaling pathways in plant senescence. *Frontiers in Plant Science* 6:552.

Zhang, G. C., W. L. Zhu, J. Y. Gai, Y. L. Zhu, and L. F. Yang. 2015. Enhanced salt tolerance of transgenic vegetable soybeans resulting from overexpression of a novel D1- pyrroline-5-carboxylate synthetase gene from *Solanum torvum.* Swartz. *Horticulture, Environment, and Biotechnology* 56(1):94–104.

Zhang, X., L. Wang, X. Xu, C. Cai, and W. Guo. 2014. Genome-wide identification of mitogen-activated protein kinase gene family in *Gossypium raimondii* and the function of their corresponding orthologs in tetraploid cultivated cotton. *BMC Plant Biology* 14(1):345.

Zheng, J. L., L. Y. Zhao, C. W. Wu, B. Shen, and A. Y. Zhu. 2015. Exogenous proline reduces NaCl-induced damage by mediating ionic and osmotic adjustment and enhancing antioxidant defense in *Eurya emarginata. Acta Physiologiae Plantarum* 37(9):181.

Zouari, M., C. B. Ahmed, W. Zorrig, et al. 2016. Exogenous proline mediates alleviation of cadmium stress by promoting photosynthetic activity, water status and antioxidative enzymes activities of young date palm (*Phoenix dactylifera* L.). *Ecotoxicology and Environmental Safety* 128:100–108.

Zouari, M., N. Elloumi, P. Labrousse, B. B. Rouina, F. B. Abdallah, and C. B. Ahmed. 2018. Olive trees response to lead stress: Exogenous proline provided better tolerance than glycine betaine. *South African Journal of Botany* 118:158–165.

11 Networking by Small Molecule Hormones during Drought Stress in Plants

Riddhi Datta, Salman Sahid, and Soumitra Paul

CONTENTS

11.1 INTRODUCTION

Drought is considered one of the most threatening abiotic stress factors. Cultivation in rain-fed land requires rainfall, inadequacy of which causes drought. Recent climate changes like insufficient rainfall and improper distribution of water during monsoon season leads to water scarcity which aggravates drought conditions and ultimately impedes crop yield. Worldwide, the percentage of drought-affected land is increasing day by day. For example, the cultivation of rice is being increasingly threatened by water scarcity since rice cultivation requires huge amounts of water. In Asia, rain-fed lowland systems for rice cultivation cover around 45 billion hectares which is roughly 30% of the total rice-cultivated area worldwide (Haefele and Bouman, 2009). In India, rice-cultivable rain-fed lowland constitutes a total area of 7.3 million hectares. In addition, 80–85% of the rice cultivars in India are considered as irrigated or rain-fed lowland cultivars and have been estimated to be drought-prone (Rao et al. 2015).

Being sessile, plants can't avoid the water scarcity by moving from the drought-prone region to a well-watered area. Instead, plants have evolved several defence mechanisms to avoid or tolerate drought stress that are reflected by their morphological, biochemical and physiological alterations. Improved water use efficiency leads to drought tolerance. Under normal conditions, plants can maintain their growth by coordinating among different physiological phenomena. However, complex signalling cascades during drought stress critically regulate a variety of metabolic processes to minimize the negative effects of drought, thus maintaining the cellular homeostasis. It is, therefore, necessary to understand the mechanism of plant response to drought stress which will ultimately help us improve crop yield under water scarcity. Different phytohormones can regulate drought stress responses in plants by modulating the expression of different signalling molecules. ABA, the major abiotic stress-related hormone, regulates drought stress responses via the SnRK2 mediated MAP kinase pathway. ABA also induces stomatal closure by generating oxidative burst and elevating Ca^{2+} ions (Drerup et al. 2013; Li et al. 2017a). Ethylene, a gaseous hormone in plants, also induces drought stress tolerance by regulating various ERF proteins in plants (Gu et al. 2017). A crosstalk between ABA and ET signalling also determines the drought stress tolerance in plants. Jasmonic acid also plays a significant role in drought stress tolerance by augmenting the osmoprotectant biosynthesis like glycine betaine in plants. JA interacts antagonistically or synergistically with ET and ABA to modulate drought stress tolerance (Gao et al. 2004). Other plant growth regulators like salicylic acid and brassinosteroid can regulate drought stress responses by modulating ROS signalling in plants (Fragniere et al. 2011; Ye et al. 2017)

In this chapter, we will emphasize how the altered phenotypic and physiological characteristics are linked to the different signalling pathways during drought stress with special reference to phytohormone signalling. We will first highlight the different morphological and physiological changes that occur in response to drought stress. Next, we will focus on the signal transduction pathways of important phytohormones which function as sophisticated adaptive strategies for drought tolerance *in planta*. We will also highlight the intricate crosstalks between these signalling pathways and other signalling molecules that enable a plant to fine-tune its defence response under water deficit conditions. Finally, we will delineate various approaches for generating drought-tolerant crop plants for survival during drought.

11.2 DROUGHT INDUCES PHENOTYPIC AND PHYSIOLOGICAL ALTERATIONS

Drought tolerance and/or avoidance are important adaptive responses in plants which trigger alterations in several morphological features and physiological processes as an attempt to survive under water scarcity. Plants sense drought stress either when the root faces water deprivation in soil or when transpiration rates are too high. Under drought stress, roots are first sensitized, which then send the appropriate signal to other tissues for stress response. Several root characteristics are altered in response to drought. The morphological alteration of roots includes changes in lignification, aerenchyma formation and enlargement of metaxylem (Redillas et al. 2012; Kulkarni et al. 2014; Kadam et al. 2015; Lee et al. 2016; Prince et al. 2017). A recent study revealed that drought tolerant genotypes exhibit a lower bleeding rate and narrow xylem diameter under stress (Henry et al. 2012). An elevated amount of root secondary metabolites is directly related to the degree of drought tolerance in plants. For example, lignin deposition in the cortex and hypodermal suberization in rice and wheat roots can be correlated with enhanced drought tolerance (Krishnamurthy et al. 2009; Shiono et al. 2011; Naseer et al. 2012; Geldner et al. 2013; Emerson et al. 2014; Weidje et al. 2017). Root length and root dry mass are also altered in response to drought (Redillas et al. 2012; Kadam et al. 2015; Dali et al. 2018). Extensive studies on roots under osmotic stress revealed several drought-tolerance responses that are thought to be associated with improved crop productivity. Deep and increased number of lateral roots and a higher root-to-shoot ratio indicate a drought-avoidance response in plants (Samson et al. 2002). These roots can penetrate deeper into the soil to accumulate water and improve drought avoidance (Lee et al. 2016; Dali et al. 2018).

Drought stress also severely affects growth and development in plants, thus hampering biomass production. During drought, plants slow down or reduce growth. The impairment of root growth, reduction of leaf surface area, wilting of leaves and delay of leaf senescence are very common when plants are exposed to drought (Blum 2011; Usman et al. 2013). Water scarcity in soil decreases the turgor pressure of the cell, leading to cessation of cell growth. Drought also impedes cell elongation as well as enlargement (Weijde et al. 2017; Dali et al. 2018). It inhibits the germination of seedlings in a number of crop plants (Jiang and Lafitte 2007), and reduces the number of tillers (Ashfaq et al. 2012; Kadam et al. 2015) as well as plant height (Ashfaq et al. 2012; Kadam et al. 2015; Lee et al. 2016). Hyponasty or leaf rolling is an important adaptive response in plants that helps to minimize transpiration rate and light perception and maintain leaf water balance (Kadioglu and Terzi 2007). The number of leaves and leaf area are also found to be reduced in different plant species in response to osmotic stress (Singh 2014; Kumar et al. 2014a; Kadam et al.2015).

Drought not only changes the morphology but also affects several physiological and biochemical parameters, such as water use efficiency, transpiration and photosynthesis rate and stomata opening and closing (Ji et al. 2012, Singh et al. 2012; Xiong et al. 2014; Li et al. 2017b) as well as the membrane stability index (Kumar et al. 2014a; Nahar et al. 2016). Plants maintain turgor pressure by reducing the solute potential as a stress-tolerance strategy (Yang et al. 2014; Nahar et al. 2016). In rice, drought severely affects PSII activity in flag leaves (Pieters and Ei Souki 2005). It degrades the D1 protein which inactivates the function of PSII in response to drought. Osmotic stress also perturbs the photosynthetic rate by preventing Rubisco activity (Bota et al. 2004). Activation of Rubisco is suppressed in response to drought (Galmes et al. 2011). Recent studies have revealed that the overexpression of C_4 enzymes like pyruvate orthophosphate kinase and phosphoenol pyruvate carboxylase showed enhanced drought tolerance in transgenic rice, suggesting that C_4 plants are more drought-tolerant than C_3 plants (Yoshimura et al. 2008; Gu et al. 2013; Qian et al. 2015; Liu et al. 2016).

Plants accumulate different osmoprotectants like glycine betaine, proline, sugar alcohol and soluble sugars in order to improve water absorption and transport when they are grown on dry soil (Shehab et al. 2010; Farahmand et al. 2014; Lee et al. 2016). These compounds decrease the solute potential thereby maintaining the turgidity of the cell (Rhodes and Samaras 1994). Reactive oxygen species (ROS) homeostasis is also considered a part of the adaptive stress response in plants. ROS can be categorized into several groups such as hydrogen peroxide, hydroxyl free radical, superoxide radical and singlet oxygen. They can cause a variety of oxidative damage in the cell by the degradation of nucleic acids and proteins as well as lipid peroxidation in membranes, thus hampering cellular homeostasis. Plant cells synthesize different ROS scavenging enzymes like superoxide dismutase (SOD), catalase (CAT), polyphenol oxidase, ascorbate peroxidase (APX), dehydroascorbate reductase (DHAR) and glutathione reductase (GR) in order to protect the cells from the harmful effect of ROS (Noctor and Foyer 1998; Nahar et al. 2016; Chen and Flurh 2018).

Emerging evidence suggests that polyamines like putrescine and spermidine (Spd) are accumulated in higher amounts in the cell during drought stress in order to maintain membrane integrity and osmotic balance and regulate drought stress by interacting with different signalling components (Nambeesan et al. 2012; Calzadilla et al. 2014; Zhou et al. 2018). The alterations in sugar metabolism, reduced and oxidied glutathione ratio and ascorbate content have also been considered as adaptive parameters under drought (Edgerton 2009; Sarker and Oba 2018).

11.3 ROLE OF PHYTOHORMONES IN DROUGHT RESPONSE

Phytohormones are essential for plants to adapt under various stress conditions. They often rapidly alter gene expression by inducing or repressing different transcriptional regulators, thereby switching between the 'normal growing mode' and the 'defence mode'. Several phytohormone biosynthesis depleted mutants confirm that these phytohormones are instrumental in regulating the plant responses under changing environmental conditions. However, hormones do not act in isolation but are interconnected by synergistic and/or antagonistic crosstalks with other signalling pathways, thus modulating the plant's response according to the type of stress encountered.

11.3.1 ABA Signalling

Abscisic acid (ABA), a sesquiterpenoid compound, regulates different developmental phenomena like vegetative growth, seed development and sprouting (Xiong and Zhu 2003). During thermal and osmotic stress, the higher abundance of ABA in the plant cell suggests its protective role against thermal and dehydration stresses (Parent et al. 2009). ABA has been considered an essential growth regulator that critically regulates stomatal opening and closing and the growth of roots and shoots during osmotic stress and communicates between roots and shoots by interacting with different signalling molecules (external and internal) during water scarcity (Vishwakarma et al. 2017). ABA also inhibits seed germination and induces dormancy and leaf senescence. It has also been reported that ABA is synthesized in guard cells and vascular tissues (Bauer et al. 2013). Despite its role in plant development, the robust stress-responsive nature of ABA has been widely reported when plants encounter different environmental stresses like cold, heat, drought, salinity, etc.

11.3.1.1 ABA Signalling by PYL-ABA-PP2C Complex

A major breakthrough in ABA research has been the progress in understanding the ABA perception and the downstream signalling machinery that operates during osmotic stress (Umezawa et al. 2010; Nakashima and Yamaguchi-Shinozaki 2013). A major plasma membrane ABA receptor includes the regulatory components of ABA receptor (RCAR)/pyrabactin resistancel (PYR)/PYR1 like (PYL) protein which is categorized under a ligand binding START domain family of protein. The *Arabidopsis* genome encodes 14 highly conserved PYR/PYL/PCAR genes of which the PYL1, PYR1 and PYL2 can directly interact with ABA. Ma et al (2009) suggested that the overexpression of PYL5, PYL8 and PYL9 in *Arabidopsis* increased ABA sensitivity and conferred drought resistance. He et al. (2018) also reported 13 members of the PYR/PYL/PCAR family in maize. The overexpression of *ZmPYL8*, *ZmPYL9* and *ZmPYL12* increased tolerance to drought stress and induced the accumulation of more proline in drought-tolerant plants. The overexpression of *Os*PYL3 in *Arabidopsis* also increased ABA sensitivity and enhanced drought tolerance (Lenka et al. 2018). The PYR/PYL/PCAR family receptors bind to ABA in the presence of co-receptors like group A protein phosphatase 2C (PP2C) (Ma et al. 2009; Park et al. 2009). Out of 9 members of group A PP2C in *Arabidopsis*, only six members are strongly induced in response to ABA signalling. The group A PP2Cs consist of ABA insensitive 1 (ABI1), ABI2, hypersensitive to ABA 1 (HAB1) and PP2CA (Razavizadeh et al.2018). When the plants are exposed to osmotic stress, accumulated ABA binds to the central hydrophobic moiety of the PYL receptor, leading to a conformational change which then induces the binding of PP2Cs to form a stable close PYL-ABA-PP2C complex (Nishimura et al. 2007; Singh et al. 2015). The protein phosphatase activity of PP2C is inhibited by the ABA-PYL complex, and this activates SNF1 related protein kinase 2 (SnRK2) including SnRK2.2, SnRK2.3 and SnRK2.6 (OST1) by autophosphorylation (Melcher et al. 2009; Soon et al. 2012). The autophosphorylation of SnRK2s switches on the downstream mitogen-activated protein kinase (MAPK) signalling cascade (Li et al. 2017a) (Figure 11.1). In the absence of ABA, PP2Cs remain associated with SnRK2s and inactivate the kinase by dephosphorylating the activation loop (Soon et al. 2012; Umezawa et al. 2013). Another study showed that group A PP2Cs are also involved in desiccation tolerance in *Physcomitrella patens*, confirming the involvement of ABA-mediated signalling during drought stress in moss (Komatsu et al. 2013).

Among the other members of the PYLs, PYL4 is known to be involved in regulating the osmotic stress response in *Arabidopsis* while PYL8 regulates auxin-mediated lateral root growth in an ABA-independent manner (Pizzio et al. 2013; Zhao et al. 2014). PYL6 interacts with MYC2, an important basic helix–loop–helix (bHLH) transcription factor (TF) that controls jasmonic acid (JA) metabolism, connecting a major link between the ABA and JA signalling pathways during abiotic stress response (Aleman et al. 2016). The SnRK2 kinase family consists of a number of drought-responsive plant-specific serine threonine protein kinases (Saruhashi et al. 2015). AAPK is the first reported protein kinase which showed ABA-mediated stomatal closure in *Vicia faba* (Li et al.

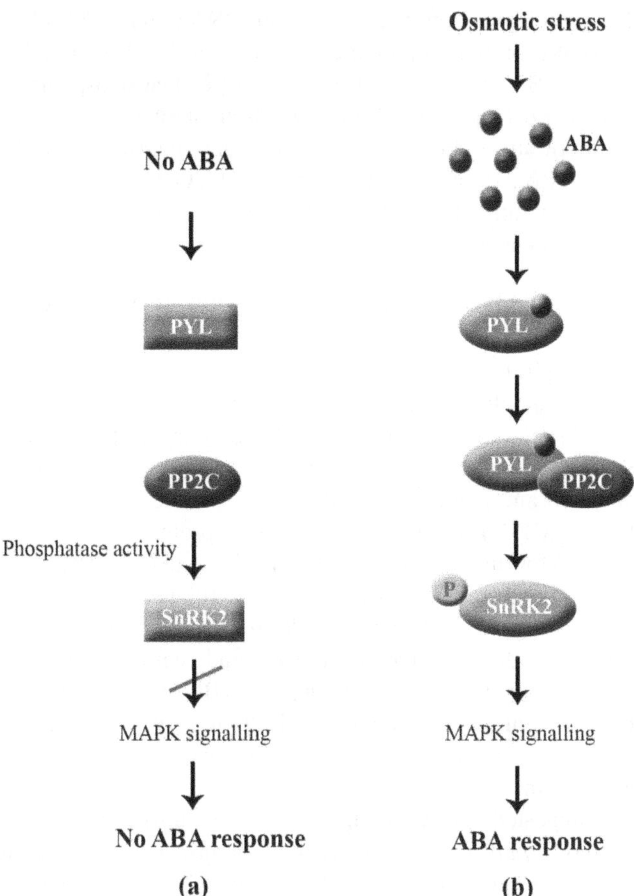

FIGURE 11.1 ABA-mediated activation of the MAPK pathway during water stress. (a) When ABA is absent, PP2C, co-receptor of ABA, dephosphorylates SnRK2 and blocks the downstream MAPK signalling pathway. (b) During osmotic stress, ABA binds to its receptor PYL and co-receptor PP2C which activates downstream SnRK2 by phosphorylation leading to the induction of the MAPK pathway for ABA-mediated response.

2000a). Subsequently, SnRK1, 2 and 6, the orthologs of AAPK in *Arabidopsis*, also regulate ABA-mediated stomatal closure (Yoshida et al. 2002). In *Arabidopsis*, out of ten members of the SnRK2 family, proteins SnRK2.1, SnRK2.2, SnRK2.3, SnRK2.6, SnRK2.7 and SnRK2.8 are activated by ABA and drought stress. Only SnRK2.9 has been found to be inactive under drought stress (Furihata et al. 2006). Boudsocq et al. (2004) further reported that only SnRK2.2, SnRK2.3 and SnRK2.6 are directly activated by ABA and play a distinct role in drought stress. The ABA-responsive element binding factor (AREB/ABF) has been found to be phosphorylated by ABA-activated SnRK2 which, in turn, activates the downstream ABA responsive genes under stress conditions (Furihata et al. 2006).

The 5′ *cis*-acting regulatory regions of ABA-responsive genes contain several conserved motifs of ABA-responsive element (ABRE; PyACGTGG/TC) which triggers gene expression upon ABA perception. It has been demonstrated that the promoter regions of most ABA-dependent genes rely on more than one ABRE or a combination of ABRE or/and a cold-responsive element (CRE) (Fujita et al. 2011, 2013; Nakashima and Yamaguchi-Shinozaki 2013). ABA-responsive element binding factors like ABA/AREB/ABF are basic leucine zipper protein (bZIP) TFs that are upregulated by ABA and water scarcity. They in turn switch on a downstream signalling cascade to modulate different physiological phenomena, leading to stress tolerance (Yoshida et al. 2015). Accumulating evidence suggests that ABRE interacts with W-box and NAM/ATAF/CUC 13 (NAC13) protein leading to the regulation of root morphology under water scarcity (Rui et al. 2018). The transcriptional activities of AREB/ABF are controlled by ABA-dependent phosphorylation during abiotic stress. The overexpression of *AREB1* showed enhanced drought tolerance in transgenic *Arabidopsis*, suggesting its role in osmotic stress (Fujita et al. 2005). Transgenic rice and soybean overexpressing *AREB1* also exhibited improved drought tolerance (Oh et al. 2005; Barbosa et al. 2013). Recently it has been shown that the overexpression of *TaAREB*3 enhanced ABA sensitivity and drought resistance in *Arabidopsis* plants (Wang et al. 2016).

11.3.1.2 ABA Signalling by Oxidative Burst and Ca^{2+} Ion Accumulation

During stomatal closure, ABA-activated SnRK2 also induces phosphorylation of the plasma membrane NADPH oxidase, respiratory burst of oxidase homolog F (RbohF), generating O_2^- and subsequently H_2O_2 within the apoplastic region of guard cells. The apoplastic ROS then enters into the cell via plasma membrane intrinsic protein 2.1 (PIP2.1) and activates two signalling components, guard cell hydrogen peroxide-resistant 1 (GHR1) and S-type anion channel (SLAC1), which aggravate the plasma membrane calcium (Ca^{2+}) channels. The elevation of Ca^{2+} activates calcium-dependent protein kinase (CPKs) and calcineurin B-like protein1/9 (CBL1/9)–CBL-interacting protein kinase 26 (CIPK26) for further phosphorylation of RbohF (Drerup et al. 2013) (Figure 11.2). In addition, the elevated amount of Ca^{2+} ions in cells upregulates ABA biosynthesis-related gene encoding 9-cis epoxy carotenoid dioxygenase (NCED), thus augmenting ABA biosynthesis. Accumulated ABA further induces different channels and regulates redox homeostasis during drought stress responses such as stomatal closure in plants (Barrero et al. 2006). On the other hand, a member of the calmodulin family of proteins, CML 20, negatively regulates ABA signalling in the guard cell, thus suggesting both positive and negative crosstalk between Ca^{2+} and ABA signalling in plants (Wu et al. 2017a)

Recent studies have suggested that ABA can also induce a novel group of lectin protein family members like *Euonymous eupatoreous* lectin (EUL) proteins, which play a vital role in osmotic stress tolerance in plants (De Schutter et al. 2017). The EUL proteins can be categorized into two groups: EULS and EULD. Several members of this family, like EULD1A and EULS2/S3,

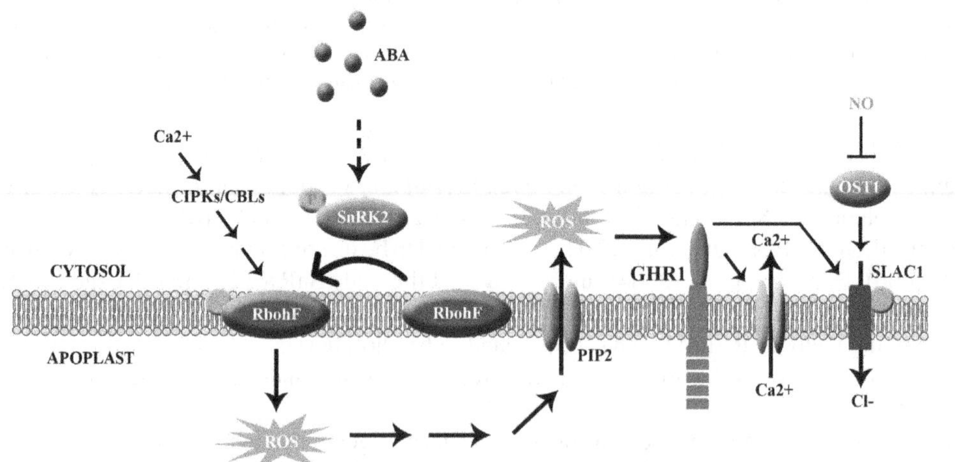

FIGURE 11.2 ABA-mediated Ca^{2+} channel activation and ROS generation. In the presence of ABA, phosphorylated SnRK2 activates a membrane channel, RbohF, by phosphorylation. This generates ROS in the apoplast. ROS, extruded out through the PIP2 membrane channel, induces the Ca^{2+out} channel, GHR1 and SLAC1 ion channel. The higher accumulation of Ca^{2+} activates downstream Ca^{2+} signalling proteins like CBL1/CIPKs to further activate the RbohF protein for ROS production. OST1, another member of the SnRK family, activates the SLAC1 protein but is inhibited by NO.

are reported to be regulated by ABA under osmotic stress (VanDamme et al. 2017). However, the detailed mechanism of ABA-mediated regulation of EUL proteins under drought stress remains unexplored to date.

11.3.2 Ethylene Signalling

Ethylene (ET), a biologically active gaseous unsaturated hydrocarbon and the simplest phytohormone, regulates different metabolic and developmental processes in plants. It regulates diverse physiological phenomena starting from seed germination to organ senescence. ET also plays a vital role in stress response and adaptation under different biotic and abiotic stress conditions (Rejeb et al. 2014). In plants, ET is synthesized in a two-step process where S-adenosyl-L-methionine is first converted to 1-aminocyclopropane-1-carboxylic acid (ACC) by ACC synthase (ACS). ACS is also considered a rate-limiting enzyme of the ET biosynthesis pathway. Finally, ET is synthesized from ACC by the enzyme ACC oxidase (ACO). Different environmental factors and endogenous signals regulate ET biosynthesis through differential expression of *ACS* and *ACO* genes (Liu et al. 2015). Under water scarcity, ACS is activated while ACC is supressed, resulting in an enhancement of ET production. This finding suggests drought-mediated activation of *ACS* gene and ET production as well. In addition, ET signalling is important for activating plant responses to flooding and water deficit (Voesenek and Bailey 2009).

ET interacts with its membrane receptor ethylene receptor 1 (ETR1), and this deactivates the copper transport protein 1 (CTR1). In the absence of ET, ETR1 remains active and it maintains CTR1 in active form. CTR1 is a negative regulator of ET signalling. In the active state, it phosphorylates and inactivates the downstream signalling proteins, like ethylene insensitive protein 2 (EIN2). This terminates the signalling cascade by the degradation of proteins like ethylene insensitive 3 (EIN3) and EIN3-like (EIL). Two EIN2-targeting proteins, EIN2-targeting protein1 (ETP1) and ETP2, interact with the C-terminal domain of EIN2 and lead to its proteolysis (Qiao et al. 2009). However, in the presence of ET, ETR1 and CTR1 are inactivated which relieves the repression of EIN2. The C-terminal domain of EIN2 is now dephosphorylated, cleaved and transported into the nucleus where it activates EIN3/EIL expression by repressing EIN3 binding F-box protein 1/2 (EBF1/2). Further, EIN3/EIL promotes the transcription of the apetala 2 (AP2)/ethylene responsive factor (ERF) (Gu et al. 2017).

Earlier studies also suggested an EIN3/EIL activation via mitogen-activated protein kinase kinase 4/5/9 (MKK4/5/9) MPK3/6 phosphorylation cascade which functions independently of the EIN2 and CTR-mediated response (Yoo et al. 2008; Hahn and Harter 2009; Li et al. 2017). In the presence of an external stimulus, MPK3/6 phosphorylates the EIN3/EIL transcription factors and reduces their interaction with the F-box protein EBF, inhibiting their degradation through the 26S proteasome (Stepanova and Alonso 2009). In addition, EIN5 shows exoribonuclease activity, decreasing EBF mRNA accumulation and thus positively regulating EIN3/EIL expression in the nucleus (Olmedo et al. 2006).

The major TFs targeted by EIN3 have been classified as ERFs. AP2/ERF family proteins characteristically possess a highly conserved AP2/ERF DNA-binding domain (Song et al. 2011; Rashid et al. 2012). This family has been categorized into four sub-groups based on the number and similarity of AP2/ERF domains. These are AP2, dehydrin responsive element binding protein (DREB), related to ABI3/VP1 (RAV) and ERF (Sharoni et al. 2011; Rashid et al. 2012). EIN3 can interact with the specific ERFs regulating different biotic and abiotic stress tolerance activities in plants. The members of the ERF family bind to the specific 5′ *cis*-acting regulatory regions which determine the downstream stress responsive pathway. For example, EIN3 interacts with ERF1, which then promotes a specific set of signalling genes by binding to the *cis*-acting regions, *GCC* box and *DRE* elements, depending on the nature of the stress encountered (Figure 11.3). ERF1 has been reported to regulate a number of osmotic stress tolerance genes like *pyrroline-5-carboxylate-synthase 1* (*P5CS1*), *early response to dehydration 7* (*ERD7*), *germin like protein 9* (*GLP9*), *similar to RCD*

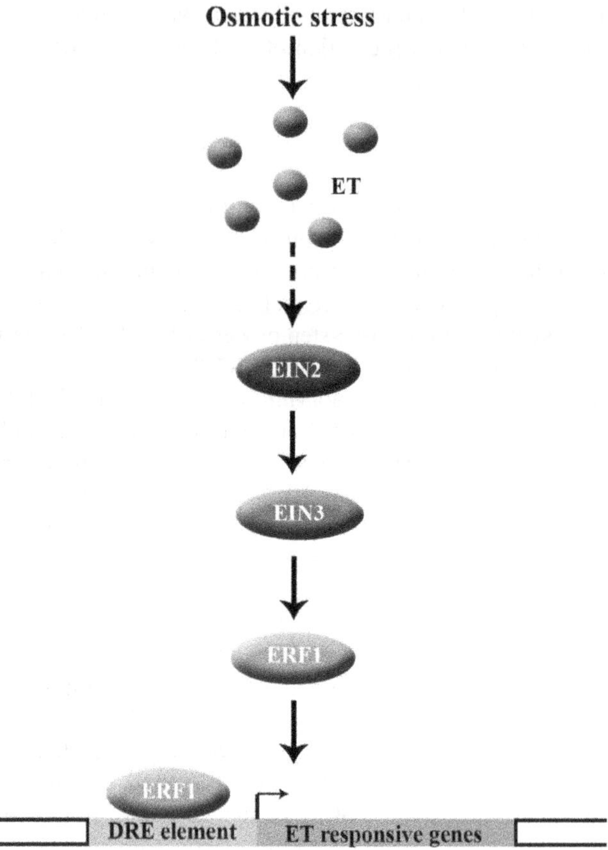

FIGURE 11.3 ET-mediated signalling during water stress. During osmotic stress, ET binds to its receptor ETR1 and then activates EIN2. Activated EIN2 moves into the nucleus and activates EIN3. EIN3 activates ERF1which then binds to DRE box to induce ET-responsive gene expression.

one 5 (*SRO5*), *response to desiccation 29A* (*RD29A*) and *RD20* (Muller and Bosch. 2015). The overexpression of the *Arabidopsis HARDY* gene (a subgroup of A-4AP2/ERF family) led to better osmotic stress tolerance in *Trifolium* (Abogadallah 2010). *OsDERF1* acts as a negative regulator during osmotic stress by interacting with the *GCC* box present in the *OsERF3* promoter (Wan et al. 2011). Moreover, *OsERF3* bears an ERF-associated amphiphilic repression (EAR) motif, which transcriptionally represses the ET production thus reducing drought tolerance. Mutant analysis of EAR lines revealed an increased ET production and drought tolerance as compared to overexpressing lines and wild-type plants (Zhang et al. 2013). However, Joo et al. (2013) reported that *OsERF4a* is a positive regulator which maintains plant growth under drought. In addition, tomato *JERF3* has also been found to render drought tolerance in rice by increasing soluble sugars and proline content (Zhang et al. 2010). Recent studies suggested the role of *OsERF71* in altering the root architecture adaptation under drought stress (Lee et al. 2016, 2017). Transgenic rice plants overexpressing *OsLG3* (an ERF family member) also displayed improved drought tolerance activity, confirming the role of ERF family members and their regulation by ET signalling during drought stress (Xiong et al. 2018).

11.3.3 JA Signalling

JA is gradually gaining importance as a major regulator of various abiotic stress responses in plants (Kazan 2015; Riemann et al. 2015). When plants are exposed to salt or dehydration stress, some of the JA biosynthesis genes like *allene oxide cyclase* 1 (*AOC*1), *AOC*2, *lipoxigenase 3* (*LOX*3) and *12-oxo phytodienoate 3* (*OPR*3) are upregulated, and the resulting JA induction can modulate root and shoot growth in plants (Kazan 2015). In response to drought stress, JA can induce the accumulation of glycine betaine, an osmoprotectant, through upregulation of its major biosynthesis-related gene *betaine aldehyde dehydrogenase* (*BADH*) (Gao et al. 2004). This suggests that JA plays a fundamental role in osmotic stress tolerance in plants by elevating the amount of osmoprotectants (Ma et al. 2006; Jiang and Deyholos 2009; Geng et al. 2013). JA-mediated signalling is regulated by the Skp1/cullin/F box protein complex (SCF) family-mediated degradation of the jasmonate zim domain (JAZ) repressor in the presence of biologically active JA compounds (JA-Isoleucine) (Chini et al. 2007; Fonseca et al. 2009). JA-isoleucine binds to the F-box protein coronatine insensitive 1 (COI1) in the SCFCOI1 complex. This leads to ubiquitination of JAZ proteins which are then degraded in the 26S proteasome. The JAZ protein interacts with bHLH TFs like MYC2, MYC3, MYC4 and MYC5 to suppress the expression of early JA-responsive genes (Chini et al. 2007; Cheng et al. 2011; Niu et al. 2011). Once JAZ is degraded, the MYC repression is relieved and switches on the transcription of downstream JA-responsive genes like *vegetative storage protein 2* (*VSP2*). JAZ degradation also activates ERF1/ORA59 proteins which in their turn induce the expression of other JA-responsive genes like *PDF1.2*.

JA can act synergistically or antagonistically with ET and ABA to impart osmotic stress tolerance activity to plants (Cheng et al. 2013). ET and JA signalling are essential in regulating the ERF proteins that play a crucial role in mitigating a number of abiotic stress responses in plants. On the other hand, the loss of functions of MYC2, an activator of the JA signalling pathway, shows less sensitivity to JA response. The *myc2* mutant plant accumulates an elevated amount of ABA, thus following the ABA-mediated response during dehydration stress. The role of MYC2 during drought stress establishes an interesting crosstalk between the ABA and JA signalling pathways in plants (Kazan and Manners 2013).

11.3.4 Salicylic Acid Signalling

Salicylic acid (SA) plays an important role in the regulation of plant growth and development, such as in plasmodesmata closure, seed germination, flowering and thermogenesis. In addition, SA has been widely reported to play a central role in imparting resistance against biotic and abiotic stresses in plants (Fragniere et al. 2011). SA has been identified as the first compound to exhibit stress tolerance activity against a wide range of abiotic stresses like drought, salinity, chilling and heat as well as heavy metal exposure (Yuan and Lin 2008; Fragniere et al. 2011). SA is synthesized by the following two pathways. In one pathway, plants produce SA from chorismate via phenylalanine, t-cinnamic acid and benzoic acid with the help of the key biosynthetic enzyme, phenylalanine ammonia lyase (PAL) (Rohde et al. 2004). The second pathway in plants is mainly confined to chorismate production from isochorismate through isochorismate synthase (ICS) in *Arabidopsis* (Wildermuth et al. 2001). SA interacts with other phytohormones, such as JA, ET, auxin, gibberellic acid (GA) and ABA, and participates in an interesting signalling crosstalk in plants (Dempsey et al. 1999, 2011; Pieterse et al. 2012).

Recent studies have revealed that transgenic plants accumulating higher SA content induces abiotic stress tolerance in plants (Rejeb et al. 2015). SA accumulated during water scarcity also maintains cellular turgidity and triggers several antioxidants similar to the ABA response in plants. For example, the SA levels in the leaves of *Phillyrea angustifolia* and roots of barley were found to be increased by five- and two-fold respectively during drought (Munne-Bosch and Penuelas 2003; Bandurska and Stroinski 2005). The levels of endogenous SA decreased significantly during modest drought stress but increased significantly after long-time re-watering. The mutants incapable of endogenous SA accumulation, like *adr1*, *myb96-1d*, *siz1*, *acd6* and *cpr5*, established that SA-dependent signalling is essential for disease resistance and drought tolerance (Bowling et al. 1997; Chini et al. 2004; Seo and Park 2010).

The application of exogenous SA can improve drought tolerance in plants, including increased antioxidant enzyme activities and decreased lipid peroxidation levels and membrane damage as well as protection of nitrate reductase activity under drought conditions (Hayat et al. 2010). However, extensive studies have revealed that the improvement of drought tolerance depends on the concentration of SA applied. For example, low concentrations of SA can increase drought tolerance while high concentrations lead to drought susceptibility. SA is involved in producing ROS in photosynthetic tissues, leading to high levels of oxidative stress which results in decreased tolerance to abiotic stress (Borsani et al. 2001). When wheat seeds were primed with 100 ppm acetyl SA, the plants showed better resistance in response to drought stress. However, plant growth and drought tolerance were suppressed when 2–3 mM SA was applied to wheat seedlings (Hamada and Al-Hakimi 2001). The application of 0.1–1 mM acetyl SA enhanced the drought tolerance in muskmelon seedlings (Korkmaz et al. 2007). MeSA application has been shown to induce drought-induced leaf senescence in *Salvia officinalis* (Abreu and Munne-Bosch 2008). The priming of barley seeds with SA reduced the damaging effect of water deficits on the cell membranes in the leaves. Interestingly, ABA levels have also been found to be increased in barley leaves after pre-treatment with SA, suggesting the crosstalk of SA and ABA during osmotic stress (Bandurska and Stroinski 2005). The pre-treatment with SA leads to the accumulation of different stress tolerance-related proteins like APX, glutathione S-transferases and 2-cysteine peroxiredoxin in plants when the plants are exposed to drought stress (Kang et al. 2012). The results suggested that the priming of plants with SA enhanced the antioxidant defence system to protect against the oxidative damage caused by drought stress. SA treatment was also thought to protect nitrate reductase activity and thus maintain the protein levels in drought-affected plants.

Drought stress generally induces a higher accumulation of SA in the plant cell, resulting in upregulation of two common SA-related genes, *pathogenesis related protein 1* (*PR1*) and *PR2*. An elevated amount of SA in the cell elicits Ca^{2+} vacuole exporter and the higher accumulation of Ca^{2+} ion thus activates Rboh-mediated redox signalling pathways (Wu et al.2012). Glutathione (GSH), an important signalling intermediate, also plays a crucial role in regulating the SA signalling pathway during drought stress responses in plants (Ghanta et al. 2014; Kumar et al. 2014b). SA binds to the non-expressor of PR 1 (NPR1) receptor, leading to its conformational change, and facilitates downstream signalling events (Wu et al. 2012). The monomeric NPR1 enters the nucleus and activates the transcriptional machinery of different osmotic stress-related genes. SA also interacts with NPR3 and NPR4 to enhance proteasomal degradation of NPR1 which maintains the optimum NPR1 level in the cell. This represents a mechanism of SA perception that determines the balance between cell death and survival (Fu et al. 2012). The changes in the cellular redox state during SA-mediated regulation under dehydration stress can induce monomerization of NPR1. It has been reported that MPK6, a member of the MAP kinase pathway, induces transcriptional activation of WRKY6 and Trxh5 which, in turn, promotes senescence in exogenous SA-treated leaves (Chai et al. 2014).

11.4 OTHERS

11.4.1 Brassinosteroid Signalling

Brassinosteroids (BRs) are a group of polyhydroxy steroidal hormones that play a pivotal role in a wide range of developmental phenomena. These include cell division and cell elongation in the stem and root, photomorphogenesis, reproductive development and leaf senescence as well as various stress responses (Sharma et al. 2013a). The BR signalling starts with the interaction of BR with a leucine-rich repeat receptor kinase brassinosteroid insensitive 1 (BRI1) and a co-receptor BRI1-associated receptor kinase 1 (BAK1), followed by a series of phosphorylation and dephosphorylation events that stimulate the various gene expressions, thus regulating varied physiological functions (Sharma et al. 2013a, b; Nakamura et al. 2017). A number of genes have been reported to be involved in cell division and cell elongation that are direct targets of BR-regulated

TF, brassinozole resistance 1(BZR1/BES1) (Sun et al. 2010; Hacham et al. 2011). BR also regulates cell cycle progression through the modulation of different TFs under stress conditions. BR interacts with R2R3 type MYB TF, MULTIPASS (MPS), which is induced under salinity stress and negatively regulates plant growth by reducing cell size (Schmidt et al. 2013). Another R2R3-MYB TF, brassinosteroids at vascular and organizing centre (BRAVO), interacts with BES1 to alter the plant stem cell quiescent centre (QC) in the primary root of *Arabidopsis* and neutralizes BR-facilitated cell division in the QC cells (Chaiwanon and Wang 2015). It has been found that BR-activated BES1 physically interacts with BRAVO to decrease its activity. This leads to regulation of the division of QC of the root stem cell (Vilarrasa et al. 2014). In addition to its primary role of cell division, BR also plays a significant role in organ differentiation and proliferation in plants (González et al. 2011; Zhiponova et al. 2013).

The BR-induced drought stress response is commonly associated with an enhanced accumulation of ROS (Jiang et al. 2012). BRs induce H_2O_2 production that can act as a signalling intermediate in different stress conditions. H_2O_2 can trigger NADPH oxidase to generate a higher amount of cellular H_2O_2 through the MAPK pathway, thus indicating a positive feedback loop for the signal. The augmented H_2O_2 level triggers the formation of several stress-related proteins, such as antioxidant enzymes, dehydrins and heat shock proteins, which scavenge ROS, thus lowering the ROS levels (Xia et al. 2009; Zhu et al. 2013). In tomato, BR-mediated stress tolerance activity is mainly facilitated by increased apoplastic H_2O_2 and subsequent activation of the MPK1/2 pathway. However, the signalling response is perturbed in RBOH1, MPK1/2 and MPK2 silenced transgenic plants but not in MPK1 silenced plants, suggesting a stronger role of MPK2 over MPK1 in BR-induced apoplastic H_2O_2 build-up (Nie et al. 2013). In addition, RD26 can inhibit BES1 activity, thus regulating drought tolerance (Ye et al. 2017). The suppression of BRI1 improves drought tolerance in plants. Recent studies have revealed that the overexpression of BR receptor BRL3 enhances drought tolerance by accumulating osmoprotectants like proline and sugars (Fabregas et al. 2018).

11.4.2 Polyamines

Polyamines like putrescine, spermine, Spd and cadaverine are widely distributed nitrogen (N)-containing organic molecules in plants. They regulate various physiological phenomena, like organogenesis, flowering, fruit development and ripening, embryogenesis and leaf senescence as well as a number of abiotic and biotic stress responses in plants (Galston et al. 1990; Kumar et al. 1997; Alcázar et al. 2010; Li et al. 2015). Recent studies have revealed that induction of the polyamine biosynthesis-related genes is closely associated with growth and development (Fuell et al. 2010; Krasuska et al. 2014; Pottosin et al. 2014) and stress tolerance in plants (Yiu et al. 2009; Do et al. 2014). It has been reported that polyamine can improve H_2O_2-induced ROS detoxification in *Medicago sativa* (Guo et al. 2014). The tolerance of drought stress is associated with the increase of endogenous polyamine content in leaves. Li et al. (2015) suggested that exogenous application of Spd induces H_2O_2 production and accumulates cytosolic free Ca^{2+} which activates the *NADPH oxidase* and *calcium dependent protein kinase (CDPK)* genes in the cell. To a greater extent, polyamine regulates the drought stress-induced H_2O_2 accumulation, free cytosolic Ca^{2+} release, activation of antioxidant enzymes and upregulation of several drought stress-responsive genes leading to the prevention of drought stress-induced oxidative damage. Several dehydrin genes like *Y2SK*, *Y2K* and *Shikimate kinase 2* (*SK*2) have been found to be highly induced in response to the application of exogenous Spd. A significant enhancement of water-soluble carbohydrates has also been reported after polyamine treatment (Li et al. 2015). This suggests that polyamine-mediated tolerance to drought stress in plants is generally associated with antioxidant defence and the regulation of dehydrins via the involvement of the Ca^{2+} messenger system and H_2O_2 signalling pathways and alteration in sugar metabolism. Recent studies have suggested that ABA can also activate the expression of polyamine biosynthesis-related genes. On the other hand, the overexpression of polyamine biosynthesis-related genes, such as *arginine decarboxylase* (*ADC*), *S-adomet decarboxylase*

(*SAMDC*) or spermine synthase (*SPMS*), can also stimulate ABA biosynthesis by the activation of *9-cis-epoxycarotenoid dioxygenase* (*NCED*), implying a positive feedback loop between the ABA and polyamine signalling pathways during drought stress. Pre-treatment with ABA also alleviates putrescine contents, but lowers Spd contents, suggesting both the synthesis and degradation of polyamines under osmotic stress (Pal et al. 2018).

11.5 PHYTOHORMONE CROSSTALK IN FINE-TUNING DROUGHT RESPONSE

Plants respond to drought stress by a coordinated interaction of different phytohormone signalling pathways. For example, the regulation of ERF1 expression in response to drought stress is regulated by an intricate ABA-ET-JA interplay (Figure 11.4). The ET-induced expression of ERF1 during drought is also regulated by JA (Cheng et al. 2013). Intriguingly, ERF1 expression has been found to be increased in the ET signalling mutant, *ctr1*, under drought stress. In addition, ABA content has also been reported to be higher in this mutant under drought stress. However, ABA can repress ERF1 activity. This implies that the JA/ET-mediated positive regulation of ERF1 under drought stress supersedes the ABA-mediated suppression of ERF1 expression (Cheng et al. 2013). On the other hand, ERF1 overexpression lines contain higher ABA content compared to wild type, indicating a probable negative feedback regulation of ERF1 by ABA. Under drought conditions, ET and JA production is commonly found in leaves where the expression of ERF1 is also detected. However, ABA production is commonly confined to the turgid tissues like vascular parenchyma in response to drought stress, thus reflecting a spatial regulation of the signalling network under drought stress (Cheng et al. 2013). ABA not only supresses ERF1, but also ERF6 in *Arabidopsis* (Dubois et al. 2013). In contrast, the ABA-mediated induction of ERFs has been reported in other plant species like *CsERF*, *GmERF3*, *LchERF*, *JERF3*, *TSRF1*, etc. (Wu et al. 2008; Zhang et al. 2009; Quan et al. 2010; Sewelam et al. 2013; Ma et al. 2014; Wu et al. 2014).

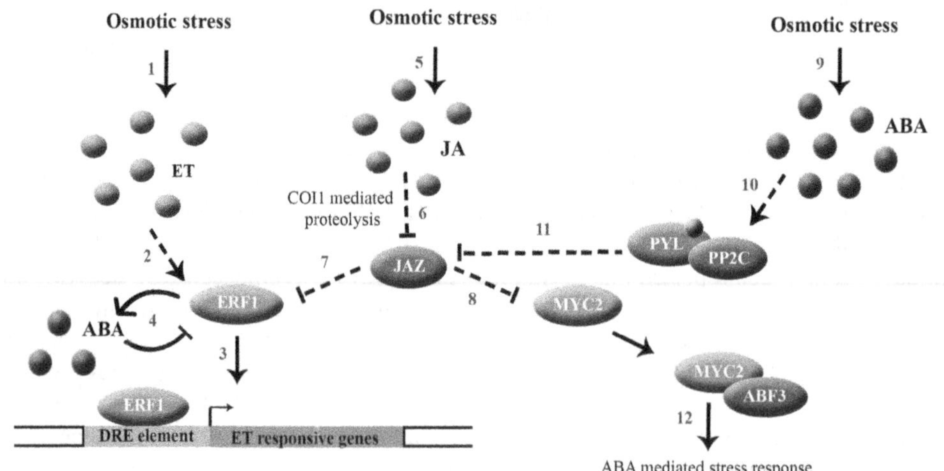

FIGURE 11.4 A complex signalling network operating during osmotic stress in plants. ET biosynthesis is induced during osmotic stress (1) which activates downstream ERF1 (2). ERF1 triggers ET-responsive gene expression (3). ERF1 also activates ABA biosynthesis; ABA can again negatively regulate the ERF1-mediated response and forms a negative feedback loop (4). JA is also accumulated during osmotic stress (5). In the presence of JA, the JAZ repressor is degraded by SCF^{COI1}-mediated proteolysis (6). JAZ is a negative regulator of ERF1 (7). JAZ degradation facilitates the ERF1-mediated signalling response. JAZ also suppresses MYC2, an important regulator of the JA signalling pathway (8). ABA accumulates during osmotic stress (9) and binds to the PYL/PP2C complex (10) to suppress the JAZ repressor (11). This frees MYC2 to interact with ABF3 and activate the ABA-mediated gene expression for rendering drought tolerance (12).

Interestingly, *JERF1*and *TSRF1* also stimulate ABA production by interacting with the 5′ *cis* acting sequence of the ABA biosynthesis-related gene *NtSDR* to increase drought tolerance in tobacco. Therefore, a major signalling network among ERF1, ET, JA and ABA has been demonstrated during drought stress in plants where ET and JA can induce the expression of ERF1 which in turn activates different downstream DRE-containing genes and also ABA production, which may negatively regulate ERF1 production. Among the different members of ERFs, ERF4 and ERF7 can act as negative regulators of ABA signalling, thus making the plant more sensitive to drought stress (Yang et al. 2005; Song et al. 2005).

Recent findings highlight the novel interaction between ABA signalling proteins and the JAZ repressor under drought stress in plants. The major ABA receptors like the PYL4, PYL5 and PP2C proteins interact with several JAZ repressor proteins like JAZ1, JAZ5 and TIFY10B to modulate drought response. MYC2, a common TF involved in JA signalling, interacts with the ABA responsive element binding factor 3 (ABF3), an important member of the ABA signalling pathway, and thus modulates the drought stress response in plants. The ABA signalling proteins not only interact with JA signalling members, but can also interact with gibberellin insensitive dwarf1B (GID1B), a GA receptor protein, suggesting an interesting ABA-JA-GA mediated crosstalk during drought stress (Liu and Xianliang 2018).

ABA also interferes with the BR signalling pathway during drought stress as a negative modulator. Brassinosteroid insensitive2 (BIN2), a repressor of the BR signalling pathway, has been found to be phosphorylated, and it interacts with ABI5 to regulate the ABA signalling pathway during seed germination and growth (Hu and Yu 2014). In rice, reproductive meristem 4.1 (REM4.1) is induced by ABA and then interacts with somatic embryogenesis receptor kinase 1 (SERK1) to impair the association with BRI1, thus establishing a crosstalk between BR and ABA (Gui et al. 2016). This also suggests the complex signalling network of different plant growth regulators as positive and negative modulators during the drought stress response in plants.

11.6 OTHER SIGNALLING MOLECULES CONNECTED TO THE PHYTOHORMONE NETWORK

11.6.1 GSH

GSH (γ-glutamyl-cysteinyl-glycine), a thiol group containing non-enzymatic signalling tripeptide, regulates various metabolic pathways in plants. During stress it protects the membranes from oxidative damage as well as preventing the oxidative denaturation of proteins. Chen et al. (2012b) reported that exogenous application of GSH in wild-type plants resulted in higher ABA content and greater drought tolerance compared to wild type. GSH can also stimulate higher levels of ABA accumulation and lead to seed germination, closing and opening of the stomatal aperture and reduction of the transpiration rate (Chen et al. 2012b). Leaf rolling in plants under drought stress and the regulation of leaf water content are also associated with the enhanced level of GSH content and reduced GSSG content. GSH also modulates the cell division of the root meristematic region to enhance root growth and is considered to be a vital adaptive drought-responsive metabolite in plants (Vernoux et al. 2000). In contrast, GSSG content has also been found to be increased in *B. napus* seedlings under drought stress and the GSH:GSSG ratio is altered (Hasanuzzaman et al. 2014). Most importantly, the higher endogenous GSH levels can confer drought stress tolerance by preventing ROS-induced damages. The levels of GSH and the enzymes of the ascorbic acid-GSH cycle have been widely reported to be involved with ROS detoxification (Meyer 2008; Koffler et al. 2014; Datta et al. 2015).

GSH and the enzymes of GSH metabolism interact with plant growth regulators to maintain cellular homeostasis under drought stress. The expression of glutathione-s-transferase (GST) has been found to be significantly increased by different phytohormones in plants such as SA, ET, cytokinin, auxin, ABA (Marrs 1996), MeJA (Moons 2003; Akter et al. 2012) and BRs (Deng et al. 2007).

During stomatal closure, the depletion of GSH content in the guard cell along with the augmentation of ABA content induce stomatal closure (Okuma et al. 2011). Furthermore, it has been revealed that GSH negatively regulates the ABA signalling pathway without affecting endogenous oscillation of Ca^{2+} ion and ROS production in the guard cell. In another study, a member of the GST gene *GST*8 has been reported to be strongly induced by 2,4-D and naphthalene acetic acid (NAA), suggesting a probable crosstalk between GSH and plant hormones like IAA and ABA during drought stress (Bianchi et al. 2002; Shi et al. 2014). Recently, translatome analysis in *Arabidopsis* revealed that exogenous GSH treatment can activate ABA, JA and IAA biosynthesis genes and also their downstream signalling genes, thus deciphering a fascinating crosstalk and complex signalling network operating under drought stress (Cheng et al. 2015).

11.6.2 Calcium

Ca^{2+} is a vital secondary messenger which plays a significant role in different signal transduction pathways including drought stress signalling. Drought stress activates signalling cascades involving several protein kinases/phosphatases such as receptor-like protein kinase 1 (RPK1), SNRK2C, CDPKs, ABI, AREBs and DREBs which are commonly associated with osmoprotectant metabolism. Most of the Ca^{2+}-regulated genes are regulated by ABA (Kaplan et al. 2006). For example, ABA activates the endogenous oscillation of Ca^{2+} ions in the guard cell during stomatal opening or closing. The accumulation of Ca^{2+} ion in the guard cell activates different members of the protein kinase family such as CPK23, CPK3/6/1 and other Ca^{2+}-binding protein family members like GHR1 and CBL1/CIPK23 which in turn regulate the ion channels like SLAC1and slow anion channel associated homolog 3 (SLAH3) by phosphorylation to induce stomatal closure. SnRK2 also plays an important role in the ABA-mediated regulation of SLAC1 channels (Maierhofer et al. 2014; Zhang et al. 2016).

Several TFs are also regulated by the Ca^{2+} ion during drought stress. For example, ABF2/AREB1, upregulated during drought stress, is a calmodulin (CaM)-binding TF which directly alters cellular Ca^{2+} levels (Popescu et al. 2007; Wu et al. 2017b). MYB2 acts as a transcriptional activator during drought stress and is also regulated by Ca^{2+} ions and subsequently by CaM-binding protein (Abe et al. 1997). The GT-2 LIKE1 (GTL1) TF, a negative regulator of drought tolerance response, also binds to CaM. A loss-of-function mutant of *AtGTL*1, *gtl*1, enhanced the ability of plants to survive under drought stress by reducing the transpiration rate (Yoo et al. 2010). In addition, the expression of the Ca^{2+}/CaM-binding NAC TF, CBNAC, is also found to be upregulated upon exposure to a combination of drought and heat stress, suggesting a regulatory role of Ca^{2+}-dependent protein kinases and different TFs in imparting drought tolerance in plants (Rizhsky et al. 2004).

11.6.3 Nitric Oxide

Nitric oxide (NO) is an important secondary messenger and a gaseous signalling molecule which is gaining much attention nowadays because of its diverse roles in a wide range of physiological phenomena and various environmental stress responses. It has been suggested that NO can activate antioxidant enzymes and has a high affinity for iron-containing enzymes, thus protecting the cell from oxidative damage during drought stress (Boyarshinov and Asafova 2011). There are several studies that support the NO-induced stimulation of major antioxidant enzymes such as APX, CAT and SOD under stress conditions (Uchida et al. 2002; Tian and Lei 2006; Sang et al. 2008). Hao et al. (2008) reported that treatment with NO prevented water loss and oxidative damage by enhancing the SOD activity in maize leaves under water deficit conditions. Recent studies highlight the NO-mediated alternate pathway of regulating photosynthetic machinery by avoiding over-reduction of the electron transport chain during drought stress (Wang et al. 2016). NO is a major component of the ABA-dependent SnRK2-mediated signalling pathway which can also regulate the different stress-inducible genes along with DREB/CBFs (Wang et al. 2013). Recently, it has been suggested that NO generated by ABA signalling causes the S-nitrosylation of cysteine residues of SnRK2.6/OST1 to suppress the kinase activity, resulting in a negative feedback loop of ABA signalling during stomatal opening and closure. The S-nitrosylation of tyrosine residue in the PYL receptor facilitates the degradation of the PYL receptor, thus negatively regulating ABA signalling in plants (Castillo et al. 2015). Therefore, a crosslink between the MAPK pathway induced by SnRK2 and RbohD/F with Ca^{2+} is crucial during ABA signalling under osmotic stress while NO remains a negative regulator critically regulating the signalling events.

11.6.4 Phosphatidylinositol

Phosphatidylinositol (PtdIns), a crucial component of the cell membrane, can act as a phospholipid signal in plants during different abiotic stress responses. It is synthesized by the enzyme, PtdIns synthase (PIS), which when overexpressed in maize increased drought stress tolerance in the transgenic lines. The resultant drought-tolerant plants displayed an increase in the digalactosyl diacylglycerol:monogalactosyl diacylglycerol (DGDG:MGDG) ratio of leaf lipids and a higher accumulation of PtdIns which helped to stabilize the membrane during drought stress. Interestingly, ABA content was also found to be increased in these drought-tolerant plants (Liu et al. 2013). In numerous plant species, drought has been reported to stimulate DGDG biosynthesis and induce the accumulation of DGDG in extrachloroplastic membranes to maintain membrane integrity and confer drought tolerance in plants (Gigon et al. 2004; Torres et al. 2007; Khorsgani et al. 2018). In several studies, it has been suggested that inositol polyphosphate 5-phosphatases (5PTases) family members are also regulated by ABA in drought stress, suggesting a close relationship between phospholipid signalling and ABA-responsive pathways during drought (Lin et al. 2005; Gunesekera et al. 2007; Ananieva et al. 2008; Perera et al. 2008).

In addition to PIS, phospholipase D (PLD) has also been considered to be a vital regulator of drought stress signalling pathways. Mane et al. (2007) reported that PLDα1-derived phosphatidic acid interacts with the RboHF and RboHD proteins to activate ROS production and also with ABI1PP2C to inhibit the trafficking of ABI1 from cytosol to nucleus. This leads to a positive regulation of the ABA signalling pathway of stomatal closure in plants. In contrast, during stress recovery, PLDα1 interacts directly with Gα, the only canonical subunit of the heterotrimeric G protein in *Arabidopsis*, and represses PLDα activity and ABA signalling, thus stimulating the intrinsic GTPase activity of the Gα subunit and inducing stomatal opening (Zhang et al. 2004; Mishra et al. 2006). The Ca^{2+} ion also triggers PLDα1-mediated phosphatidic acid production which interacts with the SnRK2-mediated signalling pathway, ensuring its role in the ABA-mediated signalling pathway during drought stress.

11.7 FUTURE GATEWAY TO SUSTAINABLE AGRICULTURE: BIOTECHNOLOGICAL APPROACHES FOR ENHANCED DROUGHT TOLERANCE

Significantly, the introduction of many drought stress-inducible genes via biolistic or *Agrobacterium*-mediated transformation has resulted in improved drought tolerance in plants (Umezawa et al. 2006). Recently, advanced molecular breeding techniques have been applied in order to develop drought-tolerant crop plants and to increase the recovery of plants from drought stress. Among these different approaches, the overexpression of *DREB/CBF* is very common, showing enhanced drought resistance in tomato, chrysanthemum, potato (Iwaki et al. 2013), soybean (de Paiva Rolla et al. 2014), rice (Datta et al. 2012), peanut (Bhatnagar-Mathur et al. 2014), wheat (Shavrukov et al. 2016) and sugarcane (Augustine et al. 2015). DREB1A can induce the expression of other osmotic stress-related genes like *R40C*1 in rice when the transgenic rice plants are exposed to drought stress (Paul et al. 2015). NAC protein family members are found to be crucial for imparting drought stress. Therefore, the overexpression of different NAC proteins such as RD26, ATAF1, SNAC1,

NAC6/SNAC2, NAC5, NAC9, NAC14 and NAC10 can improve drought tolerance via modification of the root structure, redox homeostasis and ABA biosynthesis (Hu et al. 2006; Nakashima et al. 2007; Redillas et al. 2012; Nakashima et al. 2014; Lee et al. 2017; Shim et al. 2018). Recent studies reported that OsNAC14 interacts with OsRAD51A1 and enhances the drought tolerance of plants (Shim et al. 2018). The overexpression of OsNAC10 using both a root-specific (*RCc*3) and a constitutive (*GOS*2) promoter can also maintain grain yields in transgenic rice plants under normal and drought conditions (Jeong et al. 2010).

Among the several bZIP family members, the overexpression of OsZIP16 in transgenic rice plants resulted in increased drought tolerance at both the vegetative and reproductive stages and also sensitivity to ABA (Chen et al. 2012a). Recent studies revealed that the overexpression of TabZIP60 improved resistance to abiotic stress along with ABA sensitivity in *Arabidopsis* plants (Zhang et al. 2015). bZIP23, a homolog of ABF/AREB, showed greater sensitivity to ABA in response to drought and salt stress; *Os*bZIP46 and *Os*bZIP46CA1, belonging to the ABF3 subfamily, improved drought tolerance in transgenic *Arabidopsis* and rice (Tang et al. 2012; Todaka et al. 2015). Recent studies revealed that the overexpression of *DZR*1 enhances drought tolerance in rice seedlings by the accumulation of free proline and different antioxidant enzymes (De Schutter et al. 2017; Rejeb et al. 2014, 2015; Boyarshinov and Asafova 2011). Different MYB factors like MYB44, MYB73 and MYB77 play crucial roles in imparting drought stress tolerance in plants. The overexpression of SbMYB44 enhanced the growth of yeast cells in response to both ionic and osmotic stress (Shukla et al. 2015). A couple of MYB factors like MYB88 and MYB124 were overexpressed in apple, showing enhanced drought tolerance by modulating the cell walls and root vessels (Geng et al. 2018).

Recently, the advancement of molecular breeding techniques like genome editing through CRISPR/Cas9 has allowed drought tolerance in plants to be improved without introducing any foreign gene. For example, the targeted mutation of SIMAPK3 and ARGOS8 can increase drought tolerance in tomato and maize (Wang et al. 2017; Shi et al. 2017). In rice, the development of different CRISPR/Cas-mediated knock-out mutants for several osmotic stress-related genes such as MPK2, PMS3, DERF1, EPSPS, MSH1 and MYB5 can improve drought tolerance and increase the productivity under drought which is considered a major breakthrough for sustainable agriculture (Shan et al. 2013; Zhang et al. 2014). In addition, the development of different abiotic stress-tolerant knock-out/knock-in plants without compromising growth and yield will be a boost for food security in developing countries in the future. Furthermore, RNA editing through CRISPR/Cas13 (Cox et al. 2017) can be applied in the future in plants to enhance drought or abiotic stress tolerance without inducing any mutation in the genomic DNA.

REFERENCES

Abe, H., Yamaguchi, S. K., Urao, T., Iwasaki, T., Hosokawa, D., and Shinozaki, K. 1997. Role of *Arabidopsis* MYC and MYB homologs in drought- and abscisic acid-regulated gene expression. *The Plant Cell* 9: 1859–68.

Abogadallah, G. M. 2010. Antioxidative defence under salt stress. *Plant Signalling and Behaviour* 5: 369–74.

Abreu, M. E., and Munne-Bosch, S. 2008. Salicylic acid may be involved in the regulation of drought-induced leaf senescence in perennials: A case study in field-grown *Salvia officinalis* L. plants. *Environmental and Experimental Botany* 64: 105–12.

Akter, N., Sobahan, M. A., and Uraji, M. 2012. Effects of depletion of glutathione on abscisic acid- and methyl jasmonate-induced stomatal closure in *Arabidopsis thaliana*. *Bioscience Biotechnology Biochemistry* 76: 2032–7.

Alcazar, R., Altabella, T., Marco, F., et al. 2010. Polyamines: Molecules with regulatory functions in plant abiotic stress tolerance. *Planta* 231: 1237–49.

Aleman, F., Yazaki, J., Lee, M., et al. 2016. An ABA increased interaction of PYL6 ABA receptor with MYC2 transcription factor: A putative link of ABA and JA signalling. *Scientific Reports* 6: 28941.

Ananieva, E. A., Gillaspy, G. E., Ely, A., Burnette, R. N., and Erickson, F. L. 2008. Interaction of the WD40 domain of a myoinositol polyphosphate 5-phosphatase with SnRK1 links inositol, sugar, and stress signalling. *Plant Physiology* 148(4): 1868–82.

Ashfaq, M., Haider, M. S., Khan, A. S., and Allah, S. U. 2012. Breeding potential of the basmati rice germplasm under water stress condition. *African Journal of Biotechnology* 11: 6647–57.

Augustine, S. M., Ashwin, N. J., Syamaladevi, D. P., Appunu, C., Chakravarthi, M., and Ravichandran, V. 2015. Overexpression of EaDREB2 and pyramiding of EaDREB2 with the pea DNA helicase gene (PDH45) enhance drought and salinity tolerance in sugarcane (*Saccharum* spp. hybrid). *Plant Cell Reports* 34: 247–63.

Bandurska, H., and Stroinski, A. 2005. The effect of salicylic acid on barley response to water deficit. *Acta Physiologiae Plantarum* 27: 379–86.

Barbosa, E. G. G., Lete, J. P., Martin, E. R. R., et al. 2013. Overexpression of the ABA-dependent *AREB1* transcription factor from *Arabidopsis thaliana* improves soybean tolerance to water deficit. *Plant Molecular Biology Reporter* 31: 719–30.

Barrero, J. M., Rodriguez, P. L., Quesada, V., Piqueras, P., Ponse, R. M., and Micol, J. L. 2006. Both abscisic acid (ABA)-dependent and ABA independent pathways govern the induction of NCED, AAO3 and ABA1 in response to salt stress. *Plant Cell and Environment* 29: 2000–8.

Bauer, H., Ache, P., Wohlfart, F., et al. 2013. How do stomata sense reductions in atmospheric relative humidity? *Molecular Plant* 6: 1703–6.

Bhatnagar-Mathur, P., Rao, J. S., Vadez, V., Dumbala, S. R., Rathore, A., and Yamaguchi-Shinozaki, K. 2014. Transgenic peanut overexpressing the DREB1A transcription factor has higher yields under drought stress. *Molecular Breeding* 33: 327–40.

Bianchi, M. W., Roux, C., and Vartanian, N. 2002. Drought regulation of GST8, encoding the *Arabidopsis* homologue of ParC/Nt107 glutathione transferase/peroxidase. *Plant Physiology* 116: 96–105.

Blum, A. 2011. Drought tolerance: Is it a complex trait? *Functional Plant Biology* 38: 753–7.

Borsani, O., Valpuesta, V., and Botella, M. A. 2001. Evidence for a role of salicylic acid in the oxidative damage generated by NaCl and osmotic stress in *Arabidopsis* seedlings. *Plant Physiology* 126: 1024–30.

Bota, J., Medrano, H., and Flexas, J. 2004. Is photosynthesis limited by decreased Rubisco activity and RuBP content under progressive water stress? *New Phytologist* 162: 671–81.

Boudsocq, M., Barbier-Brygoo, H., and Lauriere, C. 2004. Identification of nine sucrose non fermenting1-related protein kinases2 activated by hyperosmotic and saline stresses in *Arabidopsis thaliana*. *Journal of Biological Chemistry* 279: 41758–66.

Bowling, S. A., Clarke, J. D., Liu, Y. D., Klessig, D. F., and Dong, X. N. 1997. The cpr5 mutant of *Arabidopsis* expresses both NPR1-dependent and NPR1 independent resistance. *The Plant Cell* 9: 1573–84.

Boyarshinov, A. V. and Asafova, E. V. 2011. Stress responses of wheat leaves to dehydration: Participation of endogenous NO and effect of sodium nitroprusside. *Russian Journal of Plant Physiology* 58:1034.

Calzadilla, P. I., Gazquez, A., Maiale, S. J., et al. 2014. Polyamines as indicators and modulators of the abiotic stress in plants. In *Plant Adaptation*, eds. N. A. Anjum, S. S. Gill, and R. Gill, pp. 109–128. CABI, Wallingford, UK.

Castillo, M. C., Lozano-Juste, J., González-Guzmán, M., Rodriguez, L., Rodriguez, P. L., and León, J. 2015. Inactivation of PYR/PYL/RCAR ABA receptors by tyrosine nitration may enable rapid inhibition of ABA signalling by nitric oxide in plants. *Science Signalling* 8: 89.

Chai, J., Liu, J., Zhou, J., and Xing, D. 2014. Mitogen-activated protein kinase 6 regulates NPR1 gene expression and activation during leaf senescence induced by salicylic acid. *Journal of Experimental Botany* 65: 6513–28.

Chaiwanon, J., and Wang, Z. Y. 2015. Spatio temporal brassinosteroid signalling and antagonism with auxin pattern stem cell dynamics in *Arabidopsis* roots. *Current Biology* 20: 1031–42.

Chen, H., Chen, W., Zhou, J., et al. 2012a. Basic leucine zipper transcription factor OsbZIP16 positively regulates drought resistance in rice. *Plant Science* 194: 8–17.

Chen, J. H., Jiang, H. W., Hsieh, E. J., et al. 2012b. Drought and salt stress tolerance of an *Arabidopsis* glutathione S-transferase U17 knockout mutant are attributed to the combined effect of glutathione and abscisic acid. *Plant Physiology* 158: 340–51.

Chen, T., and Flurh, R. 2018. Singlet oxygen plays an essential role in the root's response to osmotic stress. *Plant Physiology* 177: 1717–27.

Cheng, M. C., Ko, K., Chang, W. L., et al. 2015. Increased glutathione contributes to stress tolerance and global translational changes in Arabidopsis. *Plant Journal* 83: 926–39.

Cheng, M. C., Liao, P. M., Kuo, W. W., and Lin, T. P. 2013. The *Arabidopsis* ETHY-LENE RESPONSE FACTOR1 regulates abiotic stress-responsive gene expression by binding to different *cis*-acting elements in response to different stress signals. *Plant Physiology* 162: 1566–82.

Cheng, Z., Sun, L., Qi, T., et al. 2011. The bHLH transcription factor MYC3 interacts with the jasmonate ZIM-domain proteins to mediate jasmonate response in *Arabidopsis*. *Molecular Plant* 4: 279–88.

Chini, A., Fonseca, S., Fernández, G., et al. 2007. The JAZ family of repressors is the missing link in jasmonate signaling. *Nature* 448: 666–71.

Chini, A., Grant, J. J., Seki, M., Shinozaki, K., and Loake, G. J. 2004. Drought tolerance established by enhanced expression of the CC-NBS-LRR gene, ADR1, requires salicylic acid, EDS1 and ABI1. *Plant Journal* 38: 810–22.

Cox, D. B. T., Gootenberg, J. S., Abudayyeh, O. O., Franklin, B., and Kellner, M. J. 2017. RNA editing with CRISPR-Cas13. *Science* 358: 1019–27.

Dali, G., Chen, P., Shen, X., et al. 2018. MdMYB88 and MdMYB124 enhance drought tolerance by modulating root vessels and cell walls in apple. *Plant Physiology* 178: 1296–1309.

Datta, K., Baisakh, N., Ganguly, M., Krishnan, S., Yamaguchi, S. K., and Datta, S. K. 2012. Overexpression of *Arabidopsis* and rice stress genes' inducible transcription factor confers drought and salinity tolerance to rice. *Plant Biotechnology Journal* 10: 579–86.

Datta, R., Kumar, R., Sultana, A., Hazra, S., Bhattacharyya, D., and Chattopadhyay, S. 2015. Glutathione regulates 1-aminocyclopropane-1-carboxylate synthase transcription via WRKY33 and 1-aminocyclopropane-1-carboxylate oxidase by modulating messenger RNA stability to induce ethylene synthesis during stress. *Plant Physiology* 169: 2963–81.

de Paiva Rolla, A. A., de Fátima CCarvalho, J., Fuganti-Pagliarini, R., Engels, C., do Rio, A., and Marin, S. R. 2014. Phenotyping soybean plants transformed with rd29A:AtDREB1A for drought tolerance in the greenhouse and field. *Transgenic Research* 23: 75–87.

De Schutter, K., Tsaneva, M., Kulkarni, S. R., Rougé, P., Vandepoele, K., and Van Damme, E. J. M. 2017. Evolutionary relationships and expression analysis of EUL domain proteins in rice (*Oryza sativa*). *Rice* 10: 26.

Dempsey, D. A., Shah, J., and Klessig, D. F. 1999. Salicylic acid and disease resistance in plants. *Critical Reviews in Plant Sciences* 18: 547–75.

Dempsey, D. A., Vlot, A. C., Wildermuth, M. C., and Klessig, D. F. 2011. Salicylic acid biosynthesis and metabolism. *The Arabidopsis Book* 9: e0156.

Deng, Z., Zhang, X., Tang, W., et al. 2007. A proteomics study of brassinosteroid response in *Arabidopsis*. *Molecular and Cellular Proteomics* 6: 2058–71.

Do, P. T., Drechsel, O., Heyer, A. G., Hincha, D. K., and Zuther, E. 2014. Changes in free polyamine levels, expression of polyamine biosynthesis genes, and performance of rice cultivars under salt stress: A comparison with responses to drought. *Frontier in Plant Sciences* 5: 182.

Drerup, M. M., Schlücking, K., Hashimoto, K., et al. 2013. The calcineurin B-like calcium sensors CBL1 and CBL9 together with their interacting protein kinase CIPK26 regulate the *Arabidopsis* NADPH oxidase RBOHF. *Molecular Plant* 6: 559–69.

Dubois, M., Skirycz, A., Claeys, H., Maleux, K., Dhondt, S., and De Bodt, S. 2013. ETHYLENE RESPONSE FACTOR6 acts as a central regulator of leaf growth under water limiting conditions in *Arabidopsis*. *Plant Physiology* 162: 319–32.

Edgerton, M. D. 2009. Increasing crop productivity to meet global needs for feed, food and fuel. *Plant Physiology* 149: 7–13.

Emerson, R., Hoover, A., Ray, A., Lacey, J., and Marnie, C. 2014. Drought effects on composition and yield for corn stover, mixed grasses, and Miscanthus as bioenergy feedstocks. *Biofuels* 5(3): 275–91.

Fabregas, N., Lozano-Elena, F., Blasco-Escamez, D., et al. 2018. Overexpression of the vascular brassinosteroid receptor BRL3 confers drought resistance without penalizing plant growth. *Nature Communications* 9: 4680.

Fahramand, M., Mahmoody, M., Keykha, A., Noori, M., and Rigi, K. 2014. Influence of abiotic stress on proline, photosynthetic enzymes and growth. *International Research Journal of Applied Basic Science* 8: 257–65.

Fonseca, S., Chini, A., Hamberg, M., et al. 2009. (+)-7-*iso*-Jasmonoyl-L-isoleucine is the endogenous bioactive jasmonate. *Nature Chemical Biology* 5: 344–50.

Fragniere, C., Serrano, M., Abou-Mansour, E., Metraux, J. P., and L'Haridon, F. 2011. Salicylic acid and its location in response to biotic and abiotic stress. *FEBS Letter* 585: 1847–52.

Fu, Z. Q., Yan, S. P., Saleh, A., Wang, W., Ruble, J., and Oka, N. 2012. NPR3 and NPR4 are receptors for the immune signal salicylic acid in plants. *Nature* 486: 228.

Fuell, C., Elliott, K. A., Hanfrey, C. C., Franceschetti, M., and Michael, A. J. 2010. Polyamine biosynthetic diversity in plants and algae. *Plant Physiology and Biochemistry* 48(7): 513–20.

Fujita, Y., Fujita, M., Satoh, R., et al. 2005. AREB1 is a transcription activator of novel ABRE-dependent ABA signalling that enhances drought stress tolerance in *Arabidopsis*. *Plant Cell* 17: 3470–88.

Fujita, Y., Fujita, M., Shinozaki, K., and Yamaguchi-Shinozaki, K. 2011. ABA mediated transcriptional regulation in response to osmotic stress in plants. *Journal of Plant Research* 124: 509–25.

Fujita, Y., Yoshida, T., and Yamaguchi, S. K. 2013. Pivotal role of the AREB/ABF-SnRK2 pathway in ABRE-mediated transcription in response to osmotic stress in plants. *Physiologia Plantarum* 147: 15–27.

Furihata, T., Maruyama, K., Fujita, Y., et al. 2006. Abscisic acid-dependent multisite phosphorylation regulates the activity of a transcription activator AREB1. *Proceedings of the National Academy of Sciences of the United States of America* 103: 1988–93.

Galmes, J., Carbo-Ribas, M., Medrano, H., and Flexas, J. 2011. Rubisco activity in Mediterranean species is regulated by the choloroplastic CO_2 concentration under water stress. *Journal of Experimental Botany* 62: 653–65.

Galston, W. A., and Ravindar, S. K. 1990. Polyamines in plant physiology. *Plant Physiology* 94: 406–10.

Gao, X. P., Wang, X. F., Lu, Y. F., et al. 2004. Jasmonic acid is involved in the water-stress-induced betaine accumulation in pear leaves. *Plant Cell and Environment* 27: 497–507.

Geldner, N., Lee, Y., Rubio, M. C., and Alassimone, J. 2013. A mechanism for localized lignin deposition in the endodermis. *Cell* 153: 402–12.

Geng, D., Chen, P., Shen, X., et al. 2018. MdMYB88 and MdMYB124 enhance drought tolerance by modulating root vessels and cell walls in apple. *Plant Physiology* 178: 1296–1309.

Geng, Y., Wu, R., Wee, C. W., et al. 2013. A spatio-temporal understanding of growth regulation during the salt stress response in *Arabidopsis. Plant Physiology* 25: 2132–54.

Ghanta, S., Datta, R., Bhattacharyya, D., et al. 2014. Multistep involvement of glutathione with salicylic acid and ethylene to combat environmental stress. *Journal of Plant Physiology* 171(11): 940–50.

Gigon, A., Matos, A. R., Laffray, D., Zuily, F. Y., and PhamThi, A. T. 2004. Effect of drought stress on lipid metabolism in the leaves of *Arabidopsis thaliana* (ecotype Columbia). *Annals of Botany* 94: 345–51.

González, G. M. P., Vilarrasa, B. J., Zhiponova, M., Divol, F., Mora, G. S., and Russinova, E. 2011. Brassinosteroids control meristem size by promoting cell cycle progression in *Arabidopsis* roots. *Development* 138: 849–59.

Gu, C., Guo, Z. H., Hao, P. P., Wang, G. M., Jin, Z. M., and Zhang, S. L. 2017. Multiple regulatory roles of AP2/ERF transcription factor in angiosperm. *Botanical Studies* 58: 6.

Gu, J. F., Qiu, M., and Yang, J. C. 2013. Enhanced tolerance to drought in transgenic rice plants overexpressing C4 photosynthesis enzymes. *Crop Journal* 1: 105–14.

Gui, J., Zheng, S., Liu, C., Shen, J., Li, J., and Li, L. 2016. OsREM4.1 interacts with OsSERK1 to coordinate the interlinking between abscisic acid and brassinosteroid signalling in rice. *Developmental Cell* 38: 201–13.

Gunesekera, B., Torabinejad, J., Robinson, J., and Gillaspy, G. E. 2007. Inositol polyphosphate 5-phosphatases 1 and 2 are required for regulating seedling growth. *Plant Physiology* 143: 1408–17.

Guo, Z. F., Tan, J. L., Zhuo, C. L., Wang, C. Y., Xiang, B., and Wang, Z. Y. 2014. Abscisic acid, H_2O_2 and nitric oxide interactions mediated cold-induced S- adenosyl methionine synthetase in *Medicago sativa* sub sp. falcate that confers cold tolerance through up-regulating polyamine oxidation. *Plant Biotechnology Journal* 12: 601–12.

Hacham, Y., Holland, N., Butterfield, C., Ubeda, T. S., Bennett, M. J., and Chory, J. 2011. Brassinosteroid perception in the epidermis controls root meristem size. *Development* 138: 839–48.

Haefele, S. M., and Bouman, B. A. M. 2009. Drought-prone rainfed lowland rice in Asia: Limitations and management options. In *Drought Frontiers in Rice: Crop Improvement for Increased Rainfed Production*, eds. R Serraj, et al., pp. 211–232. World Scientific Publishing, Singapore and International Rice Research Institute, Los Baños, Philippines.

Hahn, A., and Harter, K. 2009. Mitogen-activated protein kinase cascades and ethylene: Signalling, biosynthesis, or both? *Plant Physiology* 149: 1207–10.

Hamada, A. M., and Al-Hakimi, A. M. A. 2001. Salicylic acid versus salinity-drought-induced stress on wheat seedlings. *Rostlinna Vyroba* 47: 444–50.

Hao, G. P., Xing, Y., and Zhang, H. 2008. Role of nitric oxide dependence on nitric oxide synthase-like activity in the water stress signalling of maize seedling. *Journal of Integrative Plant Biology* 50: 435–42.

Hasanuzzaman, M., Nahar, K., Gill, S. S., Gill, R., and Fujita, M. 2014. Drought stress responses in plants, oxidative stress and antioxidant defence. In *Climate Change and Plant Abiotic Stress Tolerance*, eds. Gill, S. S., and Tuteja, N., pp. 209–49. Blackwell, Germany and Wiley.

Hayat, Q., Hayat, S., Irfan, M., and Ahmad, A. 2010. Effect of exogenous salicylic acid under changing environment: A review. *Environmental and Experimental Botany* 68: 14–25.

Henry, A., Cal, A. J., Batoto, T. C., Torres, R. O., and Serraj, R. 2012. Root attributes affecting water uptake of rice (*Oryza sativa*) under drought. *Journal of Experimental Botany* 63(13): 4751–63.

Hu, H. H., Dai, M. Q., Yoo, J. L., et al. 2006. Overexpressing a NAM, ATAF, and CUC (NAC) transcription factor enhances drought resistance and salt tolerance in rice. *Proceedings of the National Academy of Sciences of the United States of America* 103: 12987–92.

Hu, Y., and Yu, D. 2014. BRASSINOSTEROID INSENSITIVE2 interacts with ABSCISIC ACID INSENSITIVE5 to mediate the antagonism of brassinosteroids to abscisic acid during seed germination in *Arabidopsis. Plant Cell* 26: 4394–08.

Iwaki, T., Guo, L., Ryals, J. A., Yasuda, S., Shimazaki, T., and Kikuchi, A. 2013. Metabolic profiling of transgenic potato tubers expressing *Arabidopsis* dehydration response element-binding protein1A (DREB1A). *Journal of Agriculture and Food Chemistry* 61: 893–900.

Jeong, J. S., Kim, Y. S., Baek, K. H., et al. 2010. Root-specific expression of OsNAC10 improves drought tolerance and grain yield in rice under field drought conditions. *Plant Physiology* 153: 185–97.

Ji, K., Wanga, Y., Sun, W., et al. 2012. Drought-responsive mechanisms in rice genotypes with contrasting drought tolerance during reproductive stage. *Journal of Plant Physiology* 169: 336–44.

Jiang, W., and Lafitte, R. 2007. Ascertain the effect of PEG and exogenous ABA on rice growth at germination stage and their contribution to selecting drought tolerant genotypes. *Asian Journal of Plant Science* 6(4): 684–87.

Jiang, Y., and Deyholos, M. K. 2009. Functional characterization of *Arabidopsis* NaCl-inducible WRKY25 and WRKY33 transcription factors in abiotic stresses. *Plant Molecular Biology* 69: 91–05.

Jiang, Y. P., Cheng, F., Zhou, Y. H., Xia, X. J., Mao, W. H., and Shi, K. 2012. Brassinosteroid induced CO2 assimilation is associated with increased stability of redox-sensitive photosynthetic enzymes in the chloroplasts in cucumber plants. *Biochemical and Biophysical Research Communications* 426: 390–4.

Joo, J., Lee, Y. H., Kim, Y. K., Nahm, B. H., and Song, S. I. 2013. Abiotic stress responsive rice ASR1 and ASR3 exhibit different tissue-dependent sugar and hormone-sensitivities. *Molecular Cells* 35: 421–35.

Kadam, N. N., Yin, X., Bindraban, P. S., Struik, C. P., and Jagadish, S. V. K. 2015. Does morphological and anatomical plasticity during the vegetative stage make wheat more tolerant of water deficit stress than rice? *Plant Physiology* 167: 1389–1401.

Kadioglu, A., and Terzi, R. 2007. A dehydration avoidance mechanism: Leaf rolling. *Botanical Review* 73: 290–302.

Kang, G., Li, G., Xu, W., Peng, X., Han, Q., and Zhu, Y. 2012. Proteomics reveals effects of salicylic acid on growth and tolerance to subsequent drought stress in wheat. *Journal of Proteomic Research* 11: 6066–79.

Kaplan, B., Davydov, O., Knight, H., et al. 2006. Rapid transcriptome changes induced by cytosolic Ca^{2+} transients reveal ABRE-related sequences as Ca^{2+} responsive cis elements in *Arabidopsis. Plant Cell* 18: 2733–48.

Kazan, K. 2015. Diverse roles of jasmonates and ethylene in abiotic stress tolerance. *Trends in Plant Sciences* 20: 219–29.

Kazan, K., and Manners, J. M. 2013. MYC2: The master in action. *Molecular Plant* 6(3): 686–703.

Khorsgani, O. A., Flores, F. B., and Mohammad, P. 2018. Plant signalling pathways involved in stomatal movement under drought stress conditions. *Advances Plants Agriculture Research* 8(3): 290–97.

Koffler, B. E., Luschin-Ebengreuth, N., Stabentheiner, E., Mu¨ller, M., and Zechmann, B. 2014. Compartment specific response of antioxidants to drought stress in *Arabidopsis. Plant Science* 227: 133–44.

Komatsu, K., Suzuki, N., Kuwamura, M., et al. 2013. Group APP2C sevolvedinl and plants as key regulators of intrinsic desiccation tolerance. *Nature Communication* 4: 2219.

Korkmaz, A., Uzunlu, M., and Demirkiran, A. R. 2007. Treatment with acetyl salicylic acid protects muskmelon seedlings against drought stress. *Acta Physiologiae Plantarum* 29: 503–08.

Krasuska, U., Ciacka, K., Bogatek, R., and Gniazdowska, A. 2014. Polyamines and nitric oxide link in regulation of dormancy removal and germination of apple (*Malusdomestica* Borkh.) embryos. *Journal Plant Growth Regulation* 33: 590–601.

Krishnamurthy, P., Ranathunge, K., Franke, R., Prakash, H. S., Schreiber, L., and Mathew, M. K. 2009. The role of root apoplastic transport barriers in salt tolerance of rice (*Oryza sativa* L.). *Planta* 230: 119–34.

Kulkarni, K. P., Vishwakarma, C., Sahoo, S., et al. 2014. A substitution mutation in *OsCCD7* cosegregates with dwarf and increased tillering phenotype in rice. *Journal of Genetics* 93: 389–401.

Kumar, A., Altabella, T., Taylor, M., and Tiburcio, A. F. 1997. Recent advances in polyamine research. *Trends in Plant Science* 2: 124–30.

Kumar, S., Dwivedi, S. K., Singh, S. S., et al. 2014a. Morphophysiological traits associatejid with reproductive stage drought tolerance of rice (*Oryza sativa* L.) genotypes under rain-fed condition of eastern indo-gangetic plain. *Indian Journal of Plant Physiology* 19: 87–93.

Kumar, D., Datta, R., Sinha, R., Ghosh, A., and Chattopadhyay, S. 2014b. Proteomic profiling of γ-ECS overexpressed transgenic Nicotiana in response to drought stress. *Plant Signalling and Behavior* 9(8): e29246.

Lee, D. K., Jung, H., Jang, G., et al. 2016. Overexpression of the OsERF71 transcription factor alters rice root structure and drought resistance. *Plant Physiology* 172: 575–88.

Lee, D. K., Yoon, S., Kim, Y. S., et al. 2017. Rice OsERF71-mediated root modification affects shoot drought tolerance. *Plant Signaling and Behaviour* 12(1): e1268311.

Lenka, K. S., Muthuswamy, K. S., Chinnusamy, V., and Bansal, K. C. 2018. Ectopic expression of rice PYL3 enhances cold and drought tolerance in *Arabidopsis thaliana. Molecular Biotechnology* 60: 350–61.

Li, J., Li, Y., Yin, Z., et al. 2017b. OsASR5 enhances drought tolerance through a stomatal closure pathway associated with ABA and H_2O_2 signalling in rice. *Plant Biotechnology Journal* 15: 183–96.

Li, J., Wang, X. Q., Watson, M. B., and Assmann, S. M. 2000. Regulation of abscisic acid-induced stomatal closure and anion channels by guard cell AAPK kinase. *Science* 287: 300–303.

Li, K., Yang, F., Zhang, G., et al. 2017a. AIK1, a mitogen-activated protein kinase, modulates abscisic acid responses through the MKK5-MPK6 kinase cascade. *Plant Physiology* 173: 1391–1408.

Li, Z., Jing, W., and Peng, Y. 2015. Spermine alleviates drought stress in white clover with different resistance by influencing carbohydrate metabolism and dehydrins synthesis. *PLoS One* 10(4): e0120708.

Lin, W. H., Wang, Y., Mueller, R. B., Brearley, C. A., Xu, Z. H., and Xue, H. W. 2005. At5PTase13 modulates cotyledon vein development through regulating auxin homeostasis. *Plant Physiology* 139: 1677–91.

Liu, C., Zhao, A., Zhu, P., et al. 2015. Characterization and expression of genes involved in the ethylene biosynthesis and signal transduction during ripening of mulberry fruit. *Plos One* 10(3): e0122081.

Liu, X., Li, X., Zhang, C., et al. 2016. Phosphoenolpyruvate carboxylase regulation in C4-PEPC-expressing transgenic rice during early responses to drought stress. *Physiologia Plantarum* 159(2): 178–200.

Liu, X., and Xianliang, H. 2018. Antagonistic regulation of ABA and GA in metabolism and signalling pathways. *Frontier in Plant Science* 9: 251.

Liu, X., Zhai, X., and Y Zhao. 2013. Overexpression of the phosphatidylinositol synthase gene (*ZmPIS*) conferring drought stress tolerance by altering membrane lipid composition and increasing ABA synthesis in maize. *Plant Cell and Environment* 36: 1037–55.

Ma, S., Gong, Q., and Bohnert, H. J. 2006. Dissecting salt stress pathways. *Journal of Experimental Botany* 57: 1097–07.

Ma, Y., Szostkiewicz, I., Korte, A., et al. 2009. Regulators of PP2C phosphatase activity function as abscisic acid sensors. *Science* 324: 1064–68.

Ma, Y., Zhang, L., Zhang, J., Chen, J., Wu, T., and Zhu, S. 2014. Expressing a citrus ortholog of *Arabidopsis* ERF1 enhanced cold tolerance in tobacco. *Scientia Horticultarae* 174: 64–76.

Maierhofer, T., Diekmann, M., Offenborn, J. N., et al. 2014. Site- and kinase-specific phosphorylation-mediated activation of SLAC1, a guard cell anion channel stimulated by abscisic acid. *Science Signalling* 7: 86.

Mane, S. P., Vasquez, R. C., Sioson, A. A., Heath, L. S., and Grene, R. 2007. Early PLDa-mediated events in response to progressive drought stress in *Arabidopsis*: A transcriptome analysis. *Journal of Experimental Botany* 58: 241–52.

Marrs, K. A. 1996. The functions and regulation of glutathione S transferases in plants. *Annual Review of Plant Physiology Plant Molecular Biology* 147: 127–58.

Melcher, K., Ng, L. M., Zhou, X. E., et al. 2009. A gate-latch-lock mechanism for hormone signalling by abscisic acid receptors. *Nature* 462: 602–08.

Meyer, A. J. 2008. The integration of glutathione homeostasis and redox signalling. *Journal of Plant Physiology* 165: 1390–1403.

Mishra, G., Zhang, W., and Deng, F. 2006. A bifurcating pathway directs abscisic acid effects on stomatal closure and opening in *Arabidopsis. Science* 312(5771): 264–6.

Moons, A. 2003. Osgstu3 and osgtu4, encoding tau class glutathione S-transferases, are heavy metal- and hypoxic stress-induced and differentially salt stress responsive in rice roots. *FEBS Letters* 553: 427–32.

Muller, M., and Munne-Bosch, S. 2015. Ethylene response factors: A key regulatory hub in hormone and stress signalling. *Plant Physiology* 169: 3–41.

Munne-Bosch, S., and Penuelas, J. 2003. Photo- and antioxidative protection, and a role for salicylic acid during drought and recovery in field-grown *Phillyrea angustifolia* plants. *Planta* 217: 758–66.

Nahar, S., Kalita, J., Sahoo, L., and Tanti, B. 2016. Morphophysiological and molecular effects of drought stress in rice. *Annals of Plant Sciences* 5(9): 1409–16.

Nakamura, A., Tochio, N., Fujioka, S., Ito, S., Kigawa, T., and Shimada, Y. 2017. Molecular actions of two synthetic brassinosteroids, iso-carbaBL and 6 deoxoBL, which cause altered physiological activities between *Arabidopsis* and rice. *PLoS ONE* 12(4): e0174015.

Nakashima, K., Tran, L. S., Nguyen, D. V., et al. 2007. Functional analysis of a NAC-type transcription factor *OsNAC6* involved in abiotic and biotic stress-responsive gene expression in rice. *Plant Journal* 51: 617–30.

Nakashima, K., Yamaguchi, S. K., and Shinozaki, K. 2014. The transcriptional regulatory network in the drought response and its crosstalk in abiotic stress responses including drought, cold, and heat. *Frontier in Plant Science* 5: 1–7.

Nakashima, K., and Yamaguchi-Shinozaki, K. 2013. ABA signalling in stress response and seed development. *Plant Cell Reports* 32: 959–70.

Nambeesan, S., AbuQumar, S., NakaKristin, L., et al. 2012. Polyamines attenuate ethylene-mediated defence responses to abrogate resistance to *Botrytis cinerea* in tomato. *Plant Physiology* 158: 1034–45.

Naseer, S., Lee, Y., Lapierre, C., Franke, R., Nawrath, C., and Geldner, N. 2012. Casparian strip diffusion barrier in *Arabidopsis* is made of a lignin polymer without suberin. *Proceedings of the National Academy of Sciences of the United States of America* 109: 10101–6.

Nie, W. F., Wang, M. M., Xia, X. J., Zhou, Y. H., Shi, K., and Chen, Z. 2013. Silencing of tomato RBOH1 and MPK2 abolishes brassinosteroid induced H_2O_2 generation and stress tolerance. *Plant Cell and Environment* 36: 789–803.

Nishimura, N., Yoshida, T., Kitahata, N., Asami, T., Shinozaki, K., and Hirayama, T. 2007. ABA-hypersensitive germination 1encodes a protein phosphatase 2C, an essential component of abscisic acid signalling in *Arabidopsis* seed. *Plant Journal* 50: 935–49.

Niu, Y., Figueroa, P., and Browse, J. 2011. Characterization of JAZ interacting bHLH transcription factors that regulate jasmonate responses in *Arabidopsis*. *Journal of Experimental Botany* 62: 2143–54.

Noctor, G., and Foyer, C. H. 1998. Ascorbate and glutathione: Keeping active oxygen under control. *Annual Review Plant Physiology and Plant Molecular Biology* 49: 249–79.

Oh, S. J., Song, S. I., Kim, Y. S., Jang, H. J., Kim, S. Y., and Kim, M. 2005. *Arabidopsis* CBF3/DREB1A and ABF3 in transgenic rice increased tolerance to abiotic stress without stunting growth. *Plant Physiology* 138: 341–51.

Okuma, E., Jahan, M. S., Munemasa, S., et al. 2011. Negative regulation of abscisic acid-induced stomatal closure by glutathione in *Arabidopsis*. *Journal of Plant Physiology* 168: 2048–55.

Olmedo, G., Guo, H., Gregory, B. D., Nourizadeh, S. D., Aguilar-Henonin, L., and Li, H. 2006. Ethylene-insensitive 5 encodes a $5' \rightarrow 3'$ exoribonuclease required for regulation of the EIN3-targeting F-box proteins EBF1/2. *Proceedings of the National Academy of Sciences of the United States of America* 103(36): 13286–93.

Pal, M., Tajti, J., Szalai, G., et al. 2018. Interaction of polyamines, abscisic acid and proline under osmotic stress in the leaves of wheat plants. *Scientific Reports* 8: 12839.

Parent, B., Hachez, C., Redondo, E., Simonneau, T., Chaumont, F., and Tardieu, F. 2009. Drought and abscisic acid effects on aquaporin content translate into changes in hydraulic conductivity and leaf growth rate: A trans-scale approach. *Plant Physiology* 149: 2000–12.

Park, S. Y., Fung, P., Nishimura, N., Jensen, D. R., Fujii, H., and Zhao, Y. 2009. Abscisic acid inhibits type2C protein phosphatases via the PYR/PYL family of START proteins. *Science* 324: 1068–71.

Paul, S., Gayen, D., Datta, S. K., and Datta, K. 2015. Dissecting root proteome of transgenic rice cultivars unravels metabolic alterations and accumulation of novel stress responsive proteins under drought stress. *Plant Science* 234: 133–43.

Perera, I. Y., Hung, C. Y., Moore, C. D., Stevenson, P. J., and Boss, W. F. 2008. Transgenic *Arabidopsis* plants expressing the type 1 inositol 5-phosphatase exhibit increased drought tolerance and altered abscisic acid signalling. *Plant Cell* 20: 2876–93.

Pieters, A. J., and SoukiEi, S. 2005. Effects of drought during grain filling on PSII activity in rice. *Journal of Plant Physiology* 162: 903–11.

Pieterse, C. M., Van der Does, D., Zamioudis, C., Leon-Reyes, A., and VanWees, S. C. 2012. Hormonal modulation of plant immunity. *Annual Review of Cell and Developmental Biology* 28: 489–521.

Pizzio, G. A., Rodriguez, L., Antoni, R., Gonzalez-Guzman, M., Yunta, C., and Merilo, E. 2013. The PYL4 A194T mutant uncovers a key role of PYR1- LIKE4/PROTEIN PHOSPHATASE 2CA interaction for abscisic acid signalling and plant drought resistance. *Plant Physiology* 163: 441–55.

Popescu, S. C., Popescu, G. V., Bachan, S., et al. 2007. Differential binding of calmodulin-related proteins to their targets revealed through high-density *Arabidopsis* protein microarrays. *Proceedings of the National Academy of Sciences of the United States of America* 104: 4730–5.

Pottosin, I., Velarde-Buendía, A. M., Bose, J., et al. 2014. Cross-talk between reactive oxygen species and polyamines in regulation of ion transport across the plasma membrane: Implications for plant adaptive responses. *Journal of Experimental Botany* 65: 1271–83.

Prince, S. J., Murphy, M., Mutava, R. N., et al. 2017. Root xylem plasticity to improve water use and yield in water-stressed soybean. *Journal of Experimental Botany* 68: 2027–36.

Qian, B., Li, X., Liu, X., Chen, P., Ren, C., and Chuanchao, D. 2015. Enhanced drought tolerance in transgenic rice over-expressing of maize C4 phosphoenolpyruvate carboxylase gene via NO and Ca^{2+}. *Journal of Plant Physiology* 175: 9–20.

Qiao, H., Chang, K. N., Yazaki, J., and Ecker, J. R. 2009. Interplay between ethylene, ETP1/ETP2 F-box proteins, and degradation of EIN2 triggers ethylene responses in *Arabidopsis. Genes and Development* 23(4): 512–21.

Quan, R., Hu, S., Zhang, Z., Zhang, H., Zhang, Z., and Huang, R. 2010. Overexpression of an ERF transcription factor TSRF1 improves rice drought tolerance. *Plant Biotechnology Journal* 8: 476–88.

Rao, S. C., Lal, R., Prasad, J. V. N. S., Gopinath, K. A., et al. 2015. Potential and challenges of rainfed farming in India. *Advances in Agronomy* 133: 113–81.

Rashid, M., Guangyuan, H., Guangxiao, Y., Hussain, J., and Xu, Y. 2012. AP2/ERF transcription factor in rice: Genome-wide canvas and syntemic relationships between monocots and eudicots. *Evolutionary Bioinformatics* 8: 321–55

Redillas, M. C., Jeong, J. S., Kim, Y. S., et al. 2012. The overexpression of OsNAC9 alters the root architecture of rice plants enhancing drought resistance and grain yield under field conditions. *Plant Biotechnology Journal* 10: 792–805.

Rejeb, I. B., Pastor, V., and Mauch-Mani, B. 2014. Plant responses to simultaneous biotic and abiotic stress: Molecular mechanisms. *Plants (Basel)* 15: 458–475.

Rejeb, K. B., Vos, D. L.-D., Disquet, I. L., Leprince, A.-S., Bordenave, M., Maldiney, R., Jdey, A., Abdelly, C., and Savoure, A. 2015. Hydrogen peroxide produced by NADPH oxidases increases proline accumulation during salt or mannitol stress in *Arabidopsis thaliana. New Phytologist* 208: 1138–1148.

Rhodes, D., and Samaras, Y. 1994. Genetic control of osmoregulation in plants. In *Cellular and Molecular Physiology of Cell Volume Regulation*, ed. Strange, S. K., pp. 347–61. CRC Press, Boca Raton.

Riemann, M., Dhakarey, R., Hazman, M., Miro, B., Kohli, A., and Nick, P. 2015. Exploring jasmonates in the hormonal network of drought and salinity responses. *Frontier in Plant Sciences* 6: 1077.

Rizhsky, L., Liang, H., Shuman, J., Shulaev, V., Davletova, S., and Mittler, R. 2004. When defence pathways collide. The response of *Arabidopsis* to a combination of drought and heat stress. *Plant Physiology* 134: 1683–96.

Rohde, A., Morreel, K., Ralph, J., Goeminne, G., and Hostyn, V. 2004. Molecular phenotyping of the pal1 and pal2 mutants of *Arabidopsis thaliana* reveals far-reaching consequences on phenylpropanoid, amino acid, and carbohydrate metabolism. *Plant Cell* 16: 2749–71.

Rui, W., Lina, D., L., José, P. P., et al. 2018. The *6xABRE* synthetic promoter enables the spatiotemporal analysis of ABA-mediated transcriptional regulation. *Plant Physiology* 177: 1650–65.

Samson, B. K., Hasan, H., and Wade, L. J. 2002. Penetration of hardpans by rice lines in the rainfed lowlands. *Field Crops Research* 76: 175–88.

Sang, J., Jiang, M., Lin, F., et al. 2008. Nitric oxide reduced hydrogen peroxide accumulation involved in water stress induced subcellular antioxidant defence in maize plants. *Journal of Integrative Plant Biology* 50: 231–43.

Sarker, U., and Oba, S. 2018. Response of nutrients, minerals, antioxidant leaf pigments, vitamins, polyphenol, flavonoid and antioxidant activity in selected vegetable amaranth under four soil water content. *Food Chemistry* 252: 72–83.

Saruhashi, M., Ghosh, T. K., Arai, K., et al. 2015. Plant Raf-like kinase integrates abscisic acid and hyper osmotic stress signalling upstream of SNF1-related protein kinase2. *Proceedings of the National Academy of Sciences of the United States of America* 112(46): E6388–96.

Schmidt, R., Schippers, J. H., Mieulet, D., Obata, T., Fernie, A. R., and Guiderdoni, E. 2013. MULTIPASS, a rice R2R3-type MYB transcription factor, regulates adaptive growth by integrating multiple hormonal pathways. *Plant Journal* 76: 258–73.

Seo, P. J., and Park, C. M. 2010. MYB96-mediated abscisic acid signals induce pathogen resistance response by promoting salicylic acid biosynthesis in *Arabidopsis. New Phytologist* 186: 471–83.

Sewelam, N., Kazan, K., Thomas-Hall, S. R., Kidd, B. N., Manners, J. M., and Schenk, P. M. 2013. Ethylene response factor 6 is a regulator of reactive oxygen species signalling in *Arabidopsis. PLoS One* 8(8): e70289.

Shan, Q., Wang, Y., Li, J., Zhang, Y., Chen, K., and Liang, Z. 2013. Targeted genome modification of crop plants using a CRISPR-Cas system. *Nature Biotechnology* 31: 686–88.

Sharma, I., Bhardwaj, R., and Pati, P. K. 2013a. Stress modulation response of 24-epibrassinolide against imidacloprid in an elite indica rice variety Pusa Basmati-1. *Pesticide Biochemistry and Physiology* 105: 144–53.

Sharma, I., Ching, E., Saini, S., Bhardwaj, R., and Pati, P. K. 2013b. Exogenous application of brassinosteroid offers tolerance to salinity by altering stress responses in rice variety Pusa Basmati-1. *Plant Physiology and Biochemistry* 69: 17–26.

Sharoni, A. M., Nuruzzaman, M., Satoh, K., Shimizu, T., Kondoh, H., and Sasaya, T. 2011. Gene structures, classification and expression models of the AP2/EREBP transcription factor family in rice. *Plant and Cell Physiology* 52: 344–60.

Shavrukov, Y., Baho, M., Lopato, S., and Langridge, P. 2016. The TaDREB3 transgene transferred by conventional crossings to different genetic backgrounds of bread wheat improves drought tolerance. *Plant Biotechnology Journal* 14: 313–22.

Shehab, G. G., Ahmed, O. K., and El-Beltagi, H. S. 2010. Effects of various chemical agents for alleviation of drought stress in rice plants (*Oryza sativa* L.). *Notulae Botanicae Horti Agrobotanici Cluj-Napoca* 38: 139–48.

Shi, H., Chen, L., Ye, T., Liu, X., Ding, K., and Chan, Z. 2014. Modulation of auxin content in *Arabidopsis* confers improved drought stress resistance. *Plant Physiology and Biochemistry* 82: 209–17.

Shi, J., Gao, H., Wang, H., et al. 2017. ARGOS8 variants generated by CRISPR-Cas9 improve maize grain yield under field drought stress conditions. *Plant Biotechnology Journal* 15(2): 207–16.

Shim, J. S., Oh, N., Pil, J. C., Kim, Y. S., Choi, Y. D., and Kim, J. K. 2018. Overexpression of *OsNAC14* improves drought tolerance in rice. *Frontier in Plant Sciences* 9: 310.

Shiono, K., Ogawa, S., Yamazaki, S., et al. 2011. Contrasting dynamics of radial O2-loss barrier induction and aerenchyma formation in rice roots of two lengths. *Annals of Botany* 107: 89–99.

Shukla, P. S., Agarwal, P., Gupta, K., and Agarwal, P. K. 2015. Molecular characterization of a MYB transcription factor from a succulent halophyte involved in stress tolerance. *AOB Plants* 7: plv054.

Singh, A., Jha, S. K., Bagri, J., and Pandey, G. K. 2015. ABA inducible rice protein phosphatase 2C confers ABA insensitivity and abiotic stress tolerance in *Arabidopsis*. *PLoS ONE* 10(4): e0125168.

Singh, C. M., Kumar, B., Mehandi, S., and Chandra, K. 2012. Effect of drought stress in rice: A review on morphological and physiological characteristics. *Trends in Biosciences Journal* 54: 261–65.

Singh, O. N. 2014. Morphophysiological traits associated with reproductive stage drought tolerance of rice (*Oryza sativa* L.) genotypes under rain-fed condition of eastern Indo-Gangetic Plain. *Indian Journal of Plant Physiology* 19(2): 87–93.

Song, C. P., Agarwal, M., Ohta, M., et al. 2005. Role of an *Arabidopsis* AP2/EREBP-type transcriptional repressor in abscisic acid and drought stress responses. *Plant Cell* 17: 2384–96.

Song, S. Y., Chen, Y., Chen, J., Dai, X. Y., and Zhang, W. H. 2011. Physiological mechanisms underlying OsNAC5-dependent tolerance of rice plants to abiotic stress. *Planta* 234: 331–45.

Soon, F. F., Ng, L. M., Zhou, X. E., et al. 2012. Molecular mimicry regulates ABA signalling by SnRK2 kinases and PP2C phosphatases. *Science* 335: 85–8.

Stepanova, A. N., and Alonso, J. M. 2009. Ethylene signalling and response: Where different regulatory modules meet. *Current Opinion in Plant Biology* 12(5): 548–55.

Sun, Y., Fan, X. Y., Cao, D. M., Tang, W., He, K., and Zhu, J. Y. 2010. Integration of brassinosteroid signal transduction with the transcription network for plant growth regulation in *Arabidopsis*. *Developmental Cell* 19: 765–77.

Tang, N., Zhang, H., Li, X., Xiao, J., and Xiong, L. 2012. Constitutive activation of transcription factor *OsbZIP46* improves drought tolerance in rice. *Plant Physiology* 158: 1755–68.

Tian, X., and Lei, Y. 2006. Nitric oxide treatment alleviates drought stress in wheat seedlings. *Biologia Plantarum* 50: 775–8.

Todaka, D., Shinozaki, K., and Yamaguchi-Shinozaki, K. 2015. Recent advances in the dissection of drought-stress regulatory networks and strategies for development of drought-tolerant transgenic rice plants. *Frontier in Plant Sciences* 6: 84.

Torres, F. M. L., Gigon, A., de Melo, D. F., Zuily, F. Y., and Pham-Thi, A. T. 2007. Drought stress and rehydration affect the balance between MGDG and DGDG synthesis in cowpea leaves. *Physiologia Plantarum* 131: 201–10.

Uchida, A., Jagendorf, A. T., Hibino, T., et al. 2002. Effects of hydrogen peroxide and nitric oxide on both salt and heat stress tolerance in rice. *Plant Science* 163: 515–23.

Umezawa, T., Fujita, M., Fujita, Y., Yamaguchi, S. K., and Shinozaki, K. 2006. Engineering drought tolerance in plants: Discovering and tailoring genes to unlock the future. *Current Opinion in Biotechnology* 17: 113–22.

Umezawa, T., Nakashima, K., Miyakawa, T., Kuromori, T., Tanokura, M., and Shinozaki, K. 2010. Molecular basis of the core regulatory network in ABA responses: Sensing, signalling and transport. *Plant and Cell Physiology* 51: 1821–39.

Umezawa, T., Sugiyama, N., Takahashi, F., et al. 2013. Genetics and phosphoproteomics reveal a protein phosphorylation network in the abscisic acid signalling pathway in *Arabidopsis thaliana*. *Science Signalling* 6(270): rs8.

Usman, M., Raheem, Z. F., Ahsan, T., Iqbal, A., Sarfaraz, Z. N., and Haq, Z. 2013. Morphological, physiological and biochemical attributes as indicators for drought tolerance in rice (*Oryza sativa* L.). *European Journal of Biological Sciences* 5(1): 23–28.

Vernoux, T., Wilson, R. C., Seeley, K. A., et al. 2000. The ROOT MERISTEMLESS1/CADMIUM SENSITIVE2 gene defines a glutathione-dependent pathway involved in initiation and maintenance of cell division during postembryonic root development. *Plant Cell* 12: 97–110.

Vilarrasa, B. J., González, G. M. P., Frigola, D., Fàbregas, N., Alexiou, K. G., and Lopez, B. N. 2014. Regulation of plant stem cell quiescence by a brassinosteroid signalling module. *Developmental Cell* 30: 36–47.

Vishwakarma, K., Upadhyay, N., Nitin, K., et al. 2017. Abscisic acid signalling and abiotic stress tolerance in plants: A review on current knowledge and future prospects. *Frontier in Plant Science* 8: 161.

Voesenek, L. A. C. J., and Bailey, S. J. 2009. Plant biology: Genetics of high-rise rice. *Nature* 460: 959–60.

Wan, L., Zhang, J., Zhang, H., Zhang, Z., Quan, R., and Zhou, S. 2011. Transcriptional activation of OsDERF1 in OsERF3 and OsAP2-39 negatively modulates ethylene synthesis and drought tolerance in rice. *PLoS ONE* 6(9): e25216.

Wang, H., Wang, H., Shao, H., and Tang, X. 2016. Recent advances in utilizing transcription factors to improve plant abiotic stress tolerance by transgenic technology. *Frontier in Plant Science* 7: 67.

Wang, L., Chen, L., Rui, L., et al. 2017. Reduced drought tolerance by CRISPR/Cas9-mediated SlMAPK3 mutagenesis in tomato plants. *Journal of Agricultural and Food Chemistry* 65: 8674–82.

Wang, P., Xue, L., Batelli, G., Lee, S., Hou, Y. J., and VanOosten, M. J. 2013. Quantitative phosphorproteomics identifies SnRK2 protein kinase substrates and reveals the effectors of abscisic acid action. *Proceedings of the National Academy of Sciences of the United States of America* 110: 11205–10.

Weijde, V. T., Huxley, L. M., and Hawkins, S. 2017. Impact of drought stress on growth and quality of miscanthus for biofuel production. *Global Change Biology Bioenergy* 9: 770–82.

Wildermuth, M. C., Dewdney, J., Wu, G., and Ausubel, F. M. 2001. Isochorismate synthase is required to synthesize salicylic acid for plant defence. *Nature* 414: 562–5.

Wu, D., Ji, J., Wang, G., Guan, C., and Jin, C. 2014. *LchERF*, a novel ethylene responsive transcription factor from *Lycium chinense*, confers salt tolerance in transgenic tobacco. *Plant Cell Reports* 33: 2033–45.

Wu, L., Zhang, Z., Zhang, H., Wang, X. C., and Huang, R. 2008. Transcriptional modulation of ethylene response factor protein JERF3 in the oxidative stress response enhances tolerance of tobacco seedlings to salt, drought, and freezing. *Plant Physiology* 148: 1953–63.

Wu, X., Qiao, Z., Liu, H., Acharya, B. R., Li, C., and Zhang, W. 2017a. CML 20, an *Arabidopsis* calmodulin-like protein, negatively regulates guard cell ABA signalling and drought stress tolerance. *Frontiers in Plant Science* 8: 824.

Wu, X., Cai, K., Zhang, G., and Zeng, F. 2017b. Metabolite profiling of barley grains subjected to water stress: To explain the genotypic difference in drought-induced impacts on malting quality. *Frontier in Plant Sciences* 8: 1547.

Wu, Y., Zhang, D., Jee, C. Y., et al. 2012. The *Arabidopsis* NPR1 protein is a receptor for the plant defence hormone salicylic acid. *Cell Reports* 1: 639–47.

Xia, X. J., Wang, Y. J., Zhou, Y. H., Tao, Y., Mao, W. H., and Shi, K. 2009. Reactive oxygen species are involved in brassinosteroid induced stress tolerance in cucumber. *Plant Physiology* 150: 801–14.

Xiong, H., Li, J., Liu, P., Duan, J., Zhao, Y., Guo, X., and Li, Y. 2014. Overexpression of OsMYB48-1, a novel MYB-related transcription factor, enhances drought and salinity tolerance in rice. *PLoS ONE* 9(3): e92913.

Xiong, H., Yu, J., Miao, J., et al. 2018. Natural variation in *OsLG3* increases drought tolerance in rice by inducing ROS scavenging. *Plant Physiology* 178: 451–67.

Xiong, L., and Zhu, J. K. 2003. Molecular and genetic aspects of plant responses to osmotic stress. *Plant Cell and Environment* 25: 131–9.

Yang, A., Dai, X., and Zhang, W. H. 2014. AR2R3-type MYB gene, OsMYB2, is involved in salt, cold, and dehydration tolerance in rice. *Journal of Experimental Botany* 63: 2541–56.

Yang, Z., Tian, L., Latoszek-Green, M., Brown, D., and Wu, K. 2005. *Arabidopsis* ERF4 is a transcriptional repressor capable of modulating ethylene and abscisic acid responses. *Plant Molecular Biology* 58: 585–96.

Ye, H., Liu, S., Tang, B., Chen, J., Xie, Z., and Nolan, T. M. 2017. RD26 mediate crosstalk between drought and brassinosteroid signalling pathways. *Nature Communications* 24: 14573.

Yiu, J. C., Liu, C. W., Fang, D. Y. T., and Lai, Y. S. 2009. Waterlogging tolerance of Welsh onion (*Allium fistulosum* L.) enhanced by exogenous spermidine and spermine. *Plant Physiology and Biochemistry* 47: 710–6.

Yoo, C. Y., Pence, H. E., Jin, J. B., et al. 2010. The *Arabidopsis* GTL1 transcription factor regulates water use efficiency and drought tolerance by modulating stomatal density via trans repression of SDD1. *Plant Cell* 22: 4128–41.

Yoo, S. D., Cho, Y. H., Tena, G., Xiong, Y., and Sheen, J. 2008. Dual control of nuclear EIN3 by bifurcate MAPK cascades in C2H4 signalling. *Nature* 451: 789–95.

Yoshida, R., Hobo, T., Ichimura, K., et al. 2002. ABA-activated SnRK2 protein kinase is required for dehydration stress signalling in *Arabidopsis. Plant Cell and Physiology* 43: 1473–83.

Yoshida, T., Fujita, Y., Maruyama, K., Mogami, J., Todaka, D., and Shinozaki, K. 2015. Four *Arabidopsis* AREB/ABF transcription factors function predominantly in gene expression downstream of SnRK2 kinases in abscisic acid signalling in response to osmotic stress. *Plant Cell and Environment* 38: 35–49.

Yoshimura, K., Masuda, A., Kuwano, M., Yokota, A., and Akashi, K. 2008. Programmed proteome response for drought avoidance/tolerance in the root of a C(3) xerophyte (wild watermelon) under water deficits. *Plant and Cell Physiology* 49: 226–41.

Yuan, S., and Lin, H. H. 2008. Role of salicylic acid in plant abiotic stress. *Z Naturforsch C* 63: 313–20.

Zhang, A., Ren, H. M., Tan, Y. Q., et al. 2016. S-type anion channels SLAC1 and SLAH3 function as essential negative regulators of inward K+ channels and stomatal opening in *Arabidopsis. Plant Physiology* 28: 949–65.

Zhang, G., Chen, M., Li, L., Xu, Z., Chen, X., and Guo, J. 2009. Overexpression of the soybean *GmERF3* gene, an AP2/ERF type transcription factor for increased tolerances to salt, drought, and diseases in transgenic tobacco. *Journal of Experimental Botany* 60: 3781–96.

Zhang, H., Liu, W., Wan, L., Li, F., Dai, L., and Li, D. 2010. Functional analyses of ethylene response factor JERF3 with the aim of improving tolerance to drought and osmotic stress in transgenic rice. *Transgenic Research* 19: 809–18.

Zhang, H., Niu, X., Liu, J., Xiao, F., Cao, S., and Liu, Y. 2013. RNAi-directed down regulation of vacuoler H+-ATPase subunit A results in enhanced stomatal aperture and density in rice. *PloS ONE* 8 (7): e69046.

Zhang, H., Zhang, J., Wei, P., et al. 2014. The CRISPR/Cas9 system produces specific and homozygous targeted gene editing in rice in one generation. *Plant Biotechnology Journal* 12: 797–807.

Zhang, L., Zhang, L., Xia, C., et al. 2015. A novel wheat bZIP transcription factor, TabZIP60, confers multiple abiotic stress tolerances in transgenic *Arabidopsis. Plant Physiology* 153(4): 538–54.

Zhang, X., Wang, H., and Takemiya, A. 2004. Inhibition of blue light-dependent H+ pumping by abscisic acid through hydrogen peroxide-induced dephosphorylation of the plasma membrane H+-ATPase in guard cell protoplasts. *Plant Physiology* 136: 4150–8.

Zhao, Y., Xing, L., Wang, X., et al. 2014. The ABA receptor PYL8 promotes lateral root growth by enhancing MYB77-dependent transcription of auxin-responsive genes. *Science Signal* 7: ra53.

Zhiponova, M. K., Vanhoutte, I., Boudolf, V., Betti, C., Dhondt, S., and Coppens, F. 2013. Brassinosteroid production and signalling differentially control cell division and expansion in the leaf. *New Phytologist* 197: 490–502.

Zhou, J., Diao, X., Wang, T., Chen, G., Lin, Q., Yang, X., and Xu, J. 2018. Phylogenetic diversity and antioxidant activities of culturable fungal endophytes associated with the mangrove species *Rhizophora stylosa* and *R. mucronata* in the South China Sea. *PLoS ONE* 13(6): e0197359.

Zhu, Y., Zuo, M., Liang, Y., Jiang, M., Zhang, J., and Scheller, H. V. 2013. MAP65-1a positively regulates H_2O_2 amplification and enhances brassinosteroid-induced antioxidant defence in maize. *Journal of Experimental Botany* 64: 3787–802.

12 Aquaporins
A Promising Gene Family for Tackling Stresses for Crop Improvement

Rahul B. Nitnavare, Aishwarya R. Shankhapal, Momina Shanwaz, Pooja Bhatnagar-Mathur, and Palakolanu Sudhakar Reddy

CONTENTS

12.1 INTRODUCTION

Water is the most abundant molecule found in all living cells, and the molecular processes that define life take place largely in an aqueous environment. One of the remarkable features of life in an aqueous environment was that it facilitated the production of membrane structures that allow the compartmentation of biological processes. During the course of evolution, specific protein molecules were created that allow for the selective transport of the necessary compounds into and out of the cell to maintain cellular processes and homeostasis, thus enhancing the capabilities of the membrane. As a result of this flux of metabolites, the transport of water across membranes became more complex, and the need for osmoregulation to maintain osmotic balance and cell volume became critical. Water could penetrate the cell membranes by diffusion and by transport through water channels (pores). AQPs were first demonstrated in erythrocytes and kidney collecting ducts. Later on, AQP-like proteins homologous to animal AQPs were found in some bacteria, fungi, and plants. On the basis of sequence conservation and localization, plant AQPs are classified into several subfamilies including plasma membrane intrinsic protein (PIPs), tonoplast intrinsic protein (TIPs), nodulin intrinsic proteins (NIPs), and small intrinsic proteins (SIPs) (Postaire et al., 2008). In its turn, the subfamily PIP is subdivided into two phylogenetic groups, PIP1 and PIP2, differing mainly in the structure of their N terminal domains: it is longer in PIP1 AQPs (Chaumont et al., 2001). Moreover, the recently discovered X intrinsic proteins (XIPs) are also part of the MIPs family now and also considered as the fifth subfamily of MIPs. The XIPs have been characterized in protozoa, fungi, mosses, and dicots. Interestingly, XIP homologs were absent from monocots (Danielson and Johanson 2008). PIP and NIPs are generally localized in the plasma membrane and expressed on the entire cell surface, while TIPs are localized to the tonoplast, the membrane of the vacuole. For most plant AQPs, localization on the endoplasmic reticulum (ER) can be observed during the processes of post-transcription, translation, and modification. However, SIPs and some NIPs have been found to localize in the ER, although the mechanism of targeting and their cellular functions are still not clear. Another difference between various MIP subfamilies is their substrate selectivity. Two factors that contribute to their substrate selectivity are the conserved NPA motifs and amino acid residues including the ar/R (aromatic/arginine) region (Forrest and Bhave 2007), which are highly conserved in plant PIPs and TIPs, while alternative motifs have been found only in the NIP or SIP groups (Ishibashi, 2006; Ishibashi et al., 2000) (Figure 12.1).

The persistence of AQPs during evolution indicates their key role in water transport and common occurrence in the plant kingdom. AQP molecules form tetramers in the membrane, and each of four subunits is an individual water-transporting pore (Postaire et al., 2008, Maurel et al., 2008). AQP monomers are hydrophobic transmembrane proteins comprising six α-helical domains connected through two cytosolic and three extracytosolic loops. The N and C terminals of the AQP molecule are located in the cytoplasm. In the second and third loops connecting transmembrane domains, there are two short α-helical domains directed toward each other. Just these domains are involved in the formation of the water channel, being closely positioned within the molecule. Such a structure of the AQP molecule was called the "hour glass model". Water transport through the AQPs occurs passively along the gradient of the water potential and without energy consumption (Maurel and Chrispeels 2001). It can occur in both directions (into and from the cell), being driven by the direction of the osmotic gradient. AQPs are involved in three pathways for water inflow to plant organs, namely in the transcellular water transport, in apoplastic, and in symplastic transport. The transcellular pathway represents sequential membrane crossing during water transport.

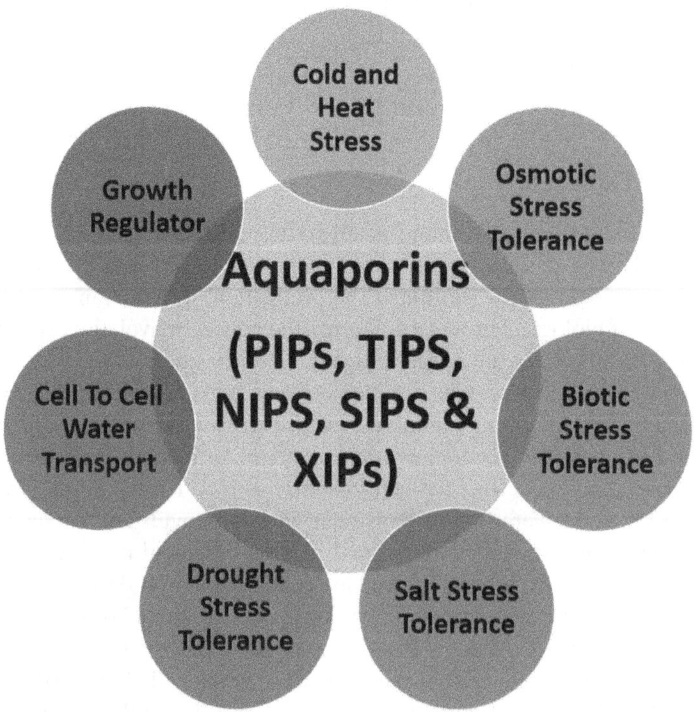

FIGURE 12.1 Summary of various functions of AQPs.

Water influx into the cell is limited only by plasmalemma AQPs, whereas the delivery of excessive water to the vacuole is limited by the tonoplast AQPs, the osmotic permeability of which for water is much higher (Javot and Maurel 2002; Trofimova et al., 2001). The rate of transmembrane transport of water depends on AQP content in the membranes and a specific permeability of each individual isoform for water.

Functional genomics and reverse genetics approaches allow a more rigorous approach and may reveal unexpected functions of AQPs. Transgenic *Arabidopsis* plants expressing an antisense copy of the *pip1b* gene showed reduced expression of several *PIP1* homologs and provided definitive evidence for the contribution of AQPs to plasma membrane water transport. Surprisingly, these antisense plants showed an increased root mass, whereas the development of the shoot was unchanged. Even though this phenotype might be related to the old observation that the root/shoot ratio of plants adjusts in response to their water status, it directly emphasizes how membrane transport can influence the developmental plasticity of plants. In the near future, analysis of single knockout aquaporin mutants will hopefully provide evidence for the multiple functions of AQPs in the growth and development of plants and in their adaptive response to stresses. It has been fascinating to observe during the last few years, how the discovery of AQPs has challenged general concepts about the role of membranes in plant–water relations. At one time it was assumed by most plant biologists that the residual water permeability of plant membrane lipids was sufficient for water flow in plants. Enthusiasm about the discovery of AQPs led to the unrealistic proposition that transmembrane water flow must be necessarily mediated by these proteins. The truth must lie somewhere in between, and we still have a long way to go to fully understand the significance of these proteins. Nevertheless, AQPs provide a unique molecular entry point into the water relations of plants and establish fascinating connections between water transport, plant development, and the adaptive responses of plants to their ever-changing environment. Abiotic stress tolerances are governed by multiple gene families involved in multiple mechanisms that may be expressed at different plant growth stages (Foolad, 1999). Functional genomics employs multiple parallel approaches including global transcriptional profiling coupled with the use of mutants and transgenic which helps high-throughput gene function studies (Vij and Tyagi, 2007). Since the area of functional genomics is very extensive, this chapter will focus on recent updates on the role of *AQP* genes in crop improvement in terms of achieving stress tolerance (Table 12.1).

12.2 DISCOVERY AND DIVERSITY OF THE AQPS

For many years it was believed that the transport of water across biological membranes occurred by simple passive diffusion through the lipid bilayer. However, some membranes exhibited such high water permeabilities that they could not be explained by simple diffusion, leading to the hypothesis of proteinaceous membrane components which could facilitate the rapid low-energy transport of water across the bilayer (Finkelstein, 1987). The conductance of water and control of transpiration steam is mainly done by vascular tissues and guard cells. Water flows through living cells while getting in or out of vascular tissues. When water flows across living tissues, it follows apoplastic or symplastic, pathways, or traverses the cell membranes. It is believed that the discovery of a class of water channel proteins named AQPs (Agre et al., 1998) has become the milestone in understanding the conductance of living cells. Because of their abundance, plant MIP homologs were identified in the late 1980s (Fortin et al., 1987), but it was recognized much later that some of them can function as highly efficient water channels and facilitate the diffusion of enormous amounts of water along transmembrane water potential gradients (Maurel et al., 1993). The discovery of AQPs establishes a conceptual advance in plants. It had been hypothesized that in certain specialized membranes, the high rate of water conductance was caused by proteinaceous membrane channels, but the molecular characterization of these channels remained elusive until the latter part of the 20th century. Aquaporin discovery allowed the understanding of a new notion about the dynamics of rapid and controlled water transport across membranes at the rate exceeding that of water diffusion.

TABLE 12.1
Summarizing a Few Illustrations of Aquaporins Transgenics Developed to Tackle Abiotic Stress Conditions such as Drought (Water), Salt (Osmotic and Salinity) and Cold

S. No.	AQP Gene	Transgenic Plant	Stress	Type of Response	Reference
1	*ScPIP1*	*Arabidopsis thaliana*	Drought	Resistance	Wang et al. (2019)
2	*ThPIP2;5*	*Tamarix hispida Arabidopsis thaliana*	Osmotic and salt	Enhanced seed germination, ROS-scavenging capability, antioxidant enzymes activities	Wang et al. (2018)
3	*PeTIP4;1*	*Arabidopsis thaliana*	Drought and salinity	Resistance	Sun et al. (2017)
4	*MdPIP1;3*	*Solanum lycopersicum* L.	Drought	Resistance	Wang et al. (2017)
5	*MaPIP1;1*	*Arabidopsis thaliana*	Drought	Resistance	Xu et al. (2014)
6	*OsPIP1;1*	*Oryza sativa*	Salinity	Resistance	Liu et al. (2013)
7	*MusaPIP1;2*	*Musa acuminata*	Cold and drought	Resistance	Sreedharan et al. (2013)
8	*TaAQP7*	*Nicotiana tabacum*	Drought and osmotic stress	Resistance	Zhou et al. (2012)
9	*VfPIP1*	*Arabidopsis thaliana*	Drought	Resistance	Uehlein et al. (2012)
10	*TdPIP1;1*	*Nicotiana tabacum*	Salinity, water stress	Resistance	Ayadi et al. (2011)
11	*SlTIP2;2*	*Solanum lycopersicum*	Salinity	Resistance	Sade et al. (2009)
12	*VvPIP2;4*	*Vitis vinifera*	Water stress	Sensitive	Vandeleur et al. (2009)
13	*StPIP1*	*Nicotiana tabacum*	Water and osmotic stress	Sensitive	Wu et al. (2009)
14	*AtPIP1;4, AtPIP2;5*	*Arabidopsis thaliana, Nicotiana tabacum*	Water, salinity, and cold stress	No effect	Jang et al. (2007)
15	*PgTIP1*	*Arabidopsis thaliana*	Salt and drought	Resistance	Peng et al. (2007)
16	*BnPIP1*	*Nicotiana tabacum*	Drought	Resistance	Yu et al. (2005)
17	*RWC3*	*Oryza sativa*	Water stress	Resistance	Lian et al. (2004)
18	*AtPIP1;b*	*Nicotiana tabacum*	Salinity	Sensitive	Aharon et al. (2003)
19	*HvPIP2;1*	*Oryza sativa*	Salinity	Sensitive	Katsuhara et al. (2002)

The genome sequencing projects have enabled many researchers to study aquaporins from different models and crops plants too, viz. in *Arabidopsis* (Johanson et al., 2001 Quigley et al., 2001), sorghum (Reddy et al., 2015), maize (Chaumont et al., 2001), populus (Gupta and Sankararamakrishnan, 2009), upland cotton (Park et al., 2010), soybean (Zhang et al., 2013), potato (Venkatesh et al., 2013), and tomato (Reuscher et al., 2013), rice (Sakurai et al., 2005) and pearl millet (Reddy et al., 2017). In addition, expressed sequence tags corresponding to 36 aquaporin isoforms have been identified in maize (Chaumont et al., 2001). Plant AQPs have been classified into their subfamilies based on substrate specificity and protein sequence homology. They can be subdivided into four subgroups which to some extent correspond to distinct sub-cellular localizations (Johanson et al., 2001; Quigley et al., 2001; Chaumont et al., 2001). The TIPs and PIPs correspond to aquaporins that are abundantly expressed in the vacuolar and plasma membranes, respectively. PIPs are further subdivided into two phylogenetic subgroups, PIP1 and PIP2. Because of their abundance, PIPs and TIPs represent central pathways for transcellular and intracellular water transport (Maurel et al., 2002 Tyerman et al., 1999; Wallace et al., 2006). A third subgroup comprises nodulin26-like intrinsic membrane proteins (NIP), i.e. aquaporins that are close homologs of GmNod26, an abundant aquaporin in the peribacteroid membrane of symbiotic nitrogen-fixing nodules of soybean roots (Wallace et al., 2006). NIPs are present in non-leguminous plants where they have been localized in plasma and intracellular membranes (Ma et al., 2006; Mizutani et al., 2006; Takano et al., 2006). The small basic intrinsic proteins (SIP) define the fourth plant aquaporin subgroup that was first uncovered from genome sequence analysis (Johanson and Gustavsson 2002). SIPs form a small class of two to three divergent aquaporin homologs and are mostly localized in the endoplasmic reticulum (ER) (Ishikawa et al., 2005). Moreover, the recently discovered X intrinsic proteins (XIPs) are also part of MIPs family now and also considered as the fifth subfamily of MIPs. They were first discovered in upland cotton (Park et al., 2010). The XIPs have been characterized in protozoa, fungi, mosses, and dicots. In grapevine, VvXIP1 was found to play a role in osmotic regulation in addition to H_2O_2 transport and metal homeostasis (Noronha et al., 2016). Interestingly, XIP homologs were absent from monocots (Danielson and Johanson 2008).

To date, the expression in *Xenopus* oocytes is still widely used because in these cells individual gene products can be tested in a well-characterized membrane environment (Zhang and Verkman, 1991). However, other systems like expression in slime mold, yeast secretory vesicles, or baculovirus-infected insect cells have been successfully used to demonstrate aquaporin activity and to determine transport rates (Laizé et al., 1995; Verkman, 1995; Chaumont et al., 1997; Yang and Verkman, 1997). It is really tough to explain the exact role of AQPs in maintaining the plant water status under different stress conditions, because different *AQP* genes may be differentially expressed or may remain unchanged under abiotic stresses. Considering that AQPs may function in transport processes of other molecules in addition to water, understanding the nature of these complex changes may prove challenging. These complex expression patterns also illustrated that the water budget levels are maintained by increased or reduced cell-to-cell water transport via AQPs under abiotic stress conditions (Javot and Maurel, 2002). Therefore, the functional validation of each *AQP* gene in the plant system is of the utmost necessity and is a bottleneck for understanding the water transport mechanism.

12.3 AQUAPORIN AS POWERFUL ARSENAL TO COMBAT VARIOUS STRESSES

12.3.1 Drought

It is a well-established fact that the water uptake and transcellular water flow in roots are largely mediated by PIPs and TIPs in most plant species. Furthermore, these two subfamilies are the most abundantly present in plant cells (Boursiac et al., 2008). Comparative transcriptome studies have shown differential expression of multiple AQP homologs in response to drought stress suggesting definite roles in stress responses. Overexpressing AQP genes to study their role in stress tolerance was believed to be a milestone in elucidating the molecular mechanisms underlying the regulation of AQP function in plants (Rizhsky et al., 2004). Elevated levels of *AtPIP2;3* under drought stress conditions are one of the earliest pieces of evidence that AQP plays central role in drought tolerance (Yamaguchi-Shinozaki, et al., 1992). Plant water retention and improving water use efficiency in the cell are considered the important goals for achieving abiotic stress tolerance. Correspondingly, the expression of *PIP1* of *Vicia faba* (*VfPIP1*) in transgenic *Arabidopsis* improved drought resistance by the reduction of transpiration rates through stomatal closure (Uehlein et al., 2012). Ranganathan et al. (2016) demonstrated that the gas exchange rates, water use efficiency, and hydraulic conductivity of the transgenic aspen lines were significantly higher than the wild-types; however, plant biomass and dry weight were unchanged. The plants exposed to a combination of heat and drought stress presented significantly higher expression of *AtPIP2;5* (Rizhsky et al., 2004). Contradictorily, substantial down-regulation of *PIP* genes under drought stress was detected in peach fruits (Sugaya et al., 2002), in the roots and twigs of olive plants (Secchi et al., 2007), and in the roots of tobacco (Mahdieh et al., 2008). Noticeably, in many other plant species, differential responses by the same aquaporin homologs have been seen between different cultivars of the same plant species. For instance, in grapevines the expression of *VvPIP1;1* in the root was up-regulated by drought stress in an isohydric cultivar but not in an anisohydric cultivar (Vandeleur et al., 2009). Furthermore, expression of the *VvPIP2;1* gene was down-regulated under drought conditions (Cramer et al., 2007). Recently, the characterization of *GoPIP1* from the legume forage *Galega orientalis* displayed its association with drought tolerance. The transcripts levels of *GoPIP1* increased significantly in roots upon exposure to the osmotic stress imposed by both high NaCl concentration and PEG. Overexpression of this gene in transgenic *Arabidopsis* made the plants more vulnerable to drought stress but not to salinity stress (Li et al., 2015). In a drought-tolerant Vitis hybrid, Richter-110 (*Vitis berlandieri* X *Vitis rupestris*) expression of five PIPs and two TIPs was checked at different levels of water stress, and it was found that the AQP genes in the leaves showed differential regulation in response to moderate and drastic water stress. A moderate decrease in water availability results in down-regulation of the AQPs. However, in roots, aquaporin expression revealed complex patterns, with no generality among different AQPs (Galmes et al., 2007). Kaldenhoff et al. (1998) showed a higher expression of *AtPIP2;2* in several root cell types including the endodermis of *Arabidopsis*. Additionally knockout mutant plants showed reduced Lpr by 14% when compared to the corresponding wild-type plants. Lately, the transcript abundance of several *PIPs* (*AtPIP1;1*, *AtPIP1;2*, *AtPIP1;4*, *AtPIP2;1*, *AtPIP2;3*, *AtPIP2;4*, and *AtPIP2;5*) in *Arabidopsis* roots was positively correlated with Lpr; this was in decent agreement with published genetic data (Lopez et al., 2013). Wang et al. (2017) screened out an ectopically expressing *MdPIP1;3* that enhanced the drought tolerance of transgenic tomatoes.

Water transport by AQPs was also found to be coupled with diurnal rhythms. The diurnal expression of PIPs in response to different intensities of drought was investigated in *Fragaria vesca*, where the PIPs were down-regulated in roots, and the expression of *FvPIP1;1* and *FvPIP2;1* was strongly correlated to the decrease in substrate moisture contents. In leaves, the amplitude of the diurnal aquaporin expression was lower in response to drought (Surbanovski et al., 2013). Many studies revealed the roles of PIPs and TIPs in drought response or water-deficient conditions. In rice, the expression of *OsTIP1;1* was up-regulated in roots and shoots in response to water stress (Liu et al., 1994). *Arabidopsis* knockout mutants of *AtPIP1;2* and *AtPIP2;2* genes were studied by Kaldenhoff et al. (1998); the study showed that there was a noteworthy reduction in the water permeability of protoplasts and a 14% decrease in Lpr, respectively, making these mutants more vulnerable to drought stress (Javot et al., 2003). On other hand, double antisense lines with lowered expression of *AtPIP1* and *AtPIP2* in *Arabidopsis* exhibited a 30-fold decrease in Lpr. Similar results were obtained in tobacco plants with antisense *NtPIP1* gene targeting, resulting in a 55% reduction in Lpr and increased sensitivity to drought. In addition to decreased Lpr, the reduced expression of *NtPIP1* also showed a significant decrease in the transpiration rate (Tr) (Siefritz et al., 2002). In moss, *Physcomitrella patens*, knockout mutants of *PpPIP2;1* and *PpPIP2;2* exhibited severe stress phenotypes when grown under water-limited conditions. These observations led researchers to believe that AQPs play a collective role as water transporters, and their reduced expressions make plants susceptible to water stress due to a lowered Lpr. Additionally, it was also postulated that decreased Tr caused reduced photosynthesis, ultimately affecting the overall survivability of the plant (Lienard et al., 2008). Consistently, transgenic plants overexpressing AQPs exhibited improved drought tolerance. Likewise, the overexpression of *BnPIP1* from *Brassica napus* in transgenic tobacco resulted in increased tolerance to drought (Yu et al., 2005). In the same way, transgenic tobacco plants overexpressing the wheat aquaporin gene *TaAQP7* (*PIP2*) were more tolerant to drought stress when compared with wild-type tobacco plants due to enhanced water retention capabilities (Zhou et al., 2012). Transgenic *Arabidopsis* plants expressing a *Vicia faba PIP1* (*VfPIP1*) showed improved drought resistance by preventing water loss through transpiration because of the induction of stomatal closure (Cui et al., 2008). In *Arabidopsis*, the overexpression of a banana PIP gene *MaPIP1;1* showed increased root growth and enhanced survival rates of transgenic plants under drought stress, when compared to wild-type plants (Xu et al., 2014). The overexpression of *PeTIP4;1–1*, a bamboo aquaporin gene, confers drought and salinity tolerance in transgenic *Arabidopsis* (Sun et al., 2017). Also, transgenic banana plants expressing banana *PIP1;2* driven by two diverse promoters exhibited higher drought tolerance. The overexpression of a tomato *SlTIP2;2* gene in transgenic tomato plants resulted in increased drought tolerance due to the ability of the plant to regulate its Tr under drought stress conditions (Sade et al., 2009). Similarly, the overexpression of Jojoba *ScPIP1* led to improved tolerance to drought stress in transgenic *Arabidopsis* by reducing membrane damage and improving osmotic adjustment (Wang et al., 2019). The above experimental evidence suggests that the overexpression of AQPs makes plants more resistant to drought stress. However, some contrasting results have also been observed because rapid water loss due to increased leaf and root hydraulic conductivity makes some plants even more vulnerable to drought stress conditions. It can be concluded that the drought stress response of AQPs is highly variable depending on stress levels, aquaporin isoform, tissue, species, presence of symbionts, and the nature of stimuli causing dehydration similar to drought stress. However, a general down-regulation of most of the *PIP* genes is thought to reduce water loss and to help prevent backflow of water to drying soil. Although TIPs are found to play a key role in controlling cell water homeostasis by rapid water transport between the vacuole and cytoplasm of plant cells, experimental evidence on their roles in response to drought stress is limited in comparison to PIPs.

12.3.2 Salt Stress

The osmatic balance of a cell is altered by salt stress which affects plant growth and development. Plant responds primarily to the salt stress by the inhibition of root-water uptake and subsequently decrease in root hydraulic conductivity (Lpr) (Boursiac et al., 2005), which is also displayed in response to drought stress. The overexpression of *PeTIP4;1–1*, a bamboo AQP gene, confers drought and salinity tolerance in transgenic *Arabidopsis* (Sun et al., 2017). The overexpression of Jojoba *ScPIP1* led to improved tolerance to salt stress in transgenic *Arabidopsis* by reducing membrane damage and improving osmotic adjustment (Wang et al., 2019). In the roots of barley, the downregulation of the *HvPIP2;1* transcript and protein product levels under osmotic stress was seen whereas up-regulation of the same transcript was recorded under salt stress (Katsuhara et al., 2002). The same results were obtained when the expression of *OsPIP1;1* was studied in leaves and roots (Liu et al., 2013). Under salt stress, *Mesembryanthemum crystallinum*, an ice plant, displayed down-regulation of *PIP* genes in roots and a *TIP* gene in leaves (Yamada et al., 1995 and Kirch et al., 2000). In maize plants, down regulation of most of the members of *ZmPIP1* and *ZmPIP2* was detected, but a transiently boosted expression of *ZmPIP1;1*, *ZmPIP1;5*, and *ZmPIP2;4* was also seen preferentially in the outer parts of the roots when maize plants were exposed to ABA-mediated salt stress, while ZmTIPs' expression was unchanged in similar conditions (Zhu et al., 2005).

The transgenic *Arabidopsis* plant, harboring the *AtTIP5;1* gene, showed tolerance to high borate concentrations, possibly suggestive of AQPs' involvement in vacuolar compartmentation of salt particles, in this case borate (Pang et al., 2010). Many genes differentially express themselves under diverse abiotic stress. The same results were obtained when Liu et al. (1994) studied the expression of *OsTIP1;1*. It showed down-regulation in response to cold stress (Sakurai et al., 2005) but up-regulation during responses to water and salinity stress. The overexpression of *PgTIP1*, *Panax ginseng* aquaporin, in *Arabidopsis* revealed improved plant growth under optimal conditions and also better tolerance to salt and drought stress (Peng et al., 2007). Similarly, in drought-tolerant, salinity-sensitive grapevine the expression of the *PIP2;1* gene was up-regulated under salt stress but down-regulated under drought (Cramer et al., 2007). Gao et al. (2010), suggested the salt stress tolerance in wheat plant is in association with ABA and other regulated pathways, by validating up-regulated expression of *TaNIP*. Overexpressed *ThPIP2;5 Tamarix hispida* aquaporin gene conferred salt and osmotic stress tolerance to transgenic *Tamarix* and *Arabidopsis* and also enhanced seed germination, ROS-scavenging capability, antioxidant enzymes activities, and seedling growth under salt and osmotic stresses (Wang et al., 2018).

12.3.3 Cold Stress

Like other stresses, cold stress is a vital abiotic stress factor that significantly confines plant growth and development. There are several research groups who have reported the early response of AQPs to cold stress. The transgenic banana plants overexpressing *MusaPIP1;2* and *MusaPIP1;2* exhibited improved tolerance to both cold and drought stress (Sreedharan et al., 2013). Cold stress tolerance was observed in transgenic tobacco plants overexpressing a wheat aquaporin *TaAQP7 (PIP2)* gene. The same plants also showed increased drought tolerance (Huang et al., 2014). The transgenic *Arabidopsis* plants overexpressing *AtPIP1;4* or *AtPIP2;5* showed enhanced tolerance to cold stress but are more susceptible to drought due to rapid water loss. On other hand, only *PIP2;5* and *PIP2;6*, which are normally among the low-expressed PIPs, are significantly up-regulated by cold stress in both the roots and aerial parts of the plant. All other *PIP* genes were found to be down regulated by cold stress, and their patterns of expression vary with the application of salt or drought stress (Jang et al., 2007). In cold-sensitive plants like rice, prolonged exposure to cold stress cause an increase in Lpr which should be regulated through root AQPs, as most, particularly *OsPIP2;5*, are found to be up-regulated (Ahamed et al., 2012). The mRNA levels of ten genes including *OsTIP1;1* and *OsTIP2;2* were significantly down-regulated, but the expression of *OsPIP1;3* increased up to 60% in roots on exposure to chilling treatment in rice (Sakurai et al., 2005). Significant up-regulation in *OsPIP1;3* was recorded in response to drought stress in a drought-tolerant rice cultivar (Lian et al., 2004). Contrastingly, *OsTIP1;1* showed up-regulation in response to water and salinity stress (Liu et al., 1994). Most abiotic stresses, including chilling, induce the production of ABA (Jang et al., 2004 and Suga et al., 2002).

12.3.4 Biotic Stress

Plant AQPs expressions are often triggered or suppressed by the symbiotic association of microorganisms. It also affects the response of aquaporin genes under various biotic and abiotic stress. This is reported in *Phaseolus vulgaris* and maize (Aroca et al., 2007 and Barzana et al., 2014). In order to elucidate the definite role of AQPs in plant disease combat, many putative members within the AQPs family need to be analyzed by using a reverse genetic-based approach. RNAseq (transcriptomic Approach) projects have provided further insight into the involvement of AQPs in host–pathogen interactions. It was assumed that in plant symbiotic relations, the re-distribution of water and nutrients takes place between the host and symbionts. The roles of NIPs in various forms of nitrogen transport (urea, ammonia) are well-known but experimental evidence is limited. Nodulin-26 was first identified in soybean root nodules and was assumed to have formed as a result of a symbiotic interaction between the plant and nitrogen-fixing bacteria, i.e. rhizobium. Many researchers are trying to elucidate the mechanism of NIP involvement in driving these nutrient exchanges between plants and symbionts or plants and pathogen interactions. The gene profiling studies done by Barzana et al. (2014) showed the roles of plant aquaporin-mediated solute transport during plant symbiosis with arbuscular mycorrhizae. These results demonstrated the aquaporin-mediated transport of glycerol from the plant to the microbe, in addition to NH_4/NH_3 from microbe to plant. Expression profiling of *Pseudomonas syringae*-infected soybean leaves showed down-regulation in a majority of the *AQP* genes (Zou et al., 2005). In citrus plants, six CsMIPs (*CsPIP1;2*, *CsPIP2;2*, *CsNIP2;2*, *CsNIP5;2*, *CsNIP6;1*, and *CsSIP1;1*) were found to be differentially expressed under the biotic stress imposed by the citrus-infecting proteobacterium, *Candidatus* Liberibacter. Comparisons of *CsPIP2;2*, *CsTIP1;2*, *CsTIP2;1*, *CsTIP2;2*, and *CsNIP5;1* expression patterns of susceptible sweet orange and tolerant rough lemon cultivars revealed that most CsMIPs are down regulated. Therefore, they could be correlated with the disease development (Martins et al., 2015 and Aritua et al., 2013). The foreseen role of AQPs in response to pathogen infection led to many studies which shed light on evidence which proves the involvement of AQPs in disease combat. Later, NIPs' involvement in biotic stress responses in maize was reported by Lawrence et al. (2013). The regulation of Si uptake through NIPs was found to be associated with plant defense against herbivory in *Festuca* spp. grasses (Hartley et al., 2015). Protein–protein interaction studies carried out by yeast two-hybrid systems revealed interactions of AQPs with bacterial and oomycete effectors (Mukhtar et al., 2011). Interactions of a cucumber mosaic virus (CMV) replication protein with TIP1 and TIP2 in the CMV1a SOS recruitment system suggested that the TIPs–CMV1a interaction potentially affects CMV replication in the host plant's tonoplasts (Kim et al., 2006).

12.4 AQUAPORIN AS TRANSPORT OF ESSENTIAL MICRONUTRIENTS

Apart from conductance of water, AQPs exhibit other functions like the transport of other substrates in various cellular processes. NIPs are known for playing a fundamental role in transporting other substrates involved in numerous cellular processes. Most NIP homologs also execute the transport of nutrients in plants. Boron is one of those minor elements, a crucial one for plant growth; however, it can be toxic when present at high concentrations. A role of AQPs in boric acid transport was first proposed by Dordas et al. (2000). B is an essential element for plant growth, development, and reproduction, and its deficiency in arable areas has drastically affected crop production worldwide. The report showing aquaporin contributing to boron transport came from *Arabidopsis. AtNIP5;1* signifies boron uptake in plants, and its gene is strikingly induced in response to boron deficiency, and boric acid transport activity of the protein was demonstrated after oocyte expression or using nip5;1 knockout plants (Takano et al., 2006). In the event of high B supply, a feedback inhibition of the *AtNIP5;1* gene is observed, thus providing strong evidence of the involvement of *NIP5;1* in B homeostasis and the adaptation of plants to B toxicity in soil. These results were further confirmed in maize, where a loss of function mutation in *ZmNIP3;1*, a maize ortholog of *AtNIP5;1*, was shown to be responsible for an abnormal phenotype caused by B deficiency. In addition, *AtNIP6;1* and *AtNIP7;1* homologs serve in B transport in *Arabidopsis*, facilitating its distribution in the shoots and anthers, respectively (Li et al., 2011). Tolerance to high B was observed in barley by a reduced expression of the *HvNIP2;1* gene (Schnurbusch et al., 2010). In addition to NIPs, an overexpression of *TIP5;1* in *Arabidopsis* suggested its involvement in the vacuolar compartmentation of borate (Pang et al., 2010).

One more key mineral is silicon (Si), which is crucial for cereal plants. It accounts for 10% of shoot dry weight in rice. Silicon plays an important role in plant defense against biotic and abiotic stresses. Si-uptake in rice plants is a well-known phenomenon. The quantitative trait loci (QTL) mapping for a Si uptake experiment in rice led to the identification of a Si transporter (OsNIP2;1) (Ma et al., 2006). Furthermore, the Si transporter NIP2;2 (Lsi6) was also identified from rice (Yamaji et al., 2008). Therefore, OsNIPs seem to play a significant role in Si transport in rice plants. Furthermore, ammonium/ammonia (NH_4+/NH_3) is an important nitrogen fertilizer for crops. Whereas NH_4+ transporters have long been identified in plants, NH_3 was initially proposed to cross the membrane by free diffusion. Yet, several TIP2 homologs of *Arabidopsis* and wheat were found to have a significant permeability to NH_3, and may therefore participate in NH_3 compartmentalization in vacuoles. This idea remains to be assessed at the whole plant level and, for instance, the overexpression of *AtTIP2;1* and *AtTIP2;3* in *Arabidopsis* failed to enhance whole-plant NH_4+/NH_3 accumulation (Loque et al., 2005). Moreover, A T-DNA knockout mutant of *Arabidopsis NIP1;1* exhibited arsenate (As) tolerance, signifying its role as an As transporter (Kamiya et al., 2009). Recently, *AtNIP3;1* has also shown to be involved in As transport. Double knockout mutants for both *AtNIP1;1* and *AtNIP3;1* show more pronounced tolerance against As stress. An additional four other isoforms, *NIP5;1*, *NIP6;1*, *NIP7;1*, and *NIP1;2*, are reported to be capable of As(III) transport on the basis of their expressions in yeast and oocytes (Bienert et al., 2008). TIP aquaporin from *P. vittata; PvTIP4;1* was shown to be involved in As(III) uptake (He et al., 2016). The transport of antimonite (Sb) by *AtNIP1;1* has also been established, and its loss of function mutant displayed an improved tolerance to high Sb stress (Kamiya et al., 2009). Plant aquaporin-mediated transport of H_2O_2 also contributed to plant defense; however a full understanding of these pathways has not been elucidated (Dynowski et al., 2008 and Hooijmaijers et al., 2012). Rodrigues et al.'s (2017) study filled the gap in our understanding of stomatal regulation and suggests a general signaling role of aquaporin in contexts involving H_2O_2.

12.5 MULTIFACETED PHYSIOLOGICAL ROLE OF AQPS

Water is a key element of all physiological processes including the growth and metabolism of the plant. The bulk flow of water through plants can take three different routes: the apoplastic route along the cell wall structure, the symplastic route from cell to cell through the plasmodesmata, and the transcellular path across the cellular membranes (Steudle and Henzler, 1995). AQPs are transmembrane water and solute transporter channels and are also considered potential regulators of plant cell–water relations, that also regulate the osmolality of cell (Wallace at al., 2006), root hydraulic conductivity (Lpr), leaf hydraulic conductivity (Siefritz et al., 2002), transpiration (Sade et al., 2010), and cell elongation (Hukin et al., 2002). Abiotic stress conditions directly influence plant–water relations and trigger an array of complicated cellular and physiological responses that lead to plant water-saving mechanisms rangin from stomatal closure to a complete cut off water loss during transpiration. To maintain balance, water-saving photosynthetic activity is reduced due to the unavailability of CO_2, ultimately leading to the production of less plant biomass. Park et al. (2010) studied the underlying mechanisms that control plant–water relations in relation to photosynthesis and their response to biotic and abiotic stresses. AQPs are explored as potential targets in developing stress-resistant crop plants as they are vital regulators of plant–water relations. A similar approach for uncoupling transpiration from light was developed in hybrid poplar (Laur and Hacke, 2013). In this particular experiment, an increased expression of *PIP1* and *PIP2* in roots was observed, and this expression was studied under low relative humidity conditions. Aquaporin expression in roots, root hydraulic conductivity, and evaporative demand (based on meteorological factors) were also determined in rice plants grown under field conditions, and were significantly correlated, in good agreement with observations made in growth chamber experiments (Hayashi et al., 2015). In these approaches, the expression of five *PIPs* and a *TIP* showed a strong positive correlation to evaporation potential, whereas the expression of a *PIP* and a *TIP* homolog, which seem to be associated with cell elongation, showed a negative correlation. The effects of low relative humidity (i.e. high transpiration) were restricted to root hydraulics. In *Arabidopsis*, a high evaporative demand resulted in an increase by > three-fold in leaf hydraulic conductance (*K* leaf) (Levin et al., 2007). In rice leaves, a coordinated up-regulation of several *PIP* and *TIP* genes could be observed as soon as 4 h after a dry air treatment (Kuwagata et al., 2012). The signaling mechanisms which link plant transpiration to aquaporin activity in shoots and roots are as yet unclear. The rapid down-regulation of root hydraulics observed after shoot topping or defoliation may pertain to the shoot-to-root signaling involved (Liu et al., 2013; Vandeleur et al., 2009). This process was more specifically investigated in soybean and grapevine. It was proposed that a xylem-mediated hydraulic signal could be responsible for the change in root aquaporin expression observed within 0.5–1 h following shoot topping (Vandeleur et al., 2009). Conversely, the negative pressure present in xylem vessels of intact, transpiring plants perceived as an activating signal for aquaporin expression in root and shoot tissues. Their significance in all facets of plant growth and development is well-established, but the mechanistic pathways behind their roles in plant defense responses are under elucidation (Forrest and Bhave, 2007). The contribution of AQPs to transpiration control goes far beyond the issue of water transport during stomatal movements and involves emerging cellular and long-distance signaling mechanisms which ultimately act on plant growth (Maurel et al., 2016). In addition to water deficiency dehydration caused by other environmental stimuli exhibited differential responses in some species. For example, osmotic stress induced by 10% polyethylene glycol (PEG) in rice revealed no effect on *OsPIP1;3*, but the expression of *OsPIP1;1* and *OsPIP1;2* was up-regulated (Guo et al., 2006). Contrastingly, the expression of *OsPIP1;1* was down-regulated in osmotic stress administrated by mannitol and also by drought stress (Malz and Sauter 1999). Similar alterations were seen in reddish AQPs in response to salt-, PEG-, and mannitol-induced osmotic stress (Suga et al., 2002).

12.6 CONCLUSIONS

The discovery of AQPs in plants has resulted in a paradigm shift in the understanding of plant–water relations. Water flux across cell membranes has been shown to occur not only through the lipid bilayer, but also through AQPs, which are members of the major intrinsic protein super-family of channel proteins. As has been found in other organisms, plant MIPs function as membrane channels permeable to water (AQPs) and in some cases to small nonelectrolytes. AQPs greatly increase the membrane permeability for water, but may also be regulated, allowing cellular control over the rate of water influx/efflux. As a result, AQPs provide a unique molecular entry point into the water relations of plants and establish fascinating connections between water transport, plant development, and the adaptive responses of plants to their ever-changing environment. Plants counteract fluctuations in water supply by regulating all AQPs in the cell plasma membrane. AQPs can serve as markers to explore the intricate flows of water and solutes that play a critical role throughout all stages of plant development. The rate of transmembrane water flux may be controlled by changing the abundance or the activity of the AQPs. Actually, there are observations showing the alteration of water permeabilities in the responses of plants to biotic or abiotic stresses such as high salinity, nutrient deprivation, and extreme temperatures. In plants, AQPs are likely to be important both at the whole plant level, for transport of water to and from the vascular tissues, and at the cellular level, for buffering osmotic fluctuations in the cytosol. By combining molecular biology with plant physiology, it should be possible to determine the role that AQPs play in water transport in the plant. There is growing evidence that suggests that AQPs play different roles throughout plant development. Therefore, aquaporin genomic information is important because assigning physiological function via transgenic reduction or removal of gene expression requires sequence information for precise targeting. Direct determination of the location of each aquaporin within tissues is still required to understand its function in the plant. A powerful tool in elucidating the aquaporin function is given by reverse genetics that can also reveal unexpected functions of water channel proteins, which benefit our understanding of sequence–structure and structure–function relationships in plants. This should be done both at the transcript and at the protein level because aquaporin turnover appears to be variable, such as when comparing constitutively expressed and inducible AQPs. The transcriptional and/or post-translational regulation of AQPs would determine changes in membrane water permeability. Both phosphorylation and translocation to/from vesicles have been reported as post-translational mechanisms. However, translocation in plants has not yet been shown. Here, the aquaporin family is a set of genes whose functions are intuitively perceived as important; much isolated information has been accumulated, yet their function is far from being understood in living plants, and we still have a long way to go to fully understand the significance of these proteins.

ACKNOWLEDGMENTS

PSR acknowledges the SERB, Department of Science and Technology, Government of India for the research grant under Core Grant Research (CRG) – File No. EMR/2016/006726. This work was undertaken as part of the CGIAR Research Program on Grain Legumes & Dryland Cereals (CRP-GLDC). ICRISAT is a member of the CGIAR Consortium.

REFERENCES

Agre, P., Bonhivers, M., & Borgnia, M. J. (1998). The aquaporins, blueprints for cellular plumbing systems. *Journal of Biological Chemistry*, *273*(24):14659–14662.

Ahamed, A., Murai-Hatano, M., Ishikawa-Sakurai, J., Hayashi, H., Kawamura, Y., & Uemura, M. (2012). Cold stress-induced acclimation in rice is mediated by root-specific aquaporins. *Plant and Cell Physiology*, *53*(8):1445–1456.

Aharon, R., Shahak, Y., Wininger, S., Bendov, R., Kapulnik, Y., & Galili, G. (2003). Overexpression of a plasma membrane aquaporin in transgenic tobacco improves plant vigor under favorable growth conditions but not under drought or salt stress. *The Plant Cell*, *15*(2):439–447.

Aritua, V., Achor, D., Gmitter, F. G., Albrigo, G., & Wang, N. (2013). Transcriptional and microscopic analyses of citrus stem and root responses to *Candidatus* Liberibacter asiaticus infection. *PloS one*, *8*(9):e73742.

Aroca, R., Porcel, R., & Ruiz-Lozano, J. M. (2007). How does arbuscular mycorrhizal symbiosis regulate root hydraulic properties and plasma membrane aquaporins in *Phaseolus vulgaris* under drought, cold or salinity stresses? *New Phytologist*, *173*(4):808–816.

Ayadi, M., Cavez, D., Miled, N., Chaumont, F., & Masmoudi, K. (2011). Identification and characterization of two plasma membrane aquaporins in durum wheat (*Triticum turgidum* L. subsp. *durum*) and their role in abiotic stress tolerance. *Plant Physiology and Biochemistry*, *49*(9):1029–1039.

Barzana, G., Aroca, R., Bienert, G. P., Chaumont, F. O., Ruiz-Lozano, J. M. (2014). New insights into the regulation of AQPs by the arbuscular mycorrhizal symbiosis in maize plants under drought stress and possible implications for plant performance. *Molecular Plant-Microbe Interactions*, *27*:349–363.

Bienert, G. P., Thorsen, M., Schüssler, M. D., Nilsson, H. R., Wagner, A., Tamás, M. J., & Jahn, T. P. (2008). A subgroup of plant aquaporins facilitate the bi-directional diffusion of As (OH) 3 and Sb (OH) 3 across membranes. *BMC Biology*, *6*(1):26.

Boursiac, Y., Boudet, J., Postaire, O., Luu, D. T., Tournaire-Roux, C., & Maurel, C. (2008). Stimulus-induced downregulation of root water transport involves reactive oxygen species-activated cell signalling and plasma membrane intrinsic protein internalization. *The Plant Journal*, *56*(2):207–218.

Boursiac, Y., Chen, S., Luu, D. T., Sorieul, M., van den Dries, N., & Maurel, C. (2005). Early effects of salinity on water transport in *Arabidopsis* roots. Molecular and cellular features of aquaporin expression. *Plant Physiology*, *139*(2):790–805.

Chaumont, F., Barrieu, F., Wojcik, E., Chrispeels, M. J., & Jung, R. (2001). Aquaporins constitute a large and highly divergent protein family in maize. *Plant Physiology*, *125*(3):1206–1215.

Chaumont, F., Loomis, W. F., & Chrispeels, M. J. (1997). Expression of an *Arabidopsis* plasma membrane aquaporin in *Dictyostelium* results in hypoosmotic sensitivity and developmental abnormalities. *Proceedings of the National Academy of Sciences*, *94*(12):6202–6209.

Cramer, G. R., Ergül, A., Grimplet, J., Tillett, R. L., Tattersall, E. A., Bohlman, M. C., et al. (2007). Water and salinity stress in grapevines: Early and late changes in transcript and metabolite profiles. *Functional and Integrative Genomics*, *7*(2):111–134.

Cui, X. H., Hao, F. S., Chen, H., Chen, J., & Wang, X. C. (2008). Expression of the *Vicia faba* VfPIP1 gene in *Arabidopsis thaliana* plants improves their drought resistance. *Journal of Plant Research*, *121*(2):207–214.

Danielson, J. Å., & Johanson, U. (2008). Unexpected complexity of the aquaporin gene family in the moss *Physcomitrella patens*. *BMC Plant Biology*, *8*(1):45.

Dordas, C., Chrispeels, M. J., & Brown, P. H. (2000). Permeability and channel-mediated transport of boric acid across membrane vesicles isolated from squash roots. *Plant Physiology*, *124*(3):1349–1362.

Dynowski, M., Schaaf, G., Loque, D., Moran, O., & Ludewig, U. (2008). Plant plasma membrane water channels conduct the signalling molecule H_2O_2. *Biochemical Journal*, *414*(1):53–61.

Finkelstein, A. (1987). *Water Movement through Lipid Bilayers, Pores, and Plasma Membranes*. John Wiley & Sons, New York.

Foolad, M. R. (1999). Comparison of salt tolerance during seed germination and vegetative growth in tomato by QTL mapping. *Genome*, *42*(4):727–734.

Forrest, K. L., & Bhave, M. (2007). Major intrinsic proteins (MIPs) in plants: A complex gene family with major impacts on plant phenotype. *Functional and Integrative Genomics*, *7*(4):263.

Fortin, M. G., Morrison, N. A., & Verma, D. P. S. (1987). Nodulin-26, a peribacteroid membrane nodulin is expressed independently of the development of the peribacteroid compartment. *Nucleic Acids Research*, *15*(2):813–824.

Galmés, J., Pou, A., Alsina, M. M., Tomas, M., Medrano, H., & Flexas, J. (2007). Aquaporin expression in response to different water stress intensities and recovery in Richter-110 (*Vitis* sp.): Relationship with ecophysiological status. *Planta, 226*(3):671–681.

Gao, Z., He, X., Zhao, B., Zhou, C., Liang, Y., Ge, R., et al. (2010). Overexpressing a putative aquaporin gene from wheat, TaNIP, enhances salt tolerance in transgenic *Arabidopsis. Plant and Cell Physiology, 51*(5):767–775.

Guo, L., Wang, Z. Y., Lin, H., Cui, W. E., Chen, J., Liu, M., et al. (2006). Expression and functional analysis of the rice plasma-membrane intrinsic protein gene family. *Cell Research, 16*(3):277.

Gupta, A. B., & Sankararamakrishnan, R. (2009). Genome-wide analysis of major intrinsic proteins in the tree plant *Populus trichocarpa*: Characterization of XIP subfamily of aquaporins from evolutionary perspective. *BMC Plant Biology, 9*(1):134.

Hartley, S. E., Fitt, R. N., McLarnon, E. L., & Wade, R. N. (2015). Defending the leaf surface: Intra-and interspecific differences in silicon deposition in grasses in response to damage and silicon supply. *Frontiers in Plant Science, 6*:35.

Hayashi, H., Ishikawa-Sakurai, J., Murai-Hatano, M., Ahamed, A., Uemura, M. (2015). AQPs in developing rice grains. *Bioscience, Biotechnology, and Biochemistry, 79*(9):1422–1429.

He, Z., Yan, H., Chen, Y., Shen, H., Xu, W., Zhang, H., et al. (2016). An aquaporin PvTIP4; 1 from *Pteris vittata* may mediate arsenite uptake. *New Phytologist, 209*(2):746–761.

Hooijmaijers, C., Rhee, J. Y., Kwak, K. J., Chung, G. C., Horie, T., Katsuhara, M., & Kang, H. (2012). Hydrogen peroxide permeability of plasma membrane aquaporins of *Arabidopsis thaliana. Journal of Plant Research, 125*(1):147–153.

Huang, C., Zhou, S., Hu, W., Deng, X., Wei, S., Yang, G., & He, G. (2014). The wheat aquaporin gene TaAQP7 confers tolerance to cold stress in transgenic tobacco. *Zeitschrift für Naturforschung C, 69*(3–4):142–148.

Hukin, D., Doering-Saad, C., Thomas, C., & Pritchard, J. (2002). Sensitivity of cell hydraulic conductivity to mercury is coincident with symplasmic isolation and expression of plasmalemma aquaporin genes in growing maize roots. *Planta, 215*(6):1047–1056.

Ishibashi, K. (2006). Aquaporin subfamily with unusual NPA boxes. *Biochimica et Biophysica Acta (BBA)-Biomembranes, 1758*(8):989–993.

Ishibashi, K., Kuwahara, M., Kageyama, Y., Sasaki, S., Suzuki, M., & Imai, M. (2000). Molecular cloning of a new aquaporin superfamily in mammals. In: Hohmann, S. & Nielsen, S. (eds), *Molecular Biology and Physiology of Water and Solute Transport* (pp. 123–126). Springer, Boston, MA.

Ishikawa, F., Suga, S., Uemura, T., Sato, M. H., & Maeshima, M. (2005). Novel type aquaporin SIPs are mainly localized to the ER membrane and show cell-specific expression in *Arabidopsis thaliana. FEBS Letters, 579*(25):5814–5820.

Jang, J. Y., Kim, D. G., Kim, Y. O., Kim, J. S., & Kang, H. (2004). An expression analysis of a gene family encoding plasma membrane aquaporins in response to abiotic stresses in *Arabidopsis thaliana. Plant Molecular Biology, 54*(5):713–725.

Jang, J. Y., Lee, S. H., Rhee, J. Y., Chung, G. C., Ahn, S. J., & Kang, H. (2007). Transgenic Arabidopsis and tobacco plants overexpressing an aquaporin respond differently to various abiotic stresses. *Plant Molecular Biology, 64*(6):621–632.

Javot, H., Lauvergeat, V., Santoni, V., Martin-Laurent, F., Güçlü, J., Vinh, J., et al. (2003). Role of a single aquaporin isoform in root water uptake. *The Plant Cell, 15*(2):509–522.

Javot, H., & Maurel, C. (2002). The role of aquaporins in root water uptake. *Annals of Botany, 90*(3):301–313.

Johanson, U., & Gustavsson, S. (2002). A new subfamily of major intrinsic proteins in plants. *Molecular Biology and Evolution, 19*(4):456–461.

Johanson, U., Karlsson, M., Johansson, I., Gustavsson, S., Sjövall, S., Fraysse, L., et al. (2001). The complete set of genes encoding major intrinsic proteins in *Arabidopsis* provides a framework for a new nomenclature for major intrinsic proteins in plants. *Plant Physiology, 126*(4):1358–1369.

Kaldenhoff, R., Grote, K., Zhu, J. J., & Zimmermann, U. (1998). Significance of plasmalemma aquaporins for water-transport in *Arabidopsis thaliana. The Plant Journal, 14*(1):121–128.

Kamiya, T., Tanaka, M., Mitani, N., Ma, J. F., Maeshima, M., & Fujiwara, T. (2009). NIP1; 1, an aquaporin homolog, determines the arsenite sensitivity of *Arabidopsis thaliana. Journal of Biological Chemistry, 284*(4):2114–2120.

Katsuhara, M., Akiyama, Y., Koshio, K., Shibasaka, M., & Kasamo, K. (2002). Functional analysis of water channels in barley roots. *Plant and Cell Physiology, 43*(8):885–893.

Kim, M. J., Kim, H. R., & Paek, K. H. (2006). *Arabidopsis* tonoplast proteins TIP1 and TIP2 interact with the cucumber mosaic virus 1a replication protein. *Journal of General Virology, 87*(11):3425–3431.

Kirch, H. H., Vera-Estrella, R., Golldack, D., Quigley, F., Michalowski, C. B., Barkla, B. J., & Bohnert, H. J. (2000). Expression of water channel proteins in *Mesembryanthemum crystallinum. Plant Physiology, 123*(1):111–124.

Kuwagata, T., Ishikawa-Sakurai, J., Hayashi, H., Nagasuga, K., Fukushi, K., Ahamed, A., et al. (2012). Influence of low air humidity and low root temperature on water uptake, growth and aquaporin expression in rice plants. *Plant and Cell Physiology, 53*(8):1418–1431.

Laizé, V., Rousselet, G., Verbavatz, J. M., Berthonaud, V., Gobin, R., Roudier, N., et al. (1995). Functional expression of the human CHIP28 water channel in a yeast secretory mutant. *FEBS Letters, 373*(3):269–274.

Laur, J., & Hacke, U. G. (2013). Transpirational demand affects aquaporin expression in poplar roots. *Journal of Experimental Botany, 64*(8):2283–2293.

Lawrence, S., Novak, N., Xu, H., & Cooke, J. (2013). Herbivory of maize by southern corn rootworm induces expression of the major intrinsic protein ZmNIP1; 1 and leads to the discovery of a novel aquaporin ZmPIP2; 8. *Plant Signaling and Behavior, 8*(8):e24937.

Levin, M., Lemcoff, J. H., Cohen, S., & Kapulnik, Y. (2007). Low air humidity increases leaf-specific hydraulic conductance of *Arabidopsis thaliana* (L.) Heynh (Brassicaceae). *Journal of Experimental Botany, 58*(13):3711–3718.

Li, J., Ban, L., Wen, H., Wang, Z., Dzyubenko, N., Chapurin, V., et al. (2015). An aquaporin protein is associated with drought stress tolerance. *Biochemical and Biophysical Research Communications, 459*(2):208–213.

Li, X., Wang, X., Yang, Y., Li, R., He, Q., Fang, X., et al. (2011). Single-molecule analysis of PIP2; 1 dynamics and partitioning reveals multiple modes of *Arabidopsis* plasma membrane aquaporin regulation. *The Plant Cell, 23*(10):3780–3797.

Lian, H. L., Yu, X., Ye, Q., Ding, X. S., Kitagawa, Y., Kwak, S. S., et al. (2004). The role of aquaporin RWC3 in drought avoidance in rice. *Plant and Cell Physiology, 45*(4):481–489.

Liénard, D., Durambur, G., Kiefer-Meyer, M. C., Nogué, F., Menu-Bouaouiche, L., Charlot, F., et al. (2008). Water transport by aquaporins in the extant plant *Physcomitrella patens. Plant Physiology, 146*(3):1207–1218.

Liu, C., Fukumoto, T., Matsumoto, T., Gena, P., Frascaria, D., Kaneko, T., et al. (2013). Aquaporin OsPIP1; 1 promotes rice salt resistance and seed germination. *Plant Physiology and Biochemistry, 63*:151–158.

Liu, Q., Umeda, M., & Uchimiya, H. (1994). Isolation and expression analysis of two rice genes encoding the major intrinsic protein. *Plant Molecular Biology, 26*(6):2003–2007.

Lopez, D., Venisse, J. S., Fumanal, B., Chaumont, F., Guillot, E., Daniels, M. J., et al. (2013). Aquaporins and leaf hydraulics: Poplar sheds new light. *Plant and Cell Physiology, 54*(12):1963–1975.

Loque, D., Ludewig, U., Yuan, L., & von Wiren, N. (2005). Tonoplast intrinsic proteins AtTIP2;1 and AtTIP2;3 facilitate NH3 transport into the vacuole. *Plant Physiology, 137*:671–680.

Ma, J. F., Tamai, K., Yamaji, N., Mitani, N., Konishi, S., Katsuhara, M., et al. (2006). A silicon transporter in rice. *Nature, 440*(7084):688.

Mahdieh, M., Mostajeran, A., Horie, T., & Katsuhara, M. (2008). Drought stress alters water relations and expression of PIP-type aquaporin genes in *Nicotiana tabacum* plants. *Plant and Cell Physiology, 49*(5):801–813.

Malz, S., & Sauter, M. (1999). Expression of two PIP genes in rapidly growing internodes of rice is not primarily controlled by meristem activity or cell expansion. *Plant Molecular Biology, 40*(6):985–995.

Martins, C. D. P. S., Pedrosa, A. M., Du, D., Gonçalves, L. P., Yu, Q., Gmitter Jr, F. G., & Costa, M. G. C. (2015). Genome-wide characterization and expression analysis of major intrinsic proteins during abiotic and biotic stresses in sweet orange (*Citrus sinensis* L. Osb.). *PLoS one, 10*(9):e0138786.

Maurel, C., & Chrispeels, M. J. (2001). Aquaporins. A molecular entry into plant water relations. *Plant Physiology, 125*(1):135–138.

Maurel, C., Javot, H., Lauvergeat, V., Gerbeau, P., Tournaire, C., Santoni, V., & Heyes, J. (2002). Molecular physiology of AQPs in plants. *International Review of Cytology, 215*:105–148.

Maurel, C., Reizer, J., Schroeder, J. I., & Chrispeels, M. J. (1993). The vacuolar membrane protein gamma-TIP creates water specific channels in Xenopus oocytes. *The EMBO Journal, 12*(6):2241–2247.

Maurel, C., Verdoucq, L., Luu, D. T., & Santoni, V. (2008). Plant aquaporins: Membrane channels with multiple integrated functions. *Annual Review of Plant Biology, 59*:595–624.

Maurel, C., Verdoucq, L., & Rodrigues, O. (2016). AQPs and plant transpiration. *Plant, Cell and Environment,* 39(11):2580–2587.

Mizutani, M., Watanabe, S., Nakagawa, T., & Maeshima, M. (2006). Aquaporin NIP2; 1 is mainly localized to the ER membrane and shows root-specific accumulation in *Arabidopsis thaliana*. *Plant and Cell Physiology*, *47*(10):1420–1426.

Mukhtar, M. S., Carvunis, A. R., Dreze, M., Epple, P., Steinbrenner, J., Moore, J., et al. (2011). Independently evolved virulence effectors converge onto hubs in a plant immune system network. *Science*, *333*(6042):596–601.

Noronha, H., Araújo, D., Conde, C., Martins, A. P., Soveral, G., Chaumont, F., et al. (2016). The grapevine uncharacterized intrinsic protein 1 (VvXIP1) is regulated by drought stress and transports glycerol, hydrogen peroxide, heavy metals but not water. *PloS one*, *11*(8):e0160976.

Pang, Y., Li, L., Ren, F., Lu, P., Wei, P., Cai, J., et al. (2010). Overexpression of the tonoplast aquaporin AtTIP5; 1 conferred tolerance to boron toxicity in *Arabidopsis*. *Journal of Genetics and Genomics*, *37*(6):389–397.

Park, W., Scheffler, B. E., Bauer, P. J., & Campbell, B. T. (2010). Identification of the family of aquaporin genes and their expression in upland cotton (*Gossypium hirsutum* L.). *BMC Plant Biology*, *10*(1):142.

Peng, Y., Lin, W., Cai, W., & Arora, R. (2007). Overexpression of a *Panax ginseng* tonoplast aquaporin alters salt tolerance, drought tolerance and cold acclimation ability in transgenic *Arabidopsis* plants. *Planta*, *226*(3):729–740.

Postaire, O., Verdoucq, L., & Maurel, C. (2008). Aquaporins in plants: From molecular structure to integrated functions. *Advances in Botanical Research*, *46*:76–137.

Quigley, F., Rosenberg, J. M., Shachar-Hill, Y., & Bohnert, H. J. (2001). From genome to function: The *Arabidopsis* aquaporins. *Genome Biology*, *3*(1):research0001-1.

Ranganathan, K., El Kayal, W., Cooke, J. E., & Zwiazek, J. J. (2016). Responses of hybrid aspen over-expressing a PIP2; 5 aquaporin to low root temperature. *Journal of Plant Physiology*, *192*:98–104.

Reddy, P. S., Divya, K., Sivasakthi, K., Tharanya, M., Lale, A., Bhatnagar-Mathur, P., et al. (2017). Role of pearl millet aquaporin genes in abiotic stress response. *InterDrought-V Conference*, Hyderabad.

Reddy, P. S., Rao, T. S. R. B., Sharma, K. K., & Vadez, V. (2015). Genome-wide identification and characterization of the aquaporin gene family in *Sorghum bicolor* (L.). *Plant Gene*, *1*:18–28.

Reuscher, S., Akiyama, M., Mori, C., Aoki, K., Shibata, D., & Shiratake, K. (2013). Genome-wide identification and expression analysis of aquaporins in tomato. *PLoS one*, *8*(11):e79052.

Rizhsky, L., Liang, H., Shuman, J., Shulaev, V., Davletova, S., & Mittler, R. (2004). When defense pathways collide. The response of *Arabidopsis* to a combination of drought and heat stress. *Plant Physiology*, *134*(4):1683–1696.

Rodrigues, O., Reshetnyak, G., Grondin, A., Saijo, Y., Leonhardt, N., Maurel, C., & Verdoucq, L. (2017). Aquaporins facilitate hydrogen peroxide entry into guard cells to mediate ABA-and pathogen-triggered stomatal closure. *Proceedings of the National Academy of Sciences*, *114*(34):9200–9205.

Sade, N., Gebretsadik, M., Seligmann, R., Schwartz, A., Wallach, R., & Moshelion, M. (2010). The role of tobacco Aquaporin1 in improving water use efficiency, hydraulic conductivity, and yield production under salt stress. *Plant Physiology*, *152*(1):245–254.

Sade, N., Vinocur, B. J., Diber, A., Shatil, A., Ronen, G., Nissan, H., et al. (2009). Improving plant stress tolerance and yield production: Is the tonoplast aquaporin SlTIP2; 2 a key to isohydric to anisohydric conversion? *New Phytologist*, *181*(3):651–661.

Sakurai, J., Ishikawa, F., Yamaguchi, T., Uemura, M., & Maeshima, M. (2005). Identification of 33 rice aquaporin genes and analysis of their expression and function. *Plant and Cell Physiology*, *46*(9):1568–1577.

Schnurbusch, T., Hayes, J., Hrmova, M., Baumann, U., Ramesh, S. A., Tyerman, S. D., et al. (2010). Boron toxicity tolerance in barley through reduced expression of the multifunctional aquaporin HvNIP2; 1. *Plant Physiology*, *153*(4):1706–1715.

Secchi, F., Lovisolo, C., & Schubert, A. (2007). Expression of OePIP2. 1 aquaporin gene and water relations of *Olea europaea* twigs during drought stress and recovery. *Annals of Applied Biology*, *150*(2):163–167.

Siefritz, F., Tyree, M. T., Lovisolo, C., Schubert, A., & Kaldenhoff, R. (2002). PIP1 plasma membrane aquaporins in tobacco: From cellular effects to function in plants. *The Plant Cell*, *14*(4):869–876.

Sreedharan, S., Shekhawat, U. K., & Ganapathi, T. R. (2013). Transgenic banana plants overexpressing a native plasma membrane aquaporin M usa PIP 1; 2 display high tolerance levels to different abiotic stresses. *Plant Biotechnology Journal*, *11*(8):942–952.

Steudle, E., & Henzler, T. (1995). Water channels in plants: Do basic concepts of water transport change? *Journal of Experimental Botany*, *46*(9):1067–1076.

Suga, S., Komatsu, S., & Maeshima, M. (2002). Aquaporin isoforms responsive to salt and water stresses and phytohormones in radish seedlings. *Plant and Cell Physiology*, *43*:1229–1237.

Sugaya, S., Ohshima, I., Gemma, H., & Iwahori, S. (2002). Expression analysis of genes encoding AQPs during the development of peach fruit. In *Proceedings of the XXVI International Horticultural Congress: Environmental Stress and Horticulture Crops 618*, Toronto, AB, Canada, pp. 363–370.

Sun, H., Li, L., Lou, Y., Zhao, H., Yang, Y., Wang, S., & Gao, Z. (2017). The bamboo aquaporin gene PeTIP4; 1-1 confers drought and salinity tolerance in transgenic *Arabidopsis*. *Plant Cell Reports*, *36*(4):597–609.

Šurbanovski, N., Sargent, D. J., Else, M. A., Simpson, D. W., Zhang, H., & Grant, O. M. (2013). Expression of *Fragaria vesca* PIP aquaporins in response to drought stress: PIP down-regulation correlates with the decline in substrate moisture content. *PloS one*, *8*(9):e74945.

Takano, J., Wada, M., Ludewig, U., Schaaf, G., Von Wirén, N., & Fujiwara, T. (2006). The *Arabidopsis* major intrinsic protein NIP5; 1 is essential for efficient boron uptake and plant development under boron limitation. *The Plant Cell*, *18*(6):1498–1509.

Trofimova, M. S., Zhestkova, I. M., Andreev, I. M., Svinov, M. M., Bobylev, Y. S., & Sorokin, E. M. (2001). Osmotic water permeability of vacuolar and plasma membranes isolated from maize roots. *Russian Journal of Plant Physiology*, *48*(3):287–293.

Tyerman, S. D., Bohnert, H. J., Maurel, C., Steudle, E., & Smith, J. A. C. (1999). Plant aquaporins: Their molecular biology, biophysics and significance for plant water relations. *Journal of Experimental Botany*, *50*:1055–1071.

Uehlein, N., Sperling, H., Heckwolf, M., & Kaldenhoff, R. (2012). The *Arabidopsis* aquaporin PIP1; 2 rules cellular CO2 uptake. *Plant, Cell and Environment*, *35*(6):1077–1083.

Vandeleur, R. K., Mayo, G., Shelden, M. C., Gilliham, M., Kaiser, B. N., & Tyerman, S. D. (2009). The role of plasma membrane intrinsic protein aquaporins in water transport through roots: Diurnal and drought stress responses reveal different strategies between isohydric and anisohydric cultivars of grapevine. *Plant Physiology*, *149*(1):445–460.

Venkatesh, J., Yu, J. W., & Park, S. W. (2013). Genome-wide analysis and expression profiling of the *Solanum tuberosum* aquaporins. *Plant Physiology and Biochemistry*, *73*:392–404.

Verkman, A. S. (1995). Optical methods to measure membrane transport processes. *The Journal of Membrane Biology*, *148*:99–110.

Vij, S., & Tyagi, A. K. (2007). Emerging trends in the functional genomics of the abiotic stress response in crop plants. *Plant Biotechnology Journal*, *5*(3):361–380.

Wang, L., Li, Q. T., Lei, Q., Feng, C., Zheng, X., Zhou, F., et al. (2017). Ectopically expressing MdPIP1; 3, an aquaporin gene, increased fruit size and enhanced drought tolerance of transgenic tomatoes. *BMC Plant Biology*, *17*(1):246.

Wang, L., Zhang, C., Wang, Y., Wang, Y., Yang, C., Lu, M., & Wang, C. (2018). *Tamarix hispida* aquaporin ThPIP2; 5 confers salt and osmotic stress tolerance to transgenic *Tamarix* and *Arabidopsis*. *Environmental and Experimental Botany*, *152*:158–166.

Wang, X., Gao, F., Bing, J., Sun, W., Feng, X., Ma, X., et al. (2019). Overexpression of the Jojoba aquaporin gene, *ScPIP1*, enhances drought and salt tolerance in transgenic *Arabidopsis*. *International Journal of Molecular Sciences*, *20*(1):153.

Wallace, I. S., Choi, W. G., & Roberts, D. M. (2006). The structure, function and regulation of the nodulin 26-like intrinsic protein family of plant aquaglyceroporins. *Biochimica et Biophysica Acta (BBA)-Biomembranes*, *1758*(8):1165–1175.

Wu, W. Z., Peng, X. L., & Wang, D. (2009). Isolation of a plasmalemma aquaporin encoding gene StPIP1 from *Solanum tuberosum* L. and its expression in transgenic tobacco. *Agricultural Sciences in China*, *8*(10):1174–1186.

Xu, Y., Hu, W., Liu, J., Zhang, J., Jia, C., Miao, H., et al. (2014). A banana aquaporin gene, MaPIP1; 1, is involved in tolerance to drought and salt stresses. *BMC Plant Biology*, *14*(1):59.

Yamada, S., Katsuhara, M., Kelly, W. B., Michalowski, C. B., & Bohnert, H. J. (1995). A family of transcripts encoding water channel proteins: Tissue-specific expression in the common ice plant. *The Plant Cell*, *7*(8):1129–1142.

Yamaguchi-Shinozaki, K., Koizumi, M., Urao, S., & Shinozaki, K. (1992). Molecular cloning and characterization of 9 c DNAs for genes that are responsive to desiccation in *Arabidopsis thaliana*: Sequence analysis of one cDNA clone that encodes a putative transmembrane channel protein. *Plant and Cell Physiology*, *33*:217–224.

Yamaji, N., Mitatni, N., & Ma, J. F. (2008). A transporter regulating silicon distribution in rice shoots. *The Plant Cell*, *20*(5):1381–1389.

Yang, B., & Verkman, A. S. (1997). Water and glycerol permeabilities of aquaporins 1–5 and MIP determined quantitatively by expression of epitope-tagged constructs in Xenopus oocytes. *Journal of Biological Chemistry*, *272*(26):16140–16146.

Yu, Q., Hu, Y., Li, J., Wu, Q., & Lin, Z. (2005). Sense and antisense expression of plasma membrane aquaporin BnPIP1 from *Brassica napus* in tobacco and its effects on plant drought resistance. *Plant Science*, *169*(4):647–656.

Zhang, D. Y., Ali, Z., Wang, C. B., Xu, L., Yi, J. X., Xu, Z. L., et al. (2013). Genome-wide sequence characterization and expression analysis of major intrinsic proteins in soybean (*Glycine max* L.). *PloS one*, *8*(2):e56312.

Zhang, R. B., & Verkman, A. S. (1991). Water and urea permeability properties of Xenopus oocytes: Expression of mRNA from toad urinary bladder. *American Journal of Physiology-Cell Physiology*, *260*(1):C26–C34.

Zhou, S., Hu, W., Deng, X., Ma, Z., Chen, L., Huang, C., et al. (2012). Overexpression of the wheat aquaporin gene, TaAQP7, enhances drought tolerance in transgenic tobacco. *PloS one*, *7*(12):e52439.

Zhu, C., Schraut, D., Hartung, W., & Schäffner, A. R. (2005). Differential responses of maize MIP genes to salt stress and ABA. *Journal of Experimental Botany*, *56*(421):2971–2981.

Zou, J., Rodriguez-Zas, S., Aldea, M., Li, M., Zhu, J., Gonzalez, D. O., et al. (2005). Expression profiling soybean response to *Pseudomonas syringae* reveals new defense-related genes and rapid HR-specific downregulation of photosynthesis. *Molecular Plant-Microbe Interactions*, *18*(11):1161–1174.

13 The Role of Extracellular ATP in Plant Abiotic Stress Signaling

Sushma Sagar and Amarjeet Singh

CONTENTS

13.1 INTRODUCTION

Adenosine 5′-triphosphate (ATP) is one of the most important molecules in living cells. ATP is majorly produced in the mitochondria by the processes of oxidative phosphorylation by the enzyme ATP synthase. It is known as the energy currency of the cell due to its ability to store energy and distribute it to various cells as and when required; thus it is crucial for diverse cellular processes. ATP is normally present inside the cell; however, in animals it was established that ATP can come out of the plasma membrane and into the extracellular spaces (Burnstock and Knight, 2004), and this ATP is referred to as extracellular ATP (eATP). In plants, the presence of ATP in extracellular space and its effect on plant processes were studied long ago (Jaffe, 1973); however, evidence for eATP acting as a signaling molecule in plants to regulate growth and development is relatively recent (Demidchik et al., 2003; Jeter et al., 2004). The identification of ATPs in the extracellular spaces posed the important question of what could be specific stimulus for the release of ATPs from within the cell to intercellular spaces in plants. To answer this question different study concluded that external stimuli like wounding, pathogen attack, touch or mechanical stimulus or abiotic stresses may cause the release of cellular ATPs to the extracellular matrix (Choi et al., 2014). In animals, ATP exodus into the extracellular matrix mainly occurs through anion channels, gap junctions, ATP binding cassettes (ABC) transporters and vesicular exocytosis (Feng et al., 2015). In addition, eATP could also be produced by a plasma membrane-localized F_0F_1-ATP synthase complex in animal cells (Mangiullo et al., 2008). On the other hand, the eATP pool in plants is accumulated mainly by the release of intracellular ATP through ABC transporters or by exocytosis, as an ATP synthase complex is not found at the plasma membrane of plant cells (Feng et al., 2015; Kim et al., 2006). In addition, a plasma membrane-localized transporter PM-ANT1 mediates ATP release from the pollen tube during its maturation (Rieder and Neuhaus, 2011), and its variable expression levels in a wide variety of plant tissues hint towards a vital role of ATP transport into extracellular space in plants (Clark and Roux, 2011). eATP is widely accepted as a signaling molecule in animal cells as it could stimulate an increase in vital signaling components, including cytosolic free calcium (Ca^{2+}), nitric oxide (NO) and reactive oxygen species (ROS) (Silva et al., 2006). Moreover, eATP is essential for several crucial physiological processes such as cell growth and cell death, neurotransmission, muscle contraction and immune response in animals (Khakh and Burnstock, 2009). Several recent studies have shown significant involvement of eATP in crucial cellular and physiological processes in plants such as root hair growth, vegetative growth, biotic and abiotic stress responses, pollen tube growth, gravitropism and cell viability (Tanaka et al., 2010, 2014; Cao et al., 2014; Chen et al., 2017; Tripathi et and Tanaka, 2018; Hou et al., 2018). The role of eATP in different processes in animal cells could be established due to knowledge of their cognitive receptors. Two types of membrane-associated purinergic receptors are known in animal cells: P2X, the ligand gated ion channels, and P2Y, G-protein coupled receptors (Khakh and Burnstock, 2009). However, the functional role of eATP as a signaling molecule in plants could be established only recently, because a membrane-associated purinoreceptor was unknown in plants until the recent identification of Does not Respond to Nucleotides (DORN1), a lectin receptor kinase, by Choi et al. (2014). The DORN1 gene was identified through a large-scale mutant screening in *Arabidopsis*, where it was shown to mediate some eATP-related processes (Choi et al., 2014). Identification of this purinoreceptor in plants has led to several interesting studies which shed the light on some important physiological roles of eATP. Thus, here we elaborate, update and provide new insights in eATP signaling, regulation and functional roles in abiotic stress signaling in plants (Figure 13.1).

13.2 eATP DETECTION IN PLANTS

The visualization of eATP in the extracellular matrix of the different tissues has been a challenging task. The lack of a suitable technique for detection of eATP is a major hindrance in assessing the dose-dependent effect of eATP *in vivo*. However, significant efforts have been made by researchers to track the changes in ATP concentrations in the extracellular matrices of different plant tissues. A luciferase reporter protein fused with cellulose binding domain peptide was used to detect eATP in *Medicago truncatula* (Kim et al., 2006). Upon application of this system at the plant roots, strong luciferase luminescence reflecting eATP accumulation in interstitial spaces between epidermal cells, mainly in the vicinity of the actively growing cells. Later on, this approach was employed for the detection of touch-induced release of eATP in *Arabidopsis* roots (Kim et al., 2006, Choi et al., 2014, Weerasinghe et al., 2009). A modified luciferase reporter with a signal peptide (for imparting a secretory property) was expressed in transgenic *Arabidopsis* plants, and eATP accumulation was monitored using a sensitive luminometer (Clark et al., 2011). These plants expressing modified luciferase reporter were used to demonstrate the release of ATP from guard cells during stomata opening or closure. Recently, different variants and commercially available kits based on these methods have

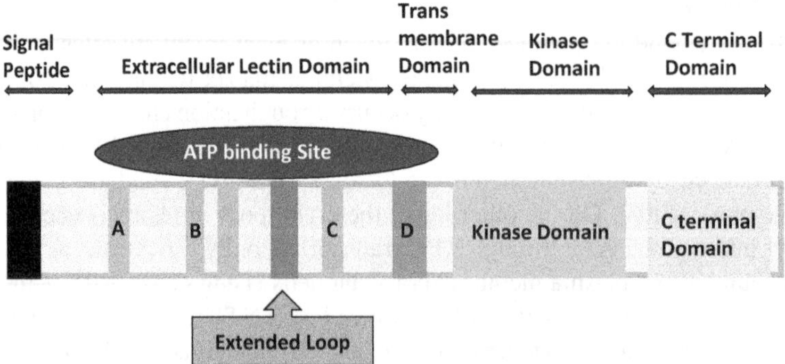

FIGURE 13.1 A schematic diagram representing various domains present in the extracellular ATP receptor DORN1. It has five regions consisting of a signal peptide followed by an extracellular lectin domain consisting of four loops, namely Loop A, B, C and D. Between loop C and D an extended loop is present. The extracellular lectin domain is followed by the transmembrane domain and a kinase domain. In the end a C terminal region is present.

been adopted in several studies to measure the eATP levels in plant tissues (Sun et al., 2012; Deng et al., 2015; Yang et al., 2015; Chen et al., 2017).

13.3 eATP RECEPTOR IN PLANTS

Though several studies reported the role of eATPs in plant growth, development and stress responses, these responses lacked a mechanistic explanation. Thus, there was a lot of skepticism until recently about the role of eATP as a signaling molecule in plants, simply because, like animals, no ATP receptor was known in plants. To end this skepticism, Choi et al. (2014) in a forward genetic screen identified an *Arabidopsis* mutant *dorn1* which was insensitive to ATP. To confirm the identity of DORN1 as an ATP receptor they analyzed the cytoplasmic calcium response of *dorn1* mutants in the presence of ATP. It was found that ATP-induced Ca^{2+} elevation was absent in EMS mutagenized alleles *dorn1–1, dorn 1-2* and T-DNA insertion line *dorn1-3*, whereas DORN1-overexpressing plants showed a 20-fold higher response to ATP as compared to the wild type (Choi et al., 2014). Moreover, it was demonstrated that the *dorn1* mutant was insensitive to ATP and other nucleotides except the pyrimidine nucleotide CTP. Based on the animal purinoreceptors nomenclature, i.e. the P2Y and P2X DORN1 receptor family was named P2K (Choi et al., 2014), and this family of lectin receptor kinases are found in plants only (Lehti-Shiu et al., 2009). The *Arabidopsis* DORN1 gene encodes a lectin receptor kinase (LecRK I.9). Structurally, DORN1 is composed of three major domains: an extra-cellular legume (L) type lectin domain, an intracellular ser/thr kinase domain and a transmembrane domain (Nguyen et al., 2016). The L-type lectin domain contains an alpha-beta sandwich fold that contain two antiparallel beta sheets connected through a loop (Nguyen et al., 2016). Generally, L-type lectins are known to bind to carbohydrates, however DORN1 does not bind to carbohydrates due to a critical substitution of Asp-79-His at its ion binding site leading to destabilization of the monosaccharide binding site (Choi et al., 2014b; Gouget et al.,2006). Importantly, DORN1 binds ATPs with high affinity (K_d – 45.7 nM) through its lectin domain as proven by *in vitro* and *in planta* binding assays (Choi et al., 2014). Due to such high affinity to ATP, even the nanomolar concentrations of eATP are enough to activate the receptor and trigger the downstream signaling response. The possibilities of identifying the specific DORN1 amino acid residues required for ATP binding were discussed and reviewed by using different EMS mutants of *dorn1* (Choi et al., 2014b). Mutations in the ser/thr kinase domain of *dorn1-1* (D572N) and *dorn1-2* (D525N) mutants led to ATP insensitivity; thus, the kinase activity of DORN1 is thought to be inevitable for eATP signaling. DORN1 has a functional kinase domain, and it exhibits autophosphorylation and transphosphorylation activities (Choi et al., 2014). The *dorn1-1* (D572N) and *dorn1-2* (D525N) mutations are crucial for autoactivation of the kinase domain and stabilization of its catalytic loop, thus abolishing the kinase function and consequently, blocking plants' response to eATP (Figure 13.2).

13.4 eATP SIGNALING IN PLANTS

One of the initial reports showed the sensitivity of the plant roots toward the auxin, upon exogenous ATP application. This effect was mediated by increased intracellular Ca^{2+} levels upon ATP addition, thereby confirming ATP's role as a signaling molecule (Demidchick et al., 2003). Similarly, ATP induced an increase in cytoplasmic Ca^{2+} levels, which in turn controlled the expression of various downstream genes (Jeter et al., 2004). Also, extraction of cellular fluid from a wound site and its analysis showed increased levels of eATP, and higher eATP levels were correlated with the formation of ROS (Song et al., 2006). Moreover, upon exogenous application of ATP, a significant amount of ROS was detected within half an hour (Clark et al., 2010). In plants, an increase in ATP levels increases cytoplasmic Ca^{2+} which subsequently induces the generation of ROS. However, the inability to generate ROS by *Arabidopsis* NADPH null mutants (rbohC, rbohD and rbohF) could not be overturned by exogenous ATP application (Tonón et al., 2010). Moreover, atrbohC mutants showed

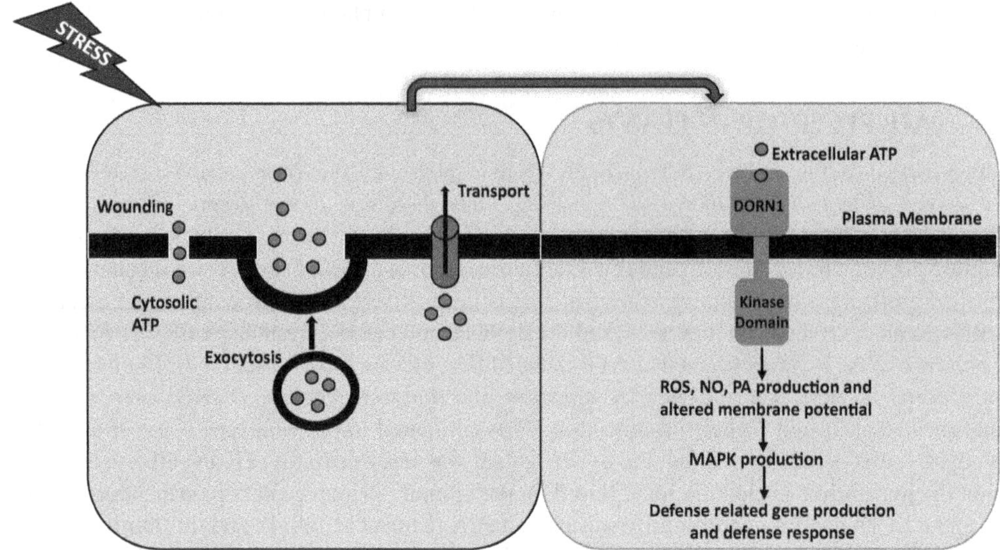

FIGURE 13.2 Mechanism of ATP perception in plants by DORN1. When plants encounter any biotic or abiotic stress or are damaged due to wounding or touch, then ATP from the cytosol is released into the extracellular matrix. This extracellular ATP is recognized by the extracellular lectin domain of the *Arabidopsis* ATP receptor DORN1. Upon recognition of eATP by the DORN1, the intracellular kinase domain of the DORN1 is activated which in turn leads to the induction of the cytoplasmic calcium gradient. It is followed by the production of reactive oxygen species (ROS), nitric oxide (NO) and phosphatidic acid (PA). The downstream protein phosphorylation events take place, and the induction of various gene expressions occurs. Many genes involved in the defense against pathogens, stress response and developmental responses are induced.

an inability to activate the stress-related gene MAPK, even upon exogenous ATP application. This finding suggests that ROS production is a downstream event and cannot be compensated by ATP application only. All these findings together suggest the role of ATP in signaling. When plants encounter any biotic or abiotic stress then ATP from the cytosol is released into the extracellular matrix (Choi et al., 2014). This extracellular ATP is recognized by the extracellular lectin domain of the ATP receptor DORN1. Upon recognition of eATP by the DORN1, the intracellular serine-threonine kinase domain of the DORN1 is activated which in turn leads to the induction of cytoplasmic calcium gradient. It is followed by the production of ROS, nitric oxide (NO) and phosphatidic acid (PA). These signaling molecules participate in further downstream signaling events, and the induction of various genes occurs. It also leads to higher intracellular mitochondrial and chloroplastic ATP biosynthesis. Additionally, DORN1 mediates adhesion between RGD motif-containing cell wall proteins and plasma membrane, thereby maintaining cell integrity (Cao et al., 2014, Tanaka et al., 2014). Also in *Arabidopsis* roots, DORN1 is involved in eATP-mediated activation of root epidermal cell plasma membrane current (Wang et al., 2018). This might control the efflux of K^+ ions under various circumstances and control growth. However further research needs to be done in this direction. Chivasa et al. (2010) showed the differential expression of various proteins involved in pathogen defense, cell wall metabolism, photosynthesis, protection against redox state and oxidative stress, etc. The expression profiles of differentially expressed proteins were distinguished into three categories. One class included proteins that underwent a change in response to AMP-PCP (an ATP analog) treatment but were unaffected by exogenous ATP application. This altered expression was explained by the hampered eATP cleavage due to the competitive exclusion of eATP from the binding sites by AMP-PCP. The second category of proteins showed differential abundance in response to both exogenous ATP and AMP-PCP application. These proteins had similar eATP binding sites as those of the P_2 receptor-mediated response. The third category of proteins exhibited differential abundance only upon exogenous ATP application and were controlled by the dissociation product of ATP (Chivasa et al., 2010). These findings shed light on the role of eATP signaling in regulating crucial cellular metabolic processes like intracellular ATP production and photosynthesis. Similarly, genes involved in programmed cell death, protein modification and carbohydrate metabolism were also found to be induced by ATP (Choi et al., 2014). These findings provide evidence for purinergic signaling in stress responses in plants.

13.5 REGULATION OF eATPS BY APYRASES

Apyrases (nucleoside triphosphate-diphosphohydrolases) are the enzymes known to hydrolyze nucleoside triphosphate (NTPs) and diphosphate and are more effective than other phosphates in removing phosphate from NTP/nucleoside diphosphate (Zimmermann et al., 2000, Chiu et al., 2015). They are present in all eukaryotes, and the majority of apyrases exist as ectoapyrases. Ectoapyrases are enzymes that are anchored in the plasma membrane while their active site faces the extracellular matrix (Zimmermann et al., 2000). Ectoapyrases are able to regulate the abundance of eATP substantially due to their high Vmax and low Km for NDPs and NTPs. These enzymes are also involved in removing the terminal phosphate from ADP and dinucleotides (Clark and Roux, 2011). In *Arabidopsis*, the highest expression of two apyrases, namely APY1 and APY2, was found in the actively growing and ATP-releasing regions (Wu et al., 2007). The expression of APY1 and APY2 was also induced by light which is known for its role in the expansion of guard cells and release of ATPs (Clark and Roux, 2011). Also, the presence of a fraction of APY1 and APY2 was detected at the plasma membrane where they limit the eATP level (Wu et al., 2007, Clark and Roux, 2009), while the major fraction was associated with the golgi vesicles. This suggests that apyrases hydrolyze the nucleotides in the lumen of golgi vesicles to regulate the eATP level, before other vesicles release their ATP content to the outside of the cells. Like any other signaling molecule, eATPs act within an optimal concentration and eATP levels are controlled by ectoapyrases. In soybean ectoapyrases are required for nodulation to take place (Tanaka et al., 2010) and they also regulate the stomatal aperture (Clark et al., 2011). In *Arabidopsis* suppression of ectoapyrases resulted in numerous growth defects and defective pollen germination (Steinberger et al., 2003). Similarly, stunted fiber growth in cotton (Clark and Roux, 2009) is re shown. These irregularities may be attributed to the increase in the eATP level and associated inhibition of auxin transport upon inhibition of ectoapyrase (Tang et al., 2003). The probable role of ectoapyrases in plant responses, such as growth and nodulation, can be established by further studies that can prove the correlation of polar auxin transport with the depleted eATP levels due to ectoapyrase activity. In another study it has been demonstrated that eATP in *Arabidopsis* is involved in regulating stomatal movement in a H_2O_2-dependent manner (Wang et al., 2015). In *Arabidopsis* and *Vicia faba*, eATP as well as ADP and GTP promoted stomatal opening (Hao et al., 2012, Wang et al., 2014). Lower levels of ATP (5–15 um) promoted stomatal opening whereas higher ATP levels (150–250 um) induce stomatal closure (Clark et al., 2011). Also, eATP is involved in DORN1-mediated NADH oxidase respiratory burst oxidase homolog D (RBOHD) phosphorylation, which in turn regulates stomatal aperture (Chen et al., 2017).

13.6 ROLE OF eATP IN ABIOTIC STRESS SIGNALING

Plants encounter numerous abiotic stress in their day-to-day life. Abiotic stress faced by plants includes heat stress, cold stress, salinity stress and many more. Plants have developed various mechanisms by which they can withstand stress conditions and reproduce. The roles of various stress hormones and signal transduction pathways have been enumerated in abiotic stress signaling and responses. eATP has also been implicated in different abiotic stress signaling pathways in diverse plant species. The release of eATP was reported in *Arabidopsis* seedlings upon salinity treatment (Jeter et al., 2004). Also, the treatment of the *Arabidopsis* seedling with $MgCl_2$, Mg $(NO_3)_2$ and

NaCl led to a several-fold increase in the concentration of eATP. Interestingly, when the seedlings were treated with a non-hydrolysable ATP analog, beta, gamma, methyladenosine-5'-triphosphate (AMP)-PLA, marked cell death was observed (Kim et al., 2009). Thus, it could be inferred that the physiological function of eATP is crucial for cell viability and the accumulation of eATP under hypertonic stress might be an important survival strategy of the plants. Furthermore, eATP was released in *Arabidopsis* roots when treated with sorbitol, salt or L-glutamate (Dark et al., 2011). In *Populus euphratica*, a transient elevation in the levels of eATP occurred upon the application of NaCl (200 mM) to the cells' suspension culture, which triggered a salinity stress response (Sun et al., 2012). On treating the *P. euphratica* cell suspension culture with suramin and H-G trap system to block the eATP signaling, the cells were unable to perform processes like ROS generation, antioxidant detoxification, vacuolar salt compartmentalization, cytosolic Na^+ exclusion and induction of salt-resistant gene expression, which are crucial for acclimation to stress (Sun et al., 2012). Moreover, the exogenous application of ATP could rescue the salt acclimation process from the effect of the H-G trap system. Also, at higher eATP levels *P. euphratica* cells accumulated Na^+ in the vacuole. The addition of ATP inhibitors like suramin diminished the tendency towards cytosolic Na^+ exclusion and vacuolar ion compartmentalization. However, the addition of exogenous ATP could restore the accumulation of Na^+ in the vacuole (Sun et al., 2012). These findings suggest that cell viability is maintained by eATP by providing salinity stress tolerance in the plants. *Arabidopsis* leaves demonstrated increased eATP levels after salt stress treatment. These increased eATP levels may be due to damage in the plasma membrane and other internal organelles due to the NaCl treatment. To understand the possible cause, the effect of eATP on photosystem II and intracellular ATP (iATP) was investigated. It was observed that exogenous AMP-PCP did not have any effect under normal conditions, however under hypertonic conditions it significantly decreased the chlorophyll fluorescence parameter and iATP production (Hou et al., 2018). Hence, the significant role of eATP was acknowledged in plant response to salt stress (Hou et al., 2018). Interestingly, it was observed that higher concentrations of eATP inhibit the cold stress tolerance in *P. euphratica*, while its lower concentration promotes cold stress tolerance (Deng et al., 2015). The cold treatment hampers the membrane integrity, resulting in a significant increase in the eATP levels in the root. In cold conditions, electrolyte leakage occurred due to inhibited vesicular trafficking. Upon perceiving this electrolyte leakage as a signal, eATP is released in the extracellular matrix. The application of ATP suppressed the root cell viability in the cold stress. Thus, eATP levels needs to be controlled and maintained under permitted concentrations by the cell to ensure cell viability (Deng et al., 2015). All these findings suggest important roles of eATP in abiotic stress signaling and tolerance in plants. However, further investigations are required to clearly understand the role of eATP in tolerating other abiotic stresses such as heat, heavy metal stress, etc.

13.7 ROLE OF eATP IN WOUNDING/MECHANICAL STRESS

Mechanical stress is known to increase eATP levels in *Arabidopsis* seedlings (Cho et al., 2017). The eATP may function as a cell–cell signaling agent in plants. Touch, osmotic stress and hypertonic stress are able to induce ATP release to a significant extent. It is surmised that wounding-induced ATP release may be a result of hampered cellular integrity (Jeter et al., 2004). ATP-dependent touch response wherein released ATP is fine-tuned by G proteins has been studied in plants. Directional root growth in response to mechano-stimulation has also been observed in *Arabidopsis*. Further, *Arabidopsis* roots showed the release of nanomolar level of extracellular ATP upon mild touch (Weerasinghe et al., 2009). It is hypothesized that eATP released upon mechanical stimulation may be crucial for the roots to detect fluctuations in pressure in the growth zone and to address such changes effectively and appropriately. Cytoskeleton rearrangement is widely associated with mechanical stress (Hardham et al., 2008). Similar effects like cytoplasmic streaming are also reported in plants upon the application of exogenous ATP (Cho et al., 2017). Furthermore, eATP has an important role in plant tolerance to herbivory-induced wounding.

ROS accumulation and the expression of defense responsive genes such as PAL1, ACS6 and LOX2, which is commonly observed in plants cells upon herbivore attack and wounding, occurred with short-term treatment with exogenous ATP (Song et al., 2006). Moreover, MAPKs that are induced upon herbivory, wounding or pathogen attack have been reported to be induced upon treatment with 100 µm ATP (Choi et al., 2014). Also, the suppression of apyrase expression contributed to increased eATP expression for 3–6 days, resulting in the induction of various genes involved in the defense response to herbivory, accumulation of ROS and biosynthesis of lignin (Lim et al., 2014). In addition to physical wounding, microbial attack can also cause mechanical stress to the plants (Tanaka et al., 2014). The penetrating peg and hyphae of the pathogenic fungi and oomycetes exert turgor pressure, causing mechanical stress to the plants (Jayaraman et al., 2014). Also, mechanical stress activates disease resistance, nuclear position and the cytoplasmic streaming response in the plants at the point of contact of the pathogen. This cytoplasmic streaming suggests that ATP released upon mechanical stress can act as a damage-associated molecular pattern (DAMP) that brings about various cellular responses in plants to fight back invading pathogens. Thus, it is assumed that wounding/herbivory causes the increase in eATP levels in plants. The eATP so produced may participate in wounding-triggered signaling and help the plant to tolerate wounding or mechanical stress.

13.8 CONCLUSION

During the past decade significant research has been done on extracellular ATP in plants. The eATP receptor has been unveiled, and the mechanism of action has also been investigated to a large extent. It is now shown how plants use extracellular ATP as a signaling molecule to combat abiotic stress. Also, the similarity and dissimilarity of the purinergic signaling between plants and animals have been shown. Current knowledge suggests that there is remarkable difference between the plant and animal receptors of extracellular ATP. The mutant screening, biochemical and molecular approaches are still being undertaken to identify and decipher the signal transduction mechanism and its crosstalk with other stress- and defense-related signal transduction pathways. Also, the mechanism by which the changes in the level of eATP induce the production of Ca^{2+}, ROS and NO and how their production translates into abiotic stress-related changes are poorly understood and demand further investigations. However, with the upcoming new knowledge and tremendous efforts in eATP research these key questions are expected to be addressed soon.

REFERENCES

Burnstock, G., and Knight, G. E. 2004. Cellular distribution and functions of P2 receptor subtypes in different systems. *International Review of Cytology* 240(1):31–304.

Cao, Y., Tanaka, K., Nguyen, C. T., and Stacey, G. 2014. Extracellular ATP is a central signaling molecule in plant stress responses. *Current Opinion in Plant Biology* 20:82–87.

Chen, D., Cao, Y., Li, H., Kim, D., Ahsan, N., Thelen, J., and Stacey, G. 2017. Extracellular ATP elicits DORN1-mediated RBOHD phosphorylation to regulate stomatal aperture. Nature Communications 8(1):2265.

Chiu, T. Y., Lao, J., Manalansan, B., Loqué, D., Roux, S. J. and Heazlewood, J. L. 2015. Biochemical characterization of *Arabidopsis* APYRASE family reveals their roles in regulating endomembrane NDP/NMP homoeostasis. *Biochemical Journal* 472(1):43–54.

Chivasa, S., Simon, W. J., Murphy, A. M., Lindsey, K., Carr, J. P., and Slabas, A. R. 2010. The effects of extracellular adenosine 5′-triphosphate on the tobacco proteome. *Proteomics* 10(2):235–244.

Cho, S. H., Choi, J., and Stacey, G. 2017. Molecular mechanism of plant recognition of extracellular ATP. In: *Protein Reviews*, ed. Atassi, M., pp. 233–253. Advances in Experimental Medicine and Biology, Springer, Singapore.

Choi, J., Tanaka, K., Cao, Y., et al. 2014. Identification of a plant receptor for extracellular ATP. *Science* 343(6168):290–294.

Clark, G., Fraley, D., Steinebrunner, I., et al. 2011. Extracellular nucleotides and apyrases regulate stomatal aperture in *Arabidopsis*. *Plant Physiology* 156(4):1740–1753.

Clark, G., and Roux, S. J. 2009. Extracellular nucleotides: Ancient signaling molecules. *Plant Science* 177(4):239–244.

Clark, G., and Roux, S. J. 2011. Apyrases, extracellular ATP and the regulation of growth. *Current Opinion in Plant Biology* 14(6):700–706.

Clark, G., Torres, J., Finlayson, S., Guan, X., Handley, C., Lee, J., and Roux, S. J. 2010. Apyrase (nucleoside triphosphate-diphosphohydrolase) and extracellular nucleotides regulate cotton fiber elongation in cultured ovules. *Plant Physiology* 152(2):1073–1083.

Dark, A., Demidchik, V., Richards, S. L., Shabala, S., and Davies, J. M. 2011. Release of extracellular purines from plant roots and effect on ion fluxes. *Plant Signaling and Behavior* 6(11):1855–1857.

Demidchik, V., Nichols, C., Oliynyk, M., Dark, A., Glover, B. J., and Davies, J. M. 2003. Is ATP a signaling agent in plants? *Plant Physiology* 133(2):456–461.

Deng, S., Sun, J., Zhao, R., et al. 2015. *Populus euphratica* APYRASE2 enhances cold tolerance by modulating vesicular trafficking and extracellular ATP in *Arabidopsis* plants. *Plant Physiology* 169(1):530–548.

Feng, H. Q., Jiao, Q. S., Sun, K., Jia, L. Y., and Tian, W. Y. 2015. Extracellular ATP affects chlorophyll fluorescence of kidney bean (*Phaseolus vulgaris*) leaves through Ca^{2+} and H_2O_2 dependent mechanism. *Photosynthetica* 53(2):201–206.

Gouget, A., Senchou, V., Govers, F., Sanson, A., Barre, A., Rougé, P., and Canut, H. 2006. Lectin receptor kinases participate in protein-protein interactions to mediate plasma membrane-cell wall adhesions in *Arabidopsis*. *Plant Physiology* 140(1):81–90.

Hardham, A. R., Takemoto, D., and White, R. G. 2008. Rapid and dynamic subcellular reorganization following mechanical stimulation of *Arabidopsis* epidermal cells mimics responses to fungal and oomycete attack. *BMC Plant Biology* 8:63.

Hao, L. H., Wang, W. X., Chen, C., Wang, Y. F., Liu, T., Li, X., and Shang, Z. L. 2012. Extracellular ATP promotes stomatal opening of *Arabidopsis thaliana* through heterotrimeric G protein α subunit and reactive oxygen species. *Molecular Plant* 5(4):852–864.

Hou, Q. Z., Sun, K., Zhang, H., Su, X., Fan, B. Q., and Feng, H. Q. 2018. The responses of photosystem II and intracellular ATP production of *Arabidopsis* leaves to salt stress are affected by extracellular ATP. *Journal of Plant Research* 131(2):331–339.

Jaffe, M. J. 1973. The role of ATP in mechanically stimulated rapid closure of the Venus's flytrap. *Plant Physiology* 51(1):17–18.

Jayaraman, D., Gilroy, S., and Ane, J. M. 2014. Staying in touch: Mechanical signals in plant–microbe interactions. *Current Opinion in Plant Biology* 20:104–109.

Jeter, C. R., Tang, W., Henaff, E., Butterfield, T., and Roux, S. J. 2004. Evidence of a novel cell signaling role for extracellular adenosine triphosphates and diphosphates in *Arabidopsis*. *The Plant Cell* 16(10):2652–2664.

Khakh, B. S., and Burnstock, G. 2009. The double life of ATP. *Scientific American* 301(6):84.

Kim, S. Y., Sivaguru, M., and Stacey, G. 2006. Extracellular ATP in plants. Visualization, localization, and analysis of physiological significance in growth and signaling. *Plant Physiology* 142(3):984–992.

Lehti-Shiu, M. D., Zou, C., Hanada, K., and Shiu, S. H. 2009. Evolutionary history and stress regulation of plant receptor-like kinase/pelle genes. *Plant Physiology* 150(1):12–26.

Lim, M. H., Wu, J., Yao, J., et al. 2014. Apyrase suppression raises extracellular ATP levels and induces gene expression and cell wall changes characteristic of stress responses. *Plant Physiology* 164(4):2054–2067.

Mangiullo, R., Gnoni, A., Leone, A., Gnoni, G. V., Papa, S., and Zanotti, F. 2008. Structural and functional characterization of FoF1-ATP synthase on the extracellular surface of rat hepatocytes. *Biochimica and Biophysica Acta (BBA)-Bioenergetics* 1777:1326–1335.

Nguyen, C. T., Tanaka, K., Cao, Y., Cho, S. H., Xu, D., and Stacey, G. 2016. Computational analysis of the ligand binding site of the extracellular ATP receptor, DORN1. *PLoS ONE* 11(9):e0161894.

Rieder, B., and Neuhaus, H. E. 2011. Identification of an *Arabidopsis* plasma membrane–located ATP transporter important for anther development. *The Plant Cell* 23(5):1932–1944.

Silva, G., Beierwaltes, W. H., and Garvin, J. L. 2006. Extracellular ATP stimulates NO production in rat thick ascending limb. *Hypertension* 47(3):563–567.

Song, C. J., Steinebrunner, I., Wang, X., Stout, S. C., and Roux, S. J. 2006. Extracellular ATP induces the accumulation of superoxide via NADPH oxidases in *Arabidopsis*. *Plant Physiology* 140(4):1222–1232.

Sun, J., Zhang, X., Deng, S., Zhang, C., Wang, M., Ding, M., and Chen, S. 2012. Extracellular ATP signaling is mediated by H_2O_2 and cytosolic Ca^{2+} in the salt response of *Populus euphratica* cells. *PLoS ONE* 7:e53136.

Tanaka, K., Choi, J., Cao, Y., and Stacey, G. 2014. Extracellular ATP acts as a damage-associated molecular pattern (DAMP) signal in plants. *Frontiers in Plant Science* 5:446.

Tanaka, K., Gilroy, S., Jones, A. M., and Stacey, G. 2010. Extracellular ATP signaling in plants. *Trends in Cell Biology* 20(10):601–608.

Tang, W., Brady, S. R., Sun, Y., Muday, G. K., and Roux, S. J. 2003. Extracellular ATP inhibits root gravitropism at concentrations that inhibit polar auxin transport. *Plant Physiology* 131(1):147–154.

Tonón, C., Terrile, M., Iglesias, M. J., Lamattina, L., and Casalongué, C. 2010. Extracellular ATP, nitric oxide and superoxide act coordinately to regulate hypocotyl growth in etiolated *Arabidopsis* seedlings. *Journal of Plant Physiology* 167(7):540–546.

Tripathi, D., and Tanaka, K. 2018. A crosstalk between extracellular ATP and jasmonate signaling pathways for plant defense. *Plant Signaling and Behavior* 13(5):511–523.

Wang, F., Jia, J., Wang, Y., Wang, W., Chen, Y., Liu, T., and Shang, Z. 2014. Hyperpolization-activated Ca^{2+} channels in guard cell plasma membrane are involved in extracellular ATP-promoted stomatal opening in *Vicia faba*. *Journal of Plant Physiology* 171(14):1241–1247.

Wang, L., Ma, X., Che, Y., Hou, L., Liu, X., and Zhang, W. 2015. Extracellular ATP mediates H2S-regulated stomatal movements and guard cell K+ current in a H_2O_2-dependent manner in *Arabidopsis*. *Science Bulletin* 60(4):419–427.

Wang, L., Wilkins, K., and Davies, J. M. 2018. *Arabidopsis* DORN1 extracellular ATP receptor; activation of plasma membrane K^+ and Ca^{2+} -permeable conductances. *The New Phytologist* 218(4):1301–1304.

Weerasinghe, R. R., Swanson, S. J., Okada, S. F., et al. 2009. Touch induces ATP release in *Arabidopsis* roots that is modulated by the heterotrimeric G-protein complex. *FEBS Letters* 583(15):2521–2526.

Wu, J., Steinebrunner, I., Sun, Y., et al. 2007. Apyrases (nucleoside triphosphate-diphosphohydrolases) play a key role in growth control in *Arabidopsis*. *Plant Physiology* 144(2):961–975.

Yang, X., Wang, B., Farris, B., Clark, G., and Roux, S. J. 2015. Modulation of root skewing in *Arabidopsis* by apyrases and extracellular ATP. *Plant and Cell Physiology* 56(11):2197–2206.

Zimmermann, H. 2000. Extracellular metabolism of ATP and other nucleotides. *Naunyn-Schmiedeberg's Archives of Pharmacology* 362(4–5):299–309.

14 Phospholipase C in Abiotic Stress-Triggered Lipid Signaling in Plants

Sushma Sagar and Amarjeet Singh

CONTENTS

14.1 INTRODUCTION

The plant cell membrane plays a crucial role in abiotic stress tolerance by acting as a physical barrier, separating the internal cellular milieu from the external surroundings. Upon perception of stimulus at the cell membrane an array of steps, such as generation of secondary messengers, activation of effector protein and modification of the cellular metabolism, takes place (Das et al., 2017). The membrane often undergoes a remodeling process in which the membrane lipid composition changes due to the action of various regulatory membrane proteins, to adapt to the changing environmental conditions. Recent advances in plant sciences have shown that lipids regulate various cellular processes, e.g. lipid remodeling, stress tolerance, hormonal response, etc. (Heilmann, 2016). Unlike lipids that have a structural role in the membrane, phosphoinositides are the regulatory lipids that are involved in membrane restructuring during stress in plants (Das et al., 2017). Phospholipids are crucial for structure development of the plant cell membrane and the synthesis of secondary messengers. Phospholipases are the enzymes that act on the phospholipids to generate secondary messengers in plants. Increasing research has shown that phospholipases are involved in a wide variety of processes in plants like growth, development and regulating abiotic and biotic stress tolerance (Heilmann and Heilmann., 2015, Singh et al., 2015). Among different classes of plant phospholipases (PLA, PLC and PLD), phospholipases C (PLCs) are the important enzymes which catalyze the hydrolysis of the phospholipids. On the basis of substrate specificity and cellular functions, PLCs in plants have been categorized as phosphoinositide phospholipase C (PI-PLC) and phosphatidylcholine-PLC (PC-PLC). PI-PLC hydrolyzes the phosphoinositides, particularly PI (4,5) P2, while PC-PLC prefers phosphatidylcholine (PC) but can also hydrolyze other lipids including phosphatidylethanolamine (PE) and phosphatidylserine (PS); therefore PC-PLC is also known as non-specific PLC (NPC) (Aloulou et al., 2018). PI-PLC hydrolyzes the glycerophosphate ester linkage on the glycerol side of phospholipid to produce diacylglycerol (DAG) and inositol 1,4,5-trisphosphate (IP_3). In animals, DAG remains bound to the membrane and activates protein kinase C (PKC) while IP_3 moves to the cytoplasm where it binds to the ligand gated calcium channels and releases calcium from the intracellular reserves (Vossen et al., 2010). However, plants lack IP_3 receptors and PKC, and the role of secondary messengers is played by the phosphorylated products of DAG and IP_3 i.e. phosphatidic acid (PA), diacylglycerol pyrophosphate (DGPP) and hexakisphosphate (IP_6). In addition to PI-PLC, DAG can also be produced by the hydrolysis activity of the NPCs that hydrolyze the phosphatidylcholine and phosphatidylethanolamine. The DAG produced by NPCs also mediates lipid signaling as a secondary messenger; thus the importance of the roles of NPCs is demonstrated in plant metabolism, plant growth and development and hormone and abiotic stress signaling (Hong et al., 2016). PI-PLCs have been reported in a wide array of plant species including nine members of *Arabidopsis thaliana* (Zhang et al., 2012), four of *Oryza sativa* (rice) (Singh et al., 2013), six of *Solanum lycopersicon* (tomato) (Vossen et al., 2010), three of *Solanum tuberosum* (potato) (Kopka et al., 1998), 12 of *Glycine max* (soybean) (Wang et al., 2015), one of *Pisum sativum* (pea) (Liu et al., 2006), three of *Vigna radiata* (mung bean) (Kim et al., 2004), one of *Avena sativa* (oat) (Huang and Crain, 2009), two of lily (Pan et al., 2005) and two of *Physcomitrella* (Repp et al., 2004). Also, NPCs have been reported in many plants. In-depth analysis revealed that six NPCs are encoded in the *Arabidopsis* genome (Hong et al., 2016), nine NPCs in soybean (Huang et al., 2010), five NPCs in rice (Singh et al., 2013) and 11 NPCs in *Gossypium hirsutum* (cotton) (Song et al., 2017). In this chapter, we present an overview of PLCs in plants and discuss their various important aspects, such as domain structure, regulation of activity and signaling mechanism, along with the recent updates on the role of PLCs in abiotic stress signaling and responses in plants.

14.2 PLC DOMAIN STRUCTURE

14.2.1 PI-PLC Domain Structure

PI-PLCs are multi-domain proteins ranging from 85 kDa to 150 kDa in various plants and animals (Rebecchi and Pentyala, 2000, Das et al., 2017). Plant PI-PLCs have various conserved structural features including catalytic X and Y domains that form a TIM barrel-like structure at the N terminus crucial for phosphoesterase activity and a Ca^{2+}/phosphoesterase binding domain at the C terminus (Tuteja and Sopory, 2008). The X and Y domains are crucial for PLC catalytic activity. A conventional plant PI-PLC, X domain is approximately 170 amino acids long while the Y domain is approximately 260 amino acids long (Chen et al., 2011). Amino acid substitution experiments have shown that even a single amino acid substitution abolishes the catalytic activity of the X domain (Dowd et al., 2006). The crystallographic structure analysis has shown that this region in its tertiary folding consists of distorted TIM barrel (Das et al., 2017). Also, eukaryotic PI-PLCs contain various conserved amino acid residues for effective substrate binding and catalysis. Two conserved His resides have been reported to be present in plants and are involved in the mixed acid/base catalysis reaction of phosphoinositide hydrolysis (Essen et al., 1996). The *Arabidopsis* AtPLC8 and AtPLC9 are widely diverse as compared to the other PI-PLCs due to the presence of large deletions in the Y region (Singh et al., 2015). Additionally, a linker region is also present between the X and Y domains. This linker region is highly hydrophilic and divergent and performs diverse functions in different plants (Pokotylo et al., 2014).

In animal PI-PLCs, this region is believed to have an autoinhibitory role. On the other hand, in plants this linker region contains various acidic residues hypothesized to be present at the surface of various folded proteins (Hicks et al., 2008). In all PI-PLCs a Cys residue is present among the first ten amino acids; however not much is known about its role in plants. In *Chlamydomonas reinhardtii* the cys7 residue helps in the formation of a disulfide bond resulting in homodimerization of its PI-PLC. The EF hand is another important domain that has been reported in various plants and animal PLCs. In animals the EF hand acts as a regulatory domain that binds Ca^{2+} and lipids (Pokotylo et al., 2014). In plants the role of the EF hand is still not known completely. The EF hand contains four helix–loop–helix motifs characteristic of calcium-binding proteins which are grouped into pairwise lobes having varied affinity to bind to Ca^{2+} (Mikami et al., 2004). It is an allosteric regulatory domain in animals since it binds both calcium and lipids, stabilizes the enzyme and helps in active site formation. However, in plants no full-length EF hands are present (Das et al., 2017). In most cases the EF hand domain in plants contains two helix–loop–helix motifs while in others a complete absence of the EF hand is observed. Surprisingly plant PI-PLCs lacking the EF hand domain are also catalytically active. Soybean PI-PLC1 has a truncated EF hand motif which is compensated by the presence of an additional EF hand motif between the X and Y domain (Otterhag et al., 2001). Apart from these, the C_2 domain is ubiquitously present in all plant PI-PLCs and is responsible for binding to the lipids (Hong et al., 2016). This domain is also present in various other proteins involved in the lipid signaling pathways (Aloulou et al., 2018). This motif is approximately 120 residues in length and contains eight antiparallel beta strands arranged as a sandwich. The C_2 domain has a Ca^{2+} binding region (CBR) which is conserved across the plant kingdom. The binding of calcium to the C_2 domain results in a change in hydrophobicity, that is responsible for targeting it to the membrane. This altered surface hydrophobicity property of PI-PLC plays a fundamental role in PLC regulation (Rupwate et al., 2012). The rice PI-PLC C_2 domain contains the polybasic stretch K-(K, R)-T-K and several hydrophobic residues that assist the binding of the C_2 domain to the anionic phospholipids (Kopka et al., 1998). Surface plasmon resonance spectroscopy and sedimentation assays have shown that in rice the C_2 domain is sufficient for the translocation of the enzyme to the membrane, in the presence of Ca^{2+} (Rupwate et al., 2012). However, in other species the EF hand domain or some effectors are crucial in targeting the enzymes to the membrane. For example, in tobacco, the C_2 domain and NtC7, a transmembrane protein, are involved in membrane targeting (Nakamura et al., 2009). In *Arabidopsis*, the N terminal region of the EF hand of AtPLC2 is important for its catalytic activity. Thus, all these domains together account for the proper functioning of the PI-PLCs in plants (Figure 14.1).

14.2.2 NPC Domain Structure

NPC/PC-PLC act upon the common phospholipid such as phosphatidylcholine and phosphatidyl ethanolamine. *Arabidopsis* NPCs are 514–538 amino acids long, weighing approximately 60 kDa (Hong et al., 2016). A lot of variation is observed in the amino acid sequences of the six *Arabidopsis* NPCs. The domain structure of the NPCs is very simple and majorly consists of a single phosphoesterase domain (Hong et al., 2016). However, a C terminal domain has also been reported, which is very divergent and known to impart functional diversity (Pokotylo et al., 2013). The bacterial NPCs contain three additional domains that are highly conserved (Nakamura et al., 2009). In an attempt to understand the structure of NPCs, homology modeling of AtNPC2 with a bacterial acid phosphatase (Acp A) revealed the presence of β sheets surrounded by seven α helixes in the NPC backbone (Pokotylo et al., 2013). Also, various active site residues were found to be conserved between NPC and Acp A. This similarity indicates a similar catalysis mechanism of the plant and bacterial NPCs and homologous acid phosphatase.

14.3 LOCALIZATION OF PLCS

Plant PI-PLCs are primarily known to be associated with the membrane; however they are also detected in the cytosol. Various techniques have been employed in different plants to study the

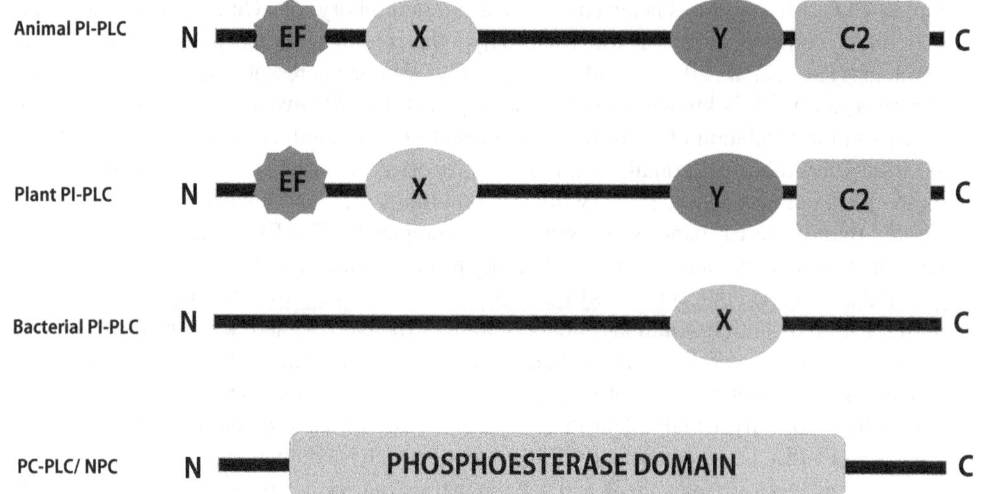

FIGURE 14.1 Schematic representation of the domain structure of PLCs in different organisms. Plant NPCs contain only a single phosphoesterase domain. Bacterial PI-PLC is the simplest, having only an X domain. Animal and plant PI-PLCs are comparatively complex and very similar in structure. They contain an EF domain, X domain, Y domain and a C2 domain. However, the presence of an EF domain in plant PI-PLC is not a universal feature since this domain is present in only a few species.

localization of PI-PLCs. In an experiment utilizing overexpression of the FLAG-tagged PI-PLC fusion protein of soybean in tobacco plants, the presence of soybean PI-PLC in both the soluble and membrane fractions was reported (Rupwate et al., 2012, Shi et al., 1995). Another study utilizing antibodies prepared from *Arabidopsis thaliana* leaf extract showed the presence of AtPLC4 in the plasma membrane and cytosol (Cao et al., 2007). Rice OsPLC1 and OsPLC4 were reported to be localized in the cytosol and nucleus, while OsNPC1 and OsNPC3 were found to be localized in the chloroplast and plastid (Singh et al., 2013). Also, *Arabidopsis* NPC3 has been reported to be localized at the tonoplast, whereas NPC4 and NPC5 were localized at the plasma membrane and in the cytosol, respectively (Hong et al., 2016). This suggested that plant PLCs are variably localized at diverse sub-cellular compartments where they may be involved in variety of plant processes.

14.4 REGULATION OF PLC ACTIVITY

Receptor tyrosine kinase and heterotrimeric G proteins activate the PI-PLCs in animals. However, to achieve such a tight regulation a defined set of domains is present in animals which is missing in the plant PI-PLCs (Munnik and Testerink, 2009). The mechanism of PLC activity regulation is still mysterious in plants (Munnik, 2014). Some of the modes of PI-PLC regulation are discussed here.

14.4.1 Calcium

Ca^{2+} is the most important element involved in the regulation of PI-PLCs in plants. PI-PLCs require calcium for catalysis, and possess an EF hand and C_2 domain for Ca^{2+} sensing and binding (Rupwate et al., 2012). In plants various cues trigger the increase of Ca^{2+}; however, only a few specific ones are known to activate PLC. The membranous PI-PLCs act upon the phosphatidyl inositol 4 phosphate (PtdIns4P) and PtdInsP2 and require calcium in the micromolar range, whereas the cytosolic PI-PLCs utilize phosphatidylinositol as a substrate and require a millimolar range of calcium (Nakamura, 2014, Dowd and Gilroy, 2010). *Cupressus lusitanica* and *Physcomitrella patens* PI-PLCs require PtdIns and a millimolar range calcium for their activity (Zhao et al., 2005; Mikami et al., 2004). Thus, it is evident that calcium is required for PI-PLC activity and regulation in plants.

Knowledge of the regulatory mechanism of plant NPCs is very scarce. Generally, PLCs are regulated by Ca^{2+}; however, most studied *Arabidopsis* NPC4 is reported to be independent of Ca^{2+}. Its activity in turn has been reported to be increased upon chelation of inhibitory divalent cations such as Zn^{2+}, Co^{2+} and Mn^{2+} by EGTA (Nakamura et al., 2005). Furthermore, NPC3 was found to be unresponsive to Mn^{2+}, Mg^{2+} and Ca^{2+} (Pokotylo et al., 2014). In *Brassica napus*, salt and a brassinosteroids hormone, 24-epibrassinolide (24-EBR), were reported to increase NPC activity (Pokotylo et al., 2014). Surprisingly, aluminum has been reported to reduce PC-derived DAG production in *Arabidopsis* (Pejchar et al., 2015). The decrease in the NPC activity upon aluminum stress was due to changes in the physical properties of the plasma membrane (Pejchar et al., 2015).

14.4.2 Phosphorylation

Phosphorylation is a post-translational modification in which a phosphate group is attached to a protein. Phosphorylation generally occurs at the tyrosine or serine/threonine in the eukaryotic protein. In animal PLCs, phosphorylation occurs at the tyrosine residue of X/Y linkers in the catalytic domain, while plant PI-PLCs do not have a single specific phosphorylation site and phosphorylation is reported to occur at multiple sites (Munnik., 2013). Thr29 of the EF hand-like motif and Ser 346 or Ser 348 of the catalytic Y domain have been found to be phosphorylated in AtPLC2 of *Arabidopsis thaliana* (Whiteman et al., 2008, Chen et al., 2010). Two additional phosphorylation sites Thr 169 and Ser 170 have been reported in the catalytic X domain of AtPLC7 (Nuhse et al., 2008).

14.4.3 SUMOylation

SUMOylation is another very common post-translational modification, and different *Arabidopsis* PI-PLCs have been reported to be SUMOylated. AtPLC2, AtPLC5, AtPLC6, AtPLC7 and AtPLC8 have been reported to contain putative SUMOylation sites (Pokotylo et al., 2104, Ren et al., 2009). Furthermore, in transgenic *Arabidopsis* plants overexpressing AtSUMO1 and AtPLC8, SUMOylation was noticed upon heat stress (Park et al., 2011).

14.5 PLC SIGNALING AND MODE OF ACTION

PLC-mediated signaling has been enormously studied in mammalian systems. The PLC activated by various stimuli, such as abiotic stress, biotic stress, wounding etc. act upon the membrane phospholipids such as PIP_2. The hydrolysis of membrane phospholipids by PLC results in the generation of IP_3 and DAG. The DAG activates protein kinase C (PKC), transducing the signal to further downstream effector molecules, and IP_3 interacts with Ca^{2+} channels located at the endoplasmic reticulum (ER), causing the release of Ca^{2+} in the cytosolic space, thereby altering the cytosolic Ca^{2+} levels (Munnik and Testerink, 2009). With the information on PI-PLC signaling in mammals, it was hypothesized that plant PI-PLCs might be regulated in a similar manner and that a similar signaling pathway existed in plants. However, the plant PI-PLC does not correspond to the mammalian paradigm, and it is still unclear whether plants contain a canonical signaling pathway. Radioactive labeling experiments and PI (4, 5) P_2 biosensors have shown that plant cell membranes have very diminished levels of PIP_2 as compared to mammalian cells (Van Leeuwen et al., 2007, Vermeer et al., 2009). Also, the absence of a pleckstrin homology (PH) domain in plant PLCs further makes it difficult to sense the low level of PI (4, 5) P_2 in the cell (Munnik and Testerink, 2009). Another hurdle in deciphering PI-PLC signaling in plants is the absence of a IP_3 receptor. Except for unicellular green algae *Chlamydomonas*, IP_3 receptors have not been reported in any other plant species, so far (Chen et al., 2011). The discovery of IP_6 receptors in plants suggests that IP_3 might convert to inositol hexakisphosphate (IP_6) by the enzyme polyinositolphosphate kinases which may aid in the release of calcium from the intracellular stores (Chen et al., 2011). Also, the potential target of

DAG, PKC, is absent in plants. The *Arabidopsis* genome encodes for seven DAG kinases (DGK) which phosphorylate DAG to PA. Importantly the presence of PA receptors in *Arabidopsis* indicates that DAG must be phosphorylated to PA to perpetuate PLC-mediated signaling (Munnik and Testerink, 2009; Arisz et al., 2009). However, it is well-known that PA is also produced by the action of PLD; therefore, a tight regulatory system in plants is essential for keeping the two PA pools separate. Reports suggest that plants have devised the mechanism for the separation of these two pools through varying degrees of unsaturation in a PA molecule and differences in the acyl chain length (Chen et al., 2011). An overview of PLC signaling in plants is provided in Figure 14.2.

14.6 PLCs IN ABIOTIC STRESS IN PLANTS

PLCs have been associated with various physiological functions in plant signal transduction in guard cells, signaling in the pollen tube, gravitropism and abiotic and biotic stress tolerance (Aloulou. et al., 2018). The role of PLCs is particularly established in abiotic stress signaling and adaptive responses in numerous plant species.

14.6.1 Osmotic Stress

A large amount of data suggests the role of PLCs in osmotic stress tolerance in plants. Munnik and Vermeer (2010) discussed the generation of phosphoinositides and their role in osmotic stress

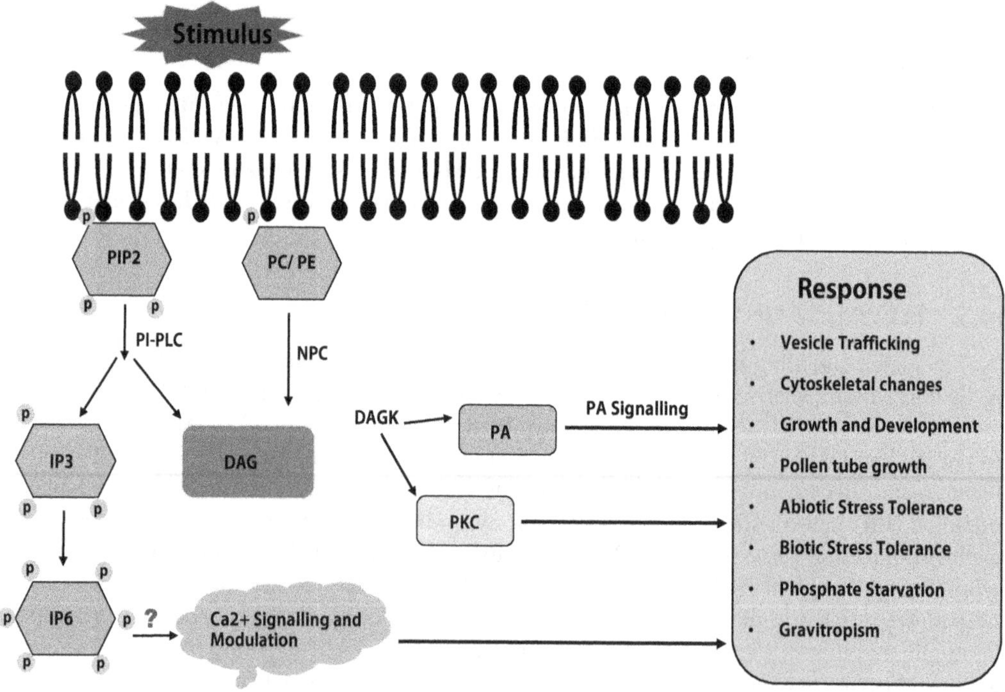

FIGURE 14.2 PLC signaling in plants. In plants, PLC hydrolyzes PI-4,5-P2 into inositol tri-phosphate (IP_3) and diacylglycerol (DAG). DAG remains bound to the membrane while IP_3 moves to the cytosol whereupon, by the action of the enzyme inositol polyphosphate kinase, it is converted to IP6. IP6 binds to an unknown receptor. This binding facilitates the release of calcium from the pool. In addition to PI-PLC, DAG is also produced by NPCs that act upon phosphatidylcholine (PC) and phosphatidylethanolamine (PE). DAG, upon phosphorylation by DAG kinase (DAGK), gets converted to phosphatidic acid (PA). PA acts as a signaling molecule and bring about various responses in plants such as biotic and abiotic stress tolerance, pollen tube growth, phosphate starvation, etc.

signaling. In plants, elevated levels of IP_3 are noticed in the presence of various salts like NaCl and KCl and osmotic stress-inducing agents such as mannitol, mannose and sorbitol (Dewald et al., 2001, Mosblech et al., 2008,). Genetic studies involving inositol polyphosphate phosphatase and kinase suggest that the increase in IP_3 levels results in osmotic stress resistance (Zhu et al., 2009). The rise in IP_3 levels was found to be associated with the increase in PI-4 and 5-P2 under osmotic stress (Dewald et al., 2001, Darwish et al., 2009). This finding suggests the parallel involvement of PI-PLC with phosphoinositide kinase during osmotic stress signaling. Additionally, PI-PLCs have been shown to affect other cellular processes triggered by osmotic stress, such as the generation of ROS, activation of mitogen-activated protein kinase (MAPK), regulation of PEP carboxylase gene expression and the release of transcription factors such as tubby-like protein from the plasma membrane (Im et al., 2012; Reitz et al., 2012; Monreal et al., 2013). In the *Arabidopsis* root tip, seedling and tobacco cells PLC-induced rapid increase in Ca^{2+} levels was observed in response to osmotic and salt stress (Dewald et al., 2001; Cessna et al., 2007; Perera et al., 2008). The role of proline as an osmolyte and its accumulation in plants to alleviate osmotic stress are well-established. In *Arabidopsis*, the PI-PLC-mediated generation of IP_3 and IP_3 gated Ca^{2+} increase have been shown to confer osmotic stress tolerance by the accumulation of proline through the regulation of various transcriptional and post-transcriptional mechanisms (Parre et al., 2007). Wheat PI-PLCs also help the plant to withstand hyperosmotic stress by forming microtubules in the plasmolyzed cells, demonstrated by defective tubulin polymerization under the influence of mannitol treatment in the wheat roots (Wang et al., 2015). Abscisic acid (ABA) is an important plant hormone known to accumulate at the onset of stress in plants. It is known to regulate various defense responses in plants (Fujita et al., 2011). The involvement of PLC in ABA-dependent signaling has been reported in plants. In *Arabidopsis* seedlings IP_3 accumulation was reported upon the application of ABA, exogenously (Xiong et al., 2001). Inhibitor studies have shown that PI-PLCs exert a regulatory role on guard cell movement by modulating the Ca^{2+} oscillations (Staxén et al., 1999). In *Commelina communis*, the addition of heparin, an IP_3 antagonist, impaired the ABA sensitivity in the guard cells that controlled the leaf transcription (Mills et al., 2004). The *Arabidopsis AtPLC3* has been implicated in lateral root elongation and ABA-dependent stomatal closure, as the guard cells of the *atplc3* mutant plants did not show ABA-dependent stomatal closure (Zhang et al., 2017). All these studies in totality suggest that PI-PLCs participate in abiotic stress signaling to enhance plant tolerance to osmotic stress.

14.6.2 Drought and Salt Stress

Drought is one of the most devastating stresses and adversely hampers plant growth and development. It is a multidimensional stress that causes various pleiotropic effects in seed germination and root growth, causes salinity stress, etc. PLCs are implicated in drought-mediated stress signaling in different plant species. Transgenic tobacco and maize plants overexpressing PLCs showed increased tolerance to drought and salt stress (Tripathi et al., 2012, Wang et al., 2008). *Arabidopsis NPC*4 knockout mutant had lower levels of DAG and was less sensitive to seed germination, root elongation and stomata movement but was susceptible to hyperosmotic stress such as high salinity and dehydration (Kocourkova et al., 2011). This suggests that under abiotic stress, *NPC*4 produces higher DAG that upon conversion to PA participates in stress and ABA signaling to enhance plant tolerance to drought and salinity stress. Under dehydration stress, gene expression changes have been reported in various plant species. Surprisingly, upon expressing inositol polyphosphate phosphatase, a decrease in the IP_3 levels brought increased drought tolerance (Perrera et al., 2008). However, the mechanism of this drought tolerance due to decreased IP_3 is not known. In the halophyte *Thellungiella halophita* PI-PLC negatively regulates proline accumulation and gene coding for proline metabolism in the absence of stress and under moderate salt levels. However, in conditions of severe osmotic and salt stress PI-PLC positively regulated proline levels (Ghars et al., 2012). This indicates that PI-PLC may act differently, depending on the severity of stress. Upon salinity stress treatment, wheat *TaPLC1* gene expression increased within half an hour with five-fold higher expression as compared to the control plant. Also, *TaPLC1* protein levels increased upon salinity treatment within half an hour and continued to increase until 48 hours (Zhang et al., 2014). Moreover, the wheat plants were treated with 20% PEG, and *TaPLC1* expression was induced within half an hour of the treatment, decreased at the sixth hour, and finally reached a maximum in 12 hours (Zhang et al., 2014). This result shows that *TaPLC1* is involved in both drought stress and salinity stress in wheat; however, there is a remarkable difference in the expression pattern. Recently, a genome-wide expression analysis of the PI-PLC members of *Glycine max* (soybean) showed differential expression of various *GmPLC* genes under different abiotic stresses. *GmPLC6, GmPLC12* transcript were induced by salinity stress treatment, *GmPLC9* was induced by PEG treatment and *GmPLC10* was induced by alkali treatment. These expression patterns suggest that various soybean PLCs are involved in abiotic stress signaling in plants (Wang et al., 2015). Similarly, in rice the differential regulation of various rice PLCs and NPCs under abiotic stress conditions was observed. *OsPLC1* and *OsPLC3* were found to be up-regulated while *OsPLC2* was down-regulated under salt and drought stresses. Additionally, *OsNPC1, OsNPC2* and *OsNPC4* were up-regulated under both drought and salt stress, while *OsNPC5* was down-regulated (Singh et al., 2013). These expression patterns provide a clue about the significant role of these PLCs and NPCs in drought and salt stress signaling in rice. In a recent study in *Arabidopsis*, the atplc4 mutants exhibited a hyposensitive response, whereas *AtPLC4* overexpression seeds were hypersensitive to the salt stress (Xia et al., 2017), suggesting that AtPLC4 negatively regulates salt stress signaling in *Arabidopsis*. Additionally, under salt stress the transcript level of salt stress-responsive genes such as *RD29B* was enhanced, whereas ZAT10 and MYB15 transcription levels were reduced in the *atplc4* mutant (Xia et al., 2017). Furthermore, *AtPLC3*-overexpression plants showed increased tolerance to drought stress (Zhang et al., 2017). These data display the involvement of PLCs in osmotic and drought stress tolerance in plants.

14.6.3 Heat Stress

A significant role of PLCs has been reported in temperature stress response in plants. Recent identification and characterization of cotton PLCs unveiled the differential regulation of GhPLC genes under various abiotic stresses. *GhPLC* 2 and *GhPLC7* transcripts were particularly highly induced under heat stress; however, weak expression of these transcripts was observed during other stresses (Zhang et al., 2017). The *Arabidopsis AtPLC9* is shown to confer heat stress tolerance to plants. *AtPLC9*-overexpression transgenic plants showed improved thermotolerance; also complementation of the *atplc9* mutant with the *AtPLC9* rescued the thermosensitive phenotype exhibited by the mutants (Zhang et al., 2012). Also, knockout mutants of *AtPLC3* had low levels of Ca^{2+} when subjected to heat stress. Additionally, *atplc3/atplc9* double mutants exhibited higher sensitivity to the heat stress as compared to the single mutant (Charng et al., 2007). This leads to the conclusion that *atplc3* and *atplc9* mutants work redundantly in heat stress signaling in *Arabidopsis*. A study in pea showed that PI-PLC activity increases in the pea membrane on the exposure to heat stress (Ruelland and Zachowski, 2010). This observation was found to be consistent with the PI-PLC protein accumulation in the heat stress-treated plants (Liu et al., 2006). A recent study in *Arabidopsis* showed that the *npc1* knockout mutant exhibited decreased chlorophyll content and survival rate upon heat stress. On the other hand, the *NPC1* overexpression line showed remarkably better heat stress tolerance, suggesting a positive role of *NPC1* in heat stress response (Krčková et al., 2015). All these observations suggest a significant role of PLCs in plant thermotolerance.

14.6.4 Cold Stress

Cold stress-induced PLC gene expression has been reported in maize (Sui et al., 2008), *Arabidopsis* (Tasma et al., 2008), winter wheat (Skinner et al., 2005) and brassica (Das et al., 2005). Cold stress-induced fleeting and transient IP_3 accumulation with a parallel decrease in PI-4-P and PI-(4, 5)-P_2 levels were found in oilseed rape leaves and wheat tissues. Also, *Arabidopsis* suspension culture, upon exposure to cold stress, showed significant losses in PI-4-P and PI -(4, 5)- P_2 levels and an increase in IP_3 levels, suggesting that under cold stress PI-PLCs play a positive role by producing more IP_3 in the cells that may initiate cold stress-induced signal transduction in plants. It is a well-known fact that under cold stress cell membranes become rigid. The rigidification of plasma membrane was observed under cold stress in *Arabidopsis* suspension cells which had mutated desaturase enzyme. The transduction of the lipid signaling pathway under cold stress is dependent on Ca^{2+}, and the substrates for PI-PLCs are supplied by type III phosphatidyl inositol 4-kinase (Delage et al., 2013). Based on these observations, a model has been proposed that emphasizes the role of membrane rigidification and the increase in cytosolic Ca^{2+} for escalating PI-PLC activity. These findings demonstrate the involvement of PI-PLC in plant cold stress tolerance.

14.6.5 Heavy Metal Stress

Toxic heavy metals affect various enzymatic reactions that grievously inhibit plant cell metabolism (Pokotylo et al., 2014). It has been shown that aluminum can block the hydrolysis of PI-4, 5-P2 by binding it or by substituting Ca^{2+} bound to the liposomal lipids (Kopka et al., 1998). In suspension cells of *Coffea arabica* (coffee), aluminum treatment led to a rapid PI-PLC activation and consequent increase in IP_3 accumulation (Estevez et al., 2003), although prolonged exposure inhibited PI-PLC activity (Ramos-Diaz et al., 2007). In coffee the Ca^{2+} and K^+ concentrations decreased in the roots upon exposure to aluminum. Detailed investigation showed the involvement of PI-PLC in this response as PI-PLC activity was decreased upon treatment with 300 μM $AlCl_3$ (Bojórquez-Quintal et al., 2014). However, the mechanism behind the rapid activation of PI-PLC by aluminum is still not clear. A study in marine alga *Ulva compressa* reported the release of intracellular Ca^{2+} in response to copper, and this was suggested to be dependent on PI-PLC activity (Gonzalez et al., 2012). Moreover, in *Catharanthus roseus* roots heavy metals such as Zn^{2+}, Cu^{2+} and Ni^{2+} decreased the *in vitro* PI-PLC activity in the membranous and soluble fractions (Piña-Chable et al., 1998). Whereas in brassica excessive copper resulted in a rapid increase in *in vivo* DAG accumulation in the root, suggesting possible activation of PI-PLC (Russo et al., 2008).

14.7 ROLE OF PLCs IN NUTRIENT STRESS

Phospholipases, especially NPCs, are implicated in the nutrient stress response in plants. NPCs are shown to be involved in the mobilization of organic phosphate from the membrane by ensuring the cleavage of the phosphate head group from phosphatidylcholine and other phospholipid, during phosphate starvation. DAG produced due to NPC activity acts as a precursor for sulfolipids and glycolipid production and, thereby, takes care of the membrane integrity (Tjellstrom et al., 2010). Under phosphate-deprived conditions the up-regulation of *NPC4* and consequent higher PC-PLC activity and DAG accumulation were detected in *Arabidopsis* (Nakamura et al., 2005). Also, *NPC5* was found to be activated in *Arabidopsis* roots during phosphate starvation (Misson et al., 2005). In addition, *NPC5* was found to be detrimental for the accumulation of galactolipid during phosphate deficiency in *Arabidopsis* roots (Gaude et al., 2008). Auxin response factors (ARF) and its inhibitors play an important role in auxin signaling (Narise et al., 2010). Interestingly, the involvement of auxin signaling was predicted in the process of membrane remodeling under phosphate starvation. Two *Arabidopsis* mutants, namely *arf7arf19* and solitary root1 (*slr1*), impaired in auxin transcription factors IAA14 and ARF7/9, showed lesser NPC activity (Narise et al., 2010). These findings suggest that auxin signaling may be upstream of the NPC activity during phosphate-deficient conditions in plants.

14.8 CONCLUSIONS

Tremendous progress has been made in understanding the various functional aspects of phospholipases C in plants, in the recent past. A huge amount of data has accumulated on the PLC structure, mode of regulation, importance in growth, development and tolerance to abiotic and biotic stress in plants. PLCs are a multi-gene family and are divided into PI-PLCs and NPC on the basis of their substrate preference. Localization experiments in plants have shown that PLCs are present at the plasma membrane and also in the cytosol. Phospholipases in animals share many common features such as structure, regulation and functions with plants, yet there are many dissimilarities as well. Various domains present in the animal PLCs are lacking in the plant PLCs. Despite tremendous efforts, the PLC-mediated abiotic stress signaling mechanism in plants is not deciphered to the accuracy required, and further investigations are needed. To understand PLCs and their role in plant abiotic stress better, it is important to understand the dynamics of IP_3 and DAG. Moreover, an important question, whether IP_3 and DAG or their phosphorylated products act as secondary signaling molecules in plants, needs to be addressed. Rigorous efforts involving novel ideas, genetic modeling experiments and newer techniques like CRISPR-Cas9 can help to fill the voids present in understanding PLC signaling in plants. Also, such techniques can be useful to uncover the previously unknown role of PLCs in plant stress tolerance mechanisms. Despite the holes present in our current knowledge, PLCs still are promising candidates for understanding abiotic stress signaling and engineering crop plants with PLCs for abiotic stress tolerance.

REFERENCES

Aloulou, A., Rahier, R., Arhab, Y., Noiriel, A., and Abousalham, A. 2018. Phospholipases: An overview. In: *Lipases and Phospholipases*, ed. G. Sandoval, 69–105, *Methods in Molecular Biology*. Humana Press, New York.

Arisz, S. A., Testerink, C., and Munnik, T. 2009. Plant PA signaling via diacylglycerol kinase. *Biochimica et Biophysica Acta (BBA)-Molecular and Cell Biology of Lipids* 1791(9):869–875.

Bojórquez-Quintal, E. A. de, Sánchez-Cach, L. A., Ku-González, A., de los Santos-Briones, C., et al. 2014. Differential effects of aluminum on in vitro primary root growth, nutrient content and phospholipase C activity in coffee seedlings (*Coffea arabica*). *Journal of Inorganic Biochemistry* 134:39–48.

Cao, Z., Zhang, J., Li, Y., et al. 2007. Preparation of polyclonal antibody specific for AtPLC4, an *Arabidopsis* phosphatidylinositol-specific phospholipase C in rabbits. *Protein Expression and Purification* 52(2):306–312.

Cessna, S. G., Matsumoto, T. K., Lamb, G. N., Rice, S. J., and Hochstedler, W. W. 2007. The externally derived portion of the hyperosmotic shock-activated cytosolic calcium pulse mediates adaptation to ionic stress in suspension-cultured tobacco cells. *Journal of Plant Physiology* 164(7):815–823.

Charng, Y. Y., Liu, H. C., Liu, N. Y., et al. 2007. A heat-inducible transcription factor, HsfA2, is required for extension of acquired thermotolerance in *Arabidopsis*. *Plant Physiology* 143(1):251–262.

Chen, G., Snyder, C. L., Greer, M. S., and Weselake, R. J. 2011. Biology and biochemistry of plant phospholipases. *Critical Reviews in Plant Sciences* 30(3):239–258.

Chen, Y., Hoehenwarter, W., and Weckwerth, W. 2010. Comparative analysis of phytohormone-responsive phosphoproteins in *Arabidopsis thaliana* using TiO_2-phosphopeptide enrichment and mass accuracy precursor alignment. *The Plant Journal* 63(1):1–17.

Darwish, E., Testerink, C., Khalil, M., El-Shihy, O., and Munnik, T. 2009. Phospholipid signaling responses in salt-stressed rice leaves. *Plant and Cell Physiology* 50(5):986-997.

Das, P., Datta, S., Samanta, M., Mukherjee, A., and Majumder, A. 2017. Phosphoinositides and phospholipase C signalling in plant stress response: A revisit. *Proceedings of the Indian National Science Academy* 83:845–863.

Das, S., Hussain, A., Bock, C., Keller, W. A., and Georges, F. 2005. Cloning of *Brassica napus* phospholipase C2 (BnPLC2), phosphatidylinositol 3-kinase (BnVPS34) and phosphatidylinositol synthasel (BnPtdIns S1)—Comparative analysis of the effect of abiotic stresses on the expression of phosphatidylinositol signal transduction-related genes in B. napus. *Planta* 220(5):777–784.

Delage, E., Puyaubert, J., Zachowski, A., and Ruelland, E. 2013. Signal transduction pathways involving phosphatidylinositol 4-phosphate and phosphatidylinositol 4,5-bisphosphate: Convergences and divergences among eukaryotic kingdoms. *Progress in Lipid Research* 52(1):1–14.

Dewald, D. B., Torabinejad, J., Jones, C. A., et al. 2001. Rapid accumulation of phosphatidylinositol 4,5-bisphosphate and inositol 1,4,5-trisphosphate correlates with calcium mobilization in salt-stressed *Arabidopsis*. *Plant Physiology* 126(2):759–769.

Dowd, P. E., Coursol, S., Skirpan, A. L., Kao, T. H., and Gilroy, S. 2006. Petunia phospholipase C1 is involved in pollen tube growth. *The Plant Cell* 18(6):1438–1453.

Dowd, P. E., and Gilroy, S. 2010. The emerging roles of phospholipase C in plant growth and development. In: *Lipid Signaling in Plants*, ed. T. Munnik, 23–37. Springer, Berlin, Heidelberg.

Essen, L., Perisic, O., Cheung, R., Katan, M., and Williams, R. L. 1996. Crystal structure of a mammalian phosphoinositide-specific phospholipase C-d. *Nature* 380:595–602.

Fujita, Y., Fujita, M., Shinozaki, K., and Yamaguchi-Shinozaki, K. 2011. ABA-mediated transcriptional regulation in response to osmotic stress in plants. *Journal of Plant Research* 124(4):509–525.

Gaude, N., Nakamura, Y., Scheible, W. R., Ohta, H., and Dörmann, P. 2008. Phospholipase C5 (NPC5) is involved in galactolipid accumulation during phosphate limitation in leaves of *Arabidopsis*. *The Plant Journal* 56(1):28–39.

Ghars, M. A., Richard, L., Lefebvre-De Vos, D., et al. 2012. Phospholipases C and D modulate proline accumulation in Thellungiella halophila/Salsuginea differently according to the severity of salt or hyperosmotic stress. *Plant and Cell Physiology* 53(1):183–192.

González, A., de los Ángeles Cabrera, M., Mellado, M., et al. 2012. Copper-induced intracellular calcium release requires extracellular calcium entry and activation of L-type voltage-dependent calcium channels in *Ulva compressa*. *Plant Signaling and Behavior* 7(7):728–732.

Heilmann, I. 2016. Plant phosphoinositide signaling: dynamics on demand. *Biochimica et Biophysica Acta (BBA)-Molecular and Cell Biology of Lipids* 1861(9 Pt B):1345–1351.

Heilmann, M., and Heilmann, I. 2015. Plant phosphoinositides-complex networks controlling growth and adaptation. *Biochimica et Biophysica Acta (BBA)-Molecular and Cell Biology of Lipids* 1851(6):759–769.

Hicks, S. N., Jezyk, M. R., Gershburg, S., Seifert, J. P., Harden, T. K., and Sondek, J. 2008. General and versatile autoinhibition of PLC isozymes. *Molecular Cell* 31(3):383–394.

Hong, Y., Zhao, J., Guo, L., et al. 2016. Plant phospholipases D and C and their diverse functions in stress responses. *Progress in Lipid Research* 62:55–74.

Huang, C. H., and Crain, R. C. 2009. Phosphoinositide-specific phospholipase C in oat roots: Association with the actin cytoskeleton. *Planta* 230(5):925–933.

Huang, X., Zheng, Y., Peters, C., and Wang, X. 2010. Characterization of non-specific phospholipase C (NPC) in soybean. *In vitro Biology Meeting and IAPB 12th World Congress*, pp. 93–211.

Im, J. H., Lee, H., Kim, J., Kim, H. B., and An, C. S. 2012. Soybean MAPK, GMK1 is dually regulated by phosphatidic acid and hydrogen peroxide and translocated to nucleus during salt stress. *Molecules and Cells* 34(3):271–278.

Kim, Y. J., Kim, J. E., Lee, J. H., et al. 2004. The Vr-PLC3 gene encodes a putative plasma membrane-localized phosphoinositide-specific phospholipase C whose expression is induced by abiotic stress in mung bean (*Vigna radiata* L.). *FEBS Letters* 556(1–3):127–136.

Kocourkova, D., Krčková, Z., Pejchar, P., et al. 2011. The phosphatidylcholine-hydrolysing phospholipase C NPC4 plays a role in response of *Arabidopsis* roots to salt stress. *Journal of Experimental Botany* 62(11):3753–3763.

Kopka, J., Pical, C., Gray, J. E., and Müller-Röber, B. 1998. Molecular and enzymatic characterization of three phosphoinositide-specific phospholipase C isoforms from potato. *Plant Physiology* 116(1):239–250.

Krčková, Z., Brouzdová, J., Daněk, M., et al. 2015. *Arabidopsis* non-specific phospholipase C1: Characterization and its involvement in response to heat stress. *Frontiers in Plant Science* 6:928.

Liu, H. T., Liu, Y. Y., Pan, Q. H., Yang, H. R., Zhan, J. C., and Huang, W. D. 2006. Novel interrelationship between salicylic acid, abscisic acid, and PIP2-specific phospholipase C in heat acclimation-induced thermotolerance in pea leaves. *Journal of Experimental Botany* 57(12):3337–3347.

Martínez-Estévez, M., Racagni-Di Palma, G., Muñoz-Sánchez, J. A., Brito-Argáez, L., Loyola-Vargas, V. M., and Hernández-Sotomayor, S. M. 2003. Aluminium differentially modifies lipid metabolism from the phosphoinositide pathway in *Coffea arabica* cells. *Journal of Plant Physiology* 160(11):1297–1303.

Mikami, K., Repp, A., Graebe-Abts, E., and Hartmann, E. 2004. Isolation of cDNAs encoding typical and novel types of phosphoinositide-specific phospholipase C from the moss *Physcomitrella patens*. *Journal of Experimental Botany* 55(401):1437–1439.

Mills, L. N., Hunt, L., Leckie, C. P., et al. 2004. The effects of manipulating phospholipase C on guard cell ABA-signalling. *Journal of Experimental Botany* 55(395):199–204.

Misson, J., Raghothama, K. G., Jain, A., *et al.* 2005. A genome-wide transcriptional analysis using *Arabidopsis thaliana* Affymetrix gene chips determined plant responses to phosphate deprivation. *Proceedings of the National Academy of Sciences of the United States of America* 102(33):11934–11939.

Monreal, J. A., Arias-Baldrich, C., Pérez-Montaño, F., Gandullo, J., Echevarría, C., and García-Mauriño, S. 2013. Factors involved in the rise of phosphoenolpyruvate carboxylase-kinase activity caused by salinity in sorghum leaves. *Planta* 237(5):1401–1413.

Mosblech, A., König, S., Stenzel, I., Grzeganek, P., Feussner, I., and Heilmann, I. 2008. Phosphoinositide and inositolpolyphosphate signalling in defense responses of *Arabidopsis thaliana* challenged by mechanical wounding. *Molecular Plant* 1(2):249–261.

Munnik, T. 2014. PI-PLC: Phosphoinositide-phospholipase C in plant signaling. In: *Phospholipases in Plant Signaling*, ed. X. Wang, 27–54. Springer, Berlin, Heidelberg.

Munnik, T., and Testerink, C. 2009. Plant phospholipid signaling: "in a nutshell". *Journal of Lipid Research* 50(Supplement):S260–S265.

Munnik, T., and Vermeer, J. E. 2010. Osmotic stress-induced phosphoinositide and inositol phosphate signalling in plants. *Plant, Cell & Environment* 33(4):655–669.

Nakamura, Y. 2014. NPC: Nonspecific phospholipase Cs in plant functions. In: *Phospholipases in Plant Signaling*, ed. X. Wang 55–67. Springer, Berlin, Heidelberg.

Nakamura, Y., Kobayashi, K., and Ohta, H. 2009. Activation of galactolipid biosynthesis in development of pistils and pollen tubes. *Plant Physiology and Biochemistry* 47(6):535–539.

Nakamura, Y., Awai, K., Masuda, T., Yoshioka, Y., Takamiya, K. I., and Ohta, H. 2005. A novel phosphatidylcholine-hydrolyzing phospholipase C induced by phosphate starvation in *Arabidopsis*. *The Journal of Biological Chemistry* 280(9):7469–7476.

Narise, T., Kobayashi, K., Baba, S., et al. 2010. Involvement of auxin signaling mediated by IAA14 and ARF7/19 in membrane lipid remodeling during phosphate starvation. *Plant Molecular Biology* 72(4–5):533–544.

Otterhag, L., Sommarin, M., and Pical, C. 2001. N-terminal EF-hand-like domain is required for phosphoinositide-specific phospholipase C activity in *Arabidopsis thaliana*. *FEBS Letters* 497(2–3):165–170.

Pan, Y. Y., Wang, X., Ma, L. G., and Sun, D. Y. 2005. Characterization of phosphatidylinositol-specific phospholipase C (PI-PLC) from *Lilium daviddi* pollen. *Plant and Cell Physiology* 46(10):1657–1665.

Park, H. C., Choi, W., Park, H. J., et al. 2011. Identification and molecular properties of SUMO-binding proteins in *Arabidopsis*. *Molecules and Cells* 32(2):143–151.

Parre, E., Ghars, M. A., Leprince, A. S., et al. 2007. Calcium signaling via phospholipase C is essential for proline accumulation upon ionic but not nonionic hyperosmotic stresses in *Arabidopsis*. *Plant Physiology* 144(1):503–512.

Pejchar, P., Potocký, M., Krčková, Z., Brouzdová, J., Daněk, M., and Martinec, J. 2015. Non-specific phospholipase C4 mediates response to aluminum toxicity in *Arabidopsis thaliana*. *Frontiers in Plant Science* 6:66.

Perera, I. Y., Hung, C. Y., Moore, C. D., Stevenson-Paulik, J., and Boss, W. F. 2008. Transgenic *Arabidopsis* plants expressing the type 1 inositol 5-phosphatase exhibit increased drought tolerance and altered abscisic acid signaling. *The Plant Cell* 20(10):2876–2893.

Piña-Chable, M. L., de los Santos-Briones, C., Muñoz-Sánchez, J. A., Machado, I. E., and Hernández-Sotomayor, S. T. 1998. Effect of different inhibitors on phospholipase C activity in *Catharanthus roseus* transformed roots. *Prostaglandins and Other Lipid Mediators* 56(1):19–31.

Pokotylo, I., Kolesnikov, Y., Kravets, V., Zachowski, A., and Ruelland, E. 2014. Plant phosphoinositide-dependent phospholipases C: Variations around a canonical theme. *Biochimie* 96:144–157

Pokotylo, I., Pejchar, P., Potocký, M., Kocourková, D., Krčková, Z., Ruelland, E., et al. 2013. The plant non-specific phospholipase C gene family. Novel competitors in lipid signalling. *Progress in Lipid Research* 52:62–79.

Ramos-Díaz, A., Brito-Argáez, L., Munnik, T., and Hernández-Sotomayor, S. T. 2007. Aluminum inhibits phosphatidic acid formation by blocking the phospholipase C pathway. *Planta* 225 (2):393–401.

Rebecchi, M. J., and Pentyala, S. N. 2000. Structure, function, and control of phosphoinositide-specific phospholipase C. *Physiological Reviews* 80(4):1291–1335.

Reitz, M. U., Bissue, J. K., Zocher, K., et al. 2012. The subcellular localization of Tubby-like proteins and participation in stress signaling and root colonization by the mutualist *Piriformospora indica*. *Plant Physiology* 160(1):349–364.

Ren, J., Gao, X., Jin, C., et al. 2009. Systematic study of protein SUMOylation: Development of a site-specific predictor of SUMOsp 2.0. *Proteomics* 9(12):3409–3412.

Repp, A., Mikami, K., Mittmann, F., and Hartmann, E. 2004. Phosphoinositide-specific phospholipase C is involved in cytokinin and gravity responses in the moss *Physcomitrella patens*. *The Plant Journal* 40(2):250–259.

Ruelland, E., and Zachowski, A. 2010. How plants sense temperature. *Environmental and Experimental Botany* 69(3):225–232.

Rupwate, S. D., and Rajasekharan, R. 2012. Plant phosphoinositide-specific phospholipase C: An insight. *Plant Signaling and Behavior* 7(10):1281–1283.

Russo, M., Sgherri, C., Izzo, R., and Navari-Izzo, F. 2008. Brassica napus subjected to copper excess: phospholipases C and D and glutathione system in signalling. *Environmental and Experimental Botany* 62(3):238–246.

Shi, J., Gonzales, R. A., and Bhattacharyya, M. K. 1995. Characterization of a plasma membrane-associated phosphoinositide-specific phospholipase C from soybean. *The Plant Journal* 8(3):381–390.

Singh, A., Bhatnagar, N., Pandey, A., and Pandey, G. K. 2015. Plant phospholipase C family: Regulation and functional role in lipid signaling. *Cell Calcium* 58(2):139–146.

Singh, A., Kanwar, P., Pandey, A., et al. 2013. Comprehensive genomic analysis and expression profiling of phospholipase C gene family during abiotic stresses and development in rice. *PLoS ONE* 8(4):e62494.

Skinner, D. Z., Bellinger, B. S., Halls, S., Baek, K., Garland-Campbell, K., and Siems, W. 2005. Phospholipid acyl chain and phospholipase dynamics during cold acclimation of winter wheat. *Crop Science* 45(5):1858e1867.

Song, J., Zhou, Y., Zhang, J., and Zhang, K. 2017. Structural, expression and evolutionary analysis of the non-specific phospholipase C gene family in *Gossypium hirsutum. BMC Genomics* 18(1):979.

Staxén, I., Pical, C., Montgomery, L. T., Gray, J. E., Hetherington, A. M., and McAinsh, M. R. 1999. Abscisic acid induces oscillations in guard-cell cytosolic free calcium that involve phosphoinositide-specific phospholipase C. *Proceedings of the National Academy of Sciences of the United States of America* 96(4):1779–1784.

Tasma, I. M., Brendel, V., Whitham, S. A., and Bhattacharyya, M. K. 2008. Expression and evolution of the phosphoinositide-specific phospholipase C gene family in *Arabidopsis thaliana. Plant Physiology and Biochemistry* 46(7):627–637.

Tjellström, H., Hellgren, L. I., Wieslander, A., and Sandelius, A. S. 2010. Lipid asymmetry in plant plasma membranes: Phosphate deficiency-induced phospholipid replacement is restricted to the cytosolic leaflet. *The FASEB Journal* 24(4):1128–1138.

Tripathy, M. K., Tyagi, W., Goswami, M., et al. 2012. Characterization and functional validation of tobacco PLC delta for abiotic stress tolerance. *Plant Molecular Biology Reporter* 30(2):488–497.

Tuteja, N., and Sopory, S. K. 2008. Plant signaling in stress: G-protein coupled receptors, heterotrimeric G-proteins and signal coupling via phospholipases. *Plant Signaling and Behavior* 3(2):79–86.

Van Leeuwen, W., Vermeer, J. E., Gadella Jr, T. W., and Munnik, T. 2007. Visualization of phosphatidylinositol 4,5-bisphosphate in the plasma membrane of suspension-cultured tobacco BY-2 cells and whole *Arabidopsis* seedlings. *The Plant Journal* 52(6):1014–1026.

Vermeer, J. E. M., Thole, J. M., Goedhart, J., Nielsen, E., Munnik, T., and Gadella, T. W. J. Jr. 2009. Imaging phosphatidylinositol 4-phosphate dynamics in living plant cells. *The Plant Journal* 57(2):356–372.

Vossen, J. H., Abd-El-Haliem, A., Fradin, E. F., et al. 2010. Identification of tomato phosphatidylinositol-specific phospholipase-C (PI-PLC) family members and the role of PLC4 and PLC6 in HR and disease resistance. *The Plant Journal* 62:224–239.

Wang, C. R., Yang, A. F., Yue, G. D., Gao, Q., Yin, H. Y., and Zhang, J. R. 2008. Enhanced expression of phospholipase C 1 (ZmPLC1) improves drought tolerance in transgenic maize. *Planta* 227(5):1127–1140.

Wang, F., Deng, Y., Zhou, Y., et al. 2015. Genome-wide analysis and expression profiling of the phospholipase C gene family in soybean (*Glycine max*). *PLoS ONE* 10(9):e0138467.

Whiteman, S. A., Serazetdinova, L., Jones, A. M., et al. 2008. Identification of novel proteins and phosphorylation sites in a tonoplast enriched membrane fraction of *Arabidopsis thaliana. Proteomics* 8(17):3536–3547.

Xia, K., Wang, B., Zhang, J., Li, Y., Yang, H., and Ren, D. 2017. *Arabidopsis* phosphoinositide-specific phospholipase C 4 negatively regulates seedling salt tolerance. *Plant, Cell and Environment* 40(8):1317–1331.

Xiong, L., Lee, B. H., Ishitani, M., Lee, H., Zhang, C., and Zhu, J. K. 2001. FIERY1 encoding an inositol polyphosphate 1-phosphatase is a negative regulator of abscisic acid and stress signaling in *Arabidopsis. Genes and Development* 15(15):1971–1984.

Zhang, K., Jin, C., Wu, L., Hou, M., Dou, S., and Pan, Y. 2014. Expression analysis of a stress-related phosphoinositide-specific phospholipase C gene in wheat (*Triticum aestivum* L.). *PLoS ONE* 9(8):e105061.

Zhang, Q., Lin, F., Mao, T., et al. 2012. Phosphatidic acid regulates microtubule organization by interacting with MAP65-1 in response to salt stress in *Arabidopsis. The Plant Cell* 24(11):4555–4576.

Zhang, Q., Van Wijk, R., Shahbaz, M., et al. 2017. *Arabidopsis* phospholipase C3 is involved in lateral root initiation and ABA responses in seed germination and stomatal closure. *Plant and Cell Physiology* 59(3):469–486.

Zhao, J., Davis, L. C., and Verpoorte, R. 2005. Elicitor signal transduction leading to production of plant secondary metabolites. *Biotechnology Advances* 23(4):283–333.

Zhu, J. Q., Zhang, J. T., Tang, R. J., Lv, Q. D., Wang, Q. Q., Yang, L., and Zhang, H. X. 2009. Molecular characterization of ThIPK2, an inositol polyphosphate kinase gene homolog from Thellungiella halophila, and its heterologous expression to improve abiotic stress tolerance in Brassica napus. *Physiologia Plantarum* 136(4):407–425.

15 Role of Silicon in Abiotic Stress Tolerance of Plants

Rakeeb Ahmad Mir, Kaisar A. Bhat, A.A. Shah, and Sajad Majeed Zargar

CONTENTS

15.1 INTRODUCTION

Agricultural productivity is adversely affected by the physiological and biochemical damage mediated by diversified classes of abiotic stresses. About 51–82% of the reduction in crop yield is due to abiotic stress (Bray et al. 2000). Plants circumvent different types of abiotic stress in conjunction with unusual weather, seasonal variations and environmental gradients (Hirt and Shinozaki, 2004). There is a dire need to adopt strategies against the adverse effects of abiotic stress to improve the crop yield for the expansion of land used for agrarian purposes (Tilman et al. 2001). The critical physiological changes mediated by abiotic stress in cells include the production of reactive oxygen species, and in turn the disruption of cellular homeostasis. The adverse effects of these toxic products include damage to the cell membrane, DNA and RNA (Mittler, 2002). Different studies have revealed that the most effective way to evade these abiotic stresses is to develop resistance in crops.

Silicon, one of the most abundant micronutrients in soil, has been found to play diversified roles in plant growth under different stress conditions (Zargar et al. 2010). It is found abundantly as the major component of the cell wall in monocot species (Inanaga and Okasaka, 1995). The diverse range of physiological roles played by Si include increases in crop yield, rate of photosynthesis and nitrogen fixation. In addition, Si helps in tolerating different biotic as well as abiotic stress like pathogens, fungal attack, UV radiation, temperature, nutrient deficiency, metal toxicity, drought and salinity (Ma, 2004; Liang et al. 2015; Zargar et al. 2010; Cooke and Leishman, 2011; Guntzer et al. 2012). In earlier studies it was assumed that Si was non-essential for plant growth (Sachs, 1860; Arnon and Stout, 1939), but several studies conducted during the last two decades confirmed that Si plays a critical role in plant growth (Liang et al. 2015). Si is considered a quasi-essential element as its benefits has been observed during stress conditions (Liang et al. 2015).

As far as its quantitative existence is concerned, silicon is the most abundant mineral in the earth's crust in the form of $Si(OH)_4$ (McGinnity, 2016). Even though limitation of silicon has no adverse physiological effect on plants, but it is observed that it increases the yield of crops. As a case study, the application of calcium and Si foliar fertilizer has been found to enhance sugar beet production and in turn sugar yields (Artyszak et al. 2015). It has been observed that the exogenous application of Si resulted in increased resistance to salinity in several plant species, like *Zea mays* (Moussa, 2006), *Lycopersicon esculentum* (Romero-Aranda et al. 2006), *Triticum aestivum* (Tuna et al. 2008; Tahir et al. 2012) and *Brassica napus* (Hashemi et al. 2010). In another experiment the exogenous application of Si ameliorated the negative effects of salinity on the growth of crops, possibly due to low content of Na^+, thus maintaining the integrity of the membrane and also helping to scavenge the ROS (Hashemi et al. 2010). In addition, studies have backed the role of Si in mitigating the negative effects of abiotic stress, in particular drought and salinity (Zhu and Gong, 2014).

15.2 SILICON IN ABIOTIC STRESS

Silicon plays an important role in reducing the adverse effects of both biotic as well as abiotic stresses. Si is believed to be the essential element for the growth and development of a wide array of plant species and in turn is linked to different morphological, physiological and molecular processes in plants (Richmond and Sussman, 2003; Ma, 2004; Liang et al., 2015). There are various types of mechanisms within plants to alleviate and mitigate different abiotic stresses (Figure 15.1). It has been found that this effect may partially be due its accumulation in intercellular spaces and cell

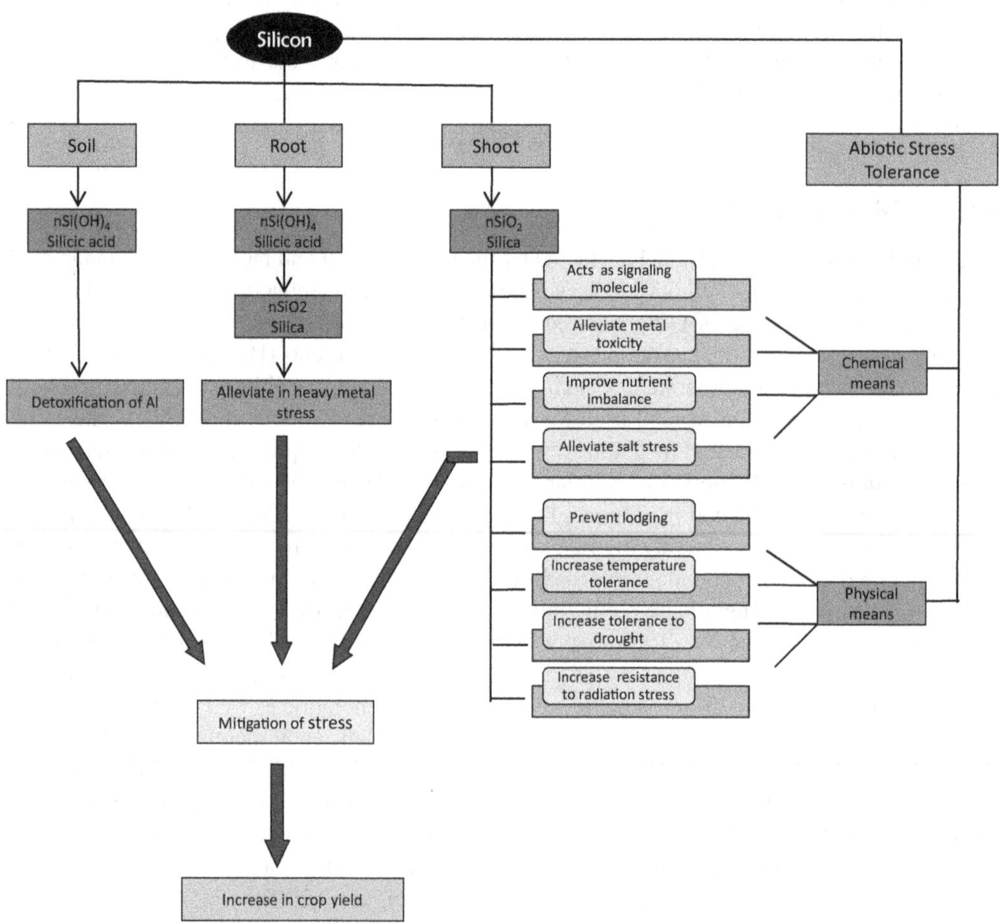

FIGURE 15.1 Schematic representation of the role of Si in mitigating different abiotic stresses by employing different mechanisms leading to increases in crop yield.

walls, thus inhibiting disease infestation in small grains. Si is acquired through roots to alleviate the negative impacts of a diverse range of abiotic stresses like metal toxicity, salinity, water stress and temperature (Ma, 2004). This positive effect of Si has led to its application in a wide range of strategies to circumvent abiotic stress for agricultural gain. An increase in the crop yield has been noticed in several crops including rice and sugarcane (Datnoff et al., 1997).

Liang et al. (2015) identified four major mechanisms employed by Si for the alleviation of abiotic stress: increased antioxidant production, clearance of toxic metal ions by binding to the latter through co-precipitation and complexation. Si can also result in varying uptake of toxic elements, metal ion storage and immobilization of metal ions. In addition, Si reduces cuticular water loss by the deposition of silicon in the form of silica beneath the cuticle (Ma & Yamaji, 2006). The deposition of silica also results in an increase in the length of the stem (Ma and Yamaji, 2008). UV-B stress has also been found to be reduced by silicon accumulation (Schaller et al. 2013). UV radiation transmission can be decreased by the silica deposited in or near the epidermis of leaves, such that very low-intensity radiation penetrates the mesophyll and sclerenchyma cells (Schaller et al. 2013). UV-B stress resulted in different physiological changes like reduced membrane damage, reduced production of superoxide radicals and enhanced biomass and chlorophyll content in Si-supplemented seedlings (Yao et al. 2011). In rice (*Oryza sativa* L.) the negative effects of UV-B radiation were reduced once supplemented with Si. This effect was accomplished by the accumulation of phenolic compounds in Si-administered rice plant (Wu et al. 2009).

It has been found that upon subjecting a single species to a single stress Si has the capacity to alleviate the abiotic stress. Estimation of the stress alleviation has been quantitatively reviewed in Si-supplemented plants (Ma et al 2001; Ma, 2004; Liang et al. 2007; Adrees et al. 2015; Pontigo et al. 2015). Most research regarding the effects of Si in alleviating the stress involves a single stress. Experimental evidence reported that Si supplementation has resulted in higher yield and biomass, enhanced tolerance to oxidative stress and enhanced rates of respiration and photosynthesis. Large numbers of experiments have revealed that the application of Si has resulted in enhancement in drought tolerance (Gong et al. 2008; Chen et al. 2011; Amin et al. 2016).

An increase in antioxidant production, the delay of leaf senescence and the maintenance of photosynthetic machinery by Si have been reported (Shen et al. 2010; Hosseini et al. 2017). The mitigation of drought by silicon in the case of tomato has been reported through the decreased synthesis of malondialdehyde (MDA) and activation of antioxidant enzymes (Shi et al. 2014). In tomatoes enhanced antioxidant activity has been shown to increase the radial hydraulic conductivity, in turn mediating drought tolerance (Cao et al. 2017). Indicators of mediation of oxidative stress include malondialdehyde, proline (PRO) and hydrogen peroxide (H_2O_2). The combating enzymes against these reactive oxygen species include peroxidase (POD), superoxide dismutase (SOD), catalase (CAT) and ascorbate peroxidase (APX). Si accumulation has enhanced the anti-oxidant properties in crop plants (Mittler, 2002; Apel and Hirt, 2004; Gill and Tuteja, 2010). In addition, several studies revealed that Si has the capability to elevate the activity of compounds like glutathione, phenolic compounds, glutathione reductases, ascorbate peroxidases, perxoidases and catalases in different plants (Shi et al. 2005; Maksimović et al. 2012; Iwasaki et al. 2002). Further Si-induced alleviation of oxidative stress has also been validated by changes in the concentration of common markers like hydrogen peroxide (H_2O_2), malondialdehyde and proline (Liang et al. 2015; Cooke and Leishman, 2016; Kim et al. 2017). Si also reduces the translocation of toxicants like Mn, Na, Cd and As from root to shoot, in turn their accumulation in leaves reducing the stress mediated by these toxicants (Yeo et al. 1999; Rogalla and Romheld, 2002; Gong et al. 2006; Sanglard et al. 2014; Che et al. 2016; Shao et al. 2017; Blamey et al. 2018; Flam-Shepherd et al. 2018).

Si has ability to alleviate both abiotic and biotic stress reported in several crop plants (Ma and Yamaji, 2008; Van Bockhaven et al. 2013). Rice is known as a strong accumulator of Si and the application of Si resulted in a prophylactic response against disease in rice (Rodrigues et al. 2003; Ma and Yamaji, 2008). Stress due to deficiencies of nutrients like copper (cu), potassium (K), phosphorus (P), iron (Fe), cadmium (cd) and manganese (Mn) is alleviated by Si (Pavlovic et al. 2016;

Kostic et al. 2017; Shao et al. 2017; Che et al. 2016; Sanglard et al. 2014; Wang et al. 2004; Mateos et al. 2015).

15.3 ACCUMULATION OF SILICON

The acquisition of Si occurs through the roots in the form of silicic acid and it is transported to the aerial parts of plant through the xylem (Raven, 1983). After uptake and transport, Si is deposited as amorphous silica in between cells and within cells and is not further remobilized (Currie and Perry, 2007). It has been reported that the Si uptake is actively regulated by intrinsic and extrinsic transporters, and in addition nodulin 26-like intrinsic proteins and other classes of aquaporins facilitate the enhanced uptake of Si in certain plant species (Ma et al. 2006; Ma and Yamaji 2015; Deshmukh et al. 2015; Belanger et al. 2016). Different species of plants accumulate Si at different rates (Hodson et al. 2005); this aspect is directly correlated with the location and expression of the aquaporin genes (Ma and Yamaji 2015). As far as the accumulation rate of Si is concerned, Bryophyta, Lycopsida and Equisetopsids (Pteridophyta) accumulated higher amounts, whereas low concentrations were accumulated by Filicopsida (Pteridophyta), Gymnospermae and most of the Angiospermae (Ma et al. 2001; Ma and Takahashi 1993; Hodson et al. 2005). Certain angiosperms accumulate above 4% of Si, including Poaceae, Cyperaceae and Balsaminaceae. A few species belonging to Solanaceae (tomato) and Fabaceace (faba bean) are considered as excluders of Si (Ma and Takahashi 1993; Hodson et al. 2005). The higher accumulator of Si has been found in the Poaceae family, which has the capability to accumulate up to 10% of Si in dry mass. The plant families Brassicaceae and Fabaceae accumulate very little Si, almost <1% (Datnoff et al. 2001; Ma, 2003).

The low water availability under drought conditions results in a decrease in the uptake of nutrients and translocation to shoots (Hosseini et al. 2016). The uptake of minerals and their translocation in crops was found to be improved by Si (Pavlovic et al. 2013). In addition, Si has resulted in the enhanced accumulation of potassium in sorghum seedlings both in shoots and roots (Miao et al. 2010). In cucumber, the iron remobilization in leaves is enhanced by Si accumulation (Pavlovic et al. 2016). The augmentation of nitrate translocation in *Brassica napus* has been demonstrated, which is linked to the elevated expression of nitrate transporter genes (Haddad et al. 2018). Even though the Si accumulation in tomatoes (*Solanum lycopersicum*) is low to moderate, salt stress, boron toxic soils and fungal infections can be tolerated by tomato plants upon the application of Si fertilizers (Al-Aghabary et al. 2004; Romero-Aranda et al. 2006; Gunes et al. 2007).

This property of Si to induce the antioxidant activity was observed in tomatoes (Shi et al. 2014). In addition, it was found that antioxidant activity induced by the Si is also dependent on the part of the plant. The latter outcome is defended by studies carried by Cocker et al. (1998a). Their findings noted that a difference in antioxidant activity was observed in the shoots and roots of a plant. These results may further help to elucidate the mechanism through which Si can induce different abiotic stresses in plants. Determination of Si accumulation was accomplished by studying the foliar Si rather than studying other organs of the plant. The differential accumulation of the Si in plants was defended by experiments in vegetables like tomato and pepper. In these plants Si accumulated more in the roots than in the shoots, revealing the capability of roots to absorb more Si as compared to the shoots. The active absorption of Si occurs through the roots of several monocots; however dicots like *Lycopersicun esculentum* have been found to be poor accumulators of Si through passive absorption (Mitani and Ma, 2005).

Monocots accumulate higher amounts of Si as compared to dicots. The overall accumulation percentage of Si in dicots is less than 0.23%, reflecting its major contribution among all the micronutrients. This property of roots has been attributed to the transporters present on roots. Si upon absorption from the roots is transported through the xylem by transporters or via transpiration towards the root endodermis, then to the membranes of the vascular bundles and epidermis of the root beneath the cuticle. Inside the cells, Si is polymerized into insoluble silica/silica gel/phytolytes.

15.4 SILICON UPTAKE AND TRANSPORT

The polymerized form of silicon, i.e. silicic acid [$Si(OH)_4$] or mono silicic acid [H_4SiO_4], has the ability to cross the plasma membrane at physiological pH (Raven, 2003). A majority of polymerized silicon has been found in the apoplast and bound to the constituents of the cell wall such as cellulose, hemi-cellulose, callose, lignin and pectin as scaffolds for silification (Guerriero et al. 2016). In certain plants Si can be polymerized in specialized cells, and in grasses certain cellular structures like long cells and silica cells, spikelet hairs and papillae are formed (Rafi et al. 1997). The pH has a great influence on Si transport which is indicated by the concentration range of 0.1 to 0.6 mM at pH 9 (Knight and Kinrade, 2001). The transported Si concentration of leaves has been found to be in the range of 0.1 to 10% on the basis of dry weight (Epstein, 1999; Ma et al. 2001). Studies related to the water uptake and Si uptake simultaneously have revealed that Si absorption and transport through lateral roots occurs possibly in three different ways including active, passive and rejective pathways. First one defines the fast uptake of Si, second, the similar uptake of Si in relation to water and third, the slower uptake of Si in comparison to water (Takahashi et al. 1990; Cornelis et al. 2011). Despite the ability of plants to accumulate Si, the passive transport of Si from soil via channels or through diffusion across the plasma membrane is energy-independent (Raven, 2003).

The uptake of silicic acid from the soil occurs via apoplastic and symplastic pathways. It must be noted that apoplastic transport of Si is prevented in the root stele (Yamaji and Ma, 2007). Si uptake through the symplastic pathway depends upon the presence of a class of the aquaporin gene family including Nod26-like intrinsic proteins (NIPs). Several classes of AQPs have been identified in monocots and dicots (Ma et al. 2006; Mitani et al. 2009a). Aquaporins are water channels which may transport solutes across the plasma membrane. They consist of an hour glass-like structure comprising six transmembrane domains. In addition, the two halves of the transmembrane domains form two NPA (asparagine-proline-alanine) domains (Murata et al. 2000). These protruding domains confer high specificity to AQPs for transporting solutes across the plasma membrane (Hove and Bhave, 2011). These transporters have also been found in plant species like horsetail, which is also called the king of silicon accumulators (Gregoire et al. 2012; Vivancos et al. 2016). As an example for the demonstration of Si uptake, the roots of rice plants have shown the ability to take up to 90% of Si available in the soil (Ma et al. 2001).

In rice mutants several classes of transporters have been identified for the uptake of Si, two of which are worth mentioning because of their significance in transporting the majority of Si available in soil: *OsLsi2* (efflux Si-transporters *OsLsi1*) and *OSLsi1* (Si-transporter AQPs, influx) (Ma et al. 2006). As far as influx Si-transporters (*OsLsi1* and *OsLsi6*) are concerned they promote the passive transport of Si through the apoplast and across the plasma membrane in the plant cells. The influx of the Si from the soil solution into the roots is mediated by the presence of *OsLsi1* (NIP-III subfamily of aquaporins) transporters. In turn the efflux transporters, in particular *OsLsi2*, help to release Si in the apoplast which is in turn translocated upwards to the shoots in conjunction with transpiration. The efflux transporter is coded by the *OsLsi2* gene; an anion channel is involved in transporting the Si out of roots into the stele (Yamaji et al., 2012). In addition, *Lsi2*, an efflux transporter, driven by gradient has been reported to transport soluble Si against the concentration gradient (Ma et al. 2007). The uptake of Si occurs comparatively more in mature roots than in root tips; this differential uptake is solely dependent on the expression of the *OsLsi1* transporter gene (Yamaji and Ma, 2007). *Lsi1* belongs to a superfamily of major intrinsic proteins and it is mainly involved in transport of $Si(OH)_4$ into root cells as compared to boric acid ($B(OH)_3$) and arsenious acid ($As(OH)_3$) (Mitani-Ueno et al. 2011). The *Lsi1* transport protein is located both in endodermal and exodermal root cells at the distal side (Yamaji and Ma, 2007) whereas, *Lsi2*, a similar class of transporter, is located in the proximal side of the endodermal and exodermal portions of cells. Xylem unloading of Si into the xylem parenchyma cells, facilitated by influx transporter (*OsLsi6*), is important to prevent its deposition. The expression of *OsLsi6* is limited to root tips, leaf blades and sheaths, and is reported to be located on the adaxial sides of xylem parenchyma cells in rice (Yamaji et al. 2008). In rice plants the excretion of Si through guttation fluid was also reported when the *Lsi6* gene was knocked out (Ma et al. 2007). Another study reported in rice plants has indicated that Si is diffused in vesicular bundles with the help of *Lsi3*, an efflux Si-transporter (Yamaji and Ma, 2007). The prototypical mechanism of Si transport was also reported in maize and barley (Mitani et al. 2009a). The unloading of silicic acid from the xylem to the leaf tissues is facilitated by NIP2 (*Lsi6*), a class of the aquaporin type of Si transporter (Yamaji et al. 2008).

Upon subjecting all the Si influx transporters to phylogenetic analysis, specific clustering within Nodulin 26-like intrinsic proteins III, a subgroup of AQPs, was found. In addition, these transporters are composed of conserved glycine-serine-glycine-arginine (GSGR) amino acid motif. Si transporters have NPA as another selective feature to transport the Si (Deshmukh et al. 2015). Other $Si(OH)_4/H^+$ antiporter transporters like *Lsi2* function as putative anion transporters having similarities with the arsenite efflux transporter ArsB from bacteria and Archaea (Ma et al. 2007). In tomato plants experiments revealed that Si impermeability due to NIP is due to differences in two NPA domains which in SiNIP2-1 form a half-helix insert. In these plants, the low accumulation of Si under stress conditions is due to passive uptake of Si. Low Si accumulators like tomato, capsicum and roses are supplied with Si to overcome the abiotic stress (Romero-Aranda et al. 2006; Jayawardana et al. 2015; Soundararajan et al. 2017b).

The cells must maintain a cytosolic concentration of $Si(OH)_4$ <2 mM; this balance is maintained by *Lsi1* and *Lsi2* in root cells. The concentration range >2 mM is reported to be toxic to cells (Iler, 1979; Montpetit et al. 2012; Exley, 2015). The vacuolar sequestration of Na^+ is promoted by Si, indicating that salinity stress is circumvented by Si, leading to the protection of important cytoplasmic functions (Liang et al. 2015). This finding is mainly due to the enhanced activity of H^+ ATPase by Si (Liang, 1999; Liang et al. 2005, 2006).

The deposition of Si in the cuticle prevents water evapo-transpiration, in turn providing resistance to water loss (Ma et al. 2001; Ma, 2004). The deposition of Si blocks the apoplastic transport of Cd^{2+}, Na^+ and Cl^-, thereby restricting the transport of these inorganic ions into the shoots. This mechanism helps to regulate the toxic levels of these minerals accumulating in shoots (Yeo et al. 1999; Ranathunge et al. 2005; Shi et al. 2014, 2013; Gong et al. 2006; Flam-Shepherd et al. 2018). In addition, Si has been reported to initiate the formation of casparian strip bands through stimulation of lignin and suberin biosynthesis, blocking the apoplastic transport of toxicants (Fleck et al. 2011, 2015). The formation of a casparian band is possibly due to interaction and cross linking of Si to phenolics or precipitation with cell walls.

15.5 PHYSIOLOGICAL EFFECTS OF SILICON

Si has been reported to be involved in number of functions like redox homeostasis, oxidative stress, nitrogen assimilation, signaling, carbohydrate metabolism, ion and water influxes, root exudation, hormone regulation, heavy metal chelation, root architecture, photosynthesis and transpiration (Manivannan and Ahn, 2017, Liang et al. 2003; Zhu et al. 2004; Farooq et al. 2016; Zhu and Gong 2014; Detmann et al. 2012; Liang et al. 2006; Liu et al. 2014; Liang et al. 2015; Markovich et al. 2017; Kidd et al. 2001; Wu et al. 2016; Wang et al. 2004; Ma et al. 2015; Gong et al. 2006; Fleck et al. 2011; Gao et al. 2006; Shen et al. 2010; Detmann et al. 2012). A few of these effects have been discussed to highlight the importance of Si in abiotic stress.

One of the major factors affecting plant growth and sustainable agriculture is low temperatures. In this class of abiotic stress Si has been shown to mitigate their adverse effects. This finding was reported in maize plant to validate the effect of Si in enhancing the tolerance to chilling stress. Reduction in plant growth and relative water content during stress can be ameliorated by Si. The Si is involved in the accumulation of free amino acids and repairing of necrosis in leaves. The accumulation of free amino acids and decline in leaf necrotic area in Si supplied plants have revealed the role of Si in increasing the metabolism of plants significantly. This physiological effect was found to be helpful in increasing the quantum yield of PSII.

Further it has been found that administration of Si resulted in upregulation of pigments like anthocyanins and carotenoid which protect PSII from damage. Si supplementation also reduces the damage caused by cold, leading to the development of excitation energy trapping and electron chain efficiency in plants.

It has been reported that resistance to cold stress is attributed to the application of Si; this results in increasing concentrations of glutathione and ascorbate. These findings have been further validated in maize plants, which showed improvement in accumulation of biomass, maintaining higher levels of ascorbic acid, glutathione and protective pigments to enhance the photosynthetic efficiency (Habibi, 2016). The enhanced metabolic rate of amino acids like glycine, methionine, serine and arginine and Si-induced uptake of ammonium and sulfur have been reported in drought-tolerant crop plants. The effects of Si on the production of amino acids like GABA and proline to lower the GSSG-to-GHS ratio for balancing the redox homeostasis were reported in drought-sensitive crop plants under osmotic stress conditions.

The increase in chlorophyll content and antioxidant activity was reported in *Lycopersicon esculentum* upon application of Si and simultaneous exposure to salt stress, even though the exact mechanism is yet to be deciphered (Al-aghabary et al. 2004). The combined effect of Si and nitrogen was found to elevate the chlorophyll 'a' content in *Oryza sativa* plants (Avila et al. 2010). Under stress conditions the Si-induced reduction of ROS leads to vast physiological changes (Liang et al. 2003, 2005, 2015; Zhu et al. 2004; Yin et al. 2018; Markovich et al. 2017). Numerous studies have revealed that Si-induced changes in the production of ROS have influenced a vast array of processes like expression of genes, development, programmed cell death and several signaling processes (Mittler, 2002; Apel and Hirt, 2004; Gill and Tuteja, 2010). In addition, many physiological aspects mediated by Si, like a decrease in ethylene signaling in salt-stressed sorghum and elevated levels of polyamines, e.g. spermine, spermidine and putrescine, back its major role in plants (Yin et al. 2014). Recent studies have revealed that Si has a role in the enhanced biosynthesis of cytokinin and may influence the delay of senescence in sorghum and *Arabidopsis* (Markovich et al. 2017). The enhanced hydraulic conductivity under hyper-osmotic stress is reported to be mediated by Si accumulation, as mentioned in the proceeding text that Si influences the expression of AQPs genes that are directly involved in water conduction (Liu et al. 2014).

The supply of Si reduces levels of ethylene and increases the synthesis of polyamines to circumvent the abiotic stress. This lowering of ethylene is mediated by the inhibition of an important precursor, i.e. 1-amino cyclopropane-1-1-carboxylic acid (ACC), upon supplementation with Si (Yin et al. 2014). Diverse functions are reported to be accomplished by polyamines during abiotic stress, for example they play a pivotal role in the stabilization of membranes, modulation of enzymatic activities, replication, transcription and protein synthesis. Water relations are greatly influenced by stomatal conductance and transpiration rate (Farooq et al. 2009). It is reported that Si directly controls transpiration in plants. It is postulated by many studies that silica forms a double layer on the epidermal tissues of leaves upon treatment with Si (Yoshida 1965; Wong et al. 1972; Matoh et al. 1991). In drought-stressed wheat the addition of Si has resulted in the thickening of leaves, resulting in a reduction of the transpiration rate; this effect is found in the cuticle rather than stomata (Gong et al. 2003; Kerstiens, 1996). In contrast in maize cuticular transpiration was not affected upon administration of Si, rather stomatal movement is regulated to control the rate and conductance of transpiration (Gao et al. 2006). Similar types of studies were reported in potted sorghum (Hattori et al. 2005). A decreased rate of leaf transpiration and water flow through xylem vessels for efficient water uptake was reported in Si-administered maize and sunflower (Gao et al. 2004). This effect is due to the deposition of silicon on the root cell wall resulting in wetting of xylem vessels, in turn resulting in efficient water transport. These findings need further investigation to give an in-depth idea about the role of Si in transpiration. Stomatal closure is the first response of plants to resist drought, and this adaption has been found to be a limiting factor for reduction in photosynthesis (Reddy et al. 2004; Farooq et al. 2009). The effects of stomatal closure include a reduced influx of the CO_2 needed for ample electrons to form reactive oxygen species (Farooq et al. 2009).

The reduction in CO_2 has been reported to result in a reduction of the photosynthetic rate in *Rosa rubiginosa* (Meyer and Genty, 1998). Stomatal and non-stomatal factors which result in improvement in photosynthesis have been shown to be due to the supplementation of Si in soil-grown rice (Chen et al. 2011). To be specific, this was found in studies carried out by Liang (1999) to show the effects of Si on improving the ultra-structure of the chloroplast and enhancing the activity of several enzymes pertaining to antioxidants like catalase and superoxide dismutase. In *Capsicum annuum* L. the role of Si is highlighted by its effect in maintaining the higher content of different photosynthetic pigments like chlorophyll a, chlorophyll b and carotenoids (Lobato et al. 2009). In rice, an increase in basal quantum yield and the efficiency of photosystem II was observed when plants were supplied with Si during drought stress conditions (Chen et al. 2011). In Si-supplied cucumber the increased activity of ribulose-bisphosphate carboxylation was observed (Adatia and Besford, 1986). As reported by Gong and Chen (2012) the Si-induced maize concentration of inorganic phosphorus was enhanced, and higher activity of phosphoenol pyruvate carboxylase was observed when subjected to drought. It may be hence concluded that Si plays an important role in controlling stomatal functions and enhancing photosynthesis under a diverse range of abiotic stresses.

15.6 SILICON-INDUCED BIOSYNTHESIS OF COMPATIBLE SOLUTES AGAINST ABIOTIC STRESS TOLERANCE

Plants synthesize a large number of compatible solutes to tolerate abiotic stress like drought, salt stress and chilling stress. During stress conditions the most commonly accumulated compatible solutes are carbohydrates (Balibrea et al. 1997), glycine betaine (Mansour, 1998), polyols (Kumar and Bandhu, 2005) and proline (Gzik, 1997). These compounds primarily stabilize the proteins and membranes under stressful conditions (Bohnert and Shen 1999; Ashraf and Foolad, 2007). Proline has been reported to be frequently involved in protecting plants against salt tolerance and osmotic stress. In addition, compatible solutes scavenge oxygen radicals (Seckin et al. 2009). To further authenticate the antioxidant capability of compatible solutes, the administration of mannitol in salt-sensitive wheat has been reported to elevate the antioxidant enzymatic activity in roots (Seckin et al. 2009). Proline has been found to be accumulated in the leaves of *Populas euphratica* in salt stress conditions (Watanabe et al. 2000). In addition, it has been found that the supply of Si induced enhanced synthesis of solutes like fructose and sucrose in sorghum plants to combat salt-induced osmotic stress (Yin et al. 2013). There is a direct effect of Si in enhancing the synthesis of compatible solutes and transport of water, so further investigation is needed to validate above findings.

15.7 ALLEVIATION OF AL STRESS BY Si

One of the major limiting factors for crop yield in acidic soils is toxicity of aluminum (Al) (Von and Mutert, 1995). The excess amount of $AlCl_3$ found in acidic soils results in a reduction of plant growth due to its severe effect on metabolic processes (Kochian et al. 2005; Mora et al. 2006; Cartes et al. 2012). Toxic levels of Al can impair the cell wall components and plasma membrane functions, signaling and nutrient homeostasis (Yamamoto et al. 2001; Horst et al. 2010; Delhaize and Ryan, 1995; Gupta et al. 2013; Singh et al. 2017; Matsumoto, 2000; Sivaguru et al. 2003; Goodwin and Sutter, 2009). Plants have adapted diverse types of mechanisms to tolerate the deleterious effects of Al; most reported tolerance is through Al exclusion and internal tolerance mechanisms (Barcelo and Poschenrieder, 2002; Kochian et al. 2005; Poschenrieder et al. 2008).

Exclusion mechanisms occur through the binding of $AlCl_3$ to organic acids and phenolic compounds through root exudation, hence limiting cytosolic uptake of Al. The internal tolerance mechanism of plants is accomplished through ROS scavenging, vacuolar compartmentalization and detoxification of Al by forming complexes with organic molecules (Barcelo and Poschenrieder, 2002; Kochian et al. 2005; Poschenrieder et al. 2008). Several plant genes encoding transporter proteins include the multidrug and toxic compound extrusion (MATE) families and aluminum-activated malate transporters (ALMT) which efflux organic acid anions (Sasaki et al. 2004; Furukawa et al. 2007; Ryan et al. 2011).

Al tolerance may also be accomplished by the antioxidant defense genes (Milla et al. 2002; Goodwin and Sutter, 2009; Du et al. 2010) and bacterial type ATP binding cassette (ABC) transporters (Huang et al. 2009). Si plays an important role in evading toxicity mediated by various metal ions like aluminum (Al), iron (Fe), manganese (Mn), chromium (Cr), arsenic (As), cadmium (Cd), copper (Cu), lead (Pb) and zinc (Zn) (Li et al. 1996; Vaculík et al. 2012; Adrees et al. 2015; Liang et al. 2015; Pontigo et al. 2015; Tripathi et al. 2015, 2016). It is evident from these studies that Si has the ability to tolerate metal toxicity by diverse classes of mechanism as briefly discussed (Cocker et al. 1998a; Liang et al. 2015; Pontigo et al. 2015; Adrees et al. 2015; Tripathi et al. 2016). Experiments have been carried out to assess the importance of Si in alleviating Al stress, including Al–Si complex formation in the growth media and in plants (Barcelo et al. 1993; Baylis et al. 1994; Cocker et al. 1998a), increase in the pH of solution by Si (Li et al. 1996; Cocker et al. 1998a), enhanced carotenoid and chlorophyll content of leaves and through exudation of phenolic compounds and organic acids (Barcelo et al. 1993; Cocker et al. 1998b; Kidd et al. 2001). Under Al stress Si supply has led to the activation of antioxidants in crop plants (Tripathi et al. 2016).

A range of mechanisms to ameliorate the toxic accumulation of metal ions, such as structural changes in plant parts, chelation of metal ions, co-precipitation of metal ions, regulation of gene expression and most importantly increasing antioxidant activity in plants, has already been discussed (Adrees M. et al. 2015; Liang et al. 2007; da Cunha and do Nascimento 2009).

The silicon detoxification of metal ions has been grouped into chemical and physical methods. In addition, as already discussed, phenolic compounds decrease Al uptake in maize by chelating or forming complexes (Adrees M. et al. 2015). In soybeans, the amelioration and toxicity of Al is mediated by silicon changing pH in a medium, leading to precipitation and unavailability of aluminum (Pontigo et al. 2017). It has been reported that detoxification of Al occurs through the formation of pH-dependent hydroxyl aluminum silicates (HAS) in roots (Kopittke et al. 2016). The pH is an important factor in the formation of HAS; this has been usually found to be formed >pH5 (Wang Y. et al. 2004). With the formation of hydroxyl aluminum silicate complexes the transportation of Si is enhanced from the root to the aerial parts of plant. It is reported that extreme Al toxicity can be reduced by Si through the apoplasm by the deposition of Si on cell walls. The cell wall is a major site in co-localizing Al-Si in epidermal and hypodermal cells (Szabela et al. 2015). The above finding is backed by studies in *Picea abies* regarding the stimulation of Al degradation by silicon in the cell wall for the protection of plant (Prabagar et al. 2011). From the above-mentioned findings we conclude that silicon significantly reduces the transport of Al through apoplastic pathways. The accumulation of Si increased in Al-stressed wheat plants, validating the role of Si in metal ion toxicity (Zsoldos et al. 2003)

15.8 SILICON-REGULATED GENES INVOLVED IN STRESS TOLERANCE AND ALTERED EXPRESSION OF REGULATORY ELEMENTS ASSOCIATED WITH STRESS RESPONSE GENES DUE TO Si

Gene expression has been found to be different during stressed conditions in plants. A large number of genes driving several processes like signal transduction, metabolic processes and stress tolerance can be upregulated in crop plants in stressful environments (Shinozaki and Yamaguchi-Shinozaki, 2000; Xiong et al. 2002; Rabbani et al. 2003; Shinozaki et al. 2003). Surprising results have been reported in rice regarding the supplementation of Si. Si has been found to downregulate the expression of housekeeping genes during normal conditions whereas during pathogenic states the expression of housekeeping genes was upregulated (Brunings et al. 2009). Among these three genes it was reported that actin cytoskeleton was found to provide basal resistance to *R. solanacearum* (Jarosch et al. 2005). Due to the low density of the Si transporter in tomato low levels of silicon, up to 0.2% dry weight, are accumulated (Ma and Yamaji, 2006). It has been reported that genes like *OsLSi1* and *OsLSi2* responsible for the transport of Si are significantly over expressed in rice plants under stress conditions (Ma et al. 2006). In addition, the upregulation of Lsi-1 gene involved in Si transport in rice and *Arabidopsis* resulted in resistance to pests and disease (Ma et al. 2006; Ma and Yamaji, 2008). Heavy metal toxicity has been mitigated by the over expression of genes related to their transport and detoxification (Brunings et al. 2009). Modulation of these genes due to metal toxicity has been ill-explored. Molecular insights into the role of Si in upregulation of several genes responsible for enhanced photosynthesis have been reported; a few of them are worth mentioning for highlighting the evolution of Si as an important mineral in combating abiotic stress and increasing crop yield. In rice a few genes have been found to be expressed upon Si supplementation, even though Si has been found to be beneficial for enhancing photosynthesis through different mechanisms. Several studies reported by Song et al. (2014) illustrated that under zinc stress the Si amendment upregulated the transcription of genes linked to photosynthesis and respiration. The protein product of the PsaH gene is a vital subunit of PSI dimer which is over expressed under the influence of Si (Pfannschmidt and Yang, 2012). It was reported that PsaH gene knockout mutants were found to damage the LHC-II complex leading to a delay in the transfer of electrons from PSII to PSI (Lunde et al. 2000). Another versatile protein called polyprotein, PsbY (Os08g02630), is an important component of photosystem II, an important part of the oxygen-evolving complex, with L-arginine metabolizing enzyme activity (Kawakami et al. 2007). An important polyprotein PsbY (Os08g02630), involved in photosystem II (PSII)-related transcripts, has been found to be over expressed upon supplementation of Si. This increase in the PsbY transcripts leads to the activation of oxidation of water, in turn increasing the electron transfer rate and efficiency of PSII (Song et al. 2014). PetC is another important gene-encoding polypeptide which binds to the Rieske Fe-S center of the cytochrome bf complex and helps in the stability of the cytochrome. This protein has been found to be upregulated once the plant is supplied with Si and downregulated upon Zn toxicity (Breyton et al. 1994). This upregulation of PetC has been found to be important for enhancing the chloroplast integrity (Song et al. 2014). Ferredoxin $NADP^+$ reductase encoding by the PetH gene is vital for NADPH synthesis through the photosynthetic electron transport chain and has an important role in reducing the glutathione content in plant cells (Song et al. 2014). Reports from several studies have revealed that Si has resulted in the upregulation of PetH gene synthesis. A series of genes, in particular Os03g57120 and Os09g26810, important for the functioning of the light harvesting complex, are upregulated by supplementation of Si. The over expression of transcription factors (TFs) has been reported to be induced against both biotic and abiotic stress (Gao et al. 2007; Lucas et al. 2011). The expressions of TFs are facilitated by cis-elements found on the promoter region of the target gene (Nakashima et al. 2009; Qin et al. 2011). A number of cis-elements called regulons, e.g. dehydration-responsive element binding protein (DREB2), are found in the promoter regions of genes in the plants which respond to stresses like temperature and drought (Mizoi et al. 2012).

Osmotic stress activates regulons like cup-shaped cotyledon (CUC), NAC regulons, no apical meristem (NAM) and *Arabidopsis thaliana* activating factor (ATAF) (Nakashima et al. 2009; Fujita et al. 2011). In rice Si supplementation mediates the over expression of TFs, in turn stimulating expression of NAC5, *Oryza sativa* RING domain containing protein (OsRDCP1), DREB2A, *Oryza sativa* choline monooxygenase (OsCMO), dehydrin OsRAB16b and signaling pathways (Chaves and Oliveira, 2004; Umezawa et al. 2006; Khattab et al. 2014). Similar experiments have also reported that OsDREB triggers stress-responsive gene expression to tolerate osmotic stress without the involvement of ABA (Dubouzet et al. 2003; Hussain et al. 2011). Another class of DREB known as OsDREB2A has also been shown to provide resistance against drought in rice plants (Chen et al. 2008; Wang et al. 2008).

NACs are a class of TFs with diverse types of functions and have been found to play a critical role during stress in plants (Tran et al. 2010). At least 140 putative NACs or NAC-like genes have been identified in the rice genome. Of these, in addition to OsNAC5, 20 genes have been found to be involved in circumventing stress through the initiation of processes like detoxification, macromolecule fortification and redox homeostasis (Hu et al. 2008). Si enhances the expression of OsNAC5 transcripts, helping to prevent the generation of excess hydrogen peroxide (H_2O_2) and lipid peroxidation; these metabolic modulations help in protecting plants from oxidative damage and dehydration stress (Takasaki et al. 2010; Song et al. 2011). The upregulation of OsNAC5 increased the stress tolerance and expression of genes like LEA3 in stressful conditions in rice (Takasaki et al. 2010). OsRAB16b is a class of genes reported to be over expressed in somatic as well as reproductive tissues during abiotic stress (Tunnacliffe and Wise, 2007; Bies-Etheve et al. 2008). In rice RING E3 Ub ligases play an important role in drought stress (Park et al. 2011; Ning et al. 2011; Bae et al. 2011). At least five homologs of OsRDCP were found to be included in the RING family (Khattab et al. 2014). Among these OsRDCP1 have been involved in combating dehydration stress in rice, and these genes are over expressed by the administration of Si (Bae et al. 2011; Khattab et al. 2014). Si has been involved in enhanced expression of OsCMO genes which encodes choline monooxygenase involved in the biosynthesis of glycine betaine (Burnet et al. 1995). The application of Si has drastically resulted in over expression of SbPIP1;6, SbPIP2;2 and SbPIP2;6, coding for plasma membrane intrinsic proteins (PIP) and important aquaporins to transport the water efficiently (Liu et al. 2015). The higher accumulation of water results in the dilution of Na^+ ions, preventing their lethal effects on plants (Gao et al. 2010).

The modulation of polyamines by Si has resulted in resistance to stress in *Sorghum bicolor* (Yin et al. 2014). The polyamine biosynthesis is elevated by the S-adenosyl-L-methionine decarboxylase (SAMDC) gene over expression augmented by a supply of Si. Polyamines are reported to mediate salt tolerance in sorghum. It may be concluded that polyamines play an important role in the regulation of several fundamental processes in crop plants. A vast range of proteins are overexpressed by Si during stress including phenylalanine-ammonia lyase (PAL), chalcone synthase (CHS), pathogenesis-related protein (PR1), peroxidase (POX), b-1, 3-glucanases and chitinases (Rodrigues et al. 2004). Most of these genes play important roles in the biosynthesis of flavonoids and secondary metabolites (Rodrigues et al. 2004). In general, the above studies strongly back the role of Si in regulating a wide range of genes responsible for maintaining the physiological functions of plants under stressful conditions.

15.9 FUTURE PROSPECTS AND CONCLUSION

The diverse range of physiological roles played by Si including increases in crop yield, rate of photosynthesis, nitrogen fixation and in particular abiotic stress shows its importance as a versatile mineral element for the growth and development of crop plants. From these studies it is evident that Si plays an important role in conjunction with its existence in the earth's crust. Extensive work has been done in this field to attract the scientific community toward utilizing Si in crop production. Its importance in biotic as well as abiotic stresses like pathogen, fungal attack, UV radiation, temperature, nutrient deficiency, metal toxicity, drought and salinity is discussed in detail in current studies. Different beneficial aspects of Si have been discussed, including its effect on the molecular modulation of genes important in the biochemical and physiological well-being of plants. Contrary studies from several reports ignore the importance of Si in crop plants. In addition, due its similar role in crop plants in comparison to other minerals, the molecular mechanism regulated by Si must be explored. Against this backdrop there is further need to validate the importance of Si through newer studies like genomic and proteomic approaches to prove its essentiality in crop plants. The present chapter emphasizes the role of Si in increased yields of plants under extreme abiotic stress conditions. This may provide a framework to further investigate the role of Si in the sustainable development of crops.

REFERENCES

Adatia MH, and Besford RT (1986) The effects of silicon on cucumber plants grown in recirculating nutrient solution. *Annals of Botany*. 58:343–351.

Adrees M, Ali S, Rizwan M, Zia-ur-Rehman M, Ibrahim M, and Abbas F (2015) Mechanisms of silicon-mediated alleviation of heavy metal toxicity in plants: A review. *Ecotoxicology and Environmental Safety*. 119:186–197.

Al-aghabary K, Zhu Z, and Shi Q (2004) Influence of silicon supply on chlorophyll content, chlorophyll fluorescence, and antioxidant enzyme activities in tomato plants under salt stress. *Journal of Plant Nutrition*. 27:2101–2115.

Amin M, Ahmad R, Ali A, Hussain I, Mahmood R, and Aslam, Lie DJ (2016) Influence of silicon fertilization on maize performance under limited water supply. *Silicon*. 10:177–183.

Apel K, and Hirt H (2004) Reactive oxygen species: Metabolism, oxidative stress, and signal transduction. *Annual Review of Plant Biology*. 55:373–399.

Arnon DI, and Stout PR (1939) The essentiality of certain elements in minute quantity for plants, with special reference to copper. *Plant Physiology*. 14:371–375.

Artyszak A, Gozdowski D, and Kucinska K (2015) The effect of silicon foliar fertilization in sugar beet – *beta vulgaris* (l.) ssp. *vulgaris* conv. *crassa* (Alef.) prov. *altissima* (Döll). *Turkish Journal of Field Crops*. 20(1):115–119.

Ashraf M, and Foolad MR (2007) Roles of glycine betaine and proline in improving plant abiotic stress resistance. *Environmental and Experimental Botany*. 59:206–216.

Avila FW, Baliza DP, Faqui V, Araújo JL, and Ramos SJ (2010) Interação entre silício e nitrogênio em arroz cultivado em solução nutritiva. *Revista Ciencia Agronomica*. 41:184–190.

Bae H, Kim SK, Cho SK, Kang BG, and Kim W T (2011) Overexpression of OsRDCP1, a rice RING domain-containing E3 ubiquitin ligase, increased tolerance to drought stress in rice (*Oryza sativa* L.). *Plant Science*. 180:775–782.

Balibrea ME, Rus-alvarez AM, Bolarfn MC, and Perez-alfocea F (1997) Fast changes in soluble carbohydrates and proline contents in tomato seedlings in response to ionic and non ionic iso-osmotic stresses. *Journal of Plant Physiology*. 151:221–226.

Barcelo J, Guevara P, and Poschenrieder CH (1993) Silicon amelioration of aluminium toxicity in teosinte (*Zea mays* L. ssp. *mexicana*). *Plant and Soil*. 154:249–255.

Barcelo J, and Poschenrieder C (2002) Fast root growth responses, root exudates, and internal detoxification as clues to the mechanisms of aluminium toxicity and resistance: A review. *Environmental and Experimental Botany*. 48:75–92.

Baylis AD, Gragopoulou C, Davidson KJ, and Birchall JD (1994) Effects of silicon on the toxicity of aluminum to soybean. *Communications in Soil Science and Plant Analysis*. 25:537–546.

Belanger R, Deshmukh R, Belzile F, Labbe C, Perumal A, and Edwards SM (2016) Plant with increased silicon uptake. *Patent No.: WO/2016/183684*.

Bies-Etheve N, Gaubier-Comella P, Debures A, Lasserre E, Jobet E, and Raynal M (2008) Inventory, evolution and expression profiling diversity of the LEA (late embryogenesis abundant) protein gene family in *Arabidopsis thaliana*. *Plant Molecular Biology*. 67:107–124.

Blamey FPC, McKenna BA, Li C, Cheng MM, Tang CX, Jiang HB, Howard DL, Paterson DJ, Kappen P, and Wang P (2018) Manganese distribution and speciation help to explain the effects of silicate and phosphate on manganese toxicity in four crop species. *New Phytologist*. 217:1146–1160.

Bohnert HJ, and Shen B (1999) Transformation and compatible solutes. *Scientia Horticulturae*. 78:237–260.

Bray EA, Bailey-Serres J, and Weretilynk E (2000) Responses to abiotic stresses. In: *Biochemistry and Molecular Biology of Plants*, Gruissem W, and Jones R (Eds.). American Society of Plant Physiologists, Rockville, pp. 1158–1203.

Breyton C, de Vitry C, and Popot JL (1994) Membrane association of cytochrome b6f subunits. The Rieske iron-sulfur protein from *Chlamydomonas reinhardtii* is an extrinsic protein. *Journal of Biological Chemistry*. 269:7597–7602.

Brunings AM, Datnoff LE, Ma JF, Mitani N, Nagamura Y, and Rathinasabapathi B (2009) Differential gene expression of rice in response to silicon and rice blast fungus *Magnaporthe oryzae*. *Annals of Applied Biology*. 155:161–170.

Burnet M, Lafontaine PJ, and Hanson AD (1995) Assay, purification, and partial characterization of choline monooxygenase from spinach. *Plant Physiology*. 108:581–588.

Cao BL, Wang L, Gao S, Xia J, and Xu K (2017) Silicon-mediated changes in radial hydraulic conductivity and cell wall stability are involved in silicon induced drought resistance in tomato. *Protoplasma*. 254:2295–2304.

Cartes P, Jara AA, Pinilla L, Rosas A, and Mora ML (2012) Selenium improves the antioxidant ability against aluminium induced oxidative stress in ryegrass roots. *Annals of Applied Biology*. 156:297–307.

Chaves MM, and Oliveira MM (2004) Mechanisms underlying plant resilience to water deficits: Prospects for water-saving agriculture. *Journal of Experimental Botany*. 55:2365–2384.

Che J, Yamaji N, Shao JF, Ma JF, and Shen RF (2016) Silicon decreases both uptake and root-to-shoot translocation of manganese in rice. *Journal of Experimental Botany*. 67:1535–1544.

Chen JQ, Meng XP, Zhang Y, Xia M, and Wang XP (2008) Over-expression of OsDREB genes lead to enhanced drought tolerance in rice. *Biotechnology Letters.* 30:2191–2198.

Chen W, Yao X, Cai K, and Chen J (2011) Silicon alleviates drought stress of rice plants by improving plant water status, photosynthesis and mineral nutrient absorption. *Biological Trace Element Research.* 142:67–76.

Cocker KM, Evans DE, and Hodson MJ (1998a) The amelioration of aluminium toxicity by silicon in higher plants: Solution chemistry or an in plants mechanism? *Physiologia Plantarum.* 104:608–614.

Cooke J, and Leishman MR (2011) Is plant ecology more siliceous than we realise? *Trends in Plant Science.* 16:61–68.

Cooke J, and Leishman MR (2016) Consistent alleviation of abiotic stress with silicon addition: A meta-analysis. *Functional Ecology.* 30:1340–1357.

Cornelis JT, Delvauz B, Georg RB, Lucas Y, Ranger J, and Opfergelt S (2011) Tracing the origin of dissolved silicon transferred from various soil-plant systems towards rivers: A review. *Biogeosciences.* 8:89–112.

Currie HA, and Perry CC (2007) Silica in plants: Biological, biochemical and chemical studies. *Annals of Botany.* 100:1383–1389.

da Cunha KPV, and do Nascimento CWA (2009) Silicon effects on metal tolerance and structural changes in maize (*Zea mays* L.) grown on a cadmium and zinc enriched soil. *Water, Air, and Soil Pollution.* 197:323–330.

Datnoff LE, Deren CW, and Snyder GH (1997) Silicon fertilization for disease management of rice in Florida. *Crop Protection.* 16:525–531.

Datnoff LE, Snyder GH, and Korndorfer GH (2001) *Silicon in Agriculture.* Elsevier, New York.

Delhaize E, and Ryan PR (1995) Aluminium toxicity and tolerance in plants. *Plant Physiology.* 107:315–321.

Deshmukh RK, Vivancos J, Ramakrishnan G, Guoerin V, Carpentier G, Sonah H, Labboe C, Isenring P, Belzile FJ, and Boelanger RR (2015) A precise spacing between the NPA domains of aquaporins is essential for silicon permeability in plants. *The Plant Journal.* 83:489–500.

Detmann KC, Araujo WL, Martins SC, Sanglard L, Reis JV, Detmann E, Rodrigues FA, Nunes-Nesi A, Fernie AR, and DaMatta FM (2012) Silicon nutrition increases grain yield, which, in turn, exerts a feed forward stimulation of photosynthetic rates via enhanced mesophyll conductance and alters primary metabolism in rice. *New Phytologist.* 196:752–762.

Du B, Nian H, Zhang Z, and Yang C (2010) Effects of aluminum on superoxide dismutase and peroxidase activities, and lipid peroxidation in the roots and calluses of soybeans differing in aluminum tolerance. *Acta Physiologiae Plantarum.* 32:883–890.

Dubouzet JG, Sakuma Y, Ito Y, Kasuga M, Dubouzet EG, and Miura S (2003) OsDREB genes in rice, *Oryza sativa* L., encode transcription activators that function in drought-, high-salt-and cold-responsive gene expression. *The Plant Journal.* 33:751–763.

Epstein E (1999) Silicon: Its manifold roles in plants. *Annals of Applied Biology.* 155:155–160.

Exley C (2015) A possible mechanism of biological silicification in plants. *Frontiers in Plant Science.* 6:853.

Farooq M, Wahid A, Kobayashi N, Fujita D, and Basra SMA (2009) Plant drought stress: Effects, mechanisms and management. *Agronomy for Sustainable Development.* 29:185–212.

Farooq MA, Detterbeck A, Clemens S, and Dietz KJ (2016) Silicon-induced reversibility of cadmium toxicity in rice. *Journal of Experimental Botany.* 67:3573–3585.

Flam-Shepherd R, Huynh WQ, Coskun D, Hamam AM, Britto DT, and Kronzucker HJ (2018) Membrane fluxes, bypass flows, and sodium stress in rice: The influence of silicon. *Journal of Experimental Botany.* 69:1679–1692.

Fleck AT, Nye T, Repenning C, Stahl F, Zahn M, and Schenk MK (2011) Silicon enhances suberization and lignification in roots of rice (*Oryza sativa*). *Journal of Experimental Botany.* 62:2001–2011.

Fleck AT, Schulze S, Hinrichs M, Specht A, Wassmann F, Schreiber L, and Schenk MK (2015) Silicon promotes exodermal casparian band formation in Si accumulating and Si-excluding species by forming phenol complexes. *PLoS ONE.* 10: e0138555.

Fujita Y, Fujita M, Shinozaki K, and Yamaguchi-Shinozaki K (2011) ABA-mediated transcriptional regulation in response to osmotic stress in plants. *Journal of Plant Research.* 124:509–525.

Furukawa J, Yamaji N, Wang H, Mitani N, Murata Y, and Sato K (2007) An Aluminum-activated citrate transporter in barley. *Plant and Cell Physiology.* 8:1081–1091.

Gao JP, Chao DY, and Lin HX (2007) Understanding abiotic stress tolerance mechanisms: Recent studies on stress response in rice. *Journal of Integrative Plant Biology.* 49:742–750.

Gao X, Zou C, Wang L, and Zhang F (2004) Silicon improves water use efficiency in maize plants. *Journal of Plant Nutrition.* 27:1457–1470.

Gao X, Zou C, Wang L, and Zhang F (2006) Silicon decreases transpiration rate and conductance from stomata of maize plants. *Journal of Plant Nutrition*. 29:1637–1647.

Gao, Z, He X, Zhao B, Zhou C, Liang Y, and Ge R (2010) Overexpressing a putative aquaporin gene from wheat, TaNIP, enhances salt tolerance in transgenic *Arabidopsis*. *Plant and Cell Physiology*. 51:767–775.

Gill SS, and Tuteja N (2010) Reactive oxygen species and antioxidant machinery in abiotic stress tolerance in crop plants. *Plant Physiology and Biochemistry*. 48:909–930.

Gong H, and Chen K (2012) The regulatory role of silicon on water relations, photosynthetic gas exchange, and carboxylation activities of wheat leaves in field drought conditions. *Acta Physiologiae Plantarum*. 34(4).

Gong HJ, Chen KM, Chen GC, Wang SM, and Zhang CL (2003) Effects of silicon on growth of wheat under drought. *Journal of Plant Nutrition*. 26:1055–1063.

Gong HJ, Chen KM, Zhao ZG, Chen GC, and Zhou WJ (2008) Effects of silicon on defense of wheat against oxidative stress under drought at different developmental stages. *Biologia Plantarum*. 52:592–596.

Gong HJ, Randall DP, and Flowers TJ (2006) Silicon deposition in the root reduces sodium uptake in rice (*Oryza sativa* L.) seedlings by reducing bypass flow. *Plant, Cell and Environment*. 29:1970–1979.

Goodwin SB, and Sutter TR (2009) Microarray analysis of *Arabidopsis* genome response to aluminum stress. *Biologia Plantarum*. 53:85–99.

Grégoire C, Rémus-Borel W, Vivancos J, Labbé C, Belzile F, and Bélanger RR (2012) Discovery of a multigene family of aquaporin silicon transporters in the primitive plant *Equisetum arvense*. *The Plant Journal*. 72:320–330.

Guerriero G, Hausman JF, and Legay S (2016) Silicon and the plant extracellular matrix. *Frontiers in Plant Science*. 7:463.

Gunes A, Inal A, Bagci EG, and Pilbeam DJ (2007) Silicon-mediated changes of some physiological and enzymatic parameters symptomatic for oxidative stress in spinach and tomato grown in sodic-B toxic soil. *Plant and Soil*. 290:103–114.

Guntzer F, Keller C, and Meunier JD (2012) Benefits of plant silicon for crops: A review. *Agronomy for Sustainable Development*. 32:201–213.

Gupta N, Gaurav SS, and Kumar A (2013) Molecular basis of aluminium toxicity in plants: A review. *American Journal of Plant Sciences*. 4:21–37.

Gzik A (1997) Accumulation of proline and pattern of α-amino acids in sugar beet plants in response to osmotic, water and salt stress. *Environmental and Experimental Botany*. 36:29–38.

Habibi G (2016) Effect of foliar-applied silicon on photochemistry, antioxidant capacity and growth in maize plants subjected to chilling stress. *Acta agriculturae Slovenica*. 107(1):33.

Haddad C, Arkoun M, Jamois F, Schwarzenberg A, Yvin JC, Etienne P, et al. (2018) Silicon promotes growth of *Brassica napus* L. and delays leaf senescence induced by nitrogen starvation. *Frontiers in Plant Science*. 9:516.

Hashemi A, Abdolzadeh A, and Sadeghipour HR (2010) Beneficial effects of silicon nutrition in alleviating salinity stress in hydroponically grown canola, *Brassica napus* L. plants. *Journal of Soil Science and Plant Nutrition*. 56:244–253.

Hattori T, Inanaga S, Araki H, An P, Morita S, Luxová M, and Lux A (2005) Application of silicon enhanced drought tolerance in *Sorghum bicolour*. *Physiologia Plantarum*. 123:459–466.

Hirt H, and Shinozaki K (2004) *Plant Responses to Abiotic Stress*. Springer, New York.

Hodson M, White PJ, Mead A, and Broadley MR (2005) Phylogenetic variation in the silicon composition of plants. *Annals of Botany*. 96:1027–1046.

Horst WJ, Wang Y, and Eticha D (2010) The role of the apoplast in Al induced inhibition of root elongation and in Al resistance of plants: A review. *Annals of Botany*. 106:185–197.

Hosseini SA, Hajirezaei, MR, Seiler C, Sreenivasulu N, and von Wirén N (2016) A potential role of flag leaf potassium in conferring tolerance to drought-induced leaf senescence in barley. *Frontiers in Plant Science*. 7:206.

Hosseini SA, Maillard A, Hajirezaei MR, Ali N, Schwarzenberg A, and Jamois F (2017) Induction of barley silicon transporter hvlsi1 and hvlsi2, increased silicon concentration in the shoot and regulated starch and ABA homeostasis under osmotic stress and concomitant potassium deficiency. *Frontiers in Plant Science*. 8:1359.

Hove RM, and Bhave M (2011) Plant aquaporins with non-aqua functions: Deciphering the signature sequences. *Plant Molecular Biology*. 75:413–430.

Hu H, You J, Fang Y, Zhu X, Qi Z, and Xiong L (2008) Characterization of transcription factor gene SNAC2 conferring cold and salt tolerance in rice. *Plant Molecular Biology*. 67:169–181.

Huang CF, Yamaji N, Mitani N, Yano M, Nagamura Y, and Ma JF (2009) A bacterial type ABC transporter is involved in aluminum tolerance in rice. *Plant Cell*. 21:655–667.

Hussain SS, Kayani MA, and Amjad M (2011) Transcription factors as tools to engineer enhanced drought stress tolerance in plants. *Biotechnology Progress.* 27:297–306.

Iler RK (1979) *The Chemistry of Silica: Solubility, Polymerization, Colloid and Surface Proteins, and Biochemistry.* Wiley Interscience, USA.

Inanaga S, and Okasaka A (1995) Calcium and silicon binding compounds in cell walls of rice shoots. *Soil Science and Plant Nutrition.* 41:103–110.

Iwasaki K, Maier P, Fecht M, and Horst WJ (2002) Leaf apoplastic silicon enhances manganese tolerance of cow-pea (*Vigna unguiculata*). *Journal of Plant Physiology.* 159(2):167–173.

Jarosch B, Collins NC, Zellerhoff N, and Schaffrath U (2005) RAR1, ROR1, and the actin cytoskeleton contribute to basal resistance to *Magnaporthe grisea* in barley. *Molecular Plant-Microbe Interactions.* 18:397–404.

Jayawardana HARK, Weerahewa HLD, and Saparamadu MDJS (2015) Enhanced resistance to anthracnose disease in chili pepper (*Capsicum annuum* L.) by amendment of the nutrient solution with silicon. *The Journal of Horticultural Science and Biotechnology.* 90:557–562.

Kawakami K, Iwai M, Ikeuchi M, Kamiya N, and Shen JR (2007) Location of PsbY in oxygen-evolving photosystem II revealed by mutagenesis and X-ray crystallography. *FEBS Letters.* 581:4983–4987.

Kerstiens G (1996) Cuticular water permeability and its physiological significance. *Journal of Experimental Botany.* 47:1813–1832.

Khattab HI, Emam MA, Emam MM, Helal NM, and Mohamed MR (2014) Effect of selenium and silicon on transcription factors NAC5 and DREB2A involved in drought responsive gene expression in rice. *Biologia Plantarum.* 58:265–273.

Kidd PS, Llugany M, Poschenrieder C, Gunse B, and Barcelo J (2001) The role of root exudates in aluminium resistance and silicon-induced amelioration of aluminium toxicity in three varieties of maize (*Zea mays* L.). *Journal of Experimental Botany.* 52:1339–1352.

Kim YH, Khan AL, Waqas M, and Lee IJ (2017) Silicon regulates antioxidant activities of crop plants under abiotic-induced oxidative stress: A review. *Frontiers in Plant Science.* 8:510.

Knight CT, and Kinrade SD (2001) A primer on the aqueous chemistry of silicon studies. *Plant Science.* 8:57–84.

Kochian LV, Pineros MA, and Hoekenga OA (2005) The physiology, genetics and molecular biology of plant aluminum resistance and toxicity. *Plant and Soil.* 274:175–195.

Kopittke PM, Menzies NW, Wang P, and Blamey FPC (2016) Kinetic and nature of aluminium rhizotoxic effects: A review. *Journal of Experimental Botany.* 67 4451–4467.

Kostic L, Nikolic N, Bosnic D, Samardzic J, and Nikolic M (2017) Silicon increases phosphorus (P) uptake by wheat under low P acid soil conditions. *Plant and Soil.* 419:447–455.

Kumar AP, and Bandhu AD (2005) Salt tolerance and salinity effects on plants: A review. *Ecotoxicology and Environmental Safety.* 60:324–349.

Li YC, Sumner ME, Miller WP, and Alva AK (1996) Mechanism of silicon induced alleviation of aluminum phytotoxicity. *Journal of Plant Nutrition.* 19:1075–1087.

Liang Y (1999) Effects of silicon on enzyme activity and sodium, potassium and calcium concentration in barley under salt stress. *Plant and Soil.* 209:217–224.

Liang Y, Chen Q, Liu Q, Zhang WH, and Ding RX (2003) Exogenous silicon (Si) increases antioxidant enzyme activity and reduces lipid peroxidation in roots of salt-stressed barley (*Hordeum vulgare* L.). *Journal of Plant Physiology.* 160:1157–1164.

Liang Y, Sun W, Zhu Y, and Christie P (2007) Mechanisms of silicon-mediated alleviation of abiotic stresses in higher plants: A review. *Environmental Pollution.* 147:422–428.

Liang Y, Sun W, Zhu YG, and Christie P (2015) Mechanisms of silicon mediated alleviation of abiotic stresses in higher plants: A review. *Environmental Pollution.* 147:422–428.

Liang Y, Zhang WH, Chen Q, and Ding RX (2005) Effects of silicon on H^+-ATPase and H^+-PPase activity, fatty acid composition and fluidity of tonoplast vesicles from roots of salt-stressed barley (*Hordeum vulgare* L.). *Environmental and Experimental Botany.* 53:29–37.

Liang Y, Zhang W, Chen Q, Liu Y, and Ding R (2006) Effect of exogenous silicon (Si) on H+-ATPase activity, phospholipids and fluidity of plasma membrane in leaves of salt-stressed barley (*Hordeum vulgare* L.). *Environmental and Experimental Botany.* 57:212–219.

Liu P, Yin L, Deng X, Wang S, Tanaka K, and Zhang S (2014) Aquaporin-mediated increase in root hydraulic conductance is involved in silicon-induced improved root water uptake under osmotic stress in *Sorghum bicolor* L. *Journal of Experimental Botany.* 65:4747–4756.

Liu P, Yin L, Wang S, Zhang M, Deng X, Zhang S, and Tanaka K (2015) Enhanced root hydraulic conductance by aquaporin regulation accounts for silicon alleviated salt-induced osmotic stress in *Sorghum bicolor* L. *Environmental and Experimental Botany.* 111:42–51.

Lobato AKS, Luz LM. Costa RCL, Santos BG, Meirelles ACS, Oliveira CF, Laughinghouse HD, Neto MAM, Alves GAR, Lopes MJS, and Neves HKB (2009) Si exercises influence on nitrogen components in pepper subjected to water deficit? *Research Journal of Biological Sciences*. 4:1048–1055.

Lucas S, Durmaz E, Akpınar B A, and Budak H (2011) The drought response displayed by a DRE-binding protein from *Triticum dicoccoides*. *Plant Physiology and Biochemistry*. 49:346–351.

Lunde C, Jensen PE, Haldrup A, Knoetzel J, and Scheller HV (2000) The PSI-H subunit of photosystem I is essential for state transitions in plant photosynthesis. *Nature*. 408:613–615.

Ma J, Cai H, He C, Zhang W, and Wang L (2015) A hemicellulose-bound form of silicon inhibits cadmium ion uptake in rice (*Oryza sativa*) cells. *New Phytologist*. 206:1063–1074.

Ma JF (2003) Functions of silicon in higher plants. *Progress in Molecular and Subcellular Biology*. 33:127–147.

Ma JF (2004) Role of silicon in enhancing the resistance of plants to biotic and abiotic stresses. *Soil Science and Plant Nutrition*. 50:11–18.

Ma JF, Miyake Y, and Takahashi E (2001) Silicon as a beneficial element for crop plants. *Studies in Plant Science*. 8(C):17–39.

Ma JF, and Takahashi E (1993) Interaction between calcium and silicon in water-cultured rice plants. *Plant Soil*. 148:107–113.

Ma JF, Tamai K, Yamaji N, Mitani N, Konishi S, Katsuhara M, Ishiguro M, Murata Y, and Yano M (2006) A silicon transporter in rice. *Nature*. 440:688–691.

Ma JF, and Yamaji N (2006) Silicon uptake and accumulation in higher plants. *Trends in Plant Science*. 11:392–397.

Ma JF, and Yamaji N (2008) Function and transport of silicon in plants. *Cellular and Molecular Life Sciences*. 65:3049–3057.

Ma JF, and Yamaji N (2015) A cooperative system of silicon transport in plants. *Trends in Plant Science*. 20:435–442.

Ma JF, Yamaji N, Mitani M, Tamai K, Konishi S, Fujiwara T, Katsuhara M, and Yano M (2007) An efflux transporter of silicon in rice. *Nature*. 448:209–12.

Maksimovic JD, Mojovic M, Maksimovic V, Romheld V, and Nikolic M (2012) Silicon ameliorates manganese toxicity in cucumber by decreasing hydroxyl radical accumulation in the leaf apoplast. *Journal of Experimental Botany*. 63:2411–2420.

Manivannan A, and Ahn YK (2017) Silicon regulates potential genes involved in major physiological processes in plants to combat stress. *Frontiers in Plant Science* 8:1346.

Mansour MMF (1998) Protection of plasma membrane of onion epidermal cells by glycine betaine and proline against NaCl stress. *Plant Physiology and Biochemistry*. 36:767–772.

Markovich O, Steiner E, Kouril S, Tarkowski P, Aharoni A, and Elbaum R (2017) Silicon promotes cytokinin biosynthesis and delays senescence in *Arabidopsis* and sorghum. *Plant, Cell and Environment*. 40:1189–1196.

Mateos NE, Galle A, Florez-SI, Perdomo JA, Galmes J, Ribas-Carbo M, and Flexas J (2015) Assessment of the role of silicon in the Cu-tolerance of the C4 grass *Spartina densiflora*. *Journal of Plant Physiology*. 178:74–83.

Matoh T, Murata S, and Takahashi E (1991) Effect of silicate application on photosynthesis of rice plants. *Japanese Journal of Soil Science and Plant Nutrition*. 62:248–251.

Matsumoto H (2000) Cell biology of aluminum toxicity and tolerance in higher plants. *International Review of Cytology*. 200:1–46.

McGinnity P (2016) Silicon and its role in crop production. Planttuff. http://www.planttuff.com/pdf (25 DEC 2016).

Meyer S, and Genty S (1998) Mapping intercellular CO_2 mole fraction (Ci) in *Rosa rubiginosa* leaves fed with abscisic acid by using chlorophyll fluorescence imaging: Significance of Ci estimated from leaf gas exchange. *Plant Physiology*. 116:947–957.

Miao BH, Xan XG, and Zhang WH (2010) the ameleriotive effects of silicon on soyabean seedlings grown on potassium deficient medium. *Annals of Botany*. 105:967–973.

Milla MA, Butler ED, Huete AR, Wilson CF, Anderson O, and Gustafson JP (2002) Expressed sequence tag-based gene expression analysis under aluminum stress in rye. *Plant Physiology*. 130:1706–1716.

Mitani N, and Ma JF (2005) Uptake system of silicon in different plant species. *Journal of Experimental Botany*. 56:1255–1261.

Mitani-Ueno N, Yamaji N, Zhao FJ, and Ma JF (2011) The aromatic/arginine selectivity filter of NIP aquaporins plays a critical role in substrate selectivity for silicon, boron, and arsenic. *Journal of Experimental Botany*. 62:4391–4398.

Mittler R (2002) Oxidative stress, antioxidants and stress tolerance. *Trends in Plant Science* 7:405–410.

Mizoi J, Shinozaki K, and Yamaguchi SK (2012) AP2/ERF family transcription factors in plant abiotic stress responses. *Biochimica et Biophysica Acta*. 1819:86–96.

Montpetit J, Vivancos J, Mitani-Ueno N, Yamaji N, Remus-Borel W, Belzile F, Ma JF, and Belanger RR (2012) Cloning, functional characterization and heterologous expression of TaLsi1, a wheat silicon transporter gene. *Plant Molecular Biology*. 79:35–46.

Mora ML, Alfaro MA, Jarvis SC, Demanet R, and Cartes P (2006) Soil aluminium availability in Andisols of southern Chile and its effect on forage production and animal metabolism. *Soil Use and Management*. 22:95–101.

Moussa HR (2006) Influence of exogenous application of silicon on physiological response of salt-stressed maize (*Zea mays* L.). *International Journal of Agriculture and Biology*. 8:293–297.

Murata K, Mitsuoka K, Hirai T, Walz T, Agre P, Heymann JB, Engel A, and Fujiyoshi Y (2000) Structural determinants of water permeation through aquaporin-1. *Nature*. 407:599–605.

Nakashima K, Ito Y, and Yamaguchi SK (2009) Transcriptional regulatory networks in response to abiotic stresses in *Arabidopsis* and grasses. *Plant Physiology*. 149:88–95.

Ning Y, Jantasuriyarat C, Zhao Q, Zhang H, Chen S, Liu J, Liu L, Tang S, Park CH, Wang X, Liu X, Dai L, Xie Q, and Wang GL (2011) The SINA E3 ligase OsDIS1 negatively regulates drought response in rice. *Plant Physiology*. 157:242–255.

Park JJ, Yi J, Yoon J, Cho LH, Ping J, and Jeong HJ (2011) OsPUB15, an E3 ubiquitin ligase, functions to reduce cellular oxidative stress during seedling establishment. *The Plant Journal*. 65:194–205.

Pavlovic J, Samardzic J, Kostic L, Laursen KH, Natic M, and Timotijevic G (2016) Silicon enhances leaf remobilization of iron in cucumber under limited iron conditions. *Annals of Botany*. 118:271–280.

Pavlovic J, Samardzic J, Maksimović V, Timotijevic G, Stevic N, Laursen KH, Hansen TH, Husted S, Schjoerring JK, Liang Y, and Nikolic M (2013) Silicon alleviates iron deficiency in cucumber by promoting mobilization of iron in the root apoplast. *New Phytologist*. 198:1096–1107.

Pfannschmidt T, and Yang C (2012) The hidden function of photosynthesis: A sensing system for environmental conditions that regulates plant acclimation responses. *Protoplasma*. 249:125–136.

Pontigo S, Godoy K, Jimenez H, Gutierrez-Moraga A, Mora MDLL, and Cartes P (2017) Silicon-mediated alleviation of aluminum toxicity by modulation of Al/Si uptake and antioxidant performance in ryegrass plants. *Frontiers in Plant Science*. 8:642.

Pontigo S, Ribera A, Gianfreda L, Mora ML, Nikolic M, and Cartes P (2015) Silicon in vascular plants: Uptake, transport and its influence on mineral stress under acidic conditions. *Planta*. 242:23–37.

Poschenrieder C, Gunse B, Corrales I, and Barcelo J (2008) A glance into aluminum toxicity and resistance in plants. *Science of the Total Environment*. 400:356–368.

Prabagar S, Hodson MJ, and Evans DE (2011) Silicon amelioration of aluminium toxicity and cell death in suspension cultures of Norway spruce (*Picea abies* L.) *Karst. Environmental and Experimental Botany*. 70: 266–276.

Qin F, Shinozaki K, and Yamaguchi-Shinozaki K (2011) Achievements and challenges in understanding plant abiotic stress responses and tolerance. *Plant and Cell Physiology*. 52:1569–1582.

Rabbani MA, Maruyama K, Abe H, Khan MA, Katsura K, and Ito Y (2003) Monitoring expression profiles of rice genes under cold, drought, and high-salinity stresses and abscisic acid application using cDNA microarray and RNA gel-blot analyses. *Plant Physiology*. 133:1755–1767.

Rafi MM, Epstein E, and Falk RH (1997) Silicon deprivation causes physical abnormalities in wheat (*Triticum aestivum* L.). *Journal of Plant Physiology*. 151:497–501.

Ranathunge K, Steudle E, and Lafitte R (2005) Blockage of apoplastic bypass-flow of water in rice roots by insoluble salt precipitates analogous to a Pfeffer cell. *Plant, Cell and Environment*. 28:121–133.

Raven JA (1983) The transport and function of silicon in plants. *Biological Reviews of the Cambridge Philosophical Society*. 58:179–207.

Raven JA (2003) Cycling silicon: The role of accumulation in plants. *New Phytologist*. 158(3):419–421.

Reddy AR, Chaitanya KV, and Vivekanandanb M (2004) Drought-induced responses of photosynthesis and antioxidant metabolism in higher plants. *Journal of Plant Physiology*. 161:1189–1202.

Richmond KE, and Sussman M (2003) Got silicon? The non-essential beneficial plant nutrient. *Current Opinion in Plant Biology*. 6:268–272.

Rodrigues FÁ, McNally DJ, Datnoff LE, Jones JB, Labbé C, and Benhamou N (2004) Silicon enhances the accumulation of diterpenoid phytoalexins in rice: A potential mechanism for blast resistance. *Phytopathology*. 94:177–183.

Rodrigues FA, Vale FXR, Korndorfer GH, Prabhu AS, Datnoff LE, Oliveira AMA, and Zambolim L (2003) Influence of silicon on sheath blight of rice in Brazil. *Crop Protection*. 22:23–29.

Rogalla H, and Romheld V (2002) Role of leaf apoplast in silicon-mediated manganese tolerance of *Cucumis sativus* L. *Plant, Cell and Environment*. 25:549–555.

Romero-Aranda MR, Jurado O, and Cuartero J (2006) Silicon alleviates the deleterious salt effect on tomato plant growth by improving plant water status. *Journal of Plant Physiology.* 163:847–855.

Ryan PR, Tyerman SD, Sasaki T, Furuichi T, Yamamoto Y, and Zhang WH (2011) The identification of aluminium-resistance genes provides opportunities for enhancing crop production on acid soils. *Journal of Experimental Botany.* 62:9–20.

Sachs JV (1860) Vegetations versuchemitausschluss des bodensüber die nährstoffe und sonstigenernährungs-bedingungen von mais, bohnen, und anderenpflanzen. *Landw. Versuchsst.* 2:219–268.

Sanglard L, Martins SCV, Detmann KC, Silva PEM, Lavinsky AO, Silva MM, Detmann E, Araujo WL, and DaMatta FM (2014) Silicon nutrition alleviates the negative impacts of arsenic on the photosynthetic apparatus of rice leaves: An analysis of the key limitations of photosynthesis. *Physiologia Plantarum.* 152:355–366.

Sasaki T, Yamamoto Y, Ezaki B, Katsuhara M, Ahn SJ, Ryan PR, Delhaize E, and Matsumoto H (2004) A wheat gene encoding an aluminum-activated malate transporter. *The Plant Journal.* 37:645–653.

Schaller J, Brackhage C, and Baucker E (2013) UV-screening of grass by plant silica? *Journal of Biosciences.* 38(2):413–416.

Seckin B, Sekmen AH, and Turkan İ (2009) An enhancing effect of exogenous mannitol on the antioxidant enzyme activities in roots of wheat under salt stress. *Journal of Plant Growth Regulation.* 28:12–20.

Shao JF, Che J, Yamaji N, Shen RF, and Ma JF (2017) Silicon reduces cadmium accumulation by suppressing expression of transporter genes involved in cadmium uptake and translocation in rice. *Journal of Experimental Botany.* 68:5641–5651.

Shen XF, Zhou YY, Duan LS, Li ZH, Eneji AE, and Li JM (2010) Silicon effects on photosynthesis and antioxidant parameters of soybean seedlings under drought and ultraviolet-B radiation. *Journal of Plant Physiology.* 167:1248–1252.

Shi X, Zhang C, Wang H, and Zhang F (2005) Effect of Si on the distribution of Cd in rice seedlings. *Plant and Soil.* 272(1):53–60.

Shi Y, Wang Y, Flowers TJ, and Gong H (2013) Silicon decreases chloride transport in rice (*Oryza sativa* L.) in saline conditions. *Journal of Plant Physiology.* 170:847–853.

Shi Y, Zhang Y, Yao H, Wu J, Sun H, and Gong H (2014) Silicon improves seed germination and alleviates oxidative stress of bud seedlings in tomato under water deficit stress. *Plant Physiology and Biochemistry.* 78:27–36.

Shinozaki K, and Yamaguchi SK (2000) Molecular responses to dehydration and low temperature: Differences and cross-talk between two stress signaling pathways. *Current Opinion in Plant Biology.* 3:217–223.

Shinozaki K, Yamaguchi-Shinozaki K, and Seki M (2003) Regulatory network of gene expression in the drought and cold stress responses. *Current Opinion in Plant Biology.* 6:410–417.

Singh S, Tripathi DK, Singh S, Sharma S, Dubey NK, and Chauhan DK (2017) Toxicity of aluminium on various levels of plant cells and organism: A review. *Environmental and Experimental Botany.* 137:177–193.

Sivaguru M, Ezaki B, He ZH, Tong H, Osawa H, and Baluska F (2003) Aluminum induced gene expression and protein localization of a cell wall-associated receptor kinase in *Arabidopsis. Plant Physiology.* 132:2256–2266.

Song A, Li P, Fan F, Li Z, and Liang Y (2014) The effect of silicon on photosynthesis and expression of its relevant genes in rice (*Oryza sativa* L.) under high-zinc stress. *PLoS ONE* 26:9–11.

Song SY, Chen Y, Chen J, Dai XY, and Zhang WH (2011) Physiological mechanisms underlying OsNAC5-dependent tolerance of rice plants to abiotic stress. *Planta.* 234:331–345.

Soundararajan P, Manivannan A, Ko CH, and Jeong BR (2017) Silicon enhanced redox homeostasis and protein expression to mitigate the salinity stress in *rosa hybrida* 'Rock Fire'. *Journal of Plant Growth Regulation.* 37:1–19.

Szabela DA, Markiewicz J, and Wolf WM (2015) Heavy metal uptake by herbs. IV. Influence of soil pH on the content of heavy metals in *Valeriana officinalis* L. *Water, Air, and Soil Pollution.* 226:106.

Tahir MA, Aziz T, Farooq M, and Sarwar G (2012) Silicon-induced changes in growth, ionic composition, water relations, chlorophyll contents and membrane permeability in two salt-stressed wheat genotypes. *Archives of Agronomy and Soil Science.* 58:247–256.

Takahashi E, Ma JF, and Miyake Y (1990) The possibility of silicon as an essential element for higher plants. *Comments on Agricultural and Food Chemistry.* 2:99–122.

Takasaki H, Maruyama K, Kidokoro S, Ito Y, Fujita Y, and Shinozaki K (2010) The abiotic stress-responsive NAC-type transcription factor OsNAC5 regulates stress-inducible genes and stress tolerance in rice. *Molecular Genetics and Genomics.* 284:173–183.

Tilman D, Fargione J, Wolff B, Antonio CD, Dobson A, Howarth R, Schindler D, Schlesinger WH, Simberloff D, and Swackhamer D (2001) Forecasting agriculturally driven environmental change. *American Association for the Advancement of Science.* 292:281–284.

Tran LSP, Nishiyama R, Yamaguchi-Shinozaki K, and Shinozaki K (2010) Potential utilization of NAC transcription factors to enhance abiotic stress tolerance in plants by biotechnological approach. *GM Crops.* 1:32–39.

Tripathi DK, Bashri G, Singh S, Singh S, Ahmad P, and Prassad SM (eds.) (2016) Efficacy of silicon against aluminum toxicity in plants. In: *An Overview in Silicon in Plants: Advances and Future Prospects.* CRC Press, Boca Raton, FL.

Tripathi DK, Singh VP, Prasad SM, Chauhan DK, Dubey NK, and Rai, AK (2015) Silicon-mediated alleviation of Cr (VI) toxicity in wheat seedlings as evidenced by chlorophyll florescence, laser induced breakdown spectroscopy and anatomical changes. *Ecotoxicology and Environmental Safety.* 113:133–144.

Tuna AL, Kaya C, Higgs D, Murillo-Amador B, Aydemir S, and Girgin AR (2008) Silicon improves salinity tolerance in wheat plants. *Environmental and Experimental Botany.* 62:10–16.

Tunnacliffe A, and Wise MJ (2007) The continuing conundrum of the LEA proteins. *Naturwissenschaften.* 94:791–812.

Umezawa T, Fujita M, Fujita Y, Yamaguchi SK, and Shinozaki K (2006) Engineering drought tolerance in plants: Discovering and tailoring genes to unlock the future. *Current Opinion in Biotechnology.* 17:113–122.

Vaculík M, Landberg T, Greger M, Luxová M, Stoláriková M, and Lux A (2012) Silicon modifies root anatomy, and uptake and subcellular distribution of cadmium in young maize plants. *Annals of Botany.* 110:433–443.

Van Bockhaven J, De Vleesschauwer D, and Hofte M (2013) Towards establishing broad-spectrum disease resistance in plants: Silicon leads the way. *Journal of Experimental Botany.* 64:1281–1293.

Vivancos J, Deshmukh R, Grégoire C, Rémus-Borel W, Belzile F, and Bélanger RR (2016) Identification and characterization of silicon efflux transporters in horsetail (*Equisetum arvense*). *Journal of Plant Physiology.* 200:82–89.

Von U, and Mutert E (1995) Global extent, development and economic impact of acid soils. *Plant and Soil.* 171:1–15.

Wang Q, Guan Y, Wu Y, Chen H, Chen F, and Chu C (2008) Overexpression of a rice OsDREB1F gene increases salt, drought, and low temperature tolerance in both *Arabidopsis* and rice. *Plant Molecular Biology.* 67:589–602.

Wang YX, Stass A, and Horst WJ (2004) Apoplastic binding of aluminum is involved in silicon-induced amelioration of aluminum toxicity in maize. *Plant Physiology.* 136:3762–3770.

Watanabe S, Kojima K, Ide Y, and Sasaki S (2000) Effects of saline and osmotic stress on proline and sugar accumulation in *Populus euphratica* in vitro. *Plant Cell, Tissue and Organ Culture.* 63:199–206.

Wong YC, Heits A, and Ville DJ (1972) Foliar symptoms of silicon deficiency in the sugarcane plant. *Proceedings of International Society of Sugar Cane Technologists.* 14:766–776.

Wu JW, Geilfus CM, Pitann B, and Muhling KH (2016) Silicon-enhanced oxalate exudation contributes to alleviation of cadmium toxicity in wheat. *Environmental and Experimental Botany.* 131:10–18.

Wu XC, Chen YK, Li QS, Fang CX, Xiong J, and Lin WX (2009) Effects of silicon nutrition on phenolic metabolization of rice (*Oryza sativa* L.) exposed to enhanced ultraviolet-B. *Chinese Agricultural Science Bulletin.* 25:225–230.

Xiong L, Schumaker KS, and Zhu JK (2002) Cell signaling during cold, drought, and salt stress. *Plant Cell.* 14:165–183.

Yamaji N, Chiba Y, Mitani-Ueno N, and Ma F (2012) Transporter gene implicated in silicon distribution in barley. *Plant Physiology.* 160:1491–1497.

Yamaji N, and Ma JF (2007) Spatial distribution and temporal variation of the rice silicon transporter Lsi1. *Plant Physiology.* 143:1306–1313.

Yamaji N, Mitatni N, and Ma JF (2008) A transporter regulating silicon distribution in rice shoots. *Plant Cell.* 20:1381–1389.

Yamamoto Y, Kobayashi Y, and Matsumoto H (2001) Lipid peroxidation is an early symptom triggered by aluminum, but not the primary cause of elongation inhibition in pea roots. *Plant Physiology.* 125:199–208.

Yao X, Chu J, Kunzheng C, Liu L, Shi J, and Geng W (2011) Silicon improves the tolerance of wheat seedlings to ultraviolet-B stress. *Biological Trace Element Research.* 143:507–517.

Yeo AR, Flowers SA, Rao G, Welfare K, Senanayake N, and Flowers TJ (1999) Silicon reduces sodium uptake in rice (*Oryza sativa* L.) in saline conditions and this is accounted for by a reduction in the transpirational bypass flow. *Plant, Cell and Environment.* 22:559–565.

Yin J, Zhang X, Zhang G, Wen Y, Liang G, and Chen X (2018) Aminocyclpropane-1-carboxylic acid is a key regulator of guard mother cell terminal division in *Arabidopsis thaliana. Journal of Experimental Botany.* 70:897–908.

Yin L, Shiwen W, Li J, and Tanaka K (2013) Application of silicon improves salt tolerance through ameliorating osmotic and ionic stresses in the seedling of Sorghum bicolor. *Acta Physiologiae Plantarum*. 35(11).

Yin ZP, Li S, Ren J, and Song SX (2014) Role of spermidine and spermine in alleviation of drought-induced oxidative stress and photosynthetic inhibition in Chinese dwarf cherry (*Cerasus humilis*) seedlings. *Plant Growth Regulation*. 74:209–218.

Yoshida S (1965) Chemical aspect of silicon in physiology of the rice plant. *Bulletin of the National Agriculture Science B*. 15:1–58.

Zargar SM, Nazir, M, Agrawal GK, Kim D, and Rakwal R (2010) Silicon in plant tolerance against environmental stressors: Towards crop improvement using omics approaches. *Current Proteomics*. 7:135–143.

Zhu Y, and Gong H (2014) Beneficial effects of silicon on salt and drought tolerance in plants. *Agronomy for Sustainable Development*. 34:455–472.

Zhu Z, Wei G, Li J, Qian Q, and Yu J (2004) Silicon alleviates salt stress and increases antioxidant enzymes activity in leaves of salt-stressed cucumber (*Cucumis sativus* L.). *Plant Science*. 167:527–533.

Zsoldos F, Vashegyi A, Pecsvaradi A, and Bona L (2003) Influence of silicon on aluminium toxicity in common and durum wheats. *Agronomie*. 23:349–354.

16 Genomics Approaches in Plant Stress Research

A.T. Vivek and Shailesh Kumar

CONTENTS

16.1 INTRODUCTION

The global climatic fluctuations and rapid population growth increase the burden to produce an adequate food supply. With the abiotic and biotic stress in plants likely to disrupt global food supplies, immediate measures must be taken to ensure food security. New plants that are tolerant to abiotic stress and pests are needed for which an interdisciplinary approach is required for understanding plant stress tolerance mechanisms, plant breeding and enhancing crop production (Lee 2014). The present landscape of genomics has been dominated by NGS technologies which provide immense sequence information (Bevan and Uauy 2013). For the transition from model plants to crop plants, it is necessary to combine various types of genomics data and NGS to meet inopportune agricultural challenges in the face of climate change. The available diversity among crops that are cultivated and their wild relatives supplies abundant resources for tracing trait/gene discovery, and needs to be utilized effectively. The application of this information to the recent genomics approaches will hasten the development of elite cultivars adapted to the constantly changing climate (Ismail and Horie 2017). In addition to that, the growth of bioinformatics tools to analyze genomic data increases the value of genomic resources available for crop improvement (Akpinar, Lucas et al. 2013).

In the past, efforts to develop stress-tolerant plant were based on genetic engineering and breeding strategies, which are difficult due to the complexity involved in regulatory mechanisms owing to stress responses. Advancement is seen through the impact of genomics approaches performed on various plants and the application of high-throughput techniques over recent years (Lee 2014). As conventional breeding efforts have their limits, the genomics era, with the assistance of a range of bioinformatics tools, presents a promising approach towards the breeding of stress-tolerant plants.

After successfully completing the sequencing of the whole genome of *Arabidopsis thaliana* in 2000, several other major crops were sequenced, providing immense sequencing data on a daily basis (Arabidopsis Genome Initiative 2000). This led to a significant rise in the development of complicated bioinformatics approaches towards the analysis of genome sequence data generated that simplify the whole-genome sequencing from data analysis, assembly, annotation and storage and release of genome data online (Thudi et al. 2012). The genomes of plants from several projects have assisted in comprehending the complexity, dynamics and architecture of the plant genome. The information gained from these sequences has been implemented to construct high-density genetic maps, allele mining, genotype-by-sequencing (GBS), genome-wide association studies, plant diversity analysis and so on. All these genomic approaches along with bioinformatics have contributed towards the identification and development of stress-tolerant crops. For instance, a golden gate SNP array was reported by Kurokawa et al. (2016) that can specifically target yield and stress-resistance genes in rice. This was possible by the intervention of bioinformatics in DNA microarray analysis and determination of SNP markers (Kurokawa et al. 2016). In this chapter, genomics approaches and some suitable bioinformatics software tools are discussed for the study of plant systems in various stress conditions, suggesting how these can be integrated to improve the genomics-assisted breeding pipeline for the effective and efficient evaluation and selection of stress-tolerant plants (Figure 16.1).

16.2 THE SIGNIFICANCE OF ADVANCED GENOME SEQUENCING TECHNOLOGIES

After the successful completion of the human genome using Sanger sequencing, there has been a rapid improvement in genomics technologies. The rise in next-generation sequencing (NGS) progressed the study of plant genomes and has contributed to the comprehensive understanding of their evolution and diversity. Nonetheless, the presence of repetitive sequences in plants poses problems such as the presence of gaps and can also lead to mis-assemblies. Also, second-generation sequencing (SGS) generates short reads and low N50 (Sedlazeck et al. 2018; Alkan, Sajjadian et al. 2011). This lowers the gene prediction quality as it might cause an increase in false-positive genes due to the presence of split genes or interrupted genes in multiple contigs (Denton et al. 2014). To address these issues, long-read sequencing and optical mapping have provided new avenues for generating high-quality assemblies. Long-read sequencing technologies (LRST) generate longer reads of mean length more than 10 kb in comparison to second-generation sequencing technologies, but LRSTs were not used commonly because of low-throughput, higher error rates and the

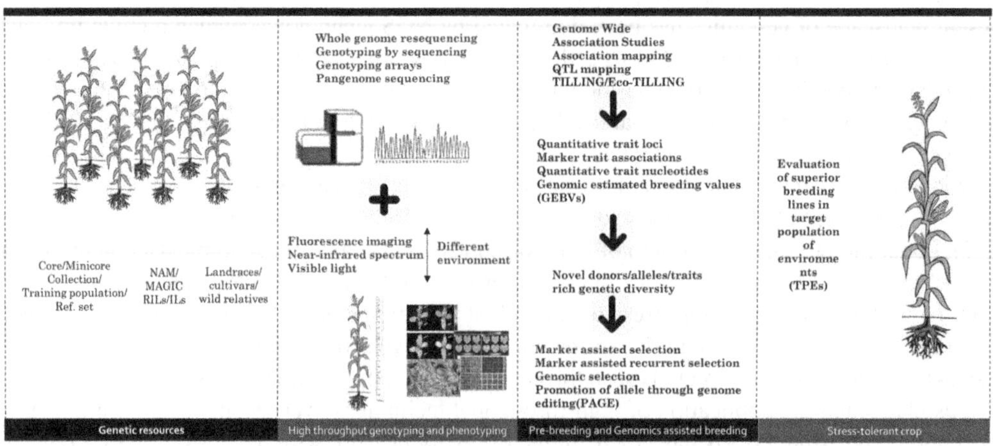

FIGURE 16.1 General pipeline for development of stress-tolerant crops using genomics-assisted breeding.

expense of data generation (Rhoads and Au 2015). But recent developments have overcome these issues, and LRSTs have been applied to a greater extent in the field of genomics. LRSTs aid in precise genome assembly by producing long reads that span regions containing repeats. Furthermore, LRSTs can help in the identification of exon connections and discover gene isoforms of corresponding mRNAs. At present, there are two type of LRSTs available, i.e. synthetic sequencing LRSTs and single molecule LRST (Yuan et al. 2017). Synthetic sequencing LRSTs include Illumina synthetic long-read sequencing technology and 10X Genomics GemCode™. Placebo single-molecule real-time (SMRT) sequencing, Oxford Nanopore sequencing and Base4 microdroplet sequencing are based on single-molecule sequencing technologies that generate long sequencing reads in real time (Eisenstein 2015). Besides DNA sequencing technologies, optical mapping is based on linkage information that assists in scaffolding genome assemblies that rely on the physical location of sites cut down by a restriction enzyme. Optical mapping produces DNA fingerprints that were earlier based on the light microscope technique which was low-throughput and less precise. This added to its negligible application to larger genomes. Howbeit, several technological improvements have allowed these limitations to be overcome and have forwarded optical mapping feasible for larger genomes (Yuan et al. 2017). Currently, OpGen Argus R and BioNano Irys R systems are two widely applied optical mapping systems (Shelton et al. 2015; Yuan et al. 2017). Precise reference assemblies promote the identification of potential candidate genes for various traits and provide enormous information to utilize in plant molecular breeding. Recent improvements in genomics technologies have been applied extensively for plant genomes, helping to expose information on the diversity, evolution and function of genes useful for the development of stress-tolerant varieties.

16.3 STRATEGIES TO DEVELOP STRESS-TOLERANT CROPS

16.3.1 Superior Allele/Haplotypes Identification

Globally, millions of accessions are preserved in seed banks, offering immense information on natural variation for plant breeders worldwide. Gene bank and other phenotype data could be used as surrogates for stresses to select genotypes that furnish essential haplotypes that can be employed in breeding programs. New sources of genetic variation, superior genes and novel genes are required to be identified from wild relatives and landraces stored in gene banks (Gur and Zamir 2004; McCouch et al. 2013). NGS technologies have been successfully employed to identify polymorphisms and associated traits of interest. Genome sequencing projects such as the 3000 Rice Genome Initiative provide possibilities to find variations for a larger set of genes based on genotypic–phenotypic associations (J.-Y. Li, Wang, and Zeigler 2014). Information on the origin of individual plant species, domestication and population structure can be attained from the resequencing of the germplasm. Such studies deliver extensive knowledge of the genetic architecture of traits relevant to stress tolerance and bolster the identification of haplotypes or alleles for improving genetic gains. On that account, it is indispensable to survey alleles adapted to adverse conditions and identify those superior alleles to incorporate into breeding programs (Varshney et al. 2018).

16.3.2 Advanced Mapping Populations

To deal with the challenges associated with the dissection of a quantitative trait, genetic architecture requires extensive genetic diversity. Generally, novel recombinants and haplotypes are generated from a bi-parental mapping population. Even though such populations are effective in detecting quantitative trait loci (QTLs), these represent the narrow genetic base of two contrasting parents (Yang Xu et al. 2017; Jannink 2007). Besides, due to a lack of information on genetic relatedness and population structure among the lines, genome-wide association mapping studies are liable to false positive detection (Lewis 2002; Zhao et al. 2007). This has led to the evolution of next-generation mapping resources referred to as multi-parent populations (MPP) to overcome the constraints of both bi-parental and genome-wide association populations. Nested association mapping (NAM) and multi-parent advanced generation inter-cross (MAGIC) populations are the two major multi-parent populations, implemented successfully in diverse crop species. Genetic diversity maximizes in such populations for the reason that more than two parents are crossed and inter-mating is structured. NAM populations are generated from a single inbred line to a sequential collection containing inbred lines. It was first developed in maize to capture the numerous recombination events (McMullen et al. 2009). Unlike NAM, MAGIC populations are generated through inter-crossing diverse parental lines to intermix the genetic background by repeated rounds of recombination. It improves mapping resolution by lowering linkage disequilibrium (B. E. Huang et al. 2015). To intensify the efficiency of genome-wide association studies, both mapping populations have been used for QTL identification in many crop species, such as *Triticum aestivum*, *Oryza sativa* and *Vigna unguiculata* (Ladejobi et al. 2016).

16.3.3 High-Throughput Genotyping and Phenotyping

Genotyping of the targeted population for trait mapping or product development is a crucial step in the identification of better alleles/superior lines for the development of stress-tolerant crops. Previously, genetic diversity assessment has been performed through cytogenetic/morphological characterization or isozyme analysis. Consequently, DNA molecular markers-based techniques have been favored to estimate genetic variation levels as types of markers, and techniques to genotype have progressed over time. The turn-around-time, information retrieval ease, reliability and genotyping assay cost are crucial to evaluate, select and advance individuals to the next generation for a plant breeder. SNPs have become a preferential marker of choice as they are abundant in the genome, bi-allelic and permit single base resolution (Mammadov et al. 2012). Advancement in high-density genotyping and SNP detection assays has promoted the adoption of large-scale genotyping. Genotyping-by-sequencing (GBS) is a rapid tool for identifying genetic variation underlying stress-responsive traits, and can be categorized into whole-genome resequencing (WGR) and reduced representation sequencing (RRS) (Scheben, Yuan, and Edwards 2016). WGRs are advantageous in high-density SNPs and can be performed at low coverages. This is because data on recombinant populations in addition to high-quality reference genome allows screening for high-confidence SNPs (X. Huang et al. 2009; Golicz, Bayer, and Edwards 2015). Nonetheless, WGRs still remain costly for sequencing populations possessing large genomes. Alternatively, the use of reduced representation sequencing lowers cost as it is based on restriction enzymes that can allow focus on a fraction of the genome. However, in comparison to WGR, RRS leads to low-density SNPs and low profiling because of polymorphisms at restriction sites (Andrews et al. 2016; Poland et al. 2012). Moreover, improved methods such as pooled mapping are being developed to avoid missing data, and more than one method is implemented in parallel to lower the costs of sequencing (Fu et al. 2016). Correspondingly, there have been major advancements in quicker genotyping using SNP array technologies of multiple markers. Illumina- and Affymetrix-based technologies are prevalent amongst all of these genotyping arrays that differ substantially other than their similar size, format and application (Rasheed et al. 2017). Nevertheless, both persist as popular SNP array technologies, considering that they permit the targeting of alleles of interest and computational analyses are straightforward. These commercial SNP arrays are implemented to develop crop-specific SNP arrays in various crops such as rice, maize and wheat as they assist in mapping, trait association studies and genomics-assisted breeding (Chen et al. 2014; Ganal et al. 2011; Winfield et al. 2016).

In order to understand the changes in phenotype for underlying genetic variations as well as multiple trait characterization in crops has been possible with the help of phenomic studies. Figure 16.2 illustrates a general schematic overview of the integration of bioinformatics to genotype data and phenotype data. Contrary to the genotyping methods, developments in technologies to assess advanced plant traits have been slow, which is referred to as a 'population bottleneck' (Furbank

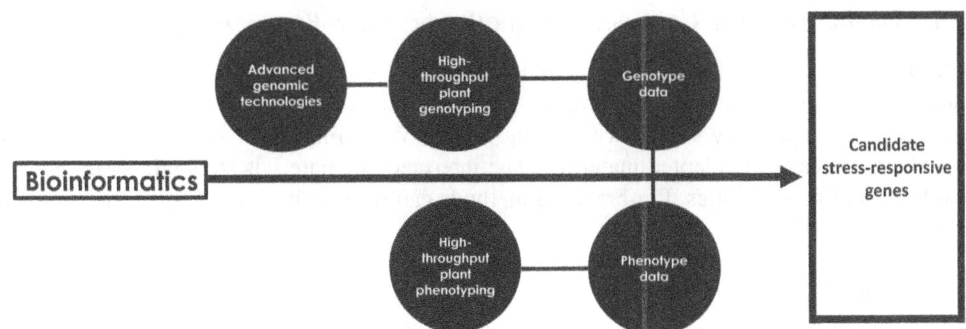

FIGURE 16.2 Integrated use of bioinformatics in candidate stress-responsive gene mining.

and Tester 2011). The main cause of phenotyping bottleneck is abiotic/biotic stress assessment and phenotyping of traits based on biochemical or metabolic approaches. The necessity to investigate many traits in bulk amounts notwithstanding its accuracy has led to the high-throughput plant phenotyping (HTTP) technologies. The application of HTTP has forwarded the complex trait dissection, examination of stress-associated traits and potential for non-destructively recording plant traits (D'Agostino et al. 2017; Coppens et al. 2017). HTTP technologies can potentially scan plant features based on analysis of images, depending on sensing methods that are non-invasive as well as automated. Thermal infrared systems, multispectral/hyperspectral remote sensing, RGB/CIR cameras, magnetic resonance imagers (MRI) and three-dimensional fluorescence imaging are some of the systems available for HTTPs. Pictures have been taken using robots and remote-controlled aerial vehicles containing sensors aggregating enormous data for analysis in a day or on a seasonal basis from germination to maturity (Perez-Sanz, Navarro, and Egea-Cortines 2017). On that account, improved phenotyping platforms have been utilized to assess and estimate traits adapted to stress conditions and further in the identification of stress-responsive genes and selection of individuals from a population that are tolerant to abiotic and biotic stresses.

16.3.4 Trait Mapping Approaches

Linkage mapping and association mapping are two supportive approaches, critical to identifying candidate loci underlying the plant response to abiotic or biotic stresses. Powerful approaches have evolved to understand interactions in complex signaling pathways, genes and regulatory regions to exploit in plant breeding. Genomic regions linked to quantitative traits called QTL investigations have successfully identified candidate genes conferring resistance to many plant stressors like heat, cold pathogens and pests in many crop species. To improve the identification of higher-resolution traits associated with genes, genome-wide association studies (GWAS) utilize the recombination of diverse association panels. Genotyping-by-sequencing (GBS) and SNP arrays have been successfully implemented in the GWAS approach for various crops such as rice, maize and soybean (Zhao et al. 2011; Hui Li et al. 2013; Hwang et al. 2014).

Progress in genomics has guided the advancement of NGS-based trait mapping approaches. Such approaches are not time-consuming as they increase the rate of trait mapping programs. Whole genome sequencing-based high-resolution mapping is a rapid method that has been improvised from the bulk-segregant analysis (BSA). It permits the mapping of target genomic regions, avoiding linkage map construction (Varshney et al. 2018). Nowadays, re-sequencing thousands of lines from different crop species is feasible in view of the fact that sequencing cost is reduced and abundant draft genome sequences of plants are available. The re-sequencing of thousands of individuals of the genetic population is advantageous in comprehending genetic relationships among individuals and provides a platform for high-resolution trait mapping as it imparts genome-wide SNPs at a large-scale (Pandey et al. 2016).

16.3.5 Pre-Breeding for Capturing Abiotic/Biotic Stress-Responsive Alleles

Pre-breeding is necessary for identifying and transferring traits and genes of interest from un-adapted to intermediate materials. Such intermediate materials are used for the production of new varieties. It is a preliminary step in associating genetic variability that has been identified from crop wild relatives and un-adapted materials. The intermediate materials so obtained are used for the production of new varieties. Pre-breeding methods can be advantageous in the following ways (Varshney et al. 2018):

- Widening the genetic base
- Identification/characterization of stress-relevant traits
- Identification/introgression of stress-responsive genes into breeding populations
- Identification and transfer of stress-responsive genes using genetic transformation experiments

It has been seen that crop wild relatives in pre-breeding pose problems because of linkage drag introduction to the new elite gene pool and crossing the barrier between wild and cultivated species. However, for the identification of markers linked to genomic regions of desired traits, genomic technologies are beneficial in minimizing these issues. Deleterious effects of alleles in crop wild relatives can be repaired using genome-editing technologies (Scheben and Edwards 2017).

16.3.6 Genomics-Assisted Breeding to Develop Stress-Tolerant Crops

Several molecular breeding approaches have been applied to the introgression of candidate genomic regions into elite lines (Varshney et al. 2012). Following the identification of markers associated with traits, they are usually employed for marker-assisted selection (MAS) or marker-assisted backcrossing (MABC). The use of marker-assisted selection allows introgression of fewer than ten loci, and it has been implemented in various crops to transfer desired traits into elite cultivars. To target around 40 loci controlling complex traits, the marker-assisted recurrent selection is adopted. By repeating intercrossing, the marker-assisted selection is better suited for the development of better lines through the incorporation of a selective combination of alleles. Recently, the forward breeding approach has been used for early generation selection that involves the application of a marker set for agronomically important traits to test early generation. Genomic selection is one of the most potentially beneficial advancements in the area of genomics-assisted breeding. It assists in the faster improvement of the crop without the need for exhaustive study of individual loci. It has been popularly used in animal breeding programs; nonetheless, there have been reports in plants in the last ten years (Varshney et al. 2018). It depends on the genomics estimated breeding values (GEBVs) for individual lines of the genotyped and phenotyped training population. Individuals can be selected to develop a breeding population that could breed over several generations, avoiding further time-consuming phenotyping (Meuwissen, Hayes, and Goddard 2001). The genomic selection approach improves the selection efficiency by reducing breeding cycles and is favorable for quantitative traits (Varshney et al. 2018). In addition to that, it can furnish a complex trait selection of abiotic/biotic stress tolerance, holding the potentiality to develop stress-tolerant crops. In many crops, genes of agronomical-relevant traits have been cloned, and detailed studies on such genes have identified quantitative trait nucleotides (QTN) that possess considerable phenotypic effects. These QTNs can be altered using genome-editing technology. Another strategy called the promotion of allele by genome editing (PAGE) is a combination of genome editing and genomic selection; it is a modern genomics approach in plant molecular breeding. On account of these, it is expected that approaches like MABC and forward breeding can be used to transfer alleles holding positive phenotypic traits, whereas lethal alleles can be rectified using genome editing (PAGE) (Varshney et al. 2018). Consequently, the evaluation of the superior lines developed in target population environments (TPEs) plays a pivotal part in the selection of lines that will perform better under abiotic/biotic stressed conditions.

16.3.7 Microarray and RNA-Seq-Based Expression Profiling

The latest genomic tools are being developed to expedite interest in global gene expression studies. Gene expression analysis allows the extraction of abundant biological information that permits breeders to decipher the molecular intricacies of complex plant processes during stress as it points to the identification of new targets for crop improvement. For any genomics experiment, it is required to prepare comparable data for identification of differentially expressed genes (DEG), biological processes/metabolic pathways and gene families under stress conditions, establish gene regulatory networks (GRN) and trace vital regulators concerned with particular biological processes and pathways (Yang and Wei 2015). For instance, comparative transcriptomic studies in tomato have shown alteration in gene expression levels due to selection pressure as most of these are stress-responsive genes involved in imparting stress tolerance (Koenig et al. 2013).

Approaches like RT-PCR, differential display and cDNA-AFLPs are limited to the capture of low-abundance transcripts, but the serial analysis of gene expression (SAGE) and massively parallel signature sequencing (MPSS) methods overcome these pitfalls (Anisimov 2008; Reinartz et al. 2002). Even then, the current genomics era is dominated by microarray and RNA-seq technologies as they can profile >10,000 of transcripts and are sensitive to lowly expressed genes. Microarray is a hybridization-based approach based on the preparation of fluorescently labelled cDNA that can be either commercial high-density microarrays or custom made. To detect and quantify specific spliced isoforms, microarrays consisting of probes covering the exon junction are used. Microarrays are less expensive except for certain approaches such as high-resolution tiling arrays that are implemented to examine larger genomes. Unlike microarray, which relies upon previous sequence information, the RNA-seq approach does not always require reference genome information. In addition to that, RNA-seq analysis yields precise locations of transcript boundaries and possesses low background signal. Thus, RNA-seq is the only method adopted to capture the entire transcriptome in both a quantitative and high-throughput manner. It has been successfully applied in various crops with distinct breeding objectives, leading to the identification of genes in various abiotic stress responses (Ong et al. 2016). RNA-seq has shown immense potential for breeding complex traits.

16.3.8 Reverse Genetics

To induce genetic variation for improved crop yield, plant breeders have been applying mutagenesis. Earlier, X-ray radiation and gamma-ray radiation were used to induce mutations, however, the use of chemical mutagens is a research routine in mutation breeding nowadays. Chemical mutagens cause substitutions rather than chromosomal mutations; therefore, ethyl methanesulfonate (EMS) is used to induce random mutations to develop mutated lines. Targeting Induced Local Lesions in Genomes (TILLING) technology is a reverse genetics, low-cost strategy that can be applied to any plant irrespective of its ploidy level and genome size. It involves the mutagenesis of plants followed by the tracking of SNPs based on mismatch detection in the mutant population generated. It is a popular plant functional genomics technique that could be used in pre-breeding as well. It is essential to detect SNP in the mutant and wild-type in the specified gene of interest. The TILLING strategy is well-suited for most plant species as seeds can be stored for long periods after self-fertilization. In a general TILLING procedure, EMS-treated seeds are grown to produce M1 plants, which are further self-fertilized to generate an M2 population. DNA isolated from the leaves of M2 plants is subject to mutational screening (Barkley and Wang 2008). Gene-specific infrared dye-labeled primers are used amplify the target fragment, where the forward and reverse primers are IRDye700 and IRDye800 respectively. Following PCR amplification, denaturation of samples and annealing is done to generate heteroduplexes between wild-type and mutant DNA strands. Sample incubation in single-strand specific nucleases such as CEL1 endonucleases promotes the digestion of mismatches and individual samples loaded onto polyacrylamide gel. Fluorescently labeled DNA is visualized using a LICOR-DNA analyzer providing two gel images, holding data extracted from the 700 nm and 800 nm channels respectively. The gel data can be analyzed using data analysis programs; however, the exact changes caused by mutation are detected from DNA sequencing (Perry et al. 2003). With an effective pooling strategy, many samples can be investigated, and high throughput is achieved by employing many thermal cyclers and LICOR analyzers (Till et al. 2006). Since the expression of natural allelic variants is more robust than induced mutations, these variants are stabilized over their period of evolution (Jiang and Ramachandran 2010). Such variants occur at an exceedingly low frequency; thus, allele mining is conducted using the EcoTILLING method in the detection of natural variations in the gene of interest in many plant individuals, similar to the conventional TILLING technology (Simsek and Kacar 2010). A combination of forward and reverse genetic screening can be used to recover alleles that would have a drastic impact on the cost compared to whole genome sequencing methods (Perry et al. 2003; Hu et al. 2018).

16.3.9 Genome Editing of Plants

Specific DNA sequences can be targeted to edit genes of interest accurately in plant genomes. This strategy may provide opportunities to improve plants by imparting stress tolerance traits. Genome editing is performed by using engineered nuclease (GEEN) to specifically target and digest DNA of choice. The use of GEEN technique causes a double-strand break (DSB), which is repaired by homologous recombination or non-homologous end-joining processes at the target site. Alterations in cleaved sites occur that include insertions/deletions, leading to gene disruption due to non-homologous recombination or exogenous sequence integration in the case of homologous recombination. Generally, engineered nucleases used in plant genome editing are as follows:

i. Zinc finger nucleases (ZFNs)
ii. Transcription activator-like effector nucleases (TALENs)
iii. Clustered interspaced short palindromic repeats (CRISPR)/CRISPR-associated protein (Cas) 9

Binding pairs of ZFNs and TALENs provide specificity of targets where each one of the pairs is responsible for the recognition of forward and reverse DNA strands, respectively. ZFNs are developed from the fusion of zinc finger transcription factors and FokI, a restriction enzyme in bacteria. Multiple domains of zinc finger nuclease can be engineered to expand the binding of 9–18 unique nucleotide sequences nearby to the target sites. However, engineering these domains is complex and expensive. Low performance has also been reported in targeting low guanine-containing sequences. However, ZFNs have been used to engineer many crops successfully. Contrarily, the TALENs method employs variable di-residue repeats to generate TAL repeat arrays in order to target sequences that allows for rapid TALEN construction. In addition, it has only a few limitations compared to ZFNs. Still, there are a few disadvantages associated such as the requirement for a larger number of candidate pairs to screen as it needs a high level of activity. In modern-day functional genomics, CRISPR/cas9 is extensively used in plant genome targeting and for basic research. The CRISPR/Cas system initiated from the understanding of bacterial and Archaea immune systems that was applied to other higher organisms. For example, viral or plasmid DNA that is foreign is eliminated based on RNA-guided cleavage. The most commonly used genome editing system is the type II CRISPR system of *Streptococcus pyogenes*. In plants, cas9 and gRNA (a chimeric guide RNA) are co-expressed, leading to a double-strand break at a specific site. This site is indicated by about a 20 nucleotide-long target sequence, and the specified target site must be adjacent to a proximal protospacer motif. Hence, to modify the double-strand break location, a template DNA is required for DNA repair. The CRISPR/Cas system is capable of editing any genomic sequence effectively in comparison to ZFNs and TALENs. Also, it is less laborious and economical in engineering nucleases for the target sequence. However, breeding programs typically rely on the mutation panels or genetic diversity to advance valuable loci of interest into elite cultivars followed by tedious backcrossing. The advantage of CRISPR/Cas9 is that it can straightforwardly deliver mutations into elite germplasm. Usually, plants are hemizygous for transgenes; however, all alleles are affected in the case of genome editing, implying that the CRISPR/Cas9 system if transformed would allow the generation of transgene-free plants after crossing or selfing. This potentially lowers the risks and requirements that may apply to transgenic crops (Scheben et al. 2017). CRISPR technology is promising for crop breeding, and plant researchers are taking efforts to focus on genome editing of major economically important crops worldwide. The CRISPR/Cas9 system has been effectively used in several plants such as tobacco, *Arabidopsis*, maize and wheat (Shan et al. 2013; Ali et al. 2015; Svitashev et al. 2015; Ma et al. 2015). Consequently, improvements in technologies relevant to genome editing are expected to lead to the development of stress-tolerant plants in the future at a rapid pace.

16.3.10 Focus on Cis-Regulatory Elements (CREs)

Substantial improvements have been made in the understanding of CREs, including promoters and enhancers. CREs are known to affect gene expression levels in terms of specific developmental stages and tissues as they modulate the expression of the unmodified sequence. As compared to mutations in coding regions, CRE modifications are most preferred owing to the fact that adverse pleiotropic effects are absent (Swinnen, Goossens, and Pauwels 2016). CRE modification studies are advantageous when the objective is not to knock out a gene but alter gene expression. Candidate CREs can be predicted by identifying open chromatin using techniques such as DNAse I, ChIP-seq and ATAC-seq experiments. Though there is not enough information on plant cis-regulatory elements, extensive efforts are being taken to screen promoters and signature sequences of such elements by exercising various computational methods and accessing cis-regulatory databases available for plants, such as Plant Care (Lescot et al. 2002; Hu et al. 2018). Otherwise, the CRE mutant library can be prepared to understand the differential expression of mutant lines. The expression data are used to identify CRE–trait associations, that are further investigated to assess variation in a target trait via genome editing (Meng et al. 2017).

16.4 BIOINFORMATICS IN CROP STRESS GENOMICS

In view of all the above, it is apparent that effective software tools and their combined use are imperative in evaluating and selecting elite plants. The deployment of pipelines in the context of next-generation sequencing for genomics-based approaches has opened new doors to intensify the development of next-generation crops, tolerant to stress. Genomics-assisted breeding involves the use of analytical and decision support tools in a sequential manner (Yunbi Xu 2010). Table 16.1 lists supportive tools commonly used in genomics-assisted breeding, and their respective utilities are explored below. Varshney et al. (2016) discussed fitting software tools and pipelines for modern molecular breeding to analyze large-scale phenotypic and genotypic data management and approaches in genomics-assisted breeding. The genetic diversity analysis relies on marker type, marker and genotype availability and heterozygosity proportion. Such analysis helps to extract information useful for the selection of contrasting lines. NTSYSSpc, MEGA and DARwin are widely used to estimate diversity, multivariate analysis and construction of phylogenetic trees from molecular data (Rohlf 1988; Tamura et al. 2011; Perrier and Jacquemoud-Collet. 2006). Likewise, population genetic analysis renders allele frequency estimates valuable to breeders. Arelequin, DNA Sequence Polymorphism (DnaSP) and GenAIEx offer a wide range of options for population genetic analysis, and numerous other tools have been developed recently such as Power Marker, specifically for SSR/SNP datasets (Excoffier, Laval, and Schneider 2007; Librado

TABLE 16.1
Software Tools Deployed in Genomics-Assisted Breeding

Method	Description	Software Tools
Genetic diversity analysis	Calculate genetic diversity estimates, evolutionary distances and construct phylogenetic trees	NTSYSpc, MEGA, DARwin, SNPphylo
Population genetic analysis	Perform equilibrium analysis, genetic diversity distance	Arlequin, DnaSP, GenAIEx, Power Maker
Linkage-mapping analysis	Small-scale experiments, <500 markers	MAPMAKER, MapDraw, JoinMap, Recombination Counting and Ordering (Record), SimpleMap
	50,000–100,000 markers	MSTMap, SEG-Map
	To develop consensus map	JoinMap, LPmerge
QTL-mapping analysis	To examine marker-trait associations and map relevant genetic loci in bi-parental/multi-parental populations	QGene, Mapmanager QTX, Mapmaker/QTL, WinQTL Cartographer, MapQTL, PLABQTL,R/qtl,R/ricalc,R/mpMap, R/mpwgaim, IciMapping
	To perform meta-QTL analysis	MetaQTL, BioMercator
Genome-wide association studies (GWAS)	To identify population genetic structure and for performing association analysis	STRUCTURE, EIGENSOFT, Bayesian analysis of population structure (BAPS), SNP analyser 2.0, Trait Analysis by aSSociation, Evolution, and Linkage (TASSEL), GenABEL, PLINK
Molecular breeding to develop stress tolerant genotypes	Selection of plants for backcrossing/ advancement	Graphical Genotypes (GGT), Flapjack, Molecular breeding design tool (MBDT), OptiMAS, solGS, ISMU 2.0

and Rozas 2009; Peakall and Smouse 2012; Liu and Muse 2005). Usually, population type and size, nature of the marker and the number of markers available are key parameters for the construction of genetic maps. Several software tools and algorithms are deployed based on marker density for the efficient generation of genetic maps such as MapDraw and JoinMap which are widely used for experiments with fewer markers; however, to develop genetic maps for large-scale genotype data, MSTMap, a mapping program, is employed (Liu and Meng. 2003; Van Ooijen 2018). The JoinMap program creates consensus maps and has been developed in multiple crops; another alternative is LPmerge which can solve the ordering of marker problems in constructing consensus maps (Endelman and Plomion 2014).

QTL analysis tools are mostly classified based on the type of mapping population and approach that need to be applied for QTL mapping. MapManager QTX, Qgene and mapMaker/QTL are software tools for single marker analysis (SMA) and PLABQTL, WinQTL cartographer is adopted for composite interval mapping respectively. Even though the majority of QTL mapping tools are available for the bi-parental population, tools have been developed to analyze marker trait associations of multi-parent populations in recent years. To map regions in multi-parent mapping populations, tools such as NAM/MAGIC, IciMapping and several R-based packages like R/mpMap, and R/qtl are available (Huihui Li et al. 2008; Broman et al. 2003; B. E. Huang and George 2011). Also, plant researchers are interested in the meta QTL study that is performed to comprehend genetic loci determining complex traits by integrating many QTLs (Wu and Hu 2012). In this context, MetaQTL and BioMercator are appropriate software packages to conduct a meta-analysis of QTLs (Veyrieras, Goffinet, and Charcosset 2007; Sosnowski, Charcosset, and Joets 2012). Genome-wide association studies demand software tools in order to conceive the distribution of linkage disequilibrium and population structure. STRUCTURE is the most popular program for inferring population genetic structure; however, EIGENSOFT and Bayesian Analysis of Population Structure (BAPS) are other important tools in GWAS (Pritchard, Stephens, and Donnelly 2000; Price et al. 2006; Corander and Martinnen. 2006). Consequently, association mapping is performed using Trait Analysis by aSSociation, Evolution, and Linkage (TASSEL) and PLINK which consist of many features and statistical approaches (Zhang et al. 2010; Purcell et al. 2007).

Marker-assisted selection/marker-assisted backcross selection has been employed in genomics-assisted breeding programs. To select plants that have attained maximum genome recovery of the recurrent parent, visualization tools such as graphical genotypes (GGT), CCSL Finder and the molecular breeding design tool (MBDT) are available (van Berloo 2008; Lorieux 2012). With respect to genomic selection, solGS, a web-based tool calculates GEBVs using a random regression best linear unbiased predictor (RR-BLUP) model (Tecle et al. 2014). Other than RR-BLUP, BayesA, BayesB and BayesCp are other frequently used models in computational tools for genomic selection. To help breeders to administer, supervise and operate everyday activities in breeding, BMS Workbench was developed. It provides tools required in breeding for the evaluation and management of germplasm, data analysis with decision support (Varshney et al. 2016). Massive data generated during the adoption of genomics-assisted strategies need to be consistently shared with different researchers to enable progress and reproducibility in a systematic procedure.

Several array platforms are utilized by plant researchers to produce expression profiles on a genome-wide scale in various crops that assist in functional genomics. For instance, microarray data on *Arabidopsis* is publicly available in ArrayExpress, NCBI Gene Expression Omnibus (GEO) and The Arabidopsis Information Resource (TAIR). Many of the above resources contain experimental data for plant species and platforms, consist of tools to perform clustering and facilitate data submission. Several databases such as PLEXdb in association with community databases provide comparative gene expression profiles in multiple plants (Dash et al. 2012). Likewise, RNA-seq experiments are submitted, stored and can be retrieved from NCBI SRA (Leinonen, Sugawara, and Shumway 2011). Unlike other transcriptomics approach, RNA-seq data integration necessitates methods to compare datasets from multiple experiments, plant species and sequencing technologies. To ensure this, Expression Atlas was established as a database of gene expression datasets under various biological conditions (Papatheodorou et al. 2018). Recently, AgriSeqDB was developed as a user-friendly RNA-seq dataset repository for major agricultural crops (Robinson et al. 2018).

Genome editing methods such as the CRISPR/Cas system do not necessitate subsequent time-consuming rounds of breeding to eliminate deleterious background mutations. Since the CRISPR/Cas system depends on guide RNA to recruit Cas protein to specific DNA binding sites, bioinformatics tools are requisite for ideal guide RNA design. Guide RNA is designed based on its binding affinity and specificity. CRISPR-P and CRISPR-Plant are plant-specific guide RNA design tools, but several features are still non-existent because guide DNA, chromatin, Cas proteins and guide RNA interactions vary in plants (Hu et al. 2018).

Many plant genome databases have become essential resource for plant researchers worldwide. It permits the extraction of diverse genetic information on gene families, regulatory elements, intronic and exonic sequences and molecular markers associated with genetic variability that serves as a significant resource for the identification of stress-responsive genes. A summary of available plant stress-dedicated database resources is provided in Table 16.2. In spite of major biological data repositories such as GenBank, Phytozome and PlantGDB, there is an urge to focus on specific databases for applied plant breeding. There is a requirement to integrate data on phenotypes and genetic variants produced by genomics and phenomics approaches in targeted environments. GrainGenes is an excellent example of such a database that consists of marker and expression data in an integrated fashion. The development of an integrative crop database is an arduous task as appropriate data are scattered in various databases in different formats and varied quality (Hu et al. 2018). KnetMiner is a data mining tool that scans for links and concepts in networks based on biological knowledge as it allows for the tracking of novel relationships between genes and traits (Hassani-Pak et al. 2016). Advancement in data mining techniques will assist breeders in better dissection of plant stress-related complex traits and associated genes.

TABLE 16.2
List of Dedicated Databases for Plant Stress Research

Database Name	URL
Plant Stress Gene Database	http://ccbb.jnu.ac.in/stressgenes/
Arabidopsis Thaliana Stress-Responsive Gene Database	http://srgdb.bicpu.edu.in/
Plant Stress Proteins Database	http://bioclues.org/pspdb/
DroughtDB: An expert-curated compilation of plant drought stress genes and their homologs in nine species	http://pgsb.helmholtz-muenchen.de/droughtdb/
STIFDB V2.0 [Stress Responsive Transcription Factor Database]	http://caps.ncbs.res.in/stifdb2/
Rice Stress-Responsive Transcription Factor Database	www.nipgr.res.in/RiceSRTFDB.html
RiceMetaSys	http://14.139.229.201/Ricemetasys/about.php
Stress2TF	http://csgenomics.ahau.edu.cn/Stress2TF/f
PlantPReS	www.proteome.ir/Default.aspx
eHALOPH – Halophytes database	www.sussex.ac.uk/affiliates/halophytes/index.php?content=plantList
QlicRice	http://cabgrid.res.in/nabg/qlicrice.html
PRGdb	http://prgdb.org/prgdb/
SolRgene	www.plantbreeding.wur.nl/SolRgenes/
Arabidopsis Resistance Genes Database	http://niblrrs.ucdavis.edu/At_RGenes/
PhytoPath	www.phytopathdb.org/
PathoPlant	www.pathoplant.de/

16.5 FUTURE PERSPECTIVES

Genomics approaches have been employed to enhance understanding of the stress-responsive biological processes that affect plants under adverse conditions. The mining and identification of genes involved in stress serve as a crucial step to manipulate the stress response in plants. Genomics technologies provide opportunities to conduct research in the understanding of stress tolerance mechanisms by providing sufficient tools to pacify the effects of abiotic and biotic stress. Eventually, the future of crop genomics will be dominated by sequencing that may replace other genotyping platforms as sequencing cost has reduced. Data overabundance in various formats from distinct genomics technologies also presents substantial problems. In such a scenario, data management and analysis are not sophisticated enough for geneticists and plant breeders with access to large-scale molecular markers and genomic tools in a systematic approach. Besides, plant genome databases and other databases specifically in plant stress genomics need to be revised and updated with the trend of growing big data and third-generation genome sequencing technologies. In the modern genomics era, it is necessary to integrate bioinformatics tools in genome sequencing, assembly, annotation, marker discovery, genomics-assisted breeding and genome editing technologies along with high-throughput phenotyping that can accelerate plant stress research and stress-tolerant plants, improving global food security.

REFERENCES

Akpinar, Bala Ani, Stuart J. Lucas, and Hikmet Budak. 2013. Genomics approaches for crop improvement against abiotic stress. *The Scientific World Journal* 2013: 361921. doi:10.1155/2013/361921.

Ali, Zahir, Aala Abul-faraj, Marek Piatek, and Magdy M Mahfouz. 2015. Activity and specificity of TRV-mediated gene editing in plants. *Plant Signaling and Behavior* 10 (10). Taylor & Francis: e1044191. doi: 10.1080/15592324.2015.1044191.

Alkan, Can, Saba Sajjadian, and Evan E. Eichler. 2011. Limitations of next-generation genome sequence assembly. *Nature Methods* 8 (1). Nature Publishing Group: 61–65. doi:10.1038/nmeth.1527.

Andrews, Kimberly R., Jeffrey M. Good, Michael R. Miller, Gordon Luikart, and Paul A. Hohenlohe. 2016. Harnessing the power of RADseq for ecological and evolutionary genomics. *Nature Reviews Genetics* 17 (2). Nature Publishing Group: 81–92. doi:10.1038/nrg.2015.28.

Anisimov, Sergey. 2008. Serial analysis of gene expression (SAGE): 13 years of application in research. *Current Pharmaceutical Biotechnology* 9 (5): 338–50. doi:10.2174/138920108785915148.

Arabidopsis Genome Initiative. 2000. Analysis of the genome sequence of the flowering plant *Arabidopsis thaliana*. *Nature* 408 (6814): 796–815. doi:10.1038/35048692.

Barkley, N A, and M. L. Wang. 2008. Application of TILLING and EcoTILLING as reverse genetic approaches to elucidate the function of genes in plants and animals. *Current Genomics* 9: 212–26. www.ncbi.nlm.nih.gov.

Berloo, Ralph van. 2008. GGT 2.0: Versatile software for visualization and analysis of genetic data. *Journal of Heredity* 99 (2). Oxford University Press: 232–36. doi:10.1093/jhered/esm109.

Bevan, Michael W., and Cristobal Uauy. 2013. Genomics reveals new landscapes for crop improvement. *Genome Biology* 14 (6). BioMed Central: 206. doi:10.1186/gb-2013-14-6-206.

Broman, K. W., H. Wu, S. Sen, and G. A. Churchill. 2003. R/Qtl: QTL mapping in experimental crosses. *Bioinformatics* 19 (7). Oxford University Press: 889–90. doi:10.1093/bioinformatics/btg112.

Chen, Haodong, Weibo Xie, Hang He, et al. 2014. A high-density SNP genotyping array for rice biology and molecular breeding. *Molecular Plant* 7 (3). Cell Press: 541–53. doi:10.1093/MP/SST135.

Coppens, Frederik, Nathalie Wuyts, Dirk Inzé, and Stijn Dhondt. 2017. Unlocking the potential of plant phenotyping data through integration and data-driven approaches. *Current Opinion in Systems Biology* 4 (August). Elsevier: 58–63. doi:10.1016/J.COISB.2017.07.002.

Corander, Jukka, and Pekka Marttinen. 2006. Bayesian identification of admixture events using multi-locus molecular markers. *Molecular Ecology* 15 (10). John Wiley & Sons, Ltd (10.1111): 2833–43. doi:10.1111/j.1365-294X.2006.02994.x.

D'Agostino, Nunzio, Pasquale Tripodi, Nunzio D'Agostino, and Pasquale Tripodi. 2017. NGS-based genotyping, high-throughput phenotyping and genome-wide association studies laid the foundations for next-generation breeding in horticultural crops. *Diversity* 9 (3). Multidisciplinary Digital Publishing Institute: 38. doi:10.3390/d9030038.

Dash, S., J. Van Hemert, L. Hong, R. P. Wise, and J. A. Dickerson. 2012. PLEXdb: Gene expression resources for plants and plant pathogens. *Nucleic Acids Research* 40 (D1). Oxford University Press: D1194–1201. doi:10.1093/nar/gkr938.

Denton, James F., Jose Lugo-Martinez, Abraham E. Tucker, Daniel R. Schrider, Wesley C. Warren, and Matthew W. Hahn. 2014. Extensive error in the number of genes inferred from draft genome assemblies. Edited by Roderic Guigo. *PLoS Computational Biology* 10 (12). Public Library of Science: e1003998. doi:10.1371/journal.pcbi.1003998.

Eisenstein, Michael. 2015. Startups use short-read data to expand long-read sequencing market. *Nature Biotechnology* 33 (5): 433–35. doi:10.1038/nbt0515-433.

Endelman, Jeffrey B., and Christophe Plomion. 2014. LPmerge: An R package for merging genetic maps by linear programming. *Bioinformatics* 30 (11). Oxford University Press: 1623–24. doi:10.1093/bioinformatics/btu091.

Excoffier, Laurent, Guillaume Laval, and Stefan Schneider. 2007. Arlequin (version 3.0): An integrated software package for population genetics data analysis. *Evolutionary Bioinformatics Online* 1 (February). SAGE Publications: 47–50. http://www.ncbi.nlm.nih.gov/pubmed/19325852.

Fu, Lixia, Chengcheng Cai, Yinan Cui, et al. 2016. Pooled mapping: An efficient method of calling variations for population samples with low-depth resequencing data. *Molecular Breeding* 36 (4): 48. doi:10.1007/s11032-016-0476-9.

Furbank, Robert T., and Mark Tester. 2011. Phenomics – Technologies to relieve the phenotyping bottleneck. *Trends in Plant Science* 16 (12). Elsevier Current Trends: 635–44. doi:10.1016/J.TPLANTS.2011.09.005.

Ganal, Martin W., Gregor Durstewitz, Andreas Polley, Aurélie Bérard, Edward S. Buckler, Alain Charcosset, Joseph D. Clarke, et al. 2011. A large maize (*Zea Mays* L.) SNP genotyping array: Development and germplasm genotyping, and genetic mapping to compare with the B73 reference genome. Edited by Lewis Lukens. *PLoS ONE* 6 (12). Public Library of Science: e28334. doi:10.1371/journal.pone.0028334.

Golicz, Agnieszka A., Philipp E. Bayer, and David Edwards. 2015. Skim-based genotyping by sequencing. In J. Batley, Ed., *Plant Genotyping*. Humana Press, New York, NY: 257–70. doi:10.1007/978-1-4939-1966-6_19.

Gur, Amit, and Dani Zamir. 2004. Unused natural variation can lift yield barriers in plant breeding. Edited by Jeffrey Dangl. *PLoS Biology* 2 (10). Public Library of Science: e245. doi:10.1371/journal.pbio.0020245.

Hassani-Pak, Keywan, Martin Castellote, M. Esch, et al. 2016. Developing integrated crop knowledge networks to advance candidate gene discovery. *Applied and Translational Genomics* 11 (December). Elsevier: 18–26. doi:10.1016/J.ATG.2016.10.003.

Hu, Haifei, Armin Scheben, David Edwards, Haifei Hu, Armin Scheben, and David Edwards. 2018. Advances in integrating genomics and bioinformatics in the plant breeding pipeline. *Agriculture* 8 (6). Multidisciplinary Digital Publishing Institute: 75. doi:10.3390/agriculture8060075.

Huang, B. Emma, and Andrew W. George. 2011. R/MpMap: A computational platform for the genetic analysis of multiparent recombinant inbred lines. *Bioinformatics* 27 (5). Oxford University Press: 727–29. doi:10.1093/bioinformatics/btq719.

Huang, B. Emma, Klara L. Verbyla, Arunas P. Verbyla, et al. 2015. MAGIC populations in crops: Current status and future prospects. *Theoretical and Applied Genetics* 128 (6). Springer, Berlin, Heidelberg: 999–1017. doi:10.1007/s00122-015-2506-0.

Huang, Xuehui, Qi Feng, Qian Qian, Qiang Zhao, Lu Wang, Ahong Wang, Jianping Guan, et al. 2009. High-throughput genotyping by whole-genome resequencing. *Genome Research* 19 (6). Cold Spring Harbor Laboratory Press: 1068–76. doi:10.1101/gr.089516.108.

Hwang, Eun-Young, Qijian Song, Gaofeng Jia, James E Specht, David L. Hyten, Jose Costa, and Perry B. Cregan. 2014. A genome-wide association study of seed protein and oil content in soybean. *BMC Genomics* 15 (1). BioMed Central: 1. doi:10.1186/1471-2164-15-1.

Ismail, Abdelbagi M., and Tomoaki Horie. 2017. Genomics, physiology, and molecular breeding approaches for improving salt tolerance. *Annual Review of Plant Biology* 68 (1): 405–34. doi:10.1146/annurev-arplant-042916-040936.

Jannink, Jean-Luc. 2007. Identifying quantitative trait locus by genetic background interactions in association studies. *Genetics* 176 (1). Genetics: 553–61. doi:10.1534/genetics.106.062992.

Jiang, Shu-Ye, and Srinivasan Ramachandran. 2010. Natural and artificial mutants as valuable resources for functional genomics and molecular breeding. *International Journal of Biological Sciences* 6 (3). Ivyspring International Publisher: 228–51. http://www.ncbi.nlm.nih.gov/pubmed/20440406.

Koenig, Daniel, José M. Jiménez-Gómez, Seisuke Kimura, Daniel Fulop, Daniel H. Chitwood, Lauren R. Headland, Ravi Kumar, et al. 2013. Comparative transcriptomics reveals patterns of selection in domesticated and wild tomato. *Proceedings of the National Academy of Sciences of the United States of America* 110 (28). National Academy of Sciences: E2655–62. doi:10.1073/pnas.1309606110.

Kurokawa, Yusuke, Tomonori Noda, Yoshiyuki Yamagata, Rosalyn Angeles-Shim, Hidehiko Sunohara, Kanako Uehara, Tomoyuki Furuta, et al. 2016. Construction of a versatile SNP array for pyramiding useful genes of rice. *Plant Science* 242 (January). Elsevier: 131–39. doi:10.1016/J.PLANTSCI.2015.09.008.

Ladejobi, Olufunmilayo, James Elderfield, Keith A. Gardner, R. Chris Gaynor, John Hickey, Julian M. Hibberd, Ian J. Mackay, and Alison R. Bentley. 2016. Maximizing the potential of multi-parental crop populations. *Applied and Translational Genomics* 11 (December). Elsevier: 9–17. doi:10.1016/J.ATG.2016.10.002.

Lee, Insuk. 2014. A showcase of future plant biology: Moving towards next-generation plant genetics assisted by genome sequencing and systems biology. *Genome Biology* 15 (5): 305. doi:10.1186/gb4176.

Leinonen, R., H. Sugawara, and M. Shumway. 2011. The sequence read archive. *Nucleic Acids Research* 39 (Database). Oxford University Press: D19–21. doi:10.1093/nar/gkq1019.

Lescot, Magali, Patrice Déhais, Gert Thijs, Kathleen Marchal, Yves Moreau, Yves Van de Peer, Pierre Rouzé, and Stephane Rombauts. 2002. PlantCARE, a database of plant cis-acting regulatory elements and a portal to tools for in silico analysis of promoter sequences. *Nucleic Acids Research* 30 (1): 325–27. http://www.ncbi.nlm.nih.gov/pubmed/11752327.

Lewis, Cathryn M. 2002. Genetic association studies: Design, analysis and interpretation. *Briefings in Bioinformatics* 3 (2): 146–53. http://www.ncbi.nlm.nih.gov/pubmed/12139434.

Li, Hui, Zhiyu Peng, Xiaohong Yang, Weidong Wang, Junjie Fu, Jianhua Wang, Yingjia Han, et al. 2013. Genome-wide association study dissects the genetic architecture of oil biosynthesis in maize kernels. *Nature Genetics* 45 (1): 43–50. doi:10.1038/ng.2484.

Li, Huihui, Jean-Marcel Ribaut, Zhonglai Li, and Jiankang Wang. 2008. Inclusive composite interval mapping (ICIM) for digenic epistasis of quantitative traits in biparental populations. *Theoretical and Applied Genetics* 116 (2). Springer-Verlag: 243–60. doi:10.1007/s00122-007-0663-5.

Li, Jia-Yang, Jun Wang, and Robert S. Zeigler. 2014. The 3,000 rice genomes project: New opportunities and challenges for future rice research. *GigaScience* 3 (1). Oxford University Press: 8. doi:10.1186/2047-217X-3-8.

Librado, P., and J. Rozas. 2009. DnaSP v5: A software for comprehensive analysis of DNA polymorphism data. *Bioinformatics* 25 (11). Oxford University Press: 1451–52. doi:10.1093/bioinformatics/btp187.

Liu, K., and S. V. Muse. 2005. PowerMarker: An integrated analysis environment for genetic marker analysis. *Bioinformatics* 21 (9). Oxford University Press: 2128–29. doi:10.1093/bioinformatics/bti282.

Liu, R. H. and J. L. Meng. 2003. MapDraw: A microsoft excel macro for drawing genetic linkage maps based on given genetic linkage data. *Yi chuan= Hereditas* 25 (3): 317–321.

Lorieux, Mathias. 2012. MapDisto: Fast and efficient computation of genetic linkage maps. *Molecular Breeding* 30 (2). Springer Netherlands: 1231–35. doi:10.1007/s11032-012-9706-y.

Ma, Xingliang, Qunyu Zhang, Qinlong Zhu, Wei Liu, Yan Chen, Rong Qiu, Bin Wang, et al. 2015. A robust CRISPR/Cas9 system for convenient, high-efficiency multiplex genome editing in monocot and dicot plants. *Molecular Plant* 8 (8): 1274–84. doi:10.1016/j.molp.2015.04.007.

Mammadov, Jafar, Rajat Aggarwal, Ramesh Buyyarapu, and Siva Kumpatla. 2012. SNP markers and their impact on plant breeding. *International Journal of Plant Genomics* 2012 (December). Hindawi: 728398. doi:10.1155/2012/728398.

McCouch, Susan, Gregory J. Baute, James Bradeen, Paula Bramel, Peter K. Bretting, Edward Buckler, John M. Burke, et al. 2013. Feeding the future. *Nature* 499 (7456): 23–24. doi:10.1038/499023a.

McMullen, Michael D., Stephen Kresovich, Hector Sanchez Villeda, Peter Bradbury, Huihui Li, Qi Sun, Sherry Flint-Garcia, et al. 2009. Genetic properties of the maize nested association mapping population. *Science (New York, N.Y.)* 325 (5941). American Association for the Advancement of Science: 737–40. doi:10.1126/science.1174320.

Meng, Xiangbing, Hong Yu, Yi Zhang, et al. 2017. Construction of a genome-wide mutant library in rice using CRISPR/Cas9. *Molecular Plant* 10 (9). Elsevier: 1238–41. doi:10.1016/j.molp.2017.06.006.

Meuwissen, T. H., B. J. Hayes, and M. E. Goddard. 2001. Prediction of total genetic value using genome-wide dense marker maps. *Genetics* 157 (4): 1819–29. http://www.ncbi.nlm.nih.gov/pubmed/11290733.

Ong, Quang, Phuc Nguyen, Nguyen Phuong Thao, and Ly Le. 2016. Bioinformatics approach in plant genomic research. *Current Genomics* 17 (4). Bentham Science Publishers: 368–78. doi:10.2174/1389202917666160331202956.

Pandey, Manish K., Manish Roorkiwal, Vikas K. Singh, et al. 2016. Emerging genomic tools for legume breeding: Current status and future prospects. *Frontiers in Plant Science* 7 (May). Frontiers: 455. doi:10.3389/fpls.2016.00455.

Papatheodorou, Irene, Nuno A. Fonseca, M. Keays, et al. 2018. Expression atlas: Gene and protein expression across multiple studies and organisms. *Nucleic Acids Research* 46 (D1). Oxford University Press: D246–51. doi:10.1093/nar/gkx1158.

Peakall, R., and P. E. Smouse. 2012. GenAlEx 6.5: Genetic analysis in excel. Population genetic software for teaching and research--an update. *Bioinformatics* 28 (19). Oxford University Press: 2537–39. doi:10.1093/bioinformatics/bts460.

Perez-Sanz, Fernando, Pedro J. Navarro, and Marcos Egea-Cortines. 2017. Plant phenomics: An overview of image acquisition technologies and image data analysis algorithms. *Giga Science* 6 (11). Oxford University Press: 1–18. doi:10.1093/gigascience/gix092.

Perrier, X. and Jacquemoud-Collet, J. P. 2006. DARwin software: Dissimilarity analysis and representation for windows. Website: http://darwin.cirad.fr/darwin [accessed 1 March 2013].

Perry, Jillian A., Trevor L. Wang, Tracey J. Welham, Sarah Gardner, Jodie M. Pike, Satoko Yoshida, and Martin Parniske. 2003. A TILLING reverse genetics tool and a web-accessible collection of mutants of the legume *Lotus Japonicus*. *Plant Physiology* 131 (3). American Society of Plant Biologists: 866–71. doi:10.1104/pp.102.017384.

Poland, Jesse A., Patrick J. Brown, Mark E. Sorrells, and Jean-Luc Jannink. 2012. Development of high-density genetic maps for barley and wheat using a novel two-enzyme genotyping-by-sequencing approach. Edited by Tongming Yin. *PLoS ONE* 7 (2). Public Library of Science: e32253. doi:10.1371/journal.pone.0032253.

Price, Alkes L., Nick J. Patterson, Robert M. Plenge, Michael E. Weinblatt, Nancy A. Shadick, and David Reich. 2006. Principal components analysis corrects for stratification in genome-wide association studies. *Nature Genetics* 38 (8). Nature Publishing Group: 904–9. doi:10.1038/ng1847.

Pritchard, Jonathan K., Matthew Stephens, and Peter Donnelly. 2000. Inference of population structure using multilocus genotype data. *Genetics* 155 (2). Genetics Society of America: 945–59.

Purcell, Shaun, Benjamin Neale, Kathe Todd-Brown, Lori Thomas, Manuel A. R. Ferreira, David Bender, Julian Maller, et al. 2007. PLINK: A tool set for whole-genome association and population-based linkage analyses. *The American Journal of Human Genetics* 81 (3). Cell Press: 559–75. doi:10.1086/519795.

Rasheed, Awais, Yuanfeng Hao, Xianchun Xia, Awais Khan, Yunbi Xu, Rajeev K. Varshney, and Zhonghu He. 2017. Crop breeding chips and genotyping platforms: Progress, challenges, and perspectives. *Molecular Plant* 10 (8): 1047–64. doi:10.1016/j.molp.2017.06.008.

Reinartz, Jeannette, Eddy Bruyns, Jing-Zhong Lin, Tim Burcham, Sydney Brenner, Ben Bowen, Michael Kramer, and Rick Woychik. 2002. Massively parallel signature sequencing (MPSS) as a tool for in-depth quantitative gene expression profiling in all organisms. *Briefings in Functional Genomics* 1 (1). Oxford University Press: 95–104. doi:10.1093/bfgp/1.1.95.

Rhoads, Anthony, and Kin Fai Au. 2015. PacBio sequencing and its applications. *Genomics, Proteomics and Bioinformatics* 13 (5). Elsevier: 278–89. doi:10.1016/J.GPB.2015.08.002.

Robinson, Andrew J., Muluneh Tamiru, Rachel Salby, et al. 2018. AgriSeqDB: An online RNA-seq database for functional studies of agriculturally relevant plant species. *BMC Plant Biology* 18 (1). BioMed Central: 200. doi:10.1186/s12870-018-1406-2.

Rohlf, F. J. 1988. *NTSYS-pc: Numerical Taxonomy and Multivariate Analysis System*. Exeter Publishing, Setauket, NY.

Scheben, Armin, and David Edwards. 2017. Genome editors take on crops. *Science* 355 (6330): 1122–23. doi:10.1126/science.aal4680.

Scheben, Armin, Felix Wolter, Jacqueline Batley, Holger Puchta, and David Edwards. 2017. Towards CRISPR/Cas crops - bringing together genomics and genome editing. *New Phytologist* 216 (3). John Wiley & Sons, Ltd (10.1111): 682–98. doi:10.1111/nph.14702.

Scheben, Armin, Yuxuan Yuan, and David Edwards. 2016. Advances in genomics for adapting crops to climate change. *Current Plant Biology* 6. Elsevier B.V.: 2–10. doi:10.1016/j.cpb.2016.09.001.

Sedlazeck, Fritz J., Hayan Lee, Charlotte A. Darby, and Michael C. Schatz. 2018. Piercing the dark matter: Bioinformatics of long-range sequencing and mapping. *Nature Reviews Genetics* 19 (6). Nature Publishing Group: 329–46. doi:10.1038/s41576-018-0003-4.

Shan, Qiwei, Yanpeng Wang, Jun Li, Yi Zhang, Kunling Chen, Zhen Liang, Kang Zhang, et al. 2013. Targeted genome modification of crop plants using a CRISPR-Cas system. *Nature Biotechnology* 31 (8): 686–88. doi:10.1038/nbt.2650.

Shelton, Jennifer M., Michelle C. Coleman, Nic Herndon, et al. 2015. Tools and pipelines for BioNano data: Molecule assembly pipeline and FASTA super scaffolding tool. *BMC Genomics* 16 (1). BioMed Central: 734. doi:10.1186/s12864-015-1911-8.

Simsek, Ozhan, and Yildiz Aka Kacar. 2010. Discovery of mutations with TILLING and ECOTILLING in plant genomes. *Scientific Research and Essays* 5 (24): 3799–3802. http://www.academicjournals.org/SRE.

Sosnowski, O., A. Charcosset, and J. Joets. 2012. BioMercator V3: An upgrade of genetic map compilation and quantitative trait loci meta-analysis algorithms. *Bioinformatics* 28 (15). Oxford University Press: 2082–3. doi:10.1093/bioinformatics/bts313.

Svitashev, Sergei, Joshua K. Young, C. Schwartz, et al. 2015. Targeted mutagenesis, precise gene editing, and site-specific gene insertion in maize using Cas9 and guide RNA. *Plant Physiology* 169 (2). American Society of Plant Biologists: 931–45. doi:10.1104/PP.15.00793.

Swinnen, Gwen, Alain Goossens, and Laurens Pauwels. 2016. Lessons from domestication: Targeting cis-regulatory elements for crop improvement. *Trends in Plant Science* 21 (6). Elsevier Current Trends: 506–15. doi:10.1016/J.TPLANTS.2016.01.014.

Tamura, K., D. Peterson, N. Peterson, G. Stecher, M. Nei, and S. Kumar. 2011. MEGA5: Molecular evolutionary genetics analysis using maximum likelihood, evolutionary distance, and maximum parsimony methods. *Molecular Biology and Evolution* 28 (10). Oxford University Press: 2731–9. doi:10.1093/molbev/msr121.

Tecle, Isaak Y., Jeremy D. Edwards, Naama Menda, Chiedozie Egesi, Ismail Y. Rabbi, Peter Kulakow, Robert Kawuki, Jean-Luc Jannink, and Lukas A. Mueller. 2014. SolGS: A web-based tool for genomic selection. *BMC Bioinformatics* 15 (1). BioMed Central: 398. doi:10.1186/s12859-014-0398-7.

Thudi, M., Y. Li, S. A. Jackson, G. D. May, and R. K. Varshney. 2012. Current state-of-art of sequencing technologies for plant genomics research. *Briefings in Functional Genomics* 11 (1). Oxford University Press: 3–11. doi:10.1093/bfgp/elr045.

Till, Bradley J., Troy Zerr, Luca Comai, and Steven Henikoff. 2006. A protocol for TILLING and ecotilling in plants and animals. *Nature Protocols* 1 (5). Nature Publishing Group: 2465–77. doi:10.1038/nprot.2006.329.

Van Ooijen, J. W. 2018. *JoinMap ® 5, Software for the Calculation of Genetic Linkage Maps in Experimental Populations of Diploid Species*. Kyazma B.V., Wageningen, Netherlands.

Varshney, Rajeev K., Jean-Marcel Ribaut, Edward S. Buckler, et al. 2012. Can genomics boost productivity of orphan crops? *Nature Biotechnology* 30 (12): 1172–6. doi:10.1038/nbt.2440.

Varshney, Rajeev K., Vikas K. Singh, J. M. Hickey, et al. 2016. Analytical and decision support tools for genomics-assisted breeding. *Trends in Plant Science* 21 (4). Elsevier Current Trends: 354–63. doi:10.1016/j.tplants.2015.10.018.

Varshney, Rajeev K, Vikas K. Singh, Arvind Kumar, Wayne Powell, and Mark E. Sorrells. 2018. Can genomics deliver climate-change ready crops? *Current Opinion in Plant Biology* 45 (October). Elsevier Current Trends: 205–11. doi:10.1016/J.PBI.2018.03.007.

Veyrieras, Jean-Baptiste, Bruno Goffinet, and Alain Charcosset. 2007. MetaQTL: A package of new computational methods for the meta-analysis of QTL mapping experiments. *BMC Bioinformatics* 8 (1). BioMed Central: 49. doi:10.1186/1471-2105-8-49.

Winfield, Mark O., Alexandra M. Allen, Amanda J. Burridge, et al. 2016. High-density SNP genotyping array for hexaploid wheat and its secondary and tertiary gene pool. *Plant Biotechnology Journal* 14 (5): 1195–1206. doi:10.1111/pbi.12485.

Wu, X.L. and Hu, Z.L., 2012. Meta-analysis of QTL mapping experiments. In Quantitative Trait Loci (QTL). Humana Press:145-171.

Xu, Yang, Pengcheng Li, Zefeng Yang, and Chenwu Xu. 2017. Genetic mapping of quantitative trait loci in crops. *The Crop Journal* 5 (2). Elsevier: 175–84. doi:10.1016/J.CJ.2016.06.003.

Xu, Yunbi. 2010. *Molecular Plant Breeding*. CABI, Wallingford, UK.

Yang, Chuanping, and Hairong Wei. 2015. Designing microarray and RNA-seq experiments for greater systems biology discovery in modern plant genomics. *Molecular Plant* 8 (2): 196–206. doi:10.1016/j.molp.2014.11.012.

Yuan, Yuxuan, Philipp E. Bayer, Jacqueline Batley, and David Edwards. 2017. Improvements in genomic technologies: Application to crop genomics. *Trends in Biotechnology* 35 (6). Elsevier Ltd: 547–58. doi:10.1016/j.tibtech.2017.02.009.

Zhang, Zhiwu, Elhan Ersoz, Chao-Qiang Lai, et al. 2010. Mixed linear model approach adapted for genome-wide association studies. *Nature Genetics* 42 (4). Nature Publishing Group: 355–60. doi:10.1038/ng.546.

Zhao, Keyan, María José Aranzana, Sung Kim, et al. 2007. An *Arabidopsis* example of association mapping in structured samples. *PLoS Genetics* 3 (1). Public Library of Science: e4. doi:10.1371/journal.pgen.0030004.

Zhao, Keyan, Chih-Wei Tung, Georgia C. Eizenga, et al. 2011. Genome-wide association mapping reveals a rich genetic architecture of complex traits in *Oryza sativa*. *Nature Communications* 2 (1). Nature Publishing Group: 467. doi:10.1038/ncomms1467.

17 High-Throughput Genomics

Application in Plant Breeding under Abiotic Stress Conditions

Aditya Banerjee and Aryadeep Roychoudhury

CONTENTS

17.1 INTRODUCTION

Abiotic stresses like salinity, drought, temperature, light, heavy metal toxicity, etc., are altogether major agricultural constraints across the world (Banerjee et al. 2018, 2019; Banerjee and Roychoudhury 2018a, b, c, d, e). In spite of exhaustive research in this field, a concise blueprint regarding the genetic regulation of abiotic stress signaling in plants has not yet been fully explored. Among the identified molecular components, the majority have been used to genetically engineer stress tolerance in model or crop plants. This has led to increased yield and stress adaptation in specific cases but with less precision (Banerjee and Roychoudhury 2018f; Banerjee and Roychoudhury 2019a, b). Hence the use of functional genomics is popularly advertised to ensure high-quality precision in genomic modification. The availability of genome databases of various model and crop plants along with efficient genome-editing tools has opened up the opportunity to incorporate targeted manipulations in the genome and record the functional aspects under various abiotic stresses (Jain 2015).

17.2 TOOLS FOR STUDYING FUNCTIONAL GENOME EDITING IN PLANTS

The availability of genome-editing tools has facilitated targeted mutation, insertion and deletion (indel) with high precision (Voytas 2013). These tools mainly include the clustered regularly interspaced short palindromic repeat (CRISPR) – CRISPR-associated nuclease 9 (Cas9), zinc finger nucleases (ZFNs) and transcriptional activator-like effector nucleases (TALENs) (Mahfouz et al. 2014; Kumar and Jain 2015). These nucleases are engineered to be sequence-specific, and they introduce double strand breaks (DSBs) at the concerned locus. The type of mutation in the targeted locus depends on the type of repair mechanism adapted by the cell to reverse the DSB. Repair by non-homologous end joining (NHEJ) introduces indel, whereas the use of homology-directed repair (HDR) results in point mutations or the incorporation of desired sequence tags through recombination events (Jain 2015).

17.3 POTENTIAL USE OF GENOME EDITING IN GENERATING ABIOTIC STRESS TOLERANCE

Genome-editing technology has been rarely used in developing stress-tolerant crops. In a recent instance, the novel allele *open stomata 2/H$^+$-ATPase 1* (*Ost2/Aha1*) has been created by optimized CRISPR/Cas9 approaches. The modified *Arabidopsis* plants exhibited altered responses to abiotic stresses (Osakabe and Osakabe 2017). In recent years, quality web- based programs have been developed to facilitate single guide RNA (sgRNA) designing for the specific targeting of genes and loci (Montague et al. 2014). The sgRNA guides the Cas9 nuclease to recognize target DNA sequences using Watson–Crick base pairing and complementarity (Kumar and Jain 2015). Different versions of Cas9 nucleases can be used to carry out variable genome editing in plants. The catalytically inactive Cas9 protein (dCas9) inhibits gene function by facilitating CRISPR interference (CRISPRi) (Qi et al. 2013). Jain (2015) proposed that the fusion of Cas9 to transcriptionally active domains of VIVIPAROUS 16 (VP16)/VP64 could lead to gain-of-function phenotypes which could then be screened for abiotic stress tolerance. Osakabe et al. (2010) designed a zinc finger nuclease fused to a heat shock promoter. This construct was used to mutate the abscisic acid (ABA)-insensitive 4 (ABI4) transcription factor, which also regulates desiccation stress responses and ABA-mediated signaling in *Arabidopsis*. Jain (2015) reviewed some reports where genome editing has been used to generate homozygous transgenic plants in the first generation itself. This will reduce the enormous time lag required for breeding or genetic transformation. CRISPR technology is often highlighted as an advanced plant breeding technology, since the nucleases can also be delivered within the plant nucleus via non-transgenic approaches (Marton et al. 2010). Hence, plants produced by this technology can qualify as non-genetically modified crops and thus aid in the scientific culmination of agricultural breeding and genomics.

17.4 NEXT-GENERATION SEQUENCING FOR SCREENING THE PLANT GENOME

Next-generation sequencing (NGS) techniques have enabled the cost-effective sequencing of large sample numbers. This technology could overcome the limitations of Sanger-based methods which were unsuitable for sequencing large number of samples and massive parallel signature sequencing (MPSS). The presence of repetitive elements due to the frequently segmented or tandem duplicated transposable elements (TEs) increases the complexity of the plant genome (Ray and Satya 2014). NGS-based whole genome sequencing has enabled the identification of single nucleotide polymorphism (SNP) in place of fragment-based polymorphism within a short time frame. This sophisticated method involves library construction prior to sequencing. Such libraries can also be partial representations of genomes if complete sequencing data is unavailable. The partially represented libraries might be complexity-reduced representation libraries formed by the use of restriction enzymes or can be sequence capture libraries lacking restriction enzyme use (Gore et al. 2009). Ray and Satya (2014) reviewed that the first group can be used for complexity reduction of polymorphic sequences, restriction-site associated DNA sequencing (RAD-seq), sequence-based polymorphic marker technology, multiplexed shotgun genotyping and genotyping-by-sequencing (GBS). The second group can be utilized for technologies involving molecular inversion probe, solution hybrid selection, microarray-based genomic selection, exome sequencing and genome region sequencing linked to a specific trait (Porreca et al. 2007; Gnirke et al. 2009; Albert et al. 2007; Teer et al. 2010). A general outline of NGS-assisted plant breeding has been illustrated in Figure 17.1.

The availability of a genomic sequence at the online public platforms enables easy development of genic molecular or functional markers. Yang et al. (2012) and Glaubitz et al. (2014) have discussed the importance of RAD-seq and GBS for efficient next-generation plant breeding. In RAD-seq, the genomic DNA is digested with a particular restriction enzyme and then ligated with barcoded adaptors with molecular identifiers. Next, the DNA from multiple plants is pooled and

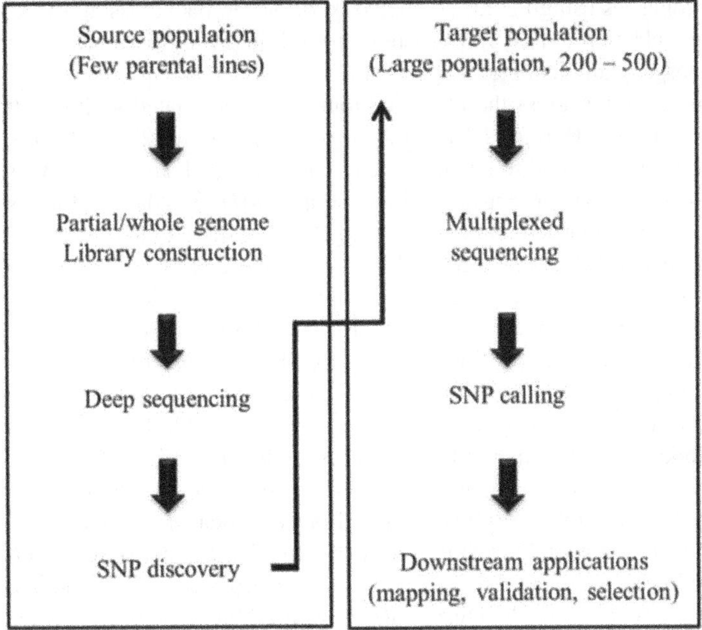

FIGURE 17.1 A generalized pipeline of NGS-assisted plant breeding [extracted from Ray and Satya (2014)].

subjected to random shear so that only a sub-set of the fragments contain the barcoded adapter. For effective PCR, another set of adapters is then ligated to the generated DNA fragments. The fragments containing both of the adapters are PCR amplified and sequenced via Illumina platforms, and the SNPs of individual plants are decoded by *in silico* analyses. This technology does not require any knowledge of the genomic sequence and has been utilized to identify QTLs for increasing the anthocyanin content in the fruit of eggplant (Barchi et al. 2012).

The constructed GBS map of wheat consists of 416,856 markers, thus indicating the robustness of the technology (Saintenac et al. 2013). Out of these, 20,000 are SNPs. About 34,000 SNPs have been identified in barley via this approach (Poland et al. 2012). GBS is performed following a modified RAD-seq protocol where the second adapter is not divergent in nature, thus allowing the generation of amplicons flanked by either of the adapter sequences. This highly multiplexed technique can be used for marker discovery and genotyping (Glaubitz et al. 2014). The ploidy character of the genome is also a challenge for sequencing projects. NGS has resolved these problems and can be used as a tool for allele mining during abiotic stress cues (Zhang et al. 2016). This field is rather young and needs exhaustive exploration.

17.5 USE OF QUANTITATIVE TRAIT LOCI (QTL) MAPPING AND GENOMICS-ASSISTED BREEDING (GAB) IN GENERATING ABIOTIC STRESS TOLERANCE

Genes and QTLs related to plant tolerance to abiotic stress are valuable resources for improving crop phenotype. Wang et al. (2016) reported that a natural variant of *vacuolar* H^+*-translocating inorganic pyrophosphatase 1* (*VPP1*) was involved in generating drought tolerance in maize. In another report, it was observed that the QTL, *COLD1*, was associated with chilling tolerance in japonica rice (Ma et al. 2015). The QTLs like *grain width 2* (*GW2*), *GW5*, *GW8*, *grain size 2* (*GS2*), *GS3* and *GS5* can be targeted for generating elite cultivars with high yield (Leng et al. 2017).

GAB uses advanced marker-assisted breeding for ensuring genome-wide genetic selection and high-density genotyping. This increases the probability of generating elite varieties with efficient stress-tackling properties (Singh et al. 2017). Marker-assisted backcrossing (MABC) and marker-assisted recurrent selection (MARS) are being currently used as strategies to introduce complex traits within the target plant genome (Singh et al. 2017).

MABC was used to introgress the identified four QTLs associated with root trait development in the upland rice cultivar Kalinga III from Azucena (Steele et al. 2007). Root size, architecture and grain yield have been found to be controlled by a major QTL in maize involved in regulating the leaf abscisic acid (ABA) concentration (Landi et al. 2007). Avoidance of leaf senescence ('stay green' trait) under drought or desiccation can be an effective strategy for generating tolerance. Four such 'stay green'-related QTLs (*Stg1–Stg4*) have been identified in *Sorghum* (Harris et al. 2007). A critical analysis of the 'stay green' phenotype and its positional cloning remains to be reported. QTLs associated with increasing the water use efficiency (WUE) have been identified in *Brassica oleracea*, rice, barley and wheat (Collins et al. 2008). The anthesis-silking interval (ASI) in maize increases during desiccation stress. Five QTL alleles associated with short ASI were introduced in a drought- sensitive line from a drought-tolerant donor using MABC. The selected lines exhibited high yield under drought compared to the unselected control plants (Ribaut and Ragot 2007). Recently, Abdelraheem et al. (2017) performed QTL mapping for abiotic stress tolerance in tetraploid cotton based on multiple morphological and physiological traits. This led to the identification of 23 QTL clusters across 15 chromosomes. Among 28 QTL hotspots, two QTL hotspots on chromosome number 24 were found to be associated with drought and salt tolerance (Abdelraheem et al. 2017). In another recent study, Diaz et al. (2018) used 95 recombinant inbred lines (RILs) of *Phaseolus vulgaris* for QTL mapping under multiple abiotic stresses. The investigation revealed a stable QTL associated with yield on chromosome number 4. Two more QTL hotspots were identified on chromosome numbers 1 and 8 (Diaz et al. 2018). QTL mapping in durum wheat led to the identification of two loci, *Nax1* and *Nax2*, which were related to Na^+ accumulation in the shoot tissue (James et al. 2006). It was observed that *Nax1* induced the retention of Na^+ in the leaf sheath instead of the leaf blade and promoted xylem unloading of Na^+ in the leaf sheath (James et al. 2006).

MABC-dependent introgression of the QTL, *Sub1*, has led to improved submergence and anoxia tolerance in the rice cultivar Swarna (Neeraja et al. 2007). Swarna was eventually converted to a submergence-tolerant variety in three backcross generations within a period of two to three years (Collins et al. 2008). Ismail et al. (2007) reported the incorporation of markers to identify the *Sub1* introgressed lines for effective breeding in flood-prone areas. *Sub1A*, *Sub1B* and *Sub1C* were identified via positional cloning as three putative ethylene- responsive factor genes. The products of these genes controlled the *Sub1* locus. Xu et al. (2006) established that *Sub1A–1* was the primary determinant of submergence tolerance in rice.

QTLs controlling pollen heat tolerance involving pollen germination and pollen tube growth were identified in maize (Frova and Sari-Gorla 1994). Hong et al. (2003) identified seven 'hot' loci in *Arabidopsis*. The compromised action of these loci led to reduced thermotolerance. Introgression of a QTL allele from wild tomato (*Solanum hirsutum*) into *S. lycopersicum* led to increased cold tolerance (Goodstal et al. 2005). The *C-repeat binding factor 2* (*CBF2*) gene was mapped to a freezing-tolerant QTL in *Arabidopsis* (Alonso-Blanco et al. 2005). Subsequently QTLs for cold tolerance have been mapped in *Lens culinaris*, *Brassica napus* and *Lolium perenne* (Collins et al. 2008). Some QTLs involved in mineral deficiency have also been mapped across plant species. These have been highlighted separately in Table 17.1.

17.6 CONCLUSION AND FUTURE PERSPECTIVE

The application of genomics in plant breeding for abiotic stress tolerance has large potential. However, exhaustive involvement of this strategy in crop genotype and phenotype improvement remains to be undertaken. Biotechnological techniques like CRISPR-Cas9 and CRISPRi can lead to the identification of modified genes that might possess the ability to confer stress-tolerant phenotypes in the successive generations. Large-scale data analysis on the whole-genome front has been

TABLE 17.1
Identified QTLs Associated with Nutrient-Related Stress in Plants

Species Where QTL Has Been Identified	Effect	Reference
Al-activated malate transporter 1 (*ALMT1*) in *Arabidopsis*, *Triticum aestivum*	Al^{3+} tolerance	Collins et al. (2008)
Medicago sativa, *Glycine max*, *Oryza sativa*, *Sorghum bicolor*, *Avena sativa*, *Secale cereale*	Tolerance against Al^{3+} toxicity	
Boron (B) toxicity tolerance (*Bo1*) in *Triticum aestivum*	Tolerance to B-mediated stress	Schnurbusch et al. (2007)
Brassica napus	Improved B efficiency	Xu et al. (2001)
Glutamine synthetase (*gln4*) locus in *Zea mays*	Improved nitrogen (N) metabolism	Gallais and Hirel (2004)
Ppd-D1, *Rht-B1* and *B1* in *Triticum aestivum*	High glutamine synthetase activity and grain N accumulation	Laperche et al. (2007)
Phosphorus uptake 1 (*Pup1*) in *Oryza sativa*	Increased phosphorus (P) uptake and utilization	Wissuwa et al. (2005)

possible by availing of NGS-based technologies. This has rapidly improved the time lag necessary for quality assessment of such a huge pool of genomic data. The availability of more genome databases has helped to sieve out potential genes for subsequent cloning and testing for stress-tolerant phenotypes in genetically modified target plant species. Stress tolerance has often been found to be guided by complex traits comprised of QTLs. The incorporation of such complex traits is now possible via QTL mapping and GAB. The field of genomics in abiotic stress tolerance is rather young, and the future perspective is based upon justified use of these highly sophisticated technologies in plant improvement and crop yield under sub-optimal regimes. The use of genomics has to be made breeder-friendly by enabling the users to properly understand the entire procedure and its advantages. Advertisement and advantages of next-generation genomics need to be properly circulated among breeders through popular science platforms and conferences. QTL cloning should be exhaustively utilized to identify superior allelic variants via EcoTILLING. Accurate phenotyping should be performed to properly dissect the functional significance of QTLs.

ACKNOWLEDGMENTS

Financial assistance from the Council of Scientific and Industrial Research (CSIR), Government of India, through the research grant [38(1387)/14/EMR-II], the Science and Engineering Research Board, Government of India through the grant [EMR/2016/004799] and the Department of Higher Education, Science and Technology and Biotechnology, Government of West Bengal, through the grant [264(Sanc.)/ST/P/S&T/1G-80/2017] to Dr. Aryadeep Roychoudhury is gratefully acknowledged. The authors are thankful to the University Grants Commission, Government of India for providing Senior Research Fellowship to Mr. Aditya Banerjee.

REFERENCES

Abdelraheem A, Liu F, Song M, Zhang JF (2017) A meta-analysis of quantitative trait loci for abiotic and biotic stress resistance in tetraploid cotton. *Mol Genet Genomics* 292: 1221–1235.

Albert TJ, Molla MN, Muzny DM, Nazareth L, Wheeler D, Song X, et al. (2007) Direct selection of human genomic loci by microarray hybridization. *Nat Methods* 4: 903–905.

Alonso-Blanco C, Gomez-Mena C, Llorente F, Koornneef M, Salinas J, Martínez-Zapater JM (2005) Genetic and molecular analyses of natural variation indicate *CBF2* as a candidate gene for underlying a freezing tolerance quantitative trait locus in *Arabidopsis*. *Plant Physiol* 139: 1304–1312.

Banerjee A, Roychoudhury A (2018a) Seed priming technology in the amelioration of salinity stress in plants. In: Rakshit A, Singh HB (Eds.) *Advances in Seed Priming*. Springer Nature, Singapore, pp. 81–93.

Banerjee A, Roychoudhury A (2018b) Role of beneficial trace elements in salt stress tolerance of plants. In: Hasanuzzaman M, Fujita M, Oku H, Nahar K, Hawrylak-Nowak B (Eds.) *Plant Nutrients and Abiotic Stress Tolerance*. Springer Nature, Singapore, pp. 377–390.

Banerjee A, Roychoudhury A (2018c) Small heat shock proteins: Structural assembly and functional responses against heat stress in plants. In: Ahmad P, Ahanger MA, Singh VP, Tripathi DK, Alam P, Alyemeni MN (Eds.) *Plant Metabolites and Regulation under Abiotic Stress*. Academic Press, Elsevier, United Kingdom and USA, pp. 367–374.

Banerjee A, Roychoudhury A (2018d) Effect of salinity stress on growth and physiology of medicinal plants. In: Ghorbanpour M, et al. (Eds.) *Medicinal Plants and Environmental Challenges*. Springer International Publishing AG, Cham, Switzerland, pp. 177–188.

Banerjee A, Roychoudhury A (2018e) Regulation of photosynthesis under salinity and drought stress. In: Singh VP, Singh S, Singh R, Prasad SM (Eds.) *Environment and Photosynthesis: A Future Prospect*. Studium Press (India) Pvt. Ltd., New Delhi, pp. 134–144.

Banerjee A, Roychoudhury A (2018f) Abiotic stress, generation of reactive oxygen species, and their consequences: An overview. In: Singh VP, Singh S, Tripathi D, Mohan Prasad S, Chauhan DK (Eds.) *Revisiting the Role of Reactive Oxygen Species (ROS) in Plants: ROS Boon or Bane for Plants?* John Wiley & Sons, Inc., USA, pp. 23–50.

Banerjee A, Roychoudhury A (2019a) Genetic engineering in plants for enhancing arsenic tolerance. In: Prasad MNV (Ed.) *Transgenic Plant Technology for Remediation of Toxic Metals and Metalloids*. Academic Press, Elsevier, United Kingdom, pp. 463–476.

Banerjee A, Roychoudhury A (2019b) Rice responses and tolerance to elevated ozone. In: Hasanuzzaman M, Fujita M, Nahar K, Biswas JK (Eds.) *Advances in Rice Research for Abiotic Stress Tolerance*. Woodhead Publishing, Elsevier, United Kingdom, pp. 399–412.

Banerjee A, Tripathi DK, Roychoudhury A (2018) Hydrogen sulphide trapeze: Environmental stress amelioration and phytohormone crosstalk. *Plant Physiol Biochem* 132: 46–53.

Banerjee A, Tripathi DK, Roychoudhury A (2019) The karrikin 'callisthenics': Can compounds derived from smoke help in stress tolerance? *Physiol Plant.* 165: 290–302.

Barchi L, Lanteri S, Portis E, Valè G, Volante A, Pulcini L, et al. (2012) A RAD tag derived marker based eggplant linkage map and the location of QTLs determining anthocyanin pigmentation. *PLoS ONE* 7: 43740.

Collins NC, Tardieu F, Tuberosa R (2008) Quantitative trait loci and crop performance under abiotic stress: Where do we stand? *J Exp Bot* 147: 469–486.

Diaz LM, Ricaurte J, Tovar E, Cajiao C, Terán H, Grajales M, et al. (2018) QTL analyses for tolerance to abiotic stresses in a common bean (*Phaseolus vulgaris* L.) population. *PLoS ONE* 13: e0202342.

Frova C, Sari-Gorla M (1994) Quantitative trait loci (QTLs) for pollen thermotolerance detected in maize. *Mol Gen Genet* 245: 424–430.

Gallais A, Hirel B (2004) An approach to the genetics of nitrogen use efficiency in maize. *J Exp Bot* 55: 295–306.

Glaubitz JC, Casstevens TM, Lu F, Harriman J, Elshire RJ, Sun Q, et al. (2014) TASSEL-GBS: A high capacity genotyping by sequencing analysis pipeline. *PLoS ONE* 9: e90346.

Gnirke A, Melnikov A, Maguire J, Rogov P, LeProust EM, Brockman W, et al. (2009) Solution hybrid selection with ultra-long oligonucleotides formassively parallel targeted sequencing. *Nat Biotechnol* 27: 182–189.

Goodstal FJ, Kohler GR, Randall LB, Bloom AJ, St Clair DA (2005) A major QTL introgressed from wild *Lycopersicon hirsutum* confers chilling tolerance to cultivated tomato (*Lycopersicon esculentum*). *Theor Appl Genet* 111: 898–905.

Gore MA, Chia JM, Elshire RJ, Sun Q, Ersoz ES, Hurwitz BL, et al. (2009) A first-generation haplotype map of maize. *Science* 326: 1115–1117.

Harris K, Subudhi PK, Borrell A, Jordan D, Rosenow D, Nguyen H, Klein P, Klein R, Mullet J (2007) Sorghum stay-green QTL individually reduce post-flowering drought-induced leaf senescence. *J Exp Bot* 58: 327–338.

Hong SW, Lee U, Vierling E (2003) *Arabidopsis* hot mutants define multiple functions required for acclimation to high temperatures. *Plant Physiol* 132: 757–767.

Ismail AM, Heuer S, Thomson MJ, Wissuwa M (2007) Genetic and genomic approaches to develop rice germplasm for problem soils. *Plant Mol Biol* 65: 547–570.

Jain M (2015) Function genomics of abiotic stress tolerance in plants: A CRISPR approach. *Front Plant Sci* 6: 375.

James RA, Davenport RJ, Munns R (2006) Physiological characterization of two genes for Na+ exclusion in durum wheat, *Nax1* and *Nax2*. *Plant Physiol* 142: 1537–1547.

Kumar V, Jain M (2015) The CRISPR-Cas system for plant genome editing: Advances and opportunities. *J Exp Bot* 66: 47–57.

Landi P, Sanguineti MC, Liu C, Li Y, Wang TY, Giuliani S, Bellotti M, Salvi S, Tuberosa R (2007) *Root-ABA1* QTL affects root lodging, grain yield, and other agronomic traits in maize grown under well-watered and water-stressed conditions. *J Exp Bot* 58: 319–326.

Laperche A, Brancourt-Hulmel M, Heumez E, Gardet O, Hanocq E, Devienne-Barret F, Le Gouis J (2007) Using genotype × nitrogen interaction variables to evaluate the QTL involved in wheat tolerance to nitrogen constraints. *Theor Appl Genet* 115: 399–415.

Leng P-f, Lubberstedt T, Xu M-l (2017) Genomics-assisted breeding - a revolutionary strategy for crop improvement. *J Integr Agric* 16: 2674–2685.

Ma Y, Dai X, Xu Y, Luo W, Zheng X, Zeng D, et al. (2015) *COLD1* confers chilling tolerance in rice. *Cell* 160: 1209–1221.

Mahfouz MM, Piatek A, Stewart CN Jr (2014) Genome engineering via TALENs and CRISPR/Cas9 systems: Challenges and perspectives. *Plant Biotechnol J* 12: 1006–1014.

Marton I, Zuker A, Shklarman E, Zeevi V, Tovkach A, Roffe S, et al. (2010) Non transgenic genome modification in plant cells. *Plant Physiol* 154: 1079–1087.

Montague TG, Cruz JM, Gagnon JA, Church GM, Valen E (2014) CHOPCHOP: A CRISPR/Cas9 and TALEN web tool for genome editing. *Nucleic Acids Res* 42: W401–W407.

Neeraja CN, Maghirang-Rodriguez R, Pamplona A, Heuer S, Collard BCY, Septiningsih EM, et al. (2007) A marker-assisted backcross approach for developing submergence-tolerant rice cultivars. *Theor Appl Genet* 115: 767–776.

Osakabe K, Osakabe Y, Toki S (2010) Site-directed mutagenesis in *Arabidopsis* using custom-designed zinc finger nucleases. *Proc Natl Acad Sci USA* 107: 12034–12039.

Osakabe Y, Osakabe K (2017) Genome editing to improve abiotic stress responses in plants. In: Weeks DP, Yang B (Eds.) *Genome Editing in Plants*. Elsevier Inc., USA, vol. 149, pp. 99–109.

Poland JA, Brown JP, Sorells ME, Jannick JL (2012) Development of high-density genetic maps for barley and wheat using a novel two-enzyme genotyping-by-sequencing approach. *PLoS ONE* 7: e32253.

Porreca GJ, Zhang K, Li JB, Xie B, Austin D, Vassallo SL, et al. (2007) Multiplex amplification of large sets of human exons. *Nat Methods* 4: 931–936.

Qi LS, Larson MH, Gilbert LA, Doudna JA, Weissman JS, Arkin AP, et al. (2013) Repurposing CRISPR as an RNA-guided platform for sequence-specific control of gene expression. *Cell* 152: 1173–1183.

Ray S, Satya P (2014) Next generation sequencing technologies for next generation plant breeding. *Front Plant Sci* 5: 367.

Ribaut JM, Ragot M (2007) Marker-assisted selection to improve drought adaptation in maize: The backcross approach, perspectives, limitations, and alternatives. *J Exp Bot* 58: 351–360.

Saintenac C, Jiang D, Wang S, Akhunov E (2013) Sequence-based mapping of the polyploid wheat genome. *G3* 3: 1105–1114.

Schnurbusch T, Collins NC, Eastwood RF, Sutton T, Jefferies SP, Langridge P (2007) Fine mapping and targeted SNP survey using rice-wheat gene colinearity in the region of the *Bo1* boron toxicity tolerance locus of bread wheat. *Theor Appl Genet* 115: 451–461.

Singh RK, Sahu PP, Muthamilarasan M, Dhaka A, Prasad M (2017) Genomics-assisted breeding for improving stress tolerance of graminaceous crops to biotic and abiotic stresses: Progress and prospects. In: Muthappa S-K (Ed.) *Plant Tolerance to Individual and Concurrent Stresses*. Springer Nature, Switzerland, pp. 59–81.

Steele KA, Virk DS, Kumar R, Prasad SC, Witcombe JR (2007) Field evaluation of upland rice lines selected for QTLs controlling root traits. *Field Crops Res* 101: 180–186.

Teer JK, Bonnycastle LL, Chines PS, Hansen NF, Aoyama N, Swift AJ, et al. (2010) Systematic comparison of three genomic enrichment methods for massively parallel DNA sequencing. *Genome Res* 20: 1420–1431.

Voytas DF (2013) Plant genome engineering with sequence-specific nucleases. *Ann Rev Plant Biol* 64: 327–350.

Wang X, Wang H, Liu S, Ferjani A, Li J, Yan J, Yang X, Qin F (2016) Genetic variation in *ZmVPP1* contributes to drought tolerance in maize seedlings. *Nat Genet* 48: 1233–1241.

Wissuwa M, Gamat G, Ismail AM (2005) Is root growth under phosphorus deficiency affected by source or sink limitations? *J Exp Bot* 56: 1943–1950.

Xu FS, Wang YH, Meng J (2001) Mapping boron efficiency gene(s) in *Brassica napus* using RFLP and AFLP markers. *Plant Breed* 120: 319–324.

Xu K, Xu X, Fukao T, Canlas P, Maghirang-Rodriguez R, Heuer S, Ismail AM, Bailey-Serres J, Ronald PC, Mackill DJ (2006) *Sub1A* is an ethylene-response-factor-like gene that confers submergence tolerance to rice. *Nature* 442: 705–708.

Yang H, Tao Y, Zheng Z, Li C, Sweetingham M, Howieson J (2012) Application of next-generation sequencing for rapid marker development in molecular plant breeding: A case study on anthracnose disease resistance in *Lupinus angustifolius* L. *BMC Genomics* 13: 318.

Zhang N, Wang S, Zhang X, Dong Z, Chen F, Cui D (2016) Transcriptome analysis of the Chinese bread wheat cultivar Yunong 201 and its ethyl methanesulfonate mutant line. *Gene* 575: 285–293.

Index